T0236321

Lecture Notes in Computer Science 10113

Commenced Publication in 1973
Founding and Former Series Editors:
Gerhard Goos, Juris Hartmanis, and Jan van Leeuwen

More information about this series at http://www.springer.com/series/7412

Shang-Hong Lai · Vincent Lepetit
Ko Nishino · Yoichi Sato (Eds.)

Computer Vision – ACCV 2016

13th Asian Conference on Computer Vision
Taipei, Taiwan, November 20–24, 2016
Revised Selected Papers, Part III

 Springer

Editors
Shang-Hong Lai
National Tsing Hua University
Hsinchu
Taiwan

Vincent Lepetit
Graz University of Technology
Graz
Austria

Ko Nishino
Drexel University
Philadelphia, PA
USA

Yoichi Sato
The University of Tokyo
Tokyo
Japan

ISSN 0302-9743 ISSN 1611-3349 (electronic)
Lecture Notes in Computer Science
ISBN 978-3-319-54186-0 ISBN 978-3-319-54187-7 (eBook)
DOI 10.1007/978-3-319-54187-7

Library of Congress Control Number: 2017932642

LNCS Sublibrary: SL6 – Image Processing, Computer Vision, Pattern Recognition, and Graphics

Printed on acid-free paper

This Springer imprint is published by Springer Nature
The registered company is Springer International Publishing AG
The registered company address is: Gewerbestrasse 11, 6330 Cham, Switzerland

Preface

Welcome to the 2016 edition of the Asian Conference on Computer Vision in Taipei. ACCV 2016 received a total number of 590 submissions, of which 479 papers went through a review process after excluding papers rejected without review because of violation of the ACCV submission guidelines or being withdrawn before review. The papers were submitted from diverse regions with 69% from Asia, 19% from Europe, and 12% from North America.

The program chairs assembled a geographically diverse team of 39 area chairs who handled nine to 15 papers each. Area chairs were selected to provide a broad range of expertise, to balance junior and senior members, and to represent a variety of geographical locations. Area chairs recommended reviewers for papers, and each paper received at least three reviews from the 631 reviewers who participated in the process. Paper decisions were finalized at an area chair meeting held in Taipei during August 13–14, 2016. At this meeting, the area chairs worked in threes to reach collective decisions about acceptance, and in panels of nine or 12 to decide on the oral/poster distinction. The total number of papers accepted was 143 (an overall acceptance rate of 24%). Of these, 33 were selected for oral presentations and 110 were selected for poster presentations.

We wish to thank all members of the local arrangements team for helping us run the area chair meeting smoothly. We also wish to extend our immense gratitude to the area chairs and reviewers for their generous participation in the process. The conference would not have been possible without this huge voluntary investment of time and effort. We acknowledge particularly the contribution of 29 reviewers designated as "Outstanding Reviewers" who were nominated by the area chairs and program chairs for having provided a large number of helpful, high-quality reviews. Last but not the least, we would like to show our deepest gratitude to all of the emergency reviewers who kindly responded to our last-minute request and provided thorough reviews for papers with missing reviews. Finally, we wish all the attendees a highly simulating, informative, and enjoyable conference.

January 2017

Shang-Hong Lai
Vincent Lepetit
Ko Nishino
Yoichi Sato

Organization

ACCV 2016 Organizers

Steering Committee

Michael Brown	National University of Singapore, Singapore
Katsu Ikeuchi	University of Tokyo, Japan
In-So Kweon	KAIST, Korea
Tieniu Tan	Chinese Academy of Sciences, China
Yasushi Yagi	Osaka University, Japan

Honorary Chairs

Thomas Huang	University of Illinois at Urbana-Champaign, USA
Wen-Hsiang Tsai	National Chiao Tung University, Taiwan, ROC

General Chairs

Yi-Ping Hung	National Taiwan University, Taiwan, ROC
Ming-Hsuan Yang	University of California at Merced, USA
Hongbin Zha	Peking University, China

Program Chairs

Shang-Hong Lai	National Tsing Hua University, Taiwan, ROC
Vincent Lepetit	TU Graz, Austria
Ko Nishino	Drexel University, USA
Yoichi Sato	University of Tokyo, Japan

Publicity Chairs

Ming-Ming Cheng	Nankai University, China
Jen-Hui Chuang	National Chiao Tung University, Taiwan, ROC
Seon Joo Kim	Yonsei University, Korea

Local Arrangements Chairs

Yung-Yu Chuang	National Taiwan University, Taiwan, ROC
Yen-Yu Lin	Academia Sinica, Taiwan, ROC
Sheng-Wen Shih	National Chi Nan University, Taiwan, ROC
Yu-Chiang Frank Wang	Academia Sinica, Taiwan, ROC

Workshops Chairs

Chu-Song Chen	Academia Sinica, Taiwan, ROC
Jiwen Lu	Tsinghua University, China
Kai-Kuang Ma	Nanyang Technological University, Singapore

Tutorial Chairs

Bernard Ghanem	King Abdullah University of Science and Technology, Saudi Arabia
Fay Huang	National Ilan University, Taiwan, ROC
Yukiko Kenmochi	Université Paris-Est, France

Exhibition and Demo Chairs

Gee-Sern Hsu	National Taiwan University of Science and Technology, Taiwan, ROC
Xue Mei	Toyota Research Institute, USA

Publication Chairs

Chih-Yi Chiu	National Chiayi University, Taiwan, ROC
Jenn-Jier (James) Lien	National Cheng Kung University, Taiwan, ROC
Huei-Yung Lin	National Chung Cheng University, Taiwan, ROC

Industry Chairs

Winston Hsu	National Taiwan University, Taiwan, ROC
Fatih Porikli	Australian National University, Australia
Li Xu	SenseTime Group Limited, Hong Kong, SAR China

Finance Chairs

Yong-Sheng Chen	National Chiao Tung University, Taiwan, ROC
Ming-Sui Lee	National Taiwan University, Taiwan, ROC

Registration Chairs

Kuan-Wen Chen	National Chiao Tung University, Taiwan, ROC
Wen-Huang Cheng	Academia Sinica, Taiwan, ROC
Min Sun	National Tsing Hua University, Taiwan, ROC

Web Chairs

Hwann-Tzong Chen	National Tsing Hua University, Taiwan, ROC
Ju-Chun Ko	National Taipei University of Technology, Taiwan, ROC
Neng-Hao Yu	National Chengchi University, Taiwan, ROC

Area Chairs

Narendra Ahuja	UIUC
Michael Brown	National University of Singapore
Yung-Yu Chuang	National Taiwan University, Taiwan, ROC
Pau-Choo Chung	National Cheng Kung University, Taiwan, ROC
Larry Davis	University of Maryland, USA

Sanja Fidler	University of Toronto, Canada
Mario Fritz	Max Planck Institute for Informatics, Germany
Yasutaka Furukawa	Washington University in St. Louis, USA
Bohyung Han	Pohang University of Science and Technology, South Korea
Hiroshi Ishikawa	Waseda University, Japan
C.V. Jawahar	IIIT Hyderabad, India
Frédéric Jurie	University of Caen, France
Iasonas Kokkinos	CentraleSupélec/Inria, France
David Kriegman	UCSD
Ivan Laptev	Inria, France
Kyoung Mu Lee	Seoul National University
Jongwoo Lim	Hanyang University, South Korea
Liang Lin	Sun Yat-Sen University, China
Tyng-Luh Liu	Academia Sinica, Taiwan, ROC
Huchuan Lu	Dalian University of Technology, China
Yasuyuki Matsushita	Osaka University, Japan
Francesc Moreno-Noguer	Institut de Robòtica i Informàtica Industrial
Greg Mori	Simon Fraser University, Canada
Srinivasa Narasimhan	CMU
Shmuel Peleg	Hebrew University of Jerusalem, Israel
Fatih Porikli	Australian National University/CSIRO, Australia
Ian Reid	University of Adelaide, Australia
Mathieu Salzmann	EPFL, Switzerland
Imari Sato	National Institute of Informatics, Japan
Shin'ichi Satoh	National Institute of Informatics, Japan
Shiguang Shan	Chinese Academy of Sciences, China
Min Sun	National Tsing Hua University, Taiwan, ROC
Raquel Urtasun	University of Toronto, Canada
Anton van den Hengel	University of Adelaide, Australia
Xiaogang Wang	Chinese University of Hong Kong, SAR China
Hanzi Wang	Xiamen University
Yu-Chiang Frank Wang	Academia Sinica, Taiwan, ROC
Jie Yang	NSF
Lei Zhang	Hong Kong Poly University, SAR China

Contents – Part III

Face and Gestures

Computational Photography

Layered Scene Reconstruction from Multiple Light Field Camera Views

Ole Johannsen, Antonin Sulc$^{(\boxtimes)}$, Nico Marniok, and Bastian Goldluecke

University of Konstanz, Konstanz, Germany
{ole.johannsen,antonin.sulc,nico.marniok,
bastian.goldluecke}@uni-konstanz.de

Abstract. We propose a framework to infer complete geometry of a scene with strong reflections or hidden by partially transparent occluders from a set of 4D light fields captured with a hand-held light field camera. For this, we first introduce a variant of bundle adjustment specifically tailored to 4D light fields to obtain improved pose parameters. Geometry is recovered in a global framework based on convex optimization for a weighted minimal surface. To allow for non-Lambertian materials and semi-transparent occluders, the point-wise costs are not based on the principle of photo-consistency. Instead, we perform a layer analysis of the light field obtained by finding superimposed oriented patterns in epipolar plane image space to obtain a set of depth hypotheses and confidence scores, which are integrated into a single functional.

1 Introduction

The reconstruction of the geometry of a non-Lambertian object from multiple views remains a tremendous challenge. As the observed color of a point on the object can strongly depend on the viewing angle, matching image patches in particular across wide-baseline views can not be done based on the information in the images alone. Thus, state-of-the art 3D reconstruction pipelines both based on sparse feature detection and matching followed by bundle adjustment as well as dense reconstruction using photo-consistency typically fail for such objects.

Recent work, as discussed in detail in Sect. 2, shows that these challenges can at least partially be overcome when working with a set of densely sampled view points, known as the 4D light field of a scene [1] or an epipolar volume [2]. Phenomena such as specular highlights or reflections manifest as patterns on epipolar plane image space, in particular, one can analyze them by working with standard pattern or statistical analysis on 2D cuts through the 4D light field, without the need for explicit matching. We believe and try to prove with our work that these ideas also form the basis to a very powerful approach to the problem of dense non-Lambertian surface reconstruction. For the plenoptic cameras now commercially available e.g. by Lytro and Raytrix, this can also open up a unique niche which they are ideally suited for.

O. Johannsen and A. Sulc contributed equally.

© Springer International Publishing AG 2017
S.-H. Lai et al. (Eds.): ACCV 2016, Part III, LNCS 10113, pp. 3–18, 2017.
DOI: 10.1007/978-3-319-54187-7_1

Fig. 1. Light field center views for one of our datasets, illustrating the type of input our method is able to handle. One can barely recognize that there is a shape behind the windows, but how exactly does it look like? The result will be revealed in Fig. 9.

Thus, the scenario we investigate in this paper is having multiple 4D light fields, which can be recorded with a hand-held plenoptic camera, instead of multiple conventional images of a given scene. The goal is to infer the geometry of the scene. The scene does not need to be Lambertian - the objects can either exhibit strong reflections, be semi-transparent, or partially occluded by other semi-transparent surfaces. In all of these cases, the epipolar plane images of the resulting light fields will exhibit overlaid patterns, whose structure is related to the depth of individual elements in the scene. This structure can be analyzed for example with a sparse coding technique pioneered in [3], or higher order structure tensors [4], resulting in a distribution of depth hypotheses for every ray in every light field. We align the light fields using linear structure-from-motion [5] and a novel bundle adjustment technique. Refined extrinsic parameters are used to reconstruct scene surface using a convex variational approach by a volumetric segmentation functional with a surface consistency data term.

Contributions. In this paper, we make the following novel contributions.

- We show how to construct non-linear bundle adjustment on top of the linear structure-from-motion framework [5] which is tailored to light fields. We show on new synthetic lightfield surround datasets with complete ground truth information available that the proposed method improves extrinsic calibration of the cameras.
- We propose an alternative model to classical photo-consistency of surface points, which is specific to 4D light fields and does not require the Lambertian color-constancy assumption. Instead, we perform multi-orientation analysis of the epipolar plane images [3,4,6] to obtain a set of depth hypotheses for every available ray, and use these to construct volumetric cost functions for the presence of object interiors and surfaces.
- Finally, we embed these cost functions into a convex variational energy minimization scheme for dense surface reconstruction to obtain the full geometry for all layers of the scene in a single pass.

All of the new building blocks are assembled into the first complete 3D scene reconstruction pipeline from multiple 4D light field camera views, which is able to deal with strongly reflective objects or partially transparent occluders.

2   Background and Related Work

Our method reconstructs an object from multiple 4D light fields. In this setting, to our knowledge only sparse structure-from-motion methods were pursued so far [5], see Sect. 3, where we construct bundle adjustment for light fields. Previous research instead focuses on computing depth from one single 4D light field, where there has been tremendous progress in the last few years. Obtaining depth from light field data is usually based on the idea of an epipolar volume [2], which can be considered as a stack of views with slowly moving view point. A two-dimensional cut through this stack is called an epipolar plane image (EPI), see Fig. 2. Every 3D scene point is projected onto a line in the EPI, which results in characteristic linear patterns. For Lambertian light fields, computing orientation of these lines yields an accurate depth estimate [7]. A theory of invariants for more general BRDF models can be found in [8].

Recent methods can deal with modest specular reflections by statistically analyzing the slight intensity variations when shifting the view point [9,10]. Notably, [11] allows the reconstruction of very accurate depth maps for arbitrary BRDF from epipolar volumes, but the BRDF needs to be known beforehand. This can not be assumed in general.

Furthermore, in our scenario, there is no unique depth due to the light field having multiple layers, see Fig. 2. In this setting, we compute center view disparities for the different layers [3,4] and decompose the scene into several layers [6]. We extend their technique to be able to integrate different depth hypotheses from different subaperture views into a distribution of likelihoods for the presence of scene points, and generate volumetric surface consistency estimates.

As approaches to depth map fusion typically require a unique depth map for each view, we also need to propose a more general reconstruction scheme. For the remainder of the section, we will briefly recap a few milestones in multi-view 3D reconstruction which led to the development of the convex approaches we employ. We will in particular focus on methods for non-Lambertian objects.

A brief history of volumetric 3D reconstruction. Purely silhouette-based approaches are the most simple and naturally independent of object BRDF, but can at best recover the visual hull [12]. Very recently, a method has been proposed to refine the visual hull based on internal occluding contours, which also yields good results on specular and transparent objects [13].

One of the earliest approaches using photo-consistency measures in volumetric 3D reconstruction was space carving [14]. Here, voxels in a volume are carved away if they are not consistent with the data in the input images. Notably, the work already considered possible consistency measures based on non-Lambertian BRDFs in the theoretical part, but did not provide concrete examples or experiments yet. Only a few years later, [15] introduced line model fitting (LMF) in RGB space as a consistency score for Phong-like BRDFs within the space carving framework to successfully deal with specular highlights.

Another line of research built upon the idea of shape reconstruction via a photo-consistent minimal surface [16]. While [16] only considered Lambertian

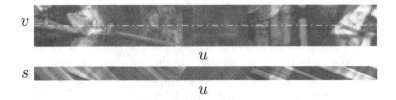

Fig. 2. *Light field with two disparity layers, epipolar volume.* The top image shows a close-up of the center view of the *Temple* dataset, see Fig. 1. The EPI corresponding to the white line is depicted below. Most parts of the center view show semi-transparent objects, with two surfaces on top of each other visible in the same pixel. These regions of the EPI thus in turn exhibit two superimposed oriented patterns, their orientations corresponding to two different disparities.

matching scores, [17] introduced a rank constraint of the radiance tensor as a non-Lambertian photo-consistency measure, which was later improved and discretized in a volumetric graph cut framework [18]. It is also possible to build surface consistency measures for transparent objects based on ray-casting, as in the approach [19] to the reconstruction of flowing liquids. Photometric stereo nowadays incorporates arbitrary BRDF models [20], and has been successfully integrated into [16] by defining a photometric cost function on a surface and performing mesh evolution [21].

Convex approaches. An efficient way to reformulate the weighted minimal surface problem in 3D reconstruction is as a convex minimization problem, as this allows to use fast solvers and can achieve globally optimal solutions independent of initialization [22,23]. The key idea is to leverage the equivalence of the weighted integral over a closed surface and the weighted total variation of the characteristic function of the enclosed volume [24], which is a convex functional. In order to avoid the trivial solution of an empty surface, additional convex constraints are necessary. These usually come from silhouette information. In the case of perfect binary silhouettes, one can make use of the observation that the set of all shapes consistent with a given set of silhouettes is convex [23]. For per-image silhouette likelihoods, one can construct volumetric data terms from probabilistic considerations [22], and perform essentially binary 3D segmentation into object interior and exterior. It is also straight-forward to formulate depth map fusion with convex functionals [25,26]. In principle this can be independent of a BRDF if a method like Helmholtz stereopsis [27] is employed to infer the depth maps.

The reconstruction method proposed in our paper builds upon the formulation in [22], and constructs both a surface-consistency term as well as a volumetric cost from the given light field data, see Sect. 4. A pre-requisite is that we have fully calibrated the cameras, which we describe in the following section.

3   Structure from Motion for Multiple 4D Light Fields

This section describes how we obtain extrinsic parameters for a set of light field cameras located around a scene. Based on the ideas for linear structure from motion [5], we model a version of bundle adjustment which is tailored to 4D light fields and achieves more robust and accurate results than the purely linear approach. Please note, that although this approach is independent of the type of surface, we still rely on Lambertian surfaces for feature detection. First, however, we briefly review the standard notation used in light field analysis, which will be used throughout the paper.

3.1   Light Field Coordinates and Projections

Our light fields are given in the absolute two-plane parametrization. Each ray is parametrized by the intersection with two parallel planes Ω and Π at a distance f of each other. The intersection coordinates are denoted $(u, v) \in \Pi$ and $(s, t) \in \Omega$, which means that each ray is described by a four-dimensional vector $\mathbf{l} = [u, v, s, t]^T \in \mathbb{R}^4$. The plane Ω is called the *image plane*, the plane Π the *focal plane*. Intuitively, coordinates in Π select a pinhole view of the scene called a *subaperture view*, with the focal point located in (s, t). Thus, (u, v) then parametrizes the image coordinates, see Fig. 2 for an illustration. We call the coordinates $\mathbf{r} = [u \; v \; s \; t \; 1]^T$ *homogeneous light field coordinates*, which are defined with respect to a local camera coordinate system.

A single scene point $\mathbf{X} = [X \; Y \; Z]^T$ given in the reference frame of the camera will be visible in multiple subaperture views. In [5], it was shown that the set of all rays $\mathbf{r}$ which pass through $\mathbf{X}$ forms a two-dimensional linear subspace defined by

$$M(\mathbf{X}, f)\mathbf{r} = 0 \text{ with } M(\mathbf{X}, f) = \begin{bmatrix} 1 & 0 & \frac{f}{Z} & 0 & -\frac{fX}{Z} \\ 0 & 1 & 0 & \frac{f}{Z} & -\frac{fY}{Z} \end{bmatrix}, \tag{1}$$

where $M(\mathbf{X}, f) \in \mathbb{R}^{2 \times 5}$ depends on $\mathbf{X}$ and the focal length f, which can in general differ between the individual light fields. In particular, this constraint needs to be satisfied for all matching features, i.e. rays in which a common point $\mathbf{X}$ is observed.

If practical images and a real-world camera are considered, we need to introduce a suitable pixel coordinate system. Here, a single ray is described within a discrete set of subaperture views indexed by (m, n), and relative pixel coordinates (x, y) within an individual subaperture view. The resulting set of coordinates $\mathbf{p} = [x \; y \; m \; n \; 1]^T$ are called the *homogeneous pixel coordinates*. These are related to the homogeneous light field coordinates by a 5×5 intrinsic matrix $\mathbf{H}$ via $\mathbf{r} = \mathbf{Hp}$. The exact structure of $\mathbf{H}$ and a way to calibrate it for a given light field camera is described in [28].

3.2   Light Field Features and Linear Structure from Motion

A light field feature F consists of a set of rays $\mathbf{r}$ which correspond to the same scene point in multiple sub-aperture images. To detect them, we employ SIFT on all subaperture views, and then search for matching features which lie approximately on a common linear subspace for a valid $M(\mathbf{X}, f)$. The estimation is performed with a RANSAC scheme to aggressively filter outliers. The result is a set of features F_{ij}, where i indexes the light field and j the feature within this light field.

In [5], they described how the relative pose can be determined from two light fields using a set of corresponding features. We fix a reference view among our input light fields, select a second light field which has the most feature correspondences with the first one, and register it to the coordinate system of the first one using [5]. The process is then iterated until an initial estimate for pose R_i, t_i relative to the reference light field is available for each input light field. In addition, we now show how to improve the initial estimate using a final global optimization pass over all views and features similar to bundle adjustment. This was not yet shown in previous work [5]. While bundle adjustment is commonly known for structure-from-motion for images, the equations are different and lead to a more rigid estimate if tailored to the light field setting, due to having multiple feature observations available from each view point.

3.3   Bundle Adjustment for Light Field Cameras

Bundle adjustment means solving a global optimization problem for the refinement of estimated 3D points and camera parameters. The initial guess, determined as in the previous subsection, is refined iteratively during the minimization of a non-linear energy. While the problem of bundle adjustment is well studied for conventional 2D imaging, it has not been specialized to light field cameras yet.

The variables to be optimized for are a set of scene points $\mathbf{X}_j^w$ in a world coordinate system and a set of pose parameters R_i, t_i for the light field camera views. The input is a set of feature observations F_{ij} in homogeneous pixel coordinates (possibly empty) for each light field and scene point.

As camera and world coordinates are related by $\mathbf{X}_{ij} = R_i \mathbf{X}_j^w + t_i$, we have by equation (1) that in the ideal noise-free case with perfect parameters,

$$M(R_i \mathbf{X}_j^w + t_i, f_i)\mathbf{H}_i \mathbf{p} = 0 \quad \text{for all } i, j, \mathbf{p} \in F_{ij}. \tag{2}$$

Thus, we minimize the residual energy

$$E(\mathbf{X}_1^w, \ldots, \mathbf{X}_M^w, R_1, \ldots, R_N, t_1, \ldots, t_N)$$
$$= \sum_{i,j} \sum_{\mathbf{p} \in F_{ij}} \left\| M(R_i \mathbf{X}_j^w + t_i, f_i)\mathbf{H}_i \mathbf{p} \right\|_2^2 \tag{3}$$

to determine the final pose parameters of the camera setup. While the structure-from-motion point cloud $\mathbf{X}_j^w$ already gives useful information about the scene, we do not yet employ it in the remainder of the pipeline except to determine an estimate of the bounding box for volumetric reconstruction. Minimization of the energy (3) is performed with Levenberg-Marquardt algorithm built into Matlab.

4   Surface Reconstruction from Multiple 4D Light Fields

In this section, we describe the volumetric reconstruction of a scene with super-imposed layers from 4D light fields, given full calibration of all extrinsics and intrinsics. The key feature is that the complete geometry of the scene is obtained in a single optimization pass. In particular, it is not necessary to separate the individual layers of a light field in advance. Rather, we work with a confidence distribution over the possible depth values for each ray, and aggregate all available information into a global cost volume.

We will explain the technique from general to specific. First, we introduce the functional to be minimized for reconstruction, then the technique how to update the individual terms in the functional given a single depth map with attached confidence values, finally, methods to obtain information about these layers given an individual 4D light field. An overview of the complete reconstruction pipeline can be found in Fig. 5.

4.1   Variational Surface Reconstruction

The aim is to reconstruct the closed scene surface Σ in a volume $\Gamma \subset \mathbb{R}^3$, which is segmented into an interior region Γ_{int} and exterior region Γ_{ext} by the surface. The key idea is to define an energy functional $E(\Sigma)$ such that the minimizer of E corresponds to a surface which is optimally consistent with all of the input data. For this, we define a surface consistency error $\rho : \Gamma \to \mathbb{R}^+$, which is large if a point is unlikely to be on the surface, and small otherwise. To simplify handling this mathematically, we follow [22] and switch to a formulation with a binary indicator function $u : \Gamma \to \{0,1\}$, which is related to the surface by setting u to the characteristic function of the interior region. The functional we use is one for segmentation in the 3D volume Γ,

$$E(u) = \lambda \int_{\Gamma} \rho \, |Du| \; + \; \int_{\Gamma} au \, dx, \tag{4}$$

where the regional cost a is negative if a point has a preference to be inside the surface, and positive otherwise. The weighted total variation corresponds to the integral of the cost function over the surface. This is a convex functional over the non-convex domain of binary functions. However, we can relax to functions valued in the unit interval, solve the resulting convex problem (we use [29]), and threshold the result to obtain the global optimum [30], all of which is straightforward.

Therefore, for the remainder of the section, it just remains to be shown how we compute the surface consistency error ρ and regional cost a for the scenario of multiple light field camera views.

4.2   Updating the Cost Volume for a Single Depth Map

Light fields naturally occupy a lot of memory. Thus, to efficiently construct the cost volume, we follow an incremental approach, where we only have to store the

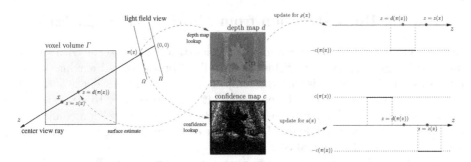

Fig. 3. *Incremental update of cost volumes from one disparity layer and confidence map.* The voxel is first projected into the image plane of the center view, where we can look up disparity and confidence. Based on these values, we update the cost functions for x. The illustration here is for the situation that the voxel x is just behind the depth sample from the point of view of the camera, so it should be inside. In this case, a negative value is added to $a(x)$, see Sect. 4.2.

data for one single light field at the same time. In a multi-layered scene, for every light field view, we will have several depth hypotheses per ray. Each hypothesis is associated with a confidence score, which estimates how sure we are that this hypothesis is the correct one. In our method, every hypothesis generates cues for the surface position within the cost volumes. In this subsection, we describe how the costs change in a single pass over the volume, given a single depth map with confidence scores from a single view. The next subsection then explains how the multiple depth hypotheses per view are generated.

The exact input to this pass is a single depth map d defined on one of the views of the light field, usually the one in the center, a confidence map c assigning a positive or zero weight to each disparity measurement, and the projection π from the cost volume region Γ onto the image plane of this view. For simplicity of notation, we assume c and d to be defined everywhere, with c being zero outside the given viewport.

The cost volumes are now updated as follows, see Fig. 3. For each $x \in \Gamma$, compute the depth $z(x)$ of the voxel. We distinguish three cases:

- If $|z(x) - d(\pi(x))| \le \epsilon$, with a small constant $\epsilon > 0$, then the surface is likely to pass through x. Thus, we *decrease* the surface consistency *error* by our confidence in this estimate,

$$\rho(x) \leftarrow \rho(x) - c(\pi(x)).$$

- If the distance of x to the camera is a bit larger than $d(\pi(x))$, i.e. $d(\pi(x)) + 2\epsilon \ge z(x) > d(\pi(x)) + \epsilon$, then x is likely to lie inside the surface. Thus, we *decrease the regional cost* a by the confidence score, i.e.

$$a(x) \leftarrow a(x) - c(\pi(x)).$$

– Conversely, if the distance of x to the camera is just a bit smaller than $d(\pi(x))$, i.e. $d(\pi(x)) + 2\epsilon \geq z(x) > d(\pi(x)) + \epsilon$, then x is likely to lie outside the surface. Thus, we *increase the regional cost a* by the confidence score, i.e.

$$a(x) \leftarrow a(x) + c(\pi(x)).$$

Our implementation discretizes Γ with a voxel volume, and the update step can be performed very efficiently on the GPU (a few milliseconds per input light field and depth hypothesis layer depending on volume resolution), as it is independent for every voxel. In a final pass, all cost volumes are filtered with a Gaussian to alleviate noise or sparsity in the disparity estimates. Examples can be observed in Fig. 4.

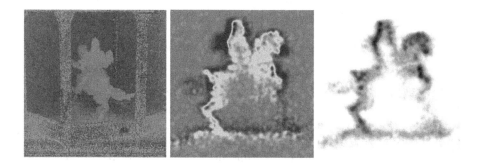

Fig. 4. *Left:* single depth map layer obtained for one of the center views in Fig. 1 using a second order structure tensor as in [4]. *Middle:* slice through the regional cost volume a, red values denote a preference for begin inside. *Right:* Slice through the surface consistency cost ρ. Darker colors mean it is more likely that a surface passes though this point. (Color figure online)

4.3   Depth Layer and Certainty Estimation

We finally describe how to obtain a set of layered depth maps $d_1, \ldots, d_N$ and associated confidence maps $c_1, \ldots, c_N$ for the center view of a given 4D light field. Obviously, any variant of multiview stereo can be used to obtain a single depth map ($N = 1$), often with a sensible way to generate a confidence score. However, a single layer is only sufficient for a Lambertian scene. Any other method which can generate a depth map for a light field is suitable if one wants to recover only a single layer, and our method works in fact fine as a replacement for depth map fusion.

For multi-layered scenes, we have two options. First, we can use second-order structure tensors for two-layered scenes, as shown in [4,6]. From these, as raw data we recover two depth layers d_{sx}^1, d_{sx}^2 for the horizontal EPIs, as well as two depth layers d_{ty}^1, d_{ty}^2 for the vertical ones. While we could just feed all of them into our pipeline, we determined experimentally that it is much better to retain a sparse number of very accurate depth cues than many inaccurate ones.

Part I: bundle adjustment	Part II: surface reconstruction
Step 1: feature detection. For each light field view i, generate sets F_{ij} of ray coordinates which (probably) observe the same scene point $\mathbf{X}_j$ (sections 3.1, 3.2).	**Step 1: build cost volumes.** Initialize ρ, a to zero. Then, for each input light field i:
	1. compute a set of depth hypothesis d_{ij} with confidence maps c_{ij}, section 4.3.
Step 2: linear SfM. From the features, compute initial pose estimates R_i, t_i using linear SfM [5], see section 3.2.	2. use each hypothesis-confidence pair (d_{ij}, c_{ij}) to update the cost volumes, section 4.2.
Step 3: bundle adjustment. Non-linear joint refinement of scene points $\mathbf{X}_j^w$ and all pose parameters R_i, t_i tailored to light fields, see section 3.3.	**Step 2: optimize for scene surface.** Solve energy 4 using the previously computed cost volumes, section 4.1.

Fig. 5. *Summary of the complete reconstruction pipeline.*

Thus, we filter out very conservatively: only if both estimates agree, i.e. $\left|d_{sx}^1 - d_{ty}^1\right| \leq \delta$ and $\left|d_{sx}^2 - d_{ty}^2\right| \leq \delta$ with a small constant $\delta > 0$, we keep the average of both estimates with confidence 1. Otherwise, we discard the pixel and set confidence to zero.

The second option to recover multiple depth layers is [3] based on sparse coding of the light field, which returns a probability volume $\alpha(\lambda, p)$ over multiple depth hypotheses $\lambda_1, \ldots, \lambda_N$ for each pixel p. This is equivalent to having a set of N constant depth maps $d_i := \lambda_i$ with probability $c_i(p) = \alpha(\lambda_i, p)$, which we provide to the reconstruction pipeline. While computationally overall much more expensive, it is more robust in practice and can in theory also deal with scenes with more than two layers. It turns out that for Lytro data, the structure tensor approach is insufficient, and we have to resort to this more sophisticated way of handling multiple layers.

5   Results

5.1   Synthetic Datasets

The synthetic test data used for the ground truth verification was created with a custom real-time graphics engine. The well-known Cook-Torrance BRDF was chosen as a material model. While shadow mapping and specular highlights are disabled, the diffuse lighting coefficient comes directly from the Fresnel term described by this model.

We render 24 light fields with 17×17 subaperture views and a resolution of 512×512 pixels around the scene to be reconstructed. Ground truth information for depth maps for all layers as well as camera orientation is stored for validation. The central object is a fully textured triangulation created from a 3D scan of

	Opacity	SfM		BA	
		R error	t error	R error	t error
Temple	50%	49.723	48.7	**20.310**	**13.177**
	70%	137.204	121.523	**37.550**	**35.311**
	90%	97.783	72.764	**42.681**	**40.407**
Warrior	50%	64.615	66.867	**26.411**	**27.941**
	70%	68.501	80.288	**29.202**	**30.013**
	90%	15.358	22.238	**12.311**	**18.388**

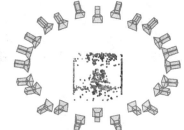

Fig. 6. *Results of SfM and Bundle adjustment. Left:* Comparison of SfM and its refinement by Bundle adjustment. Error is measured as average angular deviation of camera extrinsic parameters R and t from the ground truth in degrees (multiplied by 100). *Right:* Example result of Bundle adjustment on the *Temple* synthetic dataset.

a bronze statue depicting a Chinese horseman done by Artec 3D, https://www.artec3d.com/3dmodel/bronze-statue.

For the first scene *Temple*, a dome consisting of planar semi-transparent surfaces is built around this object. The surface textures have been obtained

Fig. 7. Results for synthetic dataset *Warrior* with a reflective surface. Note the exceptionally strong view point dependence of the appearance in the input images (top row). *Bottom left:* result with ground truth depth for reference. Voxel volume resolution is $374 \times 400 \times 224$. *Bottom center:* result using our method with depth estimated according to Sect. 4.3 using a second-order structure tensor. *Bottom right:* once depth has been estimated, the light field can be separated into its individual layers using [6], and the layer corresponding to the surface used to infer a texture via projective texture mapping. Here, a color has been assigned to each vertex based on the average color of the projections into the center views.

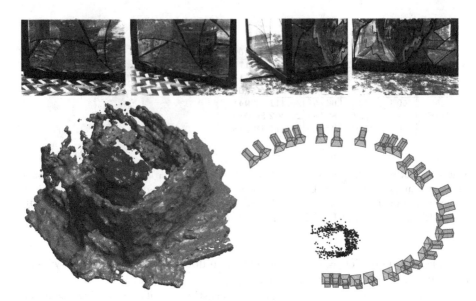

Fig. 8. Results for real-world dataset *Lantern*. *Top:* Four center views from overall 28 views captured with Lytro Illum. The green crosses depict inlier features which were detected in center views. The triangulated feature 3D points projected onto center view given pose from SfM are depicted by blue circles. Red circles show the closest 3D point refined by bundle adjustment and projected with refined extrinsics. *Bottom left:* Geometry of entire scene as result of surface reconstruction with our proposed method. Voxel volume resolution is $300 \times 300 \times 300$. *Bottom right:* Camera poses and point cloud refined with bundle adjustment.

from http://www.rgbstock.com and http://www.deviantart.com, and are reminiscent of stained glass. Due to their partial transparency, they provide a nontrivial challenge for 3D reconstruction, see Fig. 1, as the object itself cannot be seen directly but only through them. We first evaluate our novel bundle adjustment approach, see Fig. 6. It compares favorably to the results from linear structure from motion described in [5], which was shown to be more accurate for light fields than earlier methods. Afterwards, we continue surface reconstruction, see Fig. 9. Despite the difficult input data, we can recover both the geometry of the dome as well as the statue within a single pass in high detail, as can be seen by visually comparing to the ground truth.

For the second scene *Warrior*, a reflectance layer is added to the statue, which reflects the walls of an opaque Lambertian cube around the scene in a distance. Reconstruction results can be observed in Fig. 7. The scene is very difficult for traditional multi-view reconstruction due to the strong view-dependence of the observed colors of surface points. Despite this, our lightfield-based algorithm faithfully recovers the surface geometry in quite a lot of detail.

5.2   Real-World Datasets

For a real-world scene, we create a scene of a lantern with a cup inside, see Fig. 8. The faces of the lantern are painted with different semi-transparent colors, each

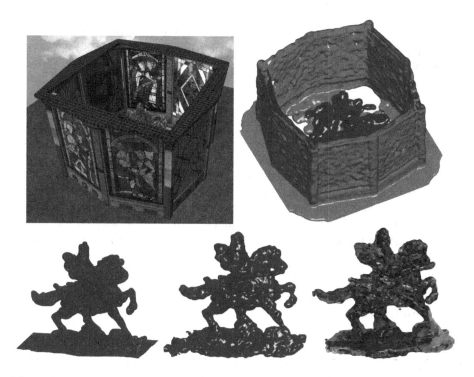

Fig. 9. Results for synthetic dataset *Temple. Top-left:* overview reference rendering of complete scene (not used as input), note the windows are partially transparent (50% in this experiment). *Top-right:* reconstruction result using our pipeline, disparity layers estimated with a second-order structure tensor [4]. Note that the full geometry including exterior and interior was recovered in a single pass, using only side view lightfields as input, see Fig. 1. The visible part of the top of the dome was also reconstructed, but cut off here for better visibility. For the same reason, the interior has been manually colored differently, it also shows the bounding box used for the closeups below. *Bottom-left:* result for the geometry of the statue within the dome if our pipeline is run with ground depth maps - this is the optimum one can hope for, but unrealistic to achieve in practice. *Bottom-center:* result in practice with estimated disparity layers. This is a close-up of the complete result above, but the voxel volume is also restricted to the area of the statue. The voxel resolution of this area is $374 \times 400 \times 224$. In our reconstruction, 93.5% of the voxels are labeled correctly compared to the ground truth. In view of the very challenging input data, we believe this is quite detailed. *Bottom-right:* Textured reconstruction, also after layer decomposition [6]. Only the light field layer corresponding to the statue has been used to compute the texture map. As layer decomposition is not perfect, some contribution of the window color is still visible. (Color figure online)

of which has a slightly different opacity. The scene was captured in a 270 degree angle. In each view, some parts of the interior object were occluded by non-transparent painting. To record our scene, we used a hand-held Lytro Illum plenoptic camera. We decoded and calibrated the images using the Light Field Toolbox 4.0 [28]. Each image consists of 15×15 subaperture views of 625×435 pixels.

There were two challenges involved in the scene in addition to conventional layer separation. First, the semi-transparent paint creates microfacets on the surface, which is thus not perfectly planar, causing refractions of the inner object. Second, there are reflections on the occluding surface which adds a third layer on the occluder.

For depth estimation, we employ the approach proposed in [3] based on sparse light field coding, where each EPI patch is encoded as a linear combination over a patch dictionary which consists of atoms with known disparity. From the estimates, we generate a cost volume according to the description in Sect. 4.3, using the expansion coefficients as a measure to describe certainty of the surface estimate at the respective depth. An initial estimate of camera extrinsics from SfM [5], which was then optimized over rotation, translation, triangulated features and focal length (assumed to be same for all captured images) with the proposed bundle adjustment framework. All results can be observed in Fig. 8. Considering the challenging data, we believe the result is acceptable as an initial demonstration that the framework can deal with real-world datasets.

6   Conclusions

We propose a method of scene reconstruction based on a variational framework, which for the first time can recover scene geometry from multi-view light fields with superimposed layers. Key ingredients are a novel light field bundle adjustment based on the framework for linear pose estimation proposed in [5], as well as multi-layered depth map reconstruction via higher-order structure tensors [4] or sparse light field coding [3]. For surface recovery, we propose a global surface reconstruction framework, which can merge multi-layered hypotheses from multiple view points and recover the complete scene in a single optimization pass. The largest limitation is that we currently still require feature matches obtained under the Lambertian assumption, so that at least some parts of scene need to consist only of a single layer. Experiments demonstrate that our algorithm works for synthetic as well as real world light fields captured with a hand-held camera under very challenging conditions.

Acknowledgement. This work was supported by the ERC Starting Grant "Light Field Imaging and Analysis" (LIA 336978, FP7-2014).

References

1. Gortler, S., Grzeszczuk, R., Szeliski, R., Cohen, M.: The lumigraph. In: Proceedings of SIGGRAPH, pp. 43–54 (1996)
2. Bolles, R., Baker, H., Marimont, D.: Epipolar-plane image analysis: an approach to determining structure from motion. Int. J. Comput. Vis. **1**, 7–55 (1987)
3. Johannsen, O., Sulc, A., Goldluecke, B.: What sparse light field coding reveals about scene structure. In: Proceedings of the International Conference on Computer Vision and Pattern Recognition (2016)
4. Wanner, S., Goldluecke, B.: Reconstructing reflective and transparent surfaces from epipolar plane images. In: Weickert, J., Hein, M., Schiele, B. (eds.) GCPR 2013. LNCS, vol. 8142, pp. 1–10. Springer, Heidelberg (2013). doi:10.1007/978-3-642-40602-7_1
5. Johannsen, O., Sulc, A., Goldluecke, B.: On linear structure from motion for light field cameras. In: Proceedings of the International Conference on Computer Vision (2015)
6. Johannsen, O., Sulc, A., Goldluecke, B.: Variational separation of light field layers. In: Vision, Modelling and Visualization (VMV) (2015)
7. Wanner, S., Goldluecke, B.: Variational light field analysis for disparity estimation and super-resolution. IEEE Trans. Pattern Anal. Mach. Intell. **36**, 606–619 (2014)
8. Chandraker, M., Reddy, D., Wang, Y., Ramamoorthi, R.: What object motion reveals about shape with unknown BRDF and lighting. In: Proceedings of the International Conference on Computer Vision and Pattern Recognition (2013)
9. Tao, M., Wang, T.C., Malik, J., Ramamoorthi, R.: Depth estimation for glossy surfaces with light-field cameras. In: Proceedings of the European Conference on Computer Vision Workshops (2014)
10. Chen, C., Lin, H., Yu, Z., Kang, S.B., J., Y.: Light field stereo matching using bilateral statistics of surface cameras. In: Proceedings of the International Conference on Computer Vision and Pattern Recognition (2014)
11. Stich, T., Tevs, A., Magnor, M.: Global depth from epipolar volumes - a general framework for reconstructing non-Lambertian surfaces. In: 3DPVT (2006)
12. Laurentini, A.: The visual hull concept for visual-based image understanding. IEEE Trans. Pattern Anal. Mach. Intell. **16**, 150–162 (1994)
13. Zuo, X., Du, C., Wang, S., Zheng, J., Yang, R.: Interactive visual hull refinement for specular and transparent object surface reconstruction. In: Proceedings of the International Conference on Computer Vision (2015)
14. Kutulakos, K.N., Seitz, S.M.: A theory of shape by space carving. Int. J. Comput. Vis. **38**, 199–218 (2000)
15. Yang, Y., Pollefeys, M., Welch, G.: Dealing with textureless regions and specular highlights - a progressive space carving scheme using a novel photo-consistency measure. In: Proceedings of the International Conference on Computer Vision (2003)
16. Faugeras, O., Keriven, R.: Variational principles, surface evolution, PDE's, level set methods, and the stereo problem. IEEE Trans. Image Process. **7**, 336–344 (1998)
17. Jin, H., Soatto, S., Yezzi, A.: Multi-view stereo reconstruction of dense shape and complex appearance. Int. J. Comput. Vis. **63**, 175–189 (2005)
18. Yu, T., Narendra, A., Chen, W.C.: SDG cut: 3D reconstruction of non-Lambertian objects using graph cuts on surface distance grid. In: Proceedings of the International Conference on Computer Vision and Pattern Recognition (2006)

19. Ihrke, I., Goldluecke, B., Magnor, M.: Reconstructing the geometry of flowing water. In: Proceedings of the International Conference on Computer Vision, pp. 1055–1060 (2005)
20. Goldman, D., Curless, B., Hertzmann, A., Seitz, S.: Shape and spatially-varying BRDFs from photometric stereo. IEEE Trans. Pattern Anal. Mach. Intell. **32**, 1060–1071 (2010)
21. Vogiatzis, G., Hernandez, C., Cipolla, R.: Reconstruction in the round using photometric normals and silhouettes. In: Proceedings of the International Conference on Computer Vision and Pattern Recognition (2006)
22. Kolev, K., Pock, T., Cremers, D.: Anisotropic minimal surfaces integrating photoconsistency and normal information for multiview stereo. In: Proceedings of the European Conference on Computer Vision (2010)
23. Cremers, D., Kolev, K.: Multiview stereo and silhouette consistency via convex functionals over convex domains. IEEE Trans. Pattern Anal. Mach. Intell. **33**, 1161–1174 (2011)
24. Federer, H.: Geometric Measure Theory. Springer, Heidelberg (1969)
25. Zach, C., Pock, T., Bischof, H.: A globally optimal algorithm for robust TV-L1 range image integration. In: Proceedings of the International Conference on Computer Vision (2007)
26. Graber, G., Pock, T., Bischof, H.: Online 3D reconstruction using convex optimization. In: Proceedings of the International Conference on Computer Vision Workshops (2011)
27. Zickler, T., Belhumeur, P., Kriegman, D., Stereopsis, H.: Exploiting reciprocity for surface reconstruction. Int. J. Comput. Vis. **49**, 215–227 (2002)
28. Dansereau, D., Pizarro, O., Williams, S.: Decoding, calibration and rectification for lenselet-based plenoptic cameras. In: Proceedings of the International Conference on Computer Vision and Pattern Recognition, pp. 1027–1034 (2013)
29. Pock, T., Chambolle, A.: Diagonal preconditioning for first order primal-dual algorithms in convex optimization. In: International Conference on Computer Vision (ICCV 2011) (2011)
30. Chan, T., Esedoglu, S., Nikolova, M.: Algorithms for finding global minimizers of image segmentation and denoising models. SIAM J. Appl. Math. **66**, 1632–1648 (2006)

A Dataset and Evaluation Methodology for Depth Estimation on 4D Light Fields

Katrin Honauer[1]([✉]), Ole Johannsen[2],
Daniel Kondermann[1], and Bastian Goldluecke[2]

[1] HCI, Heidelberg University, Heidelberg, Germany
{katrin.honauer,daniel.kondermann}@iwr.uni-heidelberg.de
[2] University of Konstanz, Konstanz, Germany
{ole.johannsen,bastian.goldluecke}@uni-konstanz.de

Abstract. In computer vision communities such as stereo, optical flow, or visual tracking, commonly accepted and widely used benchmarks have enabled objective comparison and boosted scientific progress.

In the emergent light field community, a comparable benchmark and evaluation methodology is still missing. The performance of newly proposed methods is often demonstrated qualitatively on a handful of images, making quantitative comparison and targeted progress very difficult. To overcome these difficulties, we propose a novel light field benchmark. We provide 24 carefully designed synthetic, densely sampled 4D light fields with highly accurate disparity ground truth. We thoroughly evaluate four state-of-the-art light field algorithms and one multi-view stereo algorithm using existing and novel error measures.

This consolidated state-of-the art may serve as a baseline to stimulate and guide further scientific progress. We publish the benchmark website http://www.lightfield-analysis.net, an evaluation toolkit, and our rendering setup to encourage submissions of both algorithms and further datasets.

1 Introduction

Over the last decade, light field analysis has grown from a niche topic to an established part of the computer vision community. While in its most general form, the light field captures the radiance distribution for every ray passing through every point in space-time [1], one usually simplifies this to a 4D parametrization, where one essentially considers a dense collection of pinhole views with parallel optical axes, sampled on a rectangular grid in a 2D plane. The key difference to the classical multi-view scenario is the dense and regular sampling, which allows to develop novel and highly accurate methods for depth reconstruction [2–6], which can correctly take occlusions into account to recover fine details [7].

K. Honauer and O. Johannsen contributed equally.

Electronic supplementary material The online version of this chapter (doi:10. 1007/978-3-319-54187-7_2) contains supplementary material, which is available to authorized users.

© Springer International Publishing AG 2017
S.-H. Lai et al. (Eds.): ACCV 2016, Part III, LNCS 10113, pp. 19–34, 2017.
DOI: 10.1007/978-3-319-54187-7_2

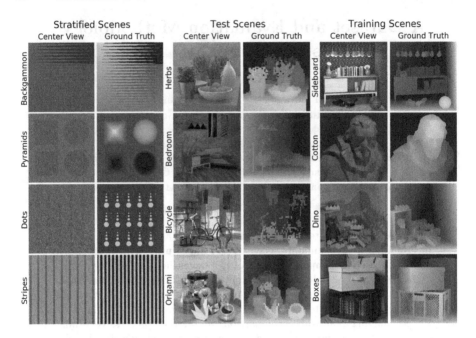

Fig. 1. We present a new light field benchmark consisting of four stratified, four test, and four training scenes. The stratified scenes are designed to pose specific, isolated challenges with spatially increasing difficulty. To warrant a deep and comprehensive understanding of algorithm performance, we quantify a variety of characteristics such as foreground fattening, texture sensitivity, and robustness to noise. For the stratified and training scenes, we provide high resolution ground truth disparity maps, normal maps and 3D depth point clouds. The same information is provided for twelve additional scenes (see Fig. 2).

In more mature vision communities such as the stereo or tracking community, standard benchmarks of sufficient variety and difficulty have proven their fundamental importance for targeted development and objective judgment of the overall progress in the respective field. Detailed evaluations and comparisons of the precise strengths and weaknesses of different methods are guiding research and thus stimulating progress. However, such a common benchmark is currently lacking in the light field community. For this reason, recent papers often resort to showing qualitative results on real-world datasets to showcase their improved results [2,6,7], but performance is very difficult to judge without ground truth. In those cases where a numeric evaluation is performed, the specific ground truth data set and/or quality metrics often vary wildly between papers, again making objective comparison hard [2–7]. Moreover, parameters might be fine-tuned towards a certain quality metric, e.g. more smoothing in general improves the mean squared error at the expense of per-pixel accuracy. Finally, there is currently no benchmarking website which offers the opportunity of a common gathering point for datasets and online performance comparison.

Contributions. The light field benchmark we present in this paper is designed to remedy the aforementioned shortcomings. In this first iteration, we focus solely on the problem of depth estimation for Lambertian scenes, although we provide some scenes with specular reflections to offer more of a challenge. Our main contributions can be summarized as follows:

– We introduce a new synthetic dataset with 24 carefully designed scenes, which overcomes technical shortcomings of previous datasets.
– We propose novel error measures and evaluation modalities enabling comprehensive and detailed characterizations of algorithm results.
– We present an initial performance analysis of four state-of-the-art light field algorithms and one multi-view stereo algorithm.
– We publish a benchmarking website and an evaluation toolkit to provide researchers with the necessary tools to facilitate algorithm evaluation.

We consider this benchmark as a first step towards a joint effort of the light field community to develop a commonly accepted benchmark suite. All researchers in the field are kindly invited to contribute existing and future algorithms, datasets, and evaluation measures.

2   Related Work

Existing Light Field Datasets. The available datasets can be grouped into synthetic light fields, real world light fields captured with a plenoptic camera (usually a Lytro Illum) and real world scenes captured with a camera array or gantry. We are aware of multiple smaller and larger collections, in particular the Stanford Light Field Archive [8], the Synthetic Light Field Archive [9], a collection of Lytro images [10], the 3D High-Resolution Disney Dataset [11], and the New Light Field Image Dataset [12]. All these datasets have in common that no ground truth data is available, making them hard to use for precise benchmarking.

To our knowledge, the only collection of light fields which comes with ground truth depth and an open benchmark is the HCI Light Field Benchmark by Wanner et al. [13]. They provide synthetic as well as real world 4D light fields including ground truth. In the past, this benchmark stimulated the growth of multiple light field algorithms, but it now reaches a point where we think it can no longer satisfy the needs of the light field community. This is due to three major drawbacks. First, their ground truth gives around 130 distinct depth labels, yielding a maximum evaluation accuracy which is already surpassed by state-of-the-art algorithms. Second, the ground truth data contains errors in the form of wrong pixels, as well as inaccuracies at occlusion boundaries, which are a key part of depth accuracy evaluation. Third, due to the way the light fields were rendered, a systematic noise pattern is present that is the same for all views.

Insights from Other Popular Benchmarks. In more mature communities such as stereo, optical flow, and visual tracking, benchmarks play a fundamental role in boosting scientific progress: they help consolidate existing research [14] and guide the community towards open challenges (e.g. [15] for large motion optical flow, [16] for automotive stereo). Building upon the experience from these successful benchmarks, we identify three key insights for the design of our benchmark. From the Visual Object Tracking Community[1], we conclude that scientific progress thrives if benchmarks are seen as a joint effort by and service for the community [17,18]. We therefore encourage researchers to not only contribute algorithms but also datasets and evaluation methods. Second, there is no single best algorithm: algorithms have different strengths and weaknesses and may be used for applications with very different requirements. We therefore use multi-dimensional performance evaluation with carefully designed metrics to reflect this diversity [19–21]. Third, as methods improve, benchmarks may no longer be suitable to differentiate algorithm performance. They may even hinder scientific progress by unintentionally stimulating overfitting. We hence designed our benchmark for a limited lifespan, focusing on those challenges where current algorithms struggle with. Together with the community, similar to [19,22], we plan to regularly maintain the dataset and add new scenes when necessary.

3   Considerations on Benchmark Design

Light field and multi-view stereo algorithms find more and more applications in real world challenges such as the movie set reconstruction and industrial optical inspection. Often, medical or automotive technologies are even safety-relevant. Designing a useful benchmark requires addressing the following four aspects:

(1) Benchmark Purpose. Test datasets should be compact to minimize dataset creation cost, maximize information gain, and reduce benchmarking efforts. Researchers across different fields of computer vision agree that a systematic, considerate compilation of imagery is necessary to allow for specific, meaningful algorithm evaluation [19,23–26]. Unintended biases can occur in dataset creation, causing e.g. an overemphasis on smooth surfaces in the scenes [23]. Using top-down approaches such as requirements engineering [25] or bottom-up methods such as HAZOP studies [23] can alleviate this risk. As shown below, state of the art algorithms often struggle with geometry and texture challenges. Hence,

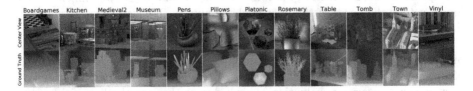

Fig. 2. We provide 12 additional scenes with ground truth. They are not part of the official benchmark but can be used for algorithm development and evaluation.

---

[1] http://www.votchallenge.net/.

we focus on five challenges, namely occlusion boundaries, fine structures, low texture, smooth surfaces and camera noise.

(2) Scene Design. Simple scenes focusing on a single challenge allow to decouple the performance analysis of each individual challenge [24]. Thus, we introduce four light fields with controlled parameters for a fixed challenge combination: Backgammon, Dots, Pyramids and Stripes (see Fig. 1). We call these scenes *stratified* since their goal is to create quantifiable challenges which can be used to re-weight performance metrics based on real-world, non-stratified data. To gradually increase each challenge, the scenes exhibit spatially increasing difficulties. This allows both for immediate visual inspection as well as quantitative comparison between algorithms. Yet, complex real-world scenes contain all the challenges in potentially statistically significant spatial combinations. We therefore create additional, photorealistically rendered scenes (see Fig. 1). This suppresses the chance of overfitting parameters to a certain challenge and helps to obtain an intuition on real-world performance.

(3) Dataset Acquisition. To date, no measurement technology exists to record real light fields with sufficiently accurate ground truth. Using computer vision algorithms to create ground truth for computer vision algorithms defeats the purpose of benchmarking. Recent research shows promising results that rendering can be a valid approach [15,27,28]. We therefore use rendered scenes, building on the advantages of near-perfect ground truth accuracy and the option to systematically vary scene parameters.

(4) Benchmarking Methodology. We adopt the approach of Scharstein et al. [19] and divide our dataset into test, training, and additional scenes (see Figs. 1 and 2). To be listed on the public benchmark table, we ask participants to submit their algorithm results and runtimes on the twelve scenes as depicted on Fig. 1. Participants may use the input data and disparity ranges as provided on the website. We further provide an evaluation toolkit which contains: (i) file IO methods for Matlab and Python (ii) a submission validation script (iii) evaluation code to compute and visualize the metrics on the stratified and training scenes. All metric scores and visualizations will be computed on our server and displayed publicly on the benchmark table. The ground truth of the training and stratified scenes may be used to optimize parameter settings. We do not publish ground truth for the four photorealistic test scenes. As in [19] algorithm results of the training scenes will be available for download in full resolution. The twelve additional scenes with full ground truth are not part of the benchmark. They are shared with the community to support further algorithm development. We refer to http://www.lightfield-analysis.net for technical submission details.

4 Description of Dataset and Metrics

In this section, we present the scenes and corresponding error metrics resulting from our theoretical considerations on scene content and performance analysis.

4.1 Technical Dataset Details

The scenes were created with Blender [29] using the internal renderer for the stratified scenes and the Cycles renderer for the photorealistic scenes. We built the light field setup in a way such that all cameras are shifted towards a common focus plane while keeping the optical axes parallel. Thus, zero disparity does not correspond to infinite depth. Most scene content lies within a range of $-1.5\,\text{px}$ and $1.5\,\text{px}$, though disparities on some scenes are up to $3\,\text{px}$.

For each scene, we provide 8 bit light fields ($9 \times 9 \times 512 \times 512 \times 3$), camera parameters, and disparity ranges. For the stratified and training scenes we further provide evaluation masks and 16 bit ground truth disparity maps in two resolutions ($512 \times 512\,\text{px}$ and $5120 \times 5120\,\text{px}$). We use the high resolution ground truth to accurately evaluate algorithm results at fine structures and depth discontinuities. The textures of the stratified scenes are generated from Gaussian noise to minimize potential unwanted interference of texture irregularities with the actual challenges in the scene. A detailed technical description of the data generation process and the source code of the Blender add-on can be found at http://www.lightfield-analysis.net.

4.2 General Evaluation Measures

Algorithms often have different strengths and weaknesses, such as overall accuracy or sensitivity to fine structures, which may be prioritized very differently depending on the application. In the spirit of [21], we quantify a variety of characteristics to warrant a deep and comprehensive understanding of individual algorithm performance. We provide the commonly used MSE * 100 and Bad-Pix(0.07) metrics as well as Bumpiness and scene specific adaptions of these metrics. The adaptations are introduced together with the respective scenes. The general MSE, BadPix, and Bumpiness metrics are defined as follows:

Given an estimated disparity map d, the ground truth disparity map gt and an evaluation mask M, MSE is quantified as

$$\text{MSE}_{\mathcal{M}} = \frac{\sum\limits_{x \in \mathcal{M}} (d(x) - gt(x))^2}{|\mathcal{M}|} * 100 \tag{1}$$

and BadPix is quantified as

$$\text{BadPix}_{\mathcal{M}}(t) = \frac{|\{x \in \mathcal{M} : |d(x) - gt(x)| > t\}|}{|\mathcal{M}|}. \tag{2}$$

To measure algorithm performance at smooth planar and curved surfaces we further define $f = d - gt$ to quantify Bumpiness as

$$\text{Bumpiness} = \frac{\sum\limits_{x \in \mathcal{M}} \min(0.05, \| \text{H}_f(x) \|_F)}{|\mathcal{M}|} * 100. \tag{3}$$

Hence, the bumpiness metric solely focuses on the smoothness of an estimation but does not assess misorientation or offset. These properties are covered by other metrics.

4.3   Scene Descriptions with Corresponding Evaluation Measures

Backgammon. This scene (see Fig. 1) is designed to assess the interplay of fine structures, occlusion boundaries and disparity differences. It consists of two parallel, slanted background planes and one foreground plane which is inversely slanted. The foreground plane is jagged to create increasingly thin foreground structures and increasingly fine background slits. On Backgammon, we quantify Foreground Fattening which is defined at occlusion boundaries on a mask M that only includes background pixels as

$$\text{FG\_Fattening} = \frac{|\{x \in \mathcal{M} : d(x) > h\}|}{|\mathcal{M}|}, \tag{4}$$

where $h = (BG + FG)/2$. Thus, Foreground Fattening calculates the fraction of pixels that are closer to the foreground than to the background and should have been estimated as background. Similarly, Foreground Thinning is defined on a mask M that only includes foreground pixels as

$$\text{FG\_Thinning} = \frac{|\{x \in \mathcal{M} : d(x) < h\}|}{|\mathcal{M}|}, \tag{5}$$

i.e. Foreground Thinning calculates the fraction of pixels that are closer to the background than to the foreground.

Pyramids. With this scene, we assess algorithm performance on convex versus concave as well as rounded versus planar geometry. The upper hemisphere and pyramid stick out of the middle plane whereas the lower hemisphere and pyramid are embedded in the plane. We quantify surface reconstruction quality by computing Bumpiness as defined in Eq. 3 on masks for the fronto-parallel plane and the slanted surfaces of the objects respectively.

Dots. This scene is designed to assess the effect of camera noise on the reconstruction of objects of varying size. The image features 15 identical grid cells. Each cell consists of 9 increasingly smaller coplanar circles. To approximate thermal and shot noise, we add Gaussian noise with variances growing linearly between 0.0 and 0.2 in row-major order. We quantify robustness against noise by computing the MSE on the background plane. We further quantify sensitivity to small geometries by computing the percentage of detected dots. A dot counts as detected if the majority of its local disparity estimates is distinguishable from the background by being in a BadPix range of 0.4 px to the ground truth dot.

Stripes. This scene is used to assess the influence of texture and contrast at occlusion boundaries. It consists of a fronto-parallel background plane and 16 coplanar stripes. The amount of texture on the background plane is gradually increasing from left to right. Likewise, the vertical stripes are increasingly textured from the bottom to the top of the image. The stripes feature alternating intensities with dark, high contrast stripes and bright, low contrast stripes.

To quantify performance, we define three types of image regions and compute BadPix(0.07) on each region individually: First, we use the no-occlusion areas on

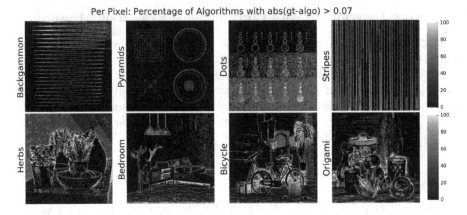

Fig. 3. The heatmaps illustrate local scene difficulty. Per pixel, they show the percentage of algorithms with a disparity error > 0.07 px. Algorithms struggle particularly with fine structures, noise, and occlusion areas.

the stripes and on the background for low texture evaluation. Second, we use the dark stripes and their occlusion areas to quantify performance at high contrast occlusion boundaries. Similarly, we use bright stripes and their occlusion areas to quantify performance at low contrast occlusion boundaries.

Photorealistic Scenes. We designed the photorealistic scenes to allow for performance evaluation on fine structures, complex occlusion areas, slanted planar surfaces, and continuous non-planar surfaces. The scenes contain various combinations of these challenges and allow to obtain an intuition of algorithm performance on real world scenes. For quantitative performance analysis, we use masks for different challenge regions. Apart from the overall MSE and Bad-Pix(0.07) scores, we compute the BadPix(0.07) score at occlusion areas. We further quantify smoothness at planar and non-planar continuous surfaces by computing Bumpiness scores on the respective image areas. Furthermore, we compute Thinning (0.15) and Fattening (-0.15) at fine structures by computing adjusted BadPix scores as follows:

$$\text{Thinning}_{\mathcal{M}}(t) = \frac{|\{x \in \mathcal{M} : gt(x) - d(x) > t\}|}{|\mathcal{M}|} \tag{6}$$

where $\mathcal{M}$ is a mask for fine structure pixels and

$$\text{Fattening}_{\mathcal{M}}(t) = \frac{|\{x \in \mathcal{M} : gt(x) - d(x) < t\}|}{|\mathcal{M}|} \tag{7}$$

where $\mathcal{M}$ is a mask for pixels surrounding fine structures.

5   Experimental Validation of Dataset and Metrics

5.1   Evaluation of Scene Content

In order to verify our reasoning on challenging scene characteristics, we analyze local scene difficulty as shown in Fig. 3. Challenging regions on the heatmaps (bright) correlate with our intended challenges as described in Sect. 4. On the stratified scenes, the fine gaps on Backgammon, low texture areas on Stripes and noisy regions on Dots feature low algorithm performance. On the photorealistic scenes, complex occlusions on Herbs, fine structures on Bedroom, and fine structure grids on Bicycle represent the most challenging image regions. By contrast, the well-textured fronto-parallel surface of Pyramids, the noise-free area on Dots as well as smooth and high-texture regions on the photorealistic scenes feature good algorithm performance.

5.2   Evaluation of Performance Measures

In this section, we examine whether our metrics appropriately quantify algorithm performance on the stratified and photorealistic scenes. We show algorithm results of one multi-view algorithm (MV) and four light field algorithms (LF, LF_OCC, EPI2, EPI1). In order to keep the focus on the metrics, we treat the algorithms as black boxes until Sect. 6. For additional results we refer to the supplemental material.

Backgammon: Fine Structures, Thin Gaps and Occlusions. The algorithm results on Fig. 4 show that algorithms do indeed struggle at gradually finer peaks and especially at thin gaps of the Backgammon scene. The depicted fattening and thinning scores quantify the respective algorithm performance appropriately. More fattening occurs at occlusion areas on the top part of the image,

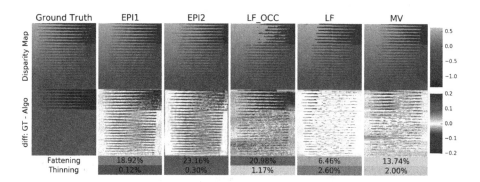

Fig. 4. All algorithms struggle with reconstructing the background depth of the narrow gaps on the left side of the image. LF_OCC and EPI2 have the strongest fattening, for LF_OCC it is concentrated between the upper bars and for EPI2 it is uniformly distributed around each bar.

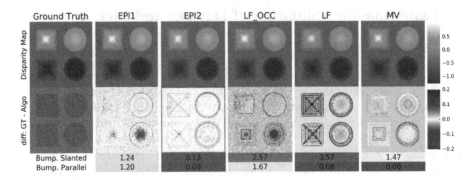

Fig. 5. The bumpiness scores correctly reflect the observation that LF produces very smooth estimates on the fronto-parallel plane but heavily staircased estimates on the slanted object surfaces. On both types of surfaces LF_OCC results are bumpy and EPI2 results are smooth.

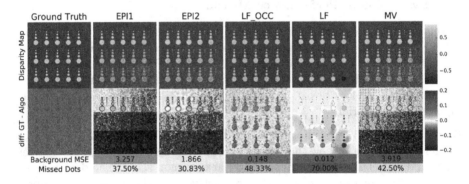

Fig. 6. The performance of most algorithms degrades by increasing levels of noise. Robustness via strong regularization is traded for low sensitivity on the smaller dots (LF) and vice versa (EPI1).

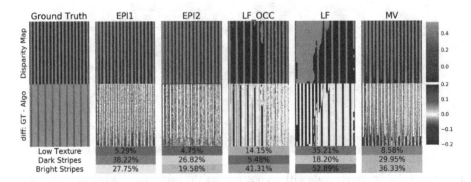

Fig. 7. Algorithms struggle with the increasingly low texture towards the bottom of the image. As reflected by our metrics, LF_OCC and LF handle dark, high contrast stripes much better than bright, low contrast stripes.

where disparity distances are large (see LF_OCC and MV). In this area, background pixels which are visible from the center view are occluded in many other views. For very thin gaps, an epipolar line belonging to a background point might then be occluded at both ends.

Pyramids: Slanted and Convex vs. Concave Surfaces. Algorithms face various difficulties on the Pyramids scene (see Fig. 5), such as systematic offset on the middle plane, bumpy surfaces and inaccurate object boundaries. The continuous disparity ranges of the slanted surfaces are particularly challenging for algorithms which estimate discrete disparity labels such as LF. The bias on the middle planes is also caused by a limited number of disparity steps where no step matches the disparity of the plane. The depicted bumpiness scores for slanted surfaces correctly identify smooth and staircased disparity maps.

Dots: Noise and Tiny Objects. Results on Fig. 6 show that algorithms struggle either with reconstructing small dots or with reconstructing smooth and accurate background planes. LF_OCC and LF robustly yield accurate results on the background, whereas EPI1 and MV are strongly affected by artifacts due to noise. In contrast, LF applies strong regularization, causing poor scores for the number of reconstructed dots; EPI1 and EPI2 perform better. These effects show that the complementary metrics of this scene nicely challenge the algorithms to find a good trade-off between regularization and fine structure sensitivity.

Stripes: Texture and Contrast at Occlusions. Algorithms struggle with correctly computing disparities at the low contrast boundaries of the bright stripes and on the low texture regions towards the bottom of the image (see Fig. 7). Our metrics quantify that algorithms such as LF, which use image gradients as priors for their occlusion handling, almost completely miss the low contrast stripes. While EPI2 shows decent performance on both types of occlusion boundaries, LF_OCC performs almost an order of magnitude better on high contrast stripes as compared to low contrast stripes.

Photorealistic Scenes. Figure 8 shows three sample algorithm results for the Herbs scene and a cutout of the Bedroom scene together with region specific challenge evaluations. MSE scores on the Herbs scene are rather similar and relatively high for all three algorithms. On this scene, high errors at the scene background and on the thyme structures reduce the expressiveness of the MSE metric. With our evaluation methods, we specifically quantify performance at smooth surfaces. The bumpiness metric is useful to show that EPI2 features smooth results, whereas the locally smooth but stepped results of LF or the noisy results of LF_OCC are not suitable in case accurate surface normals are needed per application requirements.

On the Bedroom cutout, MSE scores are much lower. Since fine structures only make up 2.8% of the total cutout, performance on these image regions is poorly reflected by MSE or BadPix scores. Hence quantifying thinning and fattening at fine structures gives additional, more specific characteristics of algorithm performance. LF may have the lowest MSE but it misses most of the fine

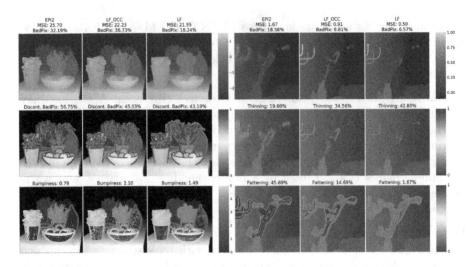

Fig. 8. Our region specific evaluation on Herbs reveals that EPI2 features the smoothest surfaces but the poorest discontinuities, whereas MSE scores for all three algorithms are close to each other. On the Bedroom cutout we quantify that LF features high fine structure thinning and low fattening whereas EPI2 and LF_OCC miss fewer structures but tend to fattening.

structures which is correctly represented by our thinning scores. By contrast, EPI2 and LF_OCC have better thinning scores but show very strong fattening.

6   Baseline Evaluation of Existing Light Field Algorithms

Experimental Setup. We evaluate four state-of-the art light field algorithms and one multi-view stereo approach. The algorithms were selected based on demonstrated state-of-the-art performance and source code availability.

LF [5] poses depth estimation as a multilabel problem which is later refined by locally fitting a quadratic function. For subpixel accurate shifting, the phase-shift theorem is used. LF_OCC [7] also poses depth estimation as a multi-label problem. As an occlusion cost, boundary orientation in the center view is compared against boundary orientation of so-called scene cam patches, which are constructed from all observed pixels for a single scene point. EPI1 [6] builds a dictionary consisting of atoms of fixed known disparity. By solving a sparse coding problem, the dictionary is employed to recover disparity for each epipolar plane image patch. EPI2 [30] employs the structure tensor to estimate the orientation of patches on the epipolar plane image. A weighted variational regularization is performed to obtain a smooth result. Finally, MV is a lab code implementation of a multi view stereo approach.

Multidimensional Algorithm Characterization. In Sect. 5, we used black-box representations of the algorithm results to show that our scenes and metrics

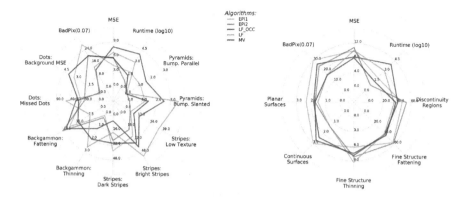

Fig. 9. The radar charts summarize all scores of the proposed metrics on the stratified (left) and photorealistic (right) test scenes. Lower scores in the center represent better performance. Neither stratified nor photorealistic scenes can be perfectly solved with a single best algorithm outperforming all others.

are capable of quantifying specific strengths and weaknesses of given algorithms. Here, we demonstrate how our scenes and metrics can be used to obtain an in-depth understanding of algorithm performance. In particular, we show how algorithms can be compared given a range of various scores instead of a single MSE score. Figure 9 summarizes all scores computed on the five algorithms, eight scenes and all associated metrics. Each radar axis represents one metric with zero in the center representing perfect performance.

Neither stratified nor photorealistic scenes can be perfectly solved with a single best algorithm outperforming all others. The radar charts illustrate that every algorithm has different strengths and weaknesses. Thus, if application data mostly contains only a subset of the challenges, optimal choice of algorithm can differ considerably. As algorithm rankings on the MSE and BadPix axes differ from rankings on other performance characteristics, our metrics indeed quantify specific properties which cannot be inferred by simply computing the MSE. For example, the multi-view stereo approach MV scores best in MSE and BadPix over all photorealistic scenes, but in no other dimension.

Furthermore, performance differences and changes in relative rankings per metric are much higher on the stratified scenes than on the photorealistic scenes. Our stratified scenes are very focused on measuring a specific algorithm characteristic, with difficulty levels ranging from feasible to almost impossible. Algorithm performance deteriorates at different levels, allowing quantification of even small differences in top performing algorithms.

Insights on Specific Algorithm Performance. The algorithms EPI1 and EPI2 are similar in that they both work on epipolar images. On the radar charts they perform similarly well on most scores of the stratified scenes, but relatively poor in the photorealistic scenes. Our metrics quantify that EPI2 outperforms all other algorithms on reconstructing smooth surfaces in the stratified scenes.

However, on the photorealistic scenes, EPI2 does not feature good scores on planar and continuous surface reconstruction. We speculate that EPI2 is very good on specific challenges but not very robust when different challenges are combined in more complex scenes.

By contrast, LF features solid performance on the photorealistic scenes, but very poor performance on many metrics of the stratified scenes. The strong regularization of LF seems to be good for scoring well on the photorealistic scenes due to the spatial distribution of the contained challenges.

LF_OCC is the only algorithm explicitly handling occlusions. Indeed, it demonstrates good performance at discontinuities and fine structure in particular on the photorealistic scenes, as well as at the high contrast stripes on the stratified scene. LF_OCC performance is much lower on the low contrast stripes since it uses image gradients for occlusion handling.

Our dataset reveals several directions for future research: based on the results shown in Figs. 3 and 9, we conclude that occlusion areas, fine structures, the reconstruction of slanted surfaces, and low texture are still unsolved challenges for light field algorithms. Additionally, while most algorithms perform well on some characteristics, there is no algorithm with solid performance on all characteristics simultaneously.

7   Conclusion

We presented and carefully justified a novel light field benchmark consisting of 4 stratified and 20 photorealistic light field scenes, a solid evaluation procedure, and a baseline evaluation to seed a public benchmark.

We thoroughly evaluated four state-of-the-art light field algorithms and one multi-view stereo algorithm using our proposed evaluation approach. Thereby, we showed that our dataset highlights open challenges for depth reconstruction algorithms. Moreover, the careful design of our dataset allowed for a structured, quantitative and specific performance analysis of the algorithms at hand. Our evaluation approach facilitated sophisticated and detailed comparisons between the strengths and weaknesses of different algorithms. The presented scenes and evaluation methods are available at http://www.lightfield-analysis.net. We encourage researchers to contribute not only algorithms but also datasets and evaluation methods to this benchmark.

In this paper we focused on the geometrical aspects of depth estimation from light fields. In future work we plan to extend the benchmark to include more non-Lambertian materials.

Acknowledgment. This work was supported by the ERC Starting Grant "Light Field Imaging and Analysis" (LIA 336978, FP7-2014), the Heidelberg Collaboratory for Image Processing (Institutional Strategy ZUK49, Measure 6.4) and the AIT Vienna, Austria.

References

1. Levoy, M.: Light fields and computational imaging. Computer **39**, 46–55 (2006)
2. Tao, M., Hadap, S., Malik, J., Ramamoorthi, R.: Depth from combining defocus and correspondence using light-field cameras. In: Proceedings of the International Conference on Computer Vision (2013)
3. Wanner, S., Goldluecke, B.: Variational light field analysis for disparity estimation and super-resolution. IEEE Trans. Pattern Anal. Mach. Intell. **36**, 606–619 (2014)
4. Heber, S., Pock, T.: Shape from light field meets robust PCA. In: Fleet, D., Pajdla, T., Schiele, B., Tuytelaars, T. (eds.) ECCV 2014. LNCS, vol. 8694, pp. 751–767. Springer, Cham (2014). doi:10.1007/978-3-319-10599-4_48
5. Jeon, H., Park, J., Choe, G., Park, J., Bok, Y., Tai, Y., Kweon, I.: Accurate depth map estimation from a lenslet light field camera. In: Proceedings of the International Conference on Computer Vision and Pattern Recognition (2015)
6. Johannsen, O., Sulc, A., Goldluecke, B.: What sparse light field coding reveals about scene structure. In: Proceedings of the IEEE Conference on Computer Vision and Pattern Recognition, pp. 3262–3270 (2016)
7. Wang, T., Efros, A., Ramamoorthi, R.: Occlusion-aware depth estimation using light-field cameras. In: Proceedings of the IEEE International Conference on Computer Vision, pp. 3487–3495 (2015)
8. Wilburn, B., Joshi, N., Vaish, V., Talvala, E.V., Antunez, E., Barth, A., Adams, A., Horowitz, M., Levoy, M.: High performance imaging using large camera arrays. ACM Trans. Graph. (TOG) **24**, 765–776 (2005). ACM. http://lightfield.stanford.edu/
9. Marwah, K., Wetzstein, G., Bando, Y., Raskar, R.: Compressive light field photography using overcomplete dictionaries and optimized projections. ACM Trans. Graph. (Proc. SIGGRAPH) **32**, 1–11 (2013). http://web.media.mit.edu/ gordonw/SyntheticLightFields/index.php
10. Mousnier, A., Vural, E., Guillemot, C.: Partial light field tomographic reconstruction from a fixed-camera focal stack. arXiv preprint arXiv:1503.01903 (2015). https://www.irisa.fr/temics/demos/lightField/index.html
11. Kim, C., Zimmer, H., Pritch, Y., Sorkine-Hornung, A., Gross, M.H.: Scene reconstruction from high spatio-angular resolution light fields. ACM Trans. Graph. **32**, 73:1–73:12 (2013). https://www.disneyresearch.com/project/lightfields/
12. Rerabek, M., Ebrahimi, T.: New light field image dataset. In: 8th International Conference on Quality of Multimedia Experience (QoMEX). Number EPFL-CONF-218363 (2016)
13. Wanner, S., Meister, S., Goldluecke, B.: Datasets and benchmarks for densely sampled 4D light fields. In: Vision, Modelling and Visualization (VMV) (2013)
14. Scharstein, D., Szeliski, R.: A taxonomy and evaluation of dense two-frame stereo correspondence algorithms. Int. J. Comput. Vis. **47**, 7–42 (2002)
15. Butler, D.J., Wulff, J., Stanley, G.B., Black, M.J.: A naturalistic open source movie for optical flow evaluation. In: Fitzgibbon, A., Lazebnik, S., Perona, P., Sato, Y., Schmid, C. (eds.) ECCV 2012. LNCS, vol. 7577, pp. 611–625. Springer, Heidelberg (2012). doi:10.1007/978-3-642-33783-3_44
16. Geiger, A., Lenz, P., Urtasun, R.: Are we ready for autonomous driving? The KITTI vision benchmark suite. In: Proceedings of the International Conference on Computer Vision and Pattern Recognition, pp. 3354–3361 (2012)
17. Kristan, M., Matas, J., Leonardis, A., Felsberg, M., Cehovin, L., Fernandez, G., Vojir, T., Hager, G., Nebehay, G., Pflugfelder, R.: The visual object tracking VOT2015 challenge results. In: Proceedings of the ICCV, pp. 1–23 (2015)

18. Kristan, M., Pflugfelder, R., Leonardis, A., Matas, J., Porikli, F., Čehovin, L., Nebehay, G., Fernandez, G., Vojir, T.: The VOT2013 challenge: overview and additional results (2014)
19. Scharstein, D., Hirschmüller, H., Kitajima, Y., Krathwohl, G., Nešić, N., Wang, X., Westling, P.: High-resolution stereo datasets with subpixel-accurate ground truth. In: Jiang, X., Hornegger, J., Koch, R. (eds.) GCPR 2014. LNCS, vol. 8753, pp. 31–42. Springer, Cham (2014). doi:10.1007/978-3-319-11752-2_3
20. Kristan, M., Matas, J., Leonardis, A., Vojir, T., Pflugfelder, R., Fernandez, G., Nebehay, G., Porikli, F., Cehovin, L.: A novel performance evaluation methodology for single-target trackers (2015)
21. Honauer, K., Maier-Hein, L., Kondermann, D.: The HCI stereo metrics: geometry-aware performance analysis of stereo algorithms. In: Proceedings of the ICCV, pp. 2120–2128 (2015)
22. Menze, M., Geiger, A.: Object scene flow for autonomous vehicles. In: Proceedings of the International Conference on Computer Vision and Pattern Recognition (2015)
23. Zendel, O., Murschitz, M., Humenberger, M., Herzner, W.: CV-HAZOP: introducing test data validation for computer vision. In: Proceedings of the ICCV (2015)
24. Haeusler, R., Kondermann, D.: Synthesizing real world stereo challenges. In: Weickert, J., Hein, M., Schiele, B. (eds.) GCPR 2013. LNCS, vol. 8142, pp. 164–173. Springer, Heidelberg (2013). doi:10.1007/978-3-642-40602-7_17
25. Kondermann, D., Nair, R., Honauer, K., Krispin, K., Andrulis, J., Brock, A., Güssefeld, B., Rahimimoghaddam, M., Hofmann, S., Brenner, C., Jähne, B.: The HCI benchmark suite: stereo and flow ground truth with uncertainties for urban autonomous driving. In: Proceedings of the International Conference on Computer Vision and Pattern Recognition Workshops (2016)
26. Perazzi, F., Pont-Tuset, J., McWilliams, B., Gool, L.V., Gross, M., Sorkine-Hornung, A.: A benchmark dataset and evaluation methodology for video object segmentation. In: Proceedings of the International Conference on Computer Vision and Pattern Recognition (2016)
27. Meister, S., Kondermann, D.: Real versus realistically rendered scenes for optical flow evaluation. In: 14th ITG Conference on Electronic Media Technology, pp. 1–6 (2011)
28. Dosovitskiy, A., Fischer, P., Ilg, E., Hausser, P., Hazirbas, C., Golkov, V., van der Smagt, P., Cremers, D., Brox, T.: Flownet: learning optical flow with convolutional networks, pp. 2758–2766 (2015)
29. Blender Online Community: Blender - a 3D modelling and rendering package (2016)
30. Wanner, S., Goldluecke, B.: Reconstructing reflective and transparent surfaces from epipolar plane images. In: Weickert, J., Hein, M., Schiele, B. (eds.) GCPR 2013. LNCS, vol. 8142, pp. 1–10. Springer, Heidelberg (2013). doi:10.1007/978-3-642-40602-7_1

Radial Lens Distortion Correction Using Convolutional Neural Networks Trained with Synthesized Images

Jiangpeng Rong, Shiyao Huang, Zeyu Shang, and Xianghua Ying[⊠]

Key Laboratory of Machine Perception (Ministry of Education),
School of Electronic Engineering and Computer Science, Peking University,
Beijing 100871, People's Republic of China
xhying@cis.pku.edu.cn

Abstract. Radial lens distortion often exists in images taken by common cameras, which violates the assumption of pinhole camera model. Estimating the radial lens distortion of an image is an important preprocessing step for many vision applications. This paper intends to employ CNNs (Convolutional Neural Networks), to achieve radial distortion correction. However, the main issue hinder its progress is the scarcity of training data with radial distortion annotations. Inspired by the growing availability of image dataset with non-radial distortion, we propose a framework to address the issue by synthesizing images with radial distortion for CNNs. We believe that a large number of images of high variation of radial distortion is generated, which can be well exploited by deep CNN with a high learning capacity. We present quantitative results that demonstrate the ability of our technique to estimate the radial distortion with comparisons against several baseline methods, including an automatic method based on Hough transforms of distorted line images.

1 Introduction

Traditional computer vision algorithms, such as 3D reconstruction and pose estimation, usually depend critically on the assumption of the ideal pinhole camera model [1]. However, most lenses in commonly used cameras suffer from lens distortion, and the radial distortion is the most significant especially in wide angle lenses [2–6]. Several kinds of distortion models [7–11] are presented to describe the radial distortion, in literature. The division model proposed by Fitzgibbon [12] is one of the most popular models.

A large number of straight lines exist in man-made scenes, and the lines will be projected to curves, no longer straight lines in the image plane due to radial distortion [13]. Under the ideal pinhole model of projection, straight line should be projected into straight line, and it is an invariant entity. Many papers have been devoted to the so-called plumb line idea based on this fact. An automatic radial estimation method using non-overlapping circular arcs is proposed by Bukhari and Dailey [14] as collections of contiguous points in the image. Hughes et al. [15] extract vanishing points from distorted image of a checkerboard, and estimate the image

© Springer International Publishing AG 2017
S.-H. Lai et al. (Eds.): ACCV 2016, Part III, LNCS 10113, pp. 35–49, 2017.
DOI: 10.1007/978-3-319-54187-7_3

center and distortion parameters. However, the method is not suitable for images of real scenes. A method based on transforming the edges of the distorted image to a 1-D angular Hough space is proposed by Rosten and Loveland [16], which optimizes the distortion correction parameters by minimizing the entropy of the corresponding normalized histogram. An automatic method for radial lens distortion correction using an extended Hough transform of image line including one radial distortion parameter is proposed by Alemán et al. [17]. However, in some cases, especially for wide angle lenses, the corrected results are not very satisfactory [17].

In this paper, we intend to employ CNNs (Convolutional Neural Networks), to achieve radial distortion correction. Convolutional neural networks, as an influential class of models for visual recognition, are high-capacity classifiers with very large numbers of parameters which are learned from enormous training dataset. CNNs have a long history, and have advocated to a wide variety of vision tasks due to good performance, such as, on image classification and object detection. With the appearance of the large-scale ImageNet dataset [18] and the rise of GPU computing. Krizhevsky et al. [19] achieve a performance leap in image classification on the ImageNet 2012 Large-Scale Visual Recognition

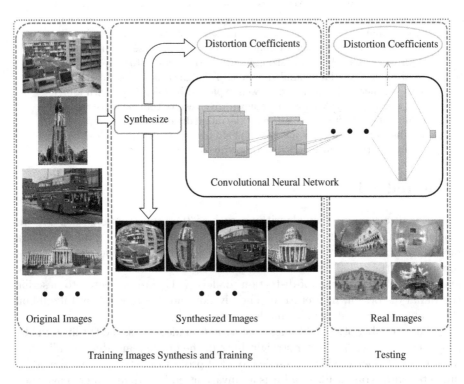

Fig. 1. System overview. We synthesize image dataset by generating radial lens distorted images simulated from real images in ImageNet. Convolutional neural network is trained to map images to the synthesized distortion coefficients. The learned framework is applied to estimate the distortion coefficients of real images.

Challenge (ILSVRC-2012), and further improve the performance by training a network on all 15 million images and 22,000 ImageNet classes.

To prepare training data for this task, in this paper, we augment real images by synthesizing thousands of distorted images which cover a high diversity in distortion coefficients. Several techniques are applied to increase the diversity of the synthesized dataset, in order to prevent the deep CNN from picking up unreliable patterns and push it to learn more robust features. To fully exploit this large-scale dataset, we modify a deep CNN specifically tailored for the distortion coefficient estimation task. We formulate a class-dependent fine-grained distortion coefficient classification problem and solve the problem with a novel loss layer adapted for this task. The results are surprising: trained on thousands of synthetic images, our CNN-based distortion coefficient estimator significantly outperforms the state-of-the-art method, tested on real image dataset. The system overview is shown in Fig. 1.

In summary, our contributions are as follows:

- We show that training CNN by massive synthetic data is an effective approach for radial distortion correction;
- Based upon existing image dataset, we propose a synthesis pipeline that generates millions of images with accurate radial distortions at negligible human cost. This pipeline is scalable, and the generated data is resistant to overfitting by CNN;
- Leveraging on the big synthesized data set, we propose a fine-grained distortion classification formulation, with a loss function encouraging strong correlation of nearby distortion coefficients.

2   Division Distortion Model of Radial Distortion

There are sufficient radial lens distortion correction methods in literature. Usually, we assume that the distortion center is known. The distance from the original distorted image point to the center of radial distortion is defined as the distorted radius r_d, and the distance corresponding to the undistorted image point is defined as the undistorted radius r_u. For a point in the distorted image $\mathbf{x}_d = [u_d, v_d, 1]^T$, and its corresponding undistorted image point $\mathbf{x}_u = [u_u, v_u, 1]^T$, we have:

$$\begin{bmatrix} u_u \\ v_u \\ 1 \end{bmatrix} \propto \begin{bmatrix} u_d \\ v_d \\ \lambda \end{bmatrix} \tag{1}$$

where λ is some unknown non-zero scale. We rewrite (1) as,

$$\begin{bmatrix} u_u \\ v_u \end{bmatrix} = \frac{1}{\lambda} \begin{bmatrix} u_d \\ v_d \end{bmatrix} \tag{2}$$

and

$$r_u = \sqrt{u_u^2 + v_u^2} = \frac{1}{\lambda} r_d = \frac{1}{\lambda} \sqrt{u_d^2 + v_d^2} \tag{3}$$

In the one-parameter division model proposed by Fitzgibbon [8], the relation between r_d and r_u is:

$$r_u = \frac{r_d}{1 + kr_d^2} \tag{4}$$

Therefore, we have:

$$\begin{bmatrix} u_u \\ v_u \end{bmatrix} = \frac{r_d}{r_u} \begin{bmatrix} u_d \\ v_d \end{bmatrix} = (1 + kr_d^2) \begin{bmatrix} u_d \\ v_d \end{bmatrix} \tag{5}$$

where k is the radial lens distortion coefficient.

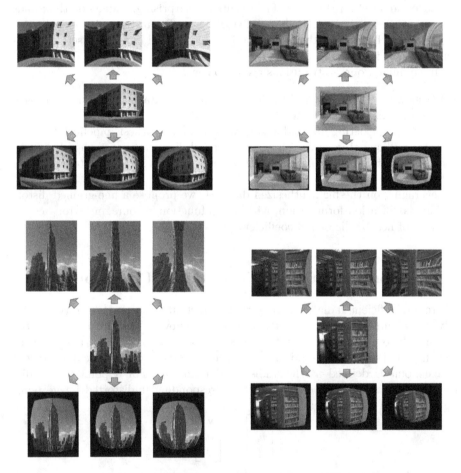

Fig. 2. Simulated distortion images with different distortion coefficients. For each group of images, three distorted images in the top row illustrate pincushion distortion with negative distortion coefficients, while the bottom three distorted images illustrate barrel distortion with positive distortion coefficients. The absolute value of distortion coefficient in each row increases from left to right for each group of images.

When the value of distortion coefficient varies, the distortion varies accordingly. The more the absolute value of k gets, the stronger radial distortion the image will express. We can easily observe that barrel distortion typically will have a negative term for k while as for pincushion distortion, the value of k will be positive, as shown in Fig. 2. For a real lens, however, the distortion coefficient k is unusually positive, thus we only focus on the negative term for the distortion coefficient k.

3   Generation of Training Distortion Image Dataset

Radial lens distortion correction can be conventionally resolved by the approaches which based on line segment extraction [7–10]. Due to such tight relationship between distortion correction and appearance of line segments in distorted images, we can intuitively infer that training on images that contain a certain number of line segments might achieve a more accurate result. Therefore, we conduct our dataset collection process with following two steps: (1) to select images which contain sufficient line segments with a specific length to form original dataset, and (2) to generate radial lens distorted images with a continuous range of distorted coefficients to establish the distorted dataset with its corresponding coefficients.

3.1   Image Selection

We construct our training images using images from ImageNet [18], a dataset of over 15 million images with various categories including animals, scenes, and artifact. In order to extract line segments in images, we can utilize Hough transform [20], one of the most widely used tools for detecting predefined shapes in the areas of image processing, pattern recognition and computer vision. Hough transform maps the coordinates (x, y) of the points in Cartesian image space to curves in the parameter space (ρ, θ) of polar presentation of lines and quantizes the parameter space into finite intervals. In this selection procedure, we employ a dataset selection factor α. Suppose that the image average pixel size is C, which defined by $C = (height + width)/2$ and the average length of the most ten longest line segments is l_{mean}, which can be obtained using Hough transform method, therefore we propose a dataset selection factor α as follows:

$$l_{mean} \geqslant \alpha \cdot C \tag{6}$$

In other words, the greater value α reaches, the more line segments will appear in a single image of dataset, and the less images will be included in the dataset. The images on ImageNet have a variety of resolutions, while our convolutional neural network requires a fixed input dimensionality. In our experiments, we choose $\alpha = 0.7$ and down-sample the images to minimal resolution of 256×256. Given a rectangular image, we rescaled the image in the first such that the shorter side was of length 256, and then cropped out the central 256×256 patch from the obtained image. Thus we build an original dataset of roughly 55,000 training images, 6,000 validation images, and 6,000 testing images.

Fig. 3. Synthetic image examples. Sample images from our Synthetic distortion dataset shows the considerable variability in terms of the content of images and degree of radial distortion. Original images are from current publicly available dataset ImageNet [18].

3.2 Distorted Image Simulation

To generate training data, we synthesize distorted images with its according distortion coefficient labelled. We note that, in application of radial lens, the order of magnitude of distortion coefficient k is usually -6 and this coefficient rarely affects the performance of distortion correction when the distortion coefficient changes within 10^{-8}. From this perspective, it is natural to replace the continuous distortion coefficient estimated in a regression problem with multiple discrete value labels. In order to label each image with a distortion coefficient, we generate a random set of integers $\{-400, -399, -398, ..., -1, 0\}$ corresponding to $-400 \times 10^{-8} \sim 0$ and this range is usually common radial lens distortion coefficients vary in. All images are of identical size 256×256. Therefore, we can simulate radial lens distortion using one-parameter division model [21] and label reshaped images with generated distortion coefficients. In this work, we consider this distortion coefficient regression problem as a 401-class classification problem. In Sect. 4.2, we will describe how to eliminate the error caused by taking a discrete classification with multiple discrete value labels as a substitution of one dimensional regression problem with a continuous range.

As shown in Fig. 3, our synthetic distortion dataset covers a realistic variability in content of images and degree of radial distortion and therefore makes the first step towards distortion image dataset establishment and represents a significant advance on distortion correction using convolutional neural networks.

4 Network Architecture and Loss Function

The proposed framework draws on recent achievements of deep convolutional neural networks (AlexNet [19], the VGG net [22], and GoogLeNet [23]) for image classification [24–26] and transfer learning [27]. The architecture of convolutional neural networks for radial lens distortion correction is as follows. Given a distorted 256×256 image, we employ a convolutional neural network to estimate

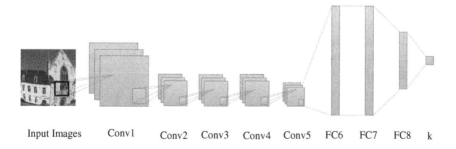

Input Images   Conv1   Conv2   Conv3   Conv4   Conv5   FC6   FC7   FC8   k

Fig. 4. Network architecture. An illustration of the architecture of our proposed convolutional neural networks, explicitly showing 5 convolutional layers and 3 fully connected layers. To compensate the difference of classification and regression tasks we add a linear regression of labels and their probabilities to estimate distortion coefficient of the input image.

the distortion coefficient for each input image. Therefore, our learned convolutional neural network is the mapping from the input distorted images to the corresponding distortion coefficients.

4.1 Network Architecture

Our convolutional neural network starts from the AlexNet [19] architecture pre-trained on the ImageNet for task of image classification. The network takes as input a square 256×256 pixel RGB image and produces a score varying in $[-400, 0]$, namely the distortion coefficient of the input image, as pointed out in Sect. 3.2. The network consists of five successive convolutional layers followed by three fully connected layers as shown in Fig. 4. We train a linear regression layer on top of the fully connected layer to estimate radial lens distortion coefficient k. The first layer is a convolutional layer which filters the input with 96 kernels and each filter is of size $11 \times 11 \times 3$ with a stride of 4 pixel. The first convolutional layer produces 96 feature maps and each one is of size 48×48. The second convolutional layer has 256 kernels of size $5 \times 5 \times 48$ connected to the max pooled outputs of the first convolutional layer. The third convolutional layer has 384 kernels of size $3 \times 3 \times 256$ connected to the normalized outputs of the second convolutional layer. For the fourth and fifth convolutional layer, the kernel size is both $3 \times 3 \times 192$, while the number of features is 384 for the fourth convolutional layer and 256 for the fifth convolutional layer. The number of neurons in the next two fully connected layer is 4096 each. The last layer is a linear regression with a one dimensional output which gives estimation of the distortion coefficient k.

4.2 Expected Value of Coefficients

Our proposed convolutional neural network for radial lens distortion correction is implemented as a classification problem of 401-class coefficients. We can increase the number of classes to gradually eliminate the error caused by this approximation. However, the smaller width of divided interval is, the more sufficient training images and computation cost will demand. Note that we must make a tradeoff between accuracy and simplicity, we decide to choose 401 classes in practice. In order to increase the accuracy of performance, we design the last loss layer with softmax loss function and calculate the expected value of distortion coefficients predicted by the last layer, just as follows:

$$k = \sum_{i=1}^{401} prob_i \cdot k_i \tag{7}$$

where k_i is the i-th class of distortion coefficient ranged in $[-400, 0]$ and $prob_i$ represents the probability of the input image belonging to the i-th class of distortion coefficient. Thus the output of our convolutional neural network is transformed to a continuous value instead of discrete classes, which leads to a more accurate experiment result without more cost of computation.

4.3 Transfer Learning of Network

Transfer learning aims to transfer knowledge between related source and target domains [28]. In computer vision, examples of transfer learning include [29], [30] which try to overcome the deficit of training samples for some categories by adapting classifiers trained for other categories. Other methods aim to cope with different data distributions in the source and target domains for the same categories. These and other related methods adapt classifiers or kernels while using standard image features. Differently to this work, we here transfer image representations trained on the source task.

The convolutional neural network architecture of [19] contains more than 60 million parameters. Directly training so many parameters from only a few training images is problematic. The key idea of this work is that the internal layers of the CNN can act as a generic extractor of mid-level image representation, which can be pre-trained on one dataset (the source task, here ImageNet) and then re-used on other target tasks (here distortion correction on real images). However, this is difficult as the labels and the distribution of images in the source and target datasets can be very different. To address these challenges we (1) design an architecture that explicitly remaps the class labels between the source and target tasks, and (2) develop training and test procedures, inspired by sliding window detectors, that explicitly deal with different distributions of object sizes, locations and scene clutter in source and target tasks. Unless otherwise specified, we use the default parameters of the open source based on Caffe implementation.

The parameters of layers Conv1, Conv2, Conv3, Conv4, Conv5, FC6 and FC7 are first trained on the source task, then transferred to the target task. We conduct a fine-tuned version of AlexNet for distortion correction in this work.

5 Experiments

In this section, before we discuss experiments details, we first introduce two evaluation metrics of distortion correction on real images.

5.1 Evaluate Distortion Correction

Score Function Related to the Mean of the Lengths of the Detected Lines. For the task of radial lens distortion correction, we observe that correction procedure is visually a transformation that maps distorted curves to previous straight line segments. Therefore, the average length of the straight line segments will generally increase after the radial lens distortion correction. In the evaluation part of our experiments, we propose a score function related to the mean of lengths of the detected lines to evaluate correction effects. Based on Hough transform, we compute the average length of detected lines in a single distorted image as l_{mean} and the average length of detected lines in a single corrected image as l'_{mean}, the score function can be defined as follows:

$$S = \frac{l'_{mean} - l_{mean}}{l_{mean}} \qquad (8)$$

Score Function Related to the Standard Variation of the Lengths of the Detected Lines. To perform distortion correction, curves should change back to straight lines. However, when the distortion correction is not correct, the original long straight lines will be divided into many short lines and their angles are close. From a statistical viewpoint to observe the lines extracted from corrected image, we may find that the more appropriate distortion correction, the greater variance of the detected lines. For example, given a single long straight line in the image (not through the distortion center), when distortion correct is suitable, the variance will be zero otherwise will be non-zero. Based on the above facts, we define the distortion correction score function as the following form:

$$T = \sqrt{\sum_{i=1}^{N}(l_i - \frac{1}{N}\sum_{j=1}^{N}l_j)^2} \qquad (9)$$

where l_i is the i-th line and N is the number of the detected lines.

5.2   Comparison with Previous Methods

Our new method outperforms previous methods since the network can be trained on distorted images which include far less line segments, while the previous methods on this kind of dataset will fail to gain a result. In Sect. 3.1, we introduce an image selection system based on Hough transform, we observe that using a larger factor $\alpha = 0.7$ leads to a better performance.

For our method, we train our network on about 55,000 synthetic distorted images and 6,000 images for validation, all of which are with distortion coefficient annotations. We compare our convolutional neural network with state-of-the-art method [21] on real distorted images, as shown in the first three rows in Fig. 5. We can see that our convolutional neural network (shown in the third column) significantly outperforms One-Parameter Division Models [21] in the second column.

A limitation of our method is that the network requires a very great number of training images to perform well, which is the same as other deep learning models. For the lack of images to train and validate, we can only synthesize distorted images with limited discrete distortion coefficients. In some unfortunate cases, the radial distortion is so drastic that coefficient may reach beyond the range covered by synthetic distorted images, which aggravates the difficulty in estimating an appropriate distortion coefficient. Figure 6 shows the correction results of this limitation that One-Parameter Division Models [21] performs a better correction result.

5.3   Effects of Architectures and Parameters

There are several parameters designed in our convolutional neural networks, and some of them can affect the accuracy of the convolutional neural networks

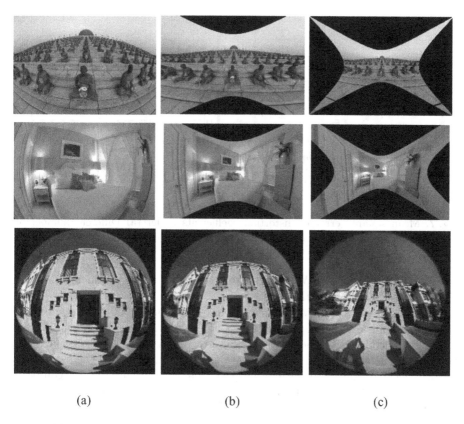

(a)   (b)   (c)

Fig. 5. (a) A real image of stone stairs, a bedroom, and a house with radial lens distortion. (b) The corrected image using One-Parameter Division Models [21]. (c) The corrected image using our method. We can observe that [21] requires more line segments in a single image to obtain a good result and our method outperforms One-Parameter Division Models [21].

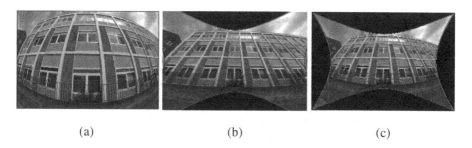

(a)   (b)   (c)

Fig. 6. (a) A real image of an apartment with radial lens distortion. (b) The corrected image using One-Parameter Division Models [21]. (c) The corrected image using our method. Since the absolute value of distortion coefficient is much greater than that of other images, which might exceed 400, [21] performs a better correction result.

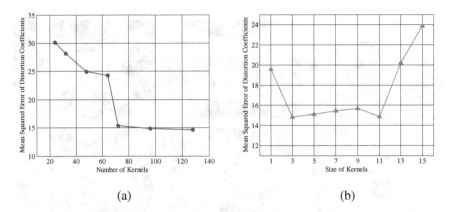

(a)　　　　　　　　　　　　　　　　　(b)

Fig. 7. Mean squared error of distortion coefficients under (a) Different number of convolutional kernels and (b) Different size of convolutional kernels.

for distortion correction. We proceed to explore different parameters for the convolutional neural network architecture.

We train the networks with different parameters on 55,000 synthetic distortion images and test the learned networks on 6,000 synthetic distortion images. Although we conduct the network training process as a 401-class classification, we note that it is not appropriate to take the label accuracy as the evaluation metrics of correction. To measure the performance of correction, we compute the mean squared error of estimated distortion coefficients, which changes with the different parameters of networks, as shown in Fig. 7. A less value of mean squared error of distortion coefficients represents a more satisfactory result.

Number of Convolutional Kernels. Figure 7(a) shows how the number of kernels in the first convolutional layer affect the performance of the network. We can generally conclude that more number of kernels using in the network can lead to a better performance of distortion correction. But the network is not sensitive when the number of kernels is more than about 70.

Size of Convolutional Kernels. To learn how the size of kernels in the first convolutional layer affects the performance of the network, we train and test the network with different sizes of kernels and fix the rest of parameters. Figure 7(b) shows the influence of size of kernels. It can be note that the tested kernel sizes within the set $\{3, 5, 7, 9, 11\}$ show similar performance, while the kernel sizes beyond the range result in a higher mean squared error.

Different Network Architectures. Our model uses the AlexNet [19] network architecture, we can also take advantage of other convolutional neural networks for distortion correction task. Table 1 shows the performance on distortion correction using LeNet [31] and AlexNet [19] with different numbers of convolutional layers (briefly written as LeNet-3, -4, -5, and AlexNet-3, -4, -5 for networks consists of 3, 4, 5 convolutional layers, respectively). We test different network architectures on real images instead of synthetic distorted images, in which way can the

Table 1. Radial lens distortion correction on real images with different architectures. The measurement is score function, and the network using AlexNet with 5 convolutional layers outperforms other architectures.

Architecture	LeNet-3	LeNet-4	LeNet-5	AlexNet-3	AlexNet-4	AlexNet-5
Score function	0.385	**0.417**	0.376	0.391	0.575	**0.634**

generality of different network architectures be tested. Since the distortion coefficients of real images are unknown, we can employ the two score functions proposed in Sect. 5.1 to measure the performance of different network architectures. As is shown in Table 1, results of the best performing network architectures are in bold. LeNet with 4 convolutional layers and the AlexNet with 5 convolutional layers perform well on real images and the latter reach a highest score of 0.634.

5.4   Implementation Details

Our convolutional neural networks are trained using the famous Caffe framework [32] by modifying the AlexNet [19]. Source code for the complete system is available provided by the authors. With Caffe we are able to easily run our framework on a GPU to speed up the process without much optimization. Our experiments are performed on a PC with 3.5 GHz CPU and NVIDIA GeForce GTX TITAN X GPU.

6   Conclusion

In this work, we demonstrated that distorted images synthesized from traditional images taken by pinhole cameras can be used to train CNN for distortion estimation on real images taken by cameras with wide-angle lenses. Our synthesis approach can leverage large collections to construct large-scale training data with fully annotated distortion information. We developed a convolutional neural network for radial lens distortion correction and introduced two score functions for evaluating the effect of correction. Furthermore we demonstrated that our network can estimate distortion coefficients on the images that include fewer line segments, which is rarely mentioned in previous literature. We will explore expanding our methods to a more complex distortion model and compare with more methods in the future.

Acknowledgement. This work was supported in part by State Key Development Program Grand No. 2016YFB1001001, NNSFC Grant No. 61322309, and NNSFC Grant No. 61273283.

References

1. Hartley, R., Zisserman, A.: Multiple View Geometry in Computer Vision. Cambridge University Press, Cambridge (2003)
2. Kuang, Y., Solem, J.E., Kahl, F., Åström, K.: Minimal solvers for relative pose with a single unknown radial distortion. In: 2014 IEEE Conference on Computer Vision and Pattern Recognition, pp. 33–40. IEEE (2014)
3. Kukelova, Z., Pajdla, T.: A minimal solution to radial distortion autocalibration. IEEE Trans. Pattern Anal. Mach. Intell. **33**, 2410–2422 (2011)
4. Ying, X., Mei, X., Yang, S., Wang, G., Rong, J., Zha, H.: Imposing differential constraints on radial distortion correction. In: Cremers, D., Reid, I., Saito, H., Yang, M.-H. (eds.) ACCV 2014. LNCS, vol. 9003, pp. 384–398. Springer, Cham (2015). doi:10.1007/978-3-319-16865-4_25
5. Ying, X., Mei, X., Yang, S., Wang, G., Zha, H.: Radial distortion correction from a single image of a planar calibration pattern using convex optimization. In: 2014 IEEE International Conference on Image Processing (ICIP), pp. 3440–3443. IEEE (2014)
6. Zhang, Z.: Flexible camera calibration by viewing a plane from unknown orientations. In: The Proceedings of the Seventh IEEE International Conference on Computer Vision, vol. 1, pp. 666–673. IEEE (1999)
7. Basu, A., Licardie, S.: Alternative models for fish-eye lenses. Pattern Recogn. Lett. **16**, 433–441 (1995)
8. Claus, D., Fitzgibbon, A.W.: A rational function lens distortion model for general cameras. In: 2005 IEEE Computer Society Conference on Computer Vision and Pattern Recognition (CVPR 2005), vol. 1, pp. 213–219. IEEE (2005)
9. Mei, X., Yang, S., Rong, J., Ying, X., Huang, S., Zha, H.: Radial lens distortion correction using cascaded one-parameter division model. In: 2015 IEEE International Conference on Image Processing (ICIP), pp. 3615–3619. IEEE (2015)
10. Ying, X., Hu, Z.: Can we consider central catadioptric cameras and fisheye cameras within a unified imaging model. In: Pajdla, T., Matas, J. (eds.) ECCV 2004. LNCS, vol. 3021, pp. 442–455. Springer, Heidelberg (2004). doi:10.1007/978-3-540-24670-1_34
11. Melo, R., Antunes, M., Barreto, J.P., Falcao, G., Goncalves, N.: Unsupervised intrinsic calibration from a single frame using a "plumb-line" approach. In: Proceedings of the IEEE International Conference on Computer Vision, pp. 537–544 (2013)
12. Fitzgibbon, A.W.: Simultaneous linear estimation of multiple view geometry and lens distortion. In: Proceedings of the 2001 IEEE Computer Society Conference on Computer Vision and Pattern Recognition (CVPR 2001), vol. 1, pp. 1–125. IEEE (2001)
13. Devernay, F., Faugeras, O.: Straight lines have to be straight. Mach. Vis. Appl. **13**, 14–24 (2001)
14. Bukhari, F., Dailey, M.N.: Automatic radial distortion estimation from a single image. J. Math. Imaging Vis. **45**, 31–45 (2013)
15. Hughes, C., Denny, P., Glavin, M., Jones, E.: Equidistant fish-eye calibration and rectification by vanishing point extraction. IEEE Trans. Pattern Anal. Mach. Intell. **32**, 2289–2296 (2010)
16. Rosten, E., Loveland, R.: Camera distortion self-calibration using the plumb-line constraint and minimal hough entropy. Mach. Vis. Appl. **22**, 77–85 (2011)

17. Alemán-Flores, M., Alvarez, L., Gomez, L., Santana-Cedrés, D.: Line detection in images showing significant lens distortion and application to distortion correction. Pattern Recognit. Lett. **36**, 261–271 (2014)
18. Deng, J., Dong, W., Socher, R., Li, L.J., Li, K., Fei-Fei, L.: Imagenet: a large-scale hierarchical image database. In: IEEE Conference on Computer Vision and Pattern Recognition (CVPR 2009), pp. 248–255. IEEE (2009)
19. Krizhevsky, A., Sutskever, I., Hinton, G.E.: Imagenet classification with deep convolutional neural networks. In: Advances in Neural Information Processing Systems, pp. 1097–1105 (2012)
20. Fernandes, L.A., Oliveira, M.M.: Real-time line detection through an improved Hough transform voting scheme. Pattern Recognit. **41**, 299–314 (2008)
21. Alemán-Flores, M., Alvarez, L., Gomez, L., Santana-Cedrés, D.: Automatic lens distortion correction using one-parameter division models. Image Process. Line **4**, 327–343 (2014)
22. Simonyan, K., Zisserman, A.: Very deep convolutional networks for large-scale image recognition. arXiv preprint arxiv:1409.1556 (2014)
23. Szegedy, C., Liu, W., Jia, Y., Sermanet, P., Reed, S., Anguelov, D., Erhan, D., Vanhoucke, V., Rabinovich, A.: Going deeper with convolutions. In: Proceedings of the IEEE Conference on Computer Vision and Pattern Recognition, pp. 1–9 (2015)
24. Juneja, M., Vedaldi, A., Jawahar, C., Zisserman, A.: Blocks that shout: distinctive parts for scene classification. In: Proceedings of the IEEE Conference on Computer Vision and Pattern Recognition, pp. 923–930 (2013)
25. Le, Q.V.: Building high-level features using large scale unsupervised learning. In: 2013 IEEE International Conference on Acoustics, Speech and Signal Processing, pp. 8595–8598. IEEE (2013)
26. Le, Q.V., Zou, W.Y., Yeung, S.Y., Ng, A.Y.: Learning hierarchical invariant spatio-temporal features for action recognition with independent subspace analysis. In: 2011 IEEE Conference on Computer Vision and Pattern Recognition (CVPR), pp. 3361–3368. IEEE (2011)
27. Donahue, J., Jia, Y., Vinyals, O., Hoffman, J., Zhang, N., Tzeng, E., Darrell, T.: DeCAF: a deep convolutional activation feature for generic visual recognition. In: ICML, pp. 647–655 (2014)
28. Pan, S.J., Yang, Q.: A survey on transfer learning. IEEE Trans. Knowl. Data Eng. **22**, 1345–1359 (2010)
29. Aytar, Y., Zisserman, A.: Tabula rasa: model transfer for object category detection. In: 2011 International Conference on Computer Vision, pp. 2252–2259. IEEE (2011)
30. Tommasi, T., Orabona, F., Caputo, B.: Safety in numbers: learning categories from few examples with multi model knowledge transfer. In: 2010 IEEE Conference on Computer Vision and Pattern Recognition (CVPR), pp. 3081–3088. IEEE (2010)
31. LeCun, Y., Bottou, L., Bengio, Y., Haffner, P.: Gradient-based learning applied to document recognition. Proc. IEEE **86**, 2278–2324 (1998)
32. Jia, Y., Shelhamer, E., Donahue, J., Karayev, S., Long, J., Girshick, R., Guadarrama, S., Darrell, T.: Caffe: convolutional architecture for fast feature embedding. In: Proceedings of the 22nd ACM International Conference on Multimedia, pp. 675–678. ACM (2014)

Ultrasound Speckle Reduction
via L_0 Minimization

Lei Zhu[1(✉)], Weiming Wang[3], Xiaomeng Li[1], Qiong Wang[3], Jing Qin[2],
Kin-Hong Wong[1], and Pheng-Ann Heng[1,3]

[1] The Chinese University of Hong Kong, Sha Tin, Hong Kong
lzhu@cse.cuhk.edu.hk
[2] The Hong Kong Polytechnic University, Kowloon, Hong Kong
[3] Guangdong Provincial Key Laboratory of Computer Vision and
Virtual Reality Technology, Shenzhen Institutes of Advanced Technology,
Chinese Academy of Science, Shenzhen, China

Abstract. Speckle reduction is a crucial prerequisite of many computer-aided ultrasound diagnosis and treatment systems. However, most of existing speckle reduction filters concentrate the blurring near features and introduced the hole artifacts, making the subsequent processing procedures complicated. Optimization-based methods can globally distribute such blurring, leading to better feature preservation. Motivated by this, we propose a novel optimization framework based on L_0 minimization for feature preserving ultrasound speckle reduction. We observed that the GAP, which integrates gradient and phase information, is extremely sparser in despeckled images than in speckled images. Based on this observation, we propose the L_0 minimization framework to remove speckle noise and simultaneously preserve features in ultrasound images. It seeks for the L_0 sparsity of the GAP values, and such sparsity is achieved by reducing small GAP values to zero in an iterative manner. Since features have larger GAP magnitudes than speckle noise, the proposed L_0 minimization is capable of effectively suppressing the speckle noise. Meanwhile, the rest of GAP values corresponding to prominent features are kept unchanged, leading to better preservation of those features. In addition, we propose an efficient and robust numerical scheme to transform the original intractable L_0 minimization into several sub-optimizations, from which we can quickly find their closed-form solutions. Experiments on synthetic and clinical ultrasound images demonstrate that our approach outperforms other state-of-the-art despeckling methods in terms of noise removal and feature preservation.

1 Introduction

Ultrasonography has become one of the most favorable imaging modalities in a wide range of clinical applications because it is safe, real-time and cost effective. However, ultrasound images are usually corrupted with granular patterns of white and dark spots that are referred as speckle [1]. Although speckle is sometimes considered as diagnostic clues on certain occasions, it has adverse effects

© Springer International Publishing AG 2017
S.-H. Lai et al. (Eds.): ACCV 2016, Part III, LNCS 10113, pp. 50–65, 2017.
DOI: 10.1007/978-3-319-54187-7_4

on precise diagnosis and treatment in most cases [2]. In addition, speckle complicates automatic processing and analysis procedures of ultrasound images [3], including detection, segmentation, registration, and so on. Therefore, speckle reduction is a crucial prerequisite of many intelligent ultrasound systems [4].

Many ultrasound speckle reduction methods have been proposed. These methods can be roughly classified into two groups: wavelet-based filters [5] and spatial filters [2,6,7]. Assuming that the multiplicative speckle noise can be transformed into additive Gaussian noise by the logarithm operation, Wavelet-based methods decomposed the content of the transformed image into multiple sub-bands at different orientations and resolution scales. Although those methods can effectively remove speckle noise, they tended to produce the ringing artifacts when preserving features [8]. By exploiting the spatial correlation, spatial filters computed a despeckling result as a weighted average on a set of candidates. According to the candidate selection, those filters are divided into the local filters [6,9] and nonlocal filters [7]. However, spatial filters would concentrate the blurring near edges and introduced the hole artifacts [10]. In order to overcome such limitation, the optimization-based approaches have been proposed with state-of-the-art results in many image processing tasks [11]. Those methods globally distributed such blurring in spatial filters into each pixel [12], causing excellent restored results without hole artifacts. In addition, the spatial filters cannot preserve sharp edges like the global optimization based filters [12].

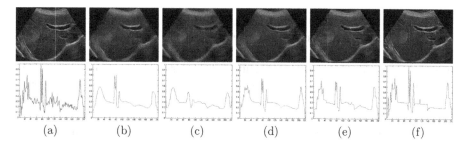

Fig. 1. Comparison of speckle reduction on a liver ultrasound image with hemangiomas, as well as the evaluation of the feature contrast preservation. (a) Original image with intensity profile. Despeckled result with intensity profile by (b) SRAD [13], (c) SBF [6], (d) OBNLM [7], (e) ADLG [2], and (f) our method. (Color figure online)

In this paper, we proposed a novel global optimization based on an L_0 minimization framework for feature-preserving speckle reduction in ultrasound images. Motivated by the sparseness prior of the image gradient and the feature detection capability of the local phase based feature asymmetry (*FA*) operator [14], we defined a new measurement (namely *GAP*) combining the gradient and the *FA* operators together. The *GAP* inherits those two properties. Consequently, we proposed a L_0 *GAP* minimization for speckle reduction in ultrasound image. It seeks for the L_0 sparsity of the *GAP* by progressively reducing low GAP values, which correspond to the speckle noise, to zero. Meanwhile, the significant features with high GAP values are kept unchanged. The proposed approach also has better performance in preserving low contrast features than existing despeckling methods, due to the intensity invariant property of the phase-based operator (FA) for feature detection. In addition, we proposed an efficient solver to minimize the proposed L_0 regularized objective function. A variable splitting strategy is employed to transfer the original optimization into the iterative optimization of a non-linear quadratic optimization and a L_0 regularized least square with a hard thresholding closed-form solution. For the non-linear quadratic optimization, the nonlinear FA measurement is linearized, producing a series of pure quadratic minimizations, for which an efficient and robust solution exists. Figure 1 compares the despeckling performance of different methods. As shown in the top row, the proposed approach effectively preserves the boundaries of important features while other methods tend to blur the edges in different degrees (see the low contrast hemangiomas of the left red rectangle). The advantages of our approach can be further verified by carefully checking the intensity profile of a vertical line in the bottom row. When smoothing out the intensity fluctuations caused by speckle noise, existing methods also remove or largely reduce the intensity peaks at some features. In contrast, our approach hardly alters the intensity values of those peaks so that the features are almost totally retained in the despeckled image.

Contributions of This Work

- We propose an observation that the GAP values in the despeckled image is highly sparser than that in the speckled image.
- We propose a L_0 norm regularized global optimization framework to seek for the GAP sparsity. During the pursuit of the sparsity, the proposed L_0 minimization can eliminate speckle noise in ultrasound images and better preserve features than previous despeckling techniques.
- We propose an efficient and robust solver to minimize the proposed objective function by first splitting the intractable problem into tractable subproblems with half-quadratic splitting method, followed by the decomposition of the non-convex subproblem into linear systems using iteratively re-weighted least squares.

2 Related Work

Many despeckling methods have been proposed in the literature, and they can be categorized into two classes. The first one is the wavelet-based filters, which assumed that multiplicative speckle noise can be transformed into the additive Gaussian noise by the logarithm operation. After the logarithm transformation, the image content was decomposed into multiple. Large coefficients corresponded to the important low frequency information (e.g. edges), and noise and image details lied in the high frequency subbands. Usually, thresholding techniques were applied to the small coefficients for noise removal. Khare et al. [15] detected strong edges using the imaginary part of the complex scaling coefficients and then performed the shrinkage on the magnitude of complex wavelet coefficients at non-edge pixels. The threshold value of the shrinkage was determined by the statistical parameters of complex wavelet coefficients of the noised image. Recently, Esakkirajan et al. [5] proposed an adaptive wavelet packet domain filtering for speckle reduction. They produced a rich set of bases with the wavelet packet decomposition technique and then the singular value decomposition was used to select the best basis. The modified NeighShrink thresholding technique was executed on all other sub-bands, except the sub-band with the largest singular value. Wavelet-based despeckling techniques can effectively preserve texture details, but they tend to produce ringing artifacts when preserving features [8].

Another direction for speckle reduction in ultrasound images is the spatial filters. By utilizing the spatial correlation existing in images, those methods computes a despeckling output as a weighted average. One is to extend the nonlinear anisotropic diffusion filtering (ADF) for the ultrasound data corrupted with speckle noise. They [2,13,16] encouraged the isotropic diffusion in the homogeneous regions for noise removal, while stop the diffusion between homogeneous regions for feature preservation. Yu and Acton [13] proposed the speckle reducing anisotropic diffusion (SRAD) by involving the edge-sensitive instantaneous coefficient of variation to determine whether a pixel should be smoothed or left intact in the ultrasound image. Recently, Flores et al. [2] proposed a ADF based despeckling method for the breast ultrasound image, where the conduction coefficient parameter involved in the ADF was adaptively selected for each pixel under the guidance of the 2D Log-Gabor filter response. Even though these diffusion based despeckling techniques can progressively smooth the speckled image, a lot of meaningful details are also discarded.

Another spatial filter group is to estimate the despeckled result of each pixel by weighted averaging on pixels in a local region or the entire image. Tay et al. [9] proposed the squeeze box filter (SBF) based upon removing outliers with a local extremum. By replacing these outliers with the local mean in each iteration, the SBF method compressed the image pixel values, so that the differences in interclass means were protected for feature preservation, while the interclass variance was reduced for speckle noise removal. Balocco et al. [6] proposed an automatic bilateral filter dedicated for the ultrasound images by embedding the noise statistics into the weighting scheme of the original bilateral filter framework. Taking the assumption that there are many repetitive patches in the image, the non-local means (NLM) method [17] averaged all pixels in the entire image, and the weights were computed by the weighted Euclidean distance between two patches. This strategy leads to a more robust denoising performance when compared to local filters. However, the original NLM was designed for suppressing the additive Gaussian noise in 2D natural images, and thus several despeckling techniques were proposed to adapt the NLM method for the multiplicative speckle noise model in ultrasound images. Couple et al. [7] utilized the Bayesian theory to define a Pearson distance for patch comparisons and implemented the NLM method in the block-wise manner to decrease the computational burden.

Recently, Yang et al. [8] combined local statistics of the ultrasound image and NLM filter to reduce speckle in ultrasound images. However, when there are not enough similar patches within the ultrasound image, those methods tend to produce severe artifacts and the performance degrades significantly.

3   Methods

Recently, many attractions in image processing has transformed from the spatial filters to the global optimization [10,11], which usually consisted of a data term and regularization terms. Compared to the spatial filters, global optimization outperformed in the sharp feature preservation [12]. In addition, spatial filters would concentrate the blurring near these edges and introduce the holes while the global optimization based methods would globally distribute such blurring to each pixel of the image. Unfortunately, existing global optimizations [10,11] employed the pixel intensity difference to formulate regularization terms. When applying those methods for despeckling, speckle noise and features will received similar penalties, since ultrasound speckle manifests itself as a form of multiplicative noise, which indicates that variances caused by speckle noise are comparable or even larger than edges [18]. In this paper, we proposed a global optimization based on L_0 minimization for speckle reduction in ultrasound images. It involves a new L_0 regularized term to seek for the sparsity of the GAP. The details of the proposed method will be introduced in the following section.

3.1   Feature Asymmetry

Local energy model developed in [19] postulates that features are perceived at points where the phase information is highly coherent. Based on this assumption, the features in the image can be located by analyzing the phase pattern of each pixel [20]. Inspired from this, Belaid et al. [14] presented a local phase-based operator, feature asymmetry (FA) [21], using Cauchy filters, to measure the significance of features in ultrasound images:

$$\begin{cases} FA = \dfrac{\lfloor |R_o| - |R_e| - \Theta \rfloor}{\sqrt{R_o^2 + R_e^2} + \varepsilon_1}, \\ R_o = g * S, R_e = (g * z_1 * S, g * z_2 * S) \end{cases} \tag{1}$$

where Θ describes the estimated noise threshold and $\lfloor \cdot \rfloor$ is the zeroing operation of negative values; ε_1 is a constant to avoid division by zero; z_1 and z_2 are the Riesz filters and g is the bandpass filter applied to the input image S. In the frequency domain, the 2D isotropic Cauchy kernel [22] is defined as:

$$G(\theta) = |\theta|^\varphi \exp(-t|\theta|) \sqrt{\dfrac{\pi 4^{\varphi+1} \sigma^{2\varphi+1}}{\Gamma(2\varphi+1)}} \tag{2}$$

where Γ is the gamma function; θ denotes the normalized coordinates of each pixel. φ and t are the bandwidth and the scaling parameter of the Cauchy kernel, respectively. The value of the FA measure ranges between 0 and 1, close to 0 in smooth regions and close to 1 near the features.

Note that the scale parameter in the Cauchy kernel is important for the quality of edge detection. Generally, details and discontinuities are maintained in fine scales while coarse scales preserve the regularities and continuities of the boundaries. Hence, finding the optimal scale that fits to the features in the image is very crucial because a finer scale might include unwanted details while a coarser scale might miss important features. We employ the following γ normalized edge strength measure Ω_γ [23] for optimal scale selection:

$$\Omega_\gamma = -t^{3\gamma}(H_x^3 H_{xxx} + 3H_x^2 H_y H_{xxy} + 3H_x H_y^2 H_{xyy} + H_y^3 H_{yyy}). \qquad (3)$$

Here, H is the filtered image obtained by convolving the input image with the Cauchy kernel at scale t and the partial derivative $H_{abc} = \frac{\partial^3 H}{\partial_a \partial_b \partial_c}$. To determine the optimal scale, we analyze the intensity distribution of Ω_γ over all a scale range and then select the one where the sum of intensities achieves the maximal value. We empirically set $\gamma = 0.5$ in our implementation, as suggested in [23].

3.2 The Proposed L_0 Minimization Framework

Given an input ultrasound image J, we define a new measurement, called GAP, by combining Gradient And Phase information:

$$GAP(J) = \partial_x J^2 + \partial_y J^2 + FA(J)^2 \qquad (4)$$

where $\partial_x J^2$ and $\partial_y J^2$ are the gradient map at the x and y direction, respectively. $FA(J)$ denotes the feature map obtained by applying the local phase based feature asymmetry on J. The proposed GAP can inherit characteristics of both the gradient information and the phase based FA, as shown in Fig. 2. First, GAP can be a good edge indicator in speckled images. Figure 2(b) is a synthetic speckled image generated from a clean image (Fig. 2(a)) by employing a theoretical speckle noise model [7] on it. The GAP map of Fig. 2(b) is shown by Fig. 2(e), in which the boundaries of different shape objects have larger GAP values than speckle noise. Second, it is well known that a clean image has a sparser gradient magnitude distribution than its corresponding speckled image. We find that the

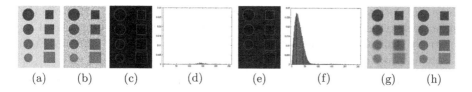

(a)   (b)   (c)   (d)   (e)   (f)   (g)   (h)

Fig. 2. Statistical analysis of GAP: (a) a clean image, (b) corresponding synthetic speckled image (noise variance $\sigma^2 = 0.08$), (c) GAP map of (a), (d) GAP magnitude histogram of (c), (e) GAP map of (b), (f) GAP magnitude histogram of (e). (g) Despeckled result of (b) by L_0 gradient minimization [11] and (h) our method.

GAP magnitude also has such sparse prior. Figure 2(c) and (d) are the *GAP* map and *GAP* histogram of Fig. 2(a), respectively. It is obvious that only pixels at edges have non-zero *GAP* values and we can observe apparent zero peaks in the *GAP* histogram distribution. In contrast, for the speckled image (Fig. 2(b)), its *GAP* distribution (Fig. 2(f)) can not be modeled by narrow peaks.

Motivated by these two properties of the *GAP*, we propose a novel L_0 minimization for feature-preserving ultrasound speckle reduction:

$$\min_{D} \left\{ \sum_{q} (D_q - I_q)^2 + \lambda \cdot P(D) \right\}, \text{ where } P(D) = \#\left\{ p \mid |GAP(D)_p| \neq 0 \right\}. \quad (5)$$

Here, I is the input image and D is the despeckled image; q is the pixel coordinate; $P(D)$ is the L_0 norm of the *GAP*. It globally counts the number of pixels whose *GAP* values are non-zero in the image D. Note that L_0 gradient minimization [11] cannot achieve satisfactory results for ultrasound speckle reduction, as the gradient magnitude is ineffective to distinguish features from speckle noise in ultrasound images [2]. Figure 2(g) shows the despeckled result of Fig. 2(b) via L_0 gradient minimization [11]. We can see that it over-smooths the shape boundaries in Fig. 2(b). On the contrary, thanks to the better edge detection capability of the *GAP*, our method can effectively eliminate speckle noise and well protect shape boundaries simultaneously, as shown in Fig. 2(h).

3.3   Numerical Solution

Solving a L_0 norm regularized objective function is usually considered as computational intractable, because the first data term models the pixel-wise difference and the second regularization term represents a global L_0 metric [11]. We propose an efficient solver for our L_0 minimization based on the half-quadratic splitting method [24,25] and iteratively re-weighted least squares framework (IRLS) [26]. First, three auxiliary variables u, v and w are introduced to represent $\partial_x D_q$, $\partial_y D_q$ and FA, leading to a new energy function:

$$\min_{D,u,v,w} \left\{ \sum_{q} (D_q - I_q)^2 + \beta \left((\partial_x D_q - u_q)^2 + (\partial_y D_q - v_q)^2 + (FA(D)_q - w_q)^2 \right) \right.$$
$$\left. + \lambda \cdot P(u,v,w) \right\} \quad (6)$$

where q is the pixel index; $P(u,v,w) = \#\{p \mid u_p^2 + v_p^2 + w_p^2 \neq 0\}$; β controls the similarity between $(\partial_x D, \partial_y D, FA)$ and (u,v,w), and its value is increased by iteratively multiplying a constant value k. Given a value of β, we minimize Eq. 6 by solving two tractable subproblems alternatively:

Subproblem 1: Updating D. Given estimated values of (u,v,w) from previous iteration, we update D by solving:

$$\min_{D} \left\{ \sum_{q} (D_q - I_q)^2 + \beta \left((\partial_x D_q - u_q)^2 + (\partial_y D_q - v_q)^2 + (FA(D)_q - w_q)^2 \right) \right\}, \quad (7)$$

Unfortunately, due to the nonlinear property of $FA(D)$, the above quadratic optimization is still highly non-convex. It is non-trivial to solve the minimization. Gradient descent methods requires tens or hundreds of iterations, and the solution is sensitive to the initialization. Inspired by the iteratively re-weighted least squares framework (IRLS) [26], we propose a numerical solver to transform the highly non-convex optimization into solving a series of sparse linear equations, for which fast and robust solutions exist. The key idea is to decompose the nonlinear FA into a linear term D_q and a nonlinear term f_q:

$$FA(D)_q \approx \frac{FA(D)_q}{D_q + \varepsilon_2} * D_q = f_q * D_q, \text{ where } f_q = \frac{FA(D)_q}{D_q + \varepsilon_2}, \tag{8}$$

where ε_2 is a small value to avoid division by zero. By incorporating f_q and D_q, we re-formulate the objective function in Eq. 7 as:

$$\min_D \left\{ \sum_q (D_q - I_q)^2 + \beta \left((\partial_x D_q - u_q)^2 + (\partial_y D_q - v_q)^2 + (f_q D_q - w_q)^2 \right) \right\} \tag{9}$$

Now, we can solve it by iteratively performing the following two steps, in a way similar to IRLS:

Step 1: Compute f_q with the estimated despeckled image D.

Step 2: Fixing f_q, we update D by solving the following sparse linear system:

$$(\Phi + \beta C_x^T C_x + \beta C_y^T C_y + \beta F^T F)V_D = V_I + \beta C_x^T V_u + \beta C_y^T V_v + \beta F^T V_w \tag{10}$$

where V_D, V_I, V_u, V_v and V_w are the vector representation of D, I, u, v, and w respectively; C_x and C_y are the Toeplitz matrices from the discrete gradient operators; F is a diagonal matrix, and we set its diagonal values as: $F[i,i] = f_i$; Φ is the identity matrix with the same size of matrix C_x. We employ the preconditioned conjugate gradient (PCG) algorithm to solve the sparse linear equation. In our experiments, we find that 3 to 5 iterations are enough to estimate the despeckled image D.

Subproblem 2: Updating (u, v, w). Given D, we estimate (u, v, w) by solving the following L_0 regularized least squares:

$$\min_{u,v,w} \left\{ \sum_q \left((\partial_x D_q - u_q)^2 + (\partial_y D_q - v_q)^2 + (FA(D)_q - w_q)^2 \right) + \frac{\lambda}{\beta} P(u,v,w) \right\}, \tag{11}$$

Similar to [11], the minimal value of Eq. 11 is obtained under the following condition:

$$(u_q, v_q, w_q) = \begin{cases} (\partial_x D_q, \partial_y D_q, FA(D)_q) & GAP(D)_q > \frac{\lambda}{\beta} \\ (0, 0, 0) & \text{otherwise.} \end{cases} \tag{12}$$

Algorithm 1. Ultrasound speckle reduction via L_0 minimization

---

Input: The input ultrasound image I, smoothness λ, the bandwidth φ, scale range Ω, similarity parameters β_0, β_{max} and increasing rate k.

1: Initialization: $D^0 \leftarrow I, \beta = \beta_0$
2: **repeat**
3:      with the estimated D, solve for u, v, w using Eq. 12,
4:      **for** $iter = 1$ to 5 **do**
5:          **for all** scale $s \in \Omega$ **do**
6:              compute FA using Eq. 1,
7:              compute the edge strength Ω_γ using Eq. 3
8:          **end for**
9:          select the optimal scale with maximal Ω_γ value,
10:          compute f_p with optimal scale using Eq. 8,
11:          update D by solving the sparse linear equation (Eq. 10),
12:      **end for**
13:      $\beta = k\beta$
14: **until** $\beta \geq \beta_{max}$
15: **return** despeckled image D

3.4   Why It Works

Here, we provide more analysis on the proposed L_0 minimization. Algorithm 1 summaries the whole optimization process. In our solver, we first introduce three auxiliary variables (u, v, w) and initialize them as: $(u, v, w) = (\partial_x D, \partial_y D, FA(D))$. Afterwards, two subproblems are alternatively updated as following:

(a) For each pixel p of the current despeckled result D, we check whether $GAP(D)_p <= \frac{\lambda}{\beta}$ is satisfied. If yes, we set $u_p = 0, v_p = 0$, and $w_p = 0$, otherwise, the value is unchanged, as described in Eq. 12.
(b) We transfer the changes happened in (u, v, w) to $(\partial_x D, \partial_y D, FA(D))$ by solving Eq. 7, so that small values in the $GAP(D)$ are also largely reduced, or even set to 0. Meanwhile, the remaining $GAP(D)$ values are unaltered.

When small $GAP(D)$ are progressively reduced to 0, their $\partial_x D$, $\partial_y D$ and $FA(D)$ values also become 0, leading to the decrease of $P(D)$. Hence, our solver can gradually minimize the proposed L_0 minimization. Once there is no pixel whose GAP value is below $\frac{\lambda}{\beta}$ in the current despeckled result D, our method reaches the convergence state. Similar to [27], our method with three auxiliary splitting variables also converges after 10–20 iterations, and it takes about 5 s to process a $300 * 255$ image. Since features have larger GAP values than speckle noise, $\partial_x D$ and $\partial_y D$ of speckle noise tend to be 0 as the iteration number increases, leading to the removal of speckle noise. At the same time, the values of $(\partial_x D, \partial_y D, FA(D))$ at features are almost unchanged, so that our method can well protect features. Furthermore, our method is also capable to preserve low contrast features, since FA is invariant to changes of brightness or contrast for feature detection. We cannot achieve the satisfied despeckling result

by employing only L_0 gradient or L_0 FA minimization. Since gradient information is unable to separate features from speckle noise in ultrasound images, the
L_0 gradient minimization [11] gives similar penalties on features and speckle
noise. Although FA has the capability to distinguish speckle noise and features,
L_0 FA minimization also cannot effectively remove speckle noise because by only
setting the FA values of the noise to 0 does not eliminate the speckle.

4   Experiments

We evaluated the performance of the proposed method on many synthetic and
clinical ultrasound images and compared its results with SRAD [13], SBF [9],
OBNLM [7], and ADLG [2]. Six parameters are included in our method, namely,
the smoothness λ, the bandwidth φ, the scale range Ω, the similarity parameters
β_0, β_{max} and the increasing rate k. In the all experiments, we set $\varphi = 1.58$,
$\beta_0 = 4\lambda$, and $\beta_{max} = 1e5$, and the scale range Ω is $[1, 20]$. The increasing rate k
balances the time efficiency and despeckling performance, and its value is in the
range of $[1.2, 2]$. λ is a critical parameter to adjust the performance of feature
preservation, and we set its value in $[1e-3, 1e-1]$.

4.1   Synthetic Images

The theoretical speckle noise model [7] is based on the equation: $Y = X + \Phi * X$, where X and Y are the noise-free and synthesized images with speckle
noise, respectively, and Φ is the zero-mean Gaussian noise with variance σ^2:
$\Phi \sim N(0, \sigma^2)$. We employed the above noise model on the clean image (Fig. 2(a))
to produce a synthetic speckled image (Fig. 3(a)) with noise variance $\sigma^2 = 0.15$.
The clean image (Fig. 2(a)) consists of shape objects with different sizes and
intensity contrasts to the background. Figure 3(b)–(f) show despeckled results of
different ultrasound speckle reduction algorithms. It is observed that all the other
methods blur the shape boundaries to some extent, especially for the low contrast
shapes in the last row. Thanks to the less sensitivity of phase information to
intensity contrasts, the proposed approach achieves the best performance in

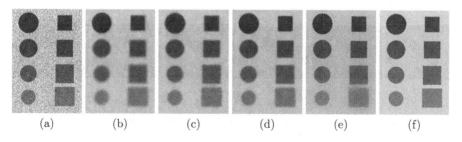

(a)   (b)   (c)   (d)   (e)   (f)

Fig. 3. (a) synthetic speckled image ($\sigma^2 = 0.15$). Despeckled result by (b) SRAD [13],
(c) SBF [9], (d) OBNLM [7], (e) ADLG [2], and (f) our method.

Table 1. Quantitative evaluation of different despeckled results.

	SRAD	SBF	OBNLM	ADLG	Our method
PSNR	22.85	23.16	23.54	23.61	**27.09**
FOM	0.3972	0.4142	0.5126	0.5291	**0.6333**
MSSIM	0.8740	0.8970	0.9155	0.9235	**0.9729**

preserving edges of all those shape objects while removing the speckle noise. In addition, three metrics are used for quantitative evaluation: peak signal-to-noise ratio $(PSNR)$, Pratt's figure of merit (FOM) [2] and structural similarity $(SSIM)$ [16]. Table 1 reports quantitative values of all the three metrics among different despeckling methods. Our method achieves the largest values for all the three metrics, which further demonstrates that our method can better preserve features during speckle reduction when compared with other techniques.

4.2   Clinical Images

We further verified the proposed method on many clinical images obtained from a public ultrasound dataset[1]. We show the results of four typical images from Figs. 4, 5, 6 and 7 and more results can be found in the supplementary material. The despeckling results of two ultrasound images with gallstones are presented in Figs. 4 and 5. As shown in the regions of red rectangles, our method effectively removes speckle noise while clearly preserving boundaries of gallbladder tissues

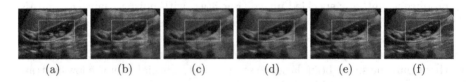

(a)   (b)   (c)   (d)   (e)   (f)

Fig. 4. Comparison of speckle reduction on a ultrasound image with floating and non-floating gallstones. (a) Original image. Despeckled result by (b) SRAD [13], (c) SBF [6], (d) OBNLM [7], (e) ADLG [2], and (f) our method.

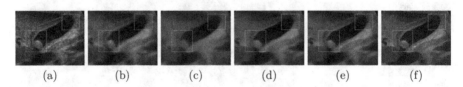

(a)   (b)   (c)   (d)   (e)   (f)

Fig. 5. Comparison of speckle reduction on a ultrasound image with a mobile gallstone. (a) Original image. Despeckled result by (b) SRAD [13], (c) SBF [6], (d) OBNLM [7], (e) ADLG [2], and (f) our method.

---

[1] http://www.ultrasoundcases.info.

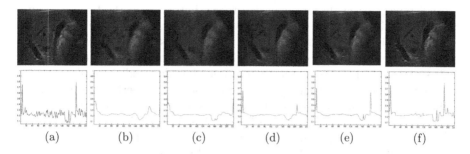

Fig. 6. Comparison of speckle reduction on a hypoechoic liver ultrasound image with metastatic melanoma, as well as the evaluation of the feature contrast preservation. (a) Original image with the intensity profile of a vertical line. Despeckled result with its intensity profile by (b) SRAD [13], (c) SBF [6], (d) OBNLM [7], (e) ADLG [2], and (f) our method.

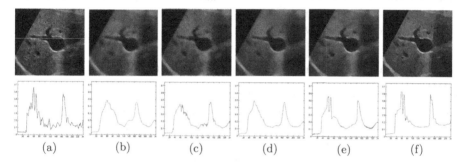

Fig. 7. Comparison of speckle reduction on a liver ultrasound image with venous dilatation of the intrahepatic portal vein, as well as the evaluation of the feature contrast preservation. (a) Original image with the intensity profile of a horizontal line. Despeckled result with its intensity profile by (b) SRAD [13], (c) SBF [6], (d) OBNLM [7], (e) ADLG [2], and (f) our method.

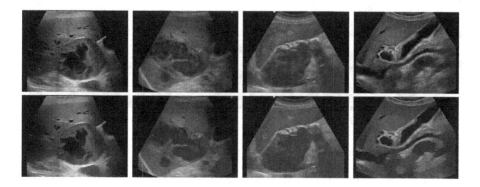

Fig. 8. More despeckled results of our method.

and gallstones, especially for small gallstones, but other methods heavily blur those boundaries and even smooth out some small gallstones. The despeckling results of two liver ultrasound images are shown in Figs. 6 and 7, in which the left column illustrates despeckled results while the right column shows the intensity profiles of a typical vertical or horizontal line. The despeckling results visually demonstrate the proposed L_0 minimization achieves best performance in maintaining features of the liver ultrasound images among all the compared despeckling techniques. The intensity profiles further demonstrate that our method does not alter peak values of prominent features in the original image, while suppressing intensity fluctuates caused by speckle noise. In contrast, other methods are all obviously reduce peak values of those features to some extent. In addition, Fig. 8 presents our despeckling results on more clinical images. Obviously, our method can effectively remove speckle noise and maintain features at the same time. In addition, many originally blurry features are much clearer in our despeckled results. We also invites some clinical experts to comment on our results, and they conclude that our results of speckle reduction improve the image quality and well protect the structure details, which can provide helps for their clinical computer-aided diagnosis systems.

4.3   Application to Ultrasound Image Segmentation

Finally, we demonstrate the proposed method can be employed as a prerequisite step in intelligent ultrasound processing systems by taking the image segmentation as an example. Breast tumor segmentation plays a vital role in the computer-aided diagnosis (CAD) system, since the clinical benign and malignant breast lesion classification relies on the shape or contour features from segmented lesions [2]. In Fig. 9, we present the segmentation results of the breast tumor on the speckled image and despeckled images by employing a famous level set based

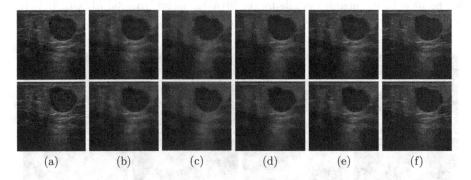

(a)   (b)   (c)   (d)   (e)   (f)

Fig. 9. Breast tumor segmentation accuracy comparison on different despeckled results. (a) The original ultrasound image and its segmentation result. Despeckled result and its segmentation result by (b) SRAD [13], (c) SBF [6], (d) OBNLM [7], (e) ADLG [2], and (f) our method. Blue color: the ground truth delineated by clinicians; Red color: the reached segmentation result. (Color figure online)

Table 2. Mean AC, HD and HM values for different segmentation results on 10 ultrasound images

	Input	SRAD	SBF	OBNLM	ADLG	Our method
AC (%)	63.46	91.00	93.09	93.44	94.03	**97.71**
HD	30.4451	10.8214	9.6129	6.5745	4.4619	**2.8892**
HM	12.785	2.4559	1.8825	1.8404	1.7014	**0.6356**

segmentation technique [16]. The red curve is the obtained segmentation result and the blue curve is the manual result delineated by a clinical doctor, which is usually regarded as the ground truth [28]. As can be seen, the segmentation performance on the original image is pretty poor, due to the adverse effect from the speckle noise. Carrying out the segmentation on those despeckled image, we find that the accuracy has been significantly improved. Among all despeckled images, the segmentation result on our despeckled result is closest to the ground truth. Since some parts of the breast tumor contour are blurred in other despeckled results, their segmentation contours leaks out the blurred edges. In addition, three metrics [16]: a combined accuracy metric of true and false positive rate (AC), Hausdorff distance (HD) and Hausdoff mean (HM), are adopted for quantitative comparison, and a better segmentation result shall have higher AC, as well as lower HD and HM. Table 2 summaries mean values of three metrics for different segmentation results of 10 ultrasound images. Obviously, our result achieves the largest AC value (97.71%), the smallest HD value (2.8892) and the smallest HM value (0.6356) among all despeckled results.

5   Conclusion

We proposed a novel L_0 minimization framework tailored for speckle reduction in ultrasound images. It involves a novel regularization term to seek for the L_0 sparsity of a new measurement GAP, which takes both gradient and phase information into consideration. In addition, we proposed an efficient and robust solver, which transfers the intractable L_0 minimization into several optimization steps with closed-form solutions. Experiments in synthetic and clinical ultrasound images verify that our approach outperforms state-of-the-art methods for speckle reduction. The proposed method has great potential to be applied to many intelligent ultrasound systems. In the future, we will accelerate our L_0 minimization using GPU, and also test our despeckling method on more clinical ultrasound images and applications.

Acknowledgement. We thank reviewers for the various valuable comments. This work was supported by the Hong Kong Research Grants Council General Research Fund (Project No. CUHK 14202514), Hong Kong Innovation and Technology Fund for Hong Kong-Shenzhen Innovation Circle Funding Program (No. GHP/002/13SZ and SGLH20131010151755080), the Natural Science Foundation of Guangdong Province

(Project No. 2014A030310381), the National Natural Science Foundation of China (Project No. 61233012 and 61305097), the Research and Development Project of Guangdong Key Laboratory of Robotics and Intelligent Systems (Grant No. ZDSYS20140509174140672), and Shenzhen Basic Research Program (Project No. JCYJ20150525092940988).

References

1. Vegas-Sanchez-Ferrero, G., Aja-Fernandez, S., Martin-Fernandez, M., Frangi, A.F., Palencia, C.: Probabilistic-driven oriented speckle reducing anisotropic diffusion with application to cardiac ultrasonic images. In: Jiang, T., Navab, N., Pluim, J.P.W., Viergever, M.A. (eds.) MICCAI 2010. LNCS, vol. 6361, pp. 518–525. Springer, Heidelberg (2010). doi:10.1007/978-3-642-15705-9_63
2. Flores, W.G., de Albuquerque Pereira, W.C., Infantosi, A.F.C.: Breast ultrasound despeckling using anisotropic diffusion guided by texture descriptors. Ultrasound Med. Biol. **40**, 2609–2621 (2014)
3. Wang, B., Cao, T., Dai, Y., Liu, D.C.: Ultrasound speckle reduction via super resolution and nonlinear diffusion. In: Zha, H., Taniguchi, R., Maybank, S. (eds.) ACCV 2009. LNCS, vol. 5996, pp. 130–139. Springer, Heidelberg (2010). doi:10.1007/978-3-642-12297-2_13
4. Cheng, H., Shan, J., Ju, W., Guo, Y., Zhang, L.: Automated breast cancer detection and classification using ultrasound images: a survey. Pattern Recogn. **43**, 299–317 (2010)
5. Esakkirajan, S., Vimalraj, C.T., Muhammed, R., Subramanian, G.: Adaptive wavelet packet-based de-speckling of ultrasound images with bilateral filter. Ultrasound Med. Biol. **39**, 2463–2476 (2013)
6. Balocco, S., Gatta, C., Pujol, O., Mauri, J., Radeva, P.: SRBF: speckle reducing bilateral filtering. Ultrasound Med. Biol. **36**, 1353–1363 (2010)
7. Coupé, P., Hellier, P., Kervrann, C., Barillot, C.: Nonlocal means-based speckle filtering for ultrasound images. IEEE Trans. Image Process. **18**, 2221–2229 (2009)
8. Yang, J., Fan, J., Ai, D., Wang, X., Zheng, Y., Tang, S., Wang, Y.: Local statistics and non-local mean filter for speckle noise reduction in medical ultrasound image. IEEE Trans. Image Process. **195**, 88–95 (2016)
9. Tay, P.C., Garson, C.D., Acton, S.T., Hossack, J.A.: Ultrasound despeckling for contrast enhancement. IEEE Trans. Image Process. **19**, 1847–1860 (2010)
10. Min, D., Choi, S., Lu, J., Ham, B., Sohn, K., Do, M.N.: Fast global image smoothing based on weighted least squares. IEEE Trans. Image Process. **23**, 5638–5653 (2014)
11. Xu, L., Lu, C., Xu, Y., Jia, J.: Image smoothing via L_0 gradient minimization. IEEE Trans. Image Process. **30**, 174 (2011)
12. Li, Z., Zheng, J., Zhu, Z., Yao, W., Wu, S.: Weighted guided image filtering. IEEE Trans. Image Process. **24**, 120–129 (2015)
13. Yu, Y., Acton, S.T.: Speckle reducing anisotropic diffusion. IEEE Trans. Image Process. **11**, 1260–1270 (2002)
14. Belaid, A., Boukerroui, D., Maingourd, Y., Lerallut, J.F.: Phase-based level set segmentation of ultrasound images. IEEE Trans. Image Process. **15**, 138–147 (2011)
15. Khare, A., Khare, M., Jeong, Y., Kim, H., Jeon, M.: Despeckling of medical ultrasound images using daubechies complex wavelet transform. Sig. Process. **90**, 428–439 (2010)

16. Cardoso, F.M., Matsumoto, M.M., Furuie, S.S.: Edge-preserving speckle texture removal by interference-based speckle filtering followed by anisotropic diffusion. Ultrasound Med. Biol. **38**, 1414–1428 (2012)
17. Buades, A., Coll, B., Morel, J.M.: A non-local algorithm for image denoising. In: CVPR, vol. 2, pp. 60–65 (2005)
18. Yu, J., Tan, J., Wang, Y.: Ultrasound speckle reduction by a Susan-controlled anisotropic diffusion method. Pattern Recogn. **43**, 3083–3092 (2010)
19. Morrone, M.C., Ross, J., Burr, D.C., Owens, R.: Mach bands are phase dependent. Nature **324**, 250–253 (1986)
20. Kovesi, P.: Symmetry and asymmetry from local phase. In: Tenth Australian Joint Conference on Artificial Intelligence, vol. 190. Citeseer (1997)
21. Kovesi, P.: Image features from phase congruency. Nature **1**, 1–26 (1999)
22. Boukerroui, D., Noble, J.A., Brady, M.: On the choice of band-pass quadrature filters. Nature **21**, 53–80 (2004)
23. Lindeberg, T.: Edge detection and ridge detection with automatic scale selection. Nature **30**, 117–156 (1998)
24. Geman, D., Yang, C.: Nonlinear image recovery with half-quadratic regularization. IEEE Trans. Image Process. **4**, 932–946 (1995)
25. Krishnan, D., Fergus, R.: Fast image deconvolution using hyper-Laplacian priors. In: Advances in Neural Information Processing Systems, pp. 1033–1041 (2009)
26. Daubechies, I., DeVore, R., Fornasier, M., Güntürk, C.S.: Iteratively reweighted least squares minimization for sparse recovery. Commun. Pure Appl. Math. **63**, 1–38 (2010)
27. Yi, S., Wang, X., Lu, C., Jia, J.: L_0 regularized stationary time estimation for crowd group analysis. In: Proceedings of the IEEE Conference on Computer Vision and Pattern Recognition, pp. 2211–2218 (2014)
28. Massoptier, L., Casciaro, S.: A new fully automatic and robust algorithm for fast segmentation of liver tissue and tumors from CT scans. Eur. Radiol. **18**, 1658–1665 (2008)

A Variational Model for Intrinsic Light Field Decomposition

Anna Alperovich$^{(\boxtimes)}$ and Bastian Goldluecke

University of Konstanz, Konstanz, Germany
anna.alperovich@uni-Konstanz.de

Abstract. We present a novel variational model for intrinsic light field decomposition, which is performed on four-dimensional ray space instead of a traditional 2D image. As most existing intrinsic image algorithms are designed for Lambertian objects, their performance suffers when considering scenes which exhibit glossy surfaces. In contrast, the rich structure of the light field with many densely sampled views allows us to cope with non-Lambertian objects by introducing an additional decomposition term that models specularity. Regularization along the epipolar plane images further encourages albedo and shading consistency across views. In evaluations of our method on real-world data sets captured with a Lytro Illum plenoptic camera, we demonstrate the advantages of our approach with respect to intrinsic image decomposition and specular removal.

1 Introduction

Intrinsic image decomposition aims at separating an illumination invariant reflectance image from an input color image. Such a decomposition has numerous applications in color enhancement, image segmentation, pattern recognition, and object tracking [1–3]. The separation of the shading component is used in BRDF estimation and shadow removal methods [4–6]. However, while intrinsic images have many applications, recovering them remains a substantial challenge for researchers. Estimation of intrinsic components is an ill-conditioned problem: a single image can be decomposed into infinitely many different combinations of reflectance and illumination. Thus, additional constraints or priors are needed to select an appropriate solution. Priors on reflectance (albedo) and shading are usually based on physical principles of light and object interaction, scene geometry, and material properties, as well as on expert knowledge of how intrinsic images should look like. Finally, decomposition into reflectance and illumination components is suitable only for diffuse (Lambertian) objects. According to the dichromatic model introduced by Shafer [7], if glossy (non-Lambertian) objects are present in a scene, a specular term should be taken into account. Many classical approaches fail when the target scene has non-Lambertian objects;

Electronic supplementary material The online version of this chapter (doi:10.1007/978-3-319-54187-7_5) contains supplementary material, which is available to authorized users.

S.-H. Lai et al. (Eds.): ACCV 2016, Part III, LNCS 10113, pp. 66–82, 2017.
DOI: 10.1007/978-3-319-54187-7_5

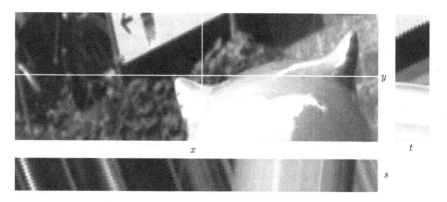

Fig. 1. The top-left image shows the center view of a light field parametrized by image coordinates x and y. On the bottom and right, the epipolar plane images (EPIs) for the white lines in the center view are shown, where s and t describe view point coordinates. As the camera moves, 3D scene points trace straight lines on the EPIs, whose slope corresponds to disparity. Any assignment of a property of a scene point to rays should be constant along these lines, which can be leveraged for consistent regularization [12].

as specularity depends on view point, it is hardly possible to estimate it from a single image.

To improve accuracy of intrinsic images, researchers use additional information, for instance, a video sequence instead of a single image, RGB-D imaging sensors, or manual labeling. This information may be incomplete, suffer from sensor noise, calibration errors, and be dependent on a human factor. Computing this information may be time consuming, require complex experiments, and special equipment. Thus, it is hardly possible to use it in for example industrial applications.

In this work, we leverage light fields for intrinsic image decomposition. 4D light fields are widely used in image analysis and computer graphics. The key idea of light field is to represent a scene not as a traditional 2D image, which contains information about accumulated intensity at each image point, but as a collection of images of the same scene from slightly different view points, see Fig. 1. The specific structure of the light field allows a wide range of applications. It is used for efficient depth estimation, virtual refocusing, automatic glare reduction as well as object insertion and removal [8–10]. Recently, the inherent structure of the light field was leveraged for shape and BRDF estimation [4,11].

Contributions. In this paper, we formulate and solve intrinsic light field decomposition by means of an optimization problem for albedo, shading, and specularity. As far as we are aware, this is the first time this problem is addressed for 4D light fields. Based on a detailed review of the state-of-the-art in intrinsic image decomposition, we propose priors for modeling all unknowns based on additional data available in the light field. Epipolar plane image constraints encourage albedo and shading to be constant for projections of the same scene point. By means of a novel term which is specific to light fields, we can also estimate specularity and highlights, and separate them from shading and albedo components. In experiments, we demonstrate that we outperform state-of-the-art intrinsic image decompostion based on RGB plus depth data [13], as well as an alternative approach to detect and remove light field specularity [10,14].

2   Related Work

Intrinsic images have been a challenging research topic for many years. First introduced by Barrow and Tenenbaum [15], they divide an observed image into the product of a reflectance and illumination image. According to Land and McCann [1], large discontinuities in pixel intensities correspond to changes in reflectance, and the remaining variation corresponds to shading. They proposed a Retinex theory that was successfully extended and implemented for intrinsic image decomposition by Tappen et al. [16], Chung et al. [17], Grosse et al. [18], Finalayson et al. [5,6], and many others.

Besides the Retinex approach, it is common to include additional regularization terms that describe certain physical properties of intrinsic components. Barron and Malik [19–21] introduce priors on reflectance, shape, and illumination to recover intrinsic images. Shen et al. [22] employ texture information. Finalyson et al. [23] search for an invariant image which is independent of lighting and shading. Gehler and Rother [24] model reflectance values drawn from a sparse set of basis colors. Bell et al. [25] also assume that reflectance values come from a predefined set which is unique for every image, then they iteratively adjust reflectance values in this set.

Recently, a significant improvement in intrinsic image decomposition was achieved by using richer types of input data. Having a sequence of images with depth information available allows to penalize albedo and shading consistency between different views, Lee et al. [26]. Depth or disparity information allows to incorporate spatial dependencies between pixels to construct shading prior, Jeon et al. [27]. Chen and Koltun [13] develop a model based on RGB-D information. They separate shading into two components: direct and indirect irradiance that significantly improved decomposition results. Barron and Malik [21] use depth to extend their SIRFS model [20] such that it is applicable for natural scenes.

Although decomposition algorithms nowadays achieve spectacular results for Lambertian scenes, their performance is suffering in the non-Lambertian case in the presence of highlights or specularity. In our paper, we will make use of the rich structure in the light field to estimate specularity for non-Lambertian objects. According to the dichromatic model introduced by Shafer [7], diffuse and specular reflections behave differently. Diffuse objects reflect incident light in multiple directions equally, thus, their color is independent of viewpoint. Specular objects reflect light in a certain direction that depends on orientation, and thus their color depends on viewpoint, light source color, and physical material properties. Blake and Bülthoff [28] made an extensive analysis of specular reflections, and propose a strategy for recovering 3D structure using specularity. Swaminathan et al. [29] study photometric properties of specular pixels, and model their motion depending on the surface geometry. Adato et al. [30] model specular flow with non-linear partial differential equations. Tao et al. [10,14] introduced depth estimation for glossy surfaces. They leverage the light field structure to cluster pixels in specular and specular-free groups, then they remove specular components from the input light field.

3   Intrinsic Light Field Model

Light Field Structure. We briefly describe the light field structure and review notation. For more detailed information, we refer to [12,31]. A light field is defined on 4D ray space $\mathcal{R} = \Pi \times \Omega$, which parametrizes rays $r = (x, y, s, t)$ by their intersection coordinates with two planes Π and Ω. Intersection with the focal plane Π gives view point coordinates (s, t), while the image plane Ω denotes image coordinates (x, y), see Fig. 1. A 4D light field is now a map $L : \mathcal{R} \to \mathbb{R}^n$ on ray space. It can be scalar or vector-valued for grey scale or color images, respectively.

Light Field Decomposition. We model an intrinsic light field as a function

$$L(r) = A(r)S(r) + H(r), \tag{1}$$

where the radiance L of every ray r is decomposed into albedo A, shading S, and specular component H. The functions $L, A, S, H : \mathcal{R} \to \mathbb{R}^3$ map ray space to RGB values. Albedo represents the color of an object independent of illumination and camera position. Shading describes intensity changes due to illumination, inter-reflections, and object geometry. Finally, specularity represents highlights that occur in case of non-Lambertian objects. They depend on illumination, object geometry, and camera position.

The common assumption in the literature related to intrinsic image decomposition is to model the shading component as mono-chromatic [16,18,24]. However, in case of multiple light sources or non-Planckian light, this modeling assumption is not sufficient. Thus, we further decompose shading into mono-chromatic shading s and trichromatic light source color C,

$$S(r) = s(r)C(r). \tag{2}$$

We directly compute the illumination component C in a pre-processing step with the illuminant estimation algorithm developed by Yang et al. [32] applied to the center view, assuming that it will be similar across views. After illumination color is computed, we exclude it from the original light field by switching to the new decomposition model

$$\frac{L(r)}{C(r)} = A(r)s(r) + \frac{H(r)}{C(r)} \tag{3}$$

which is illumination color free. Vector division is to be understood component-wise. As a further simplification, we obtain System (3) in linear form

$$L^{log}(r) = A^{log}(r) + 1s^{log}(r) + H^{log}(r, A, s, H) \tag{4}$$

by applying the logarithm. We now want to solve (4) with respect to albedo, shading, and specularity.

System (4) is ill-posed, since its number of variables is three times larger than the number of equations. To select a solution that agrees with physical meaning of intrinsic components, we pose it as an inverse problem and introduce a number of constraints or regularization terms for albedo, shading, and specularity. As usual, dependence of H^{log} on all arguments except r is ignored during optimization, and it is estimated as another independent component. We thus solve a global energy minimization problem where we weight the residual of (4) with different priors and regularization terms,

$$
\begin{aligned}
\underset{(A^{log}, s^{log}, H^{log})}{\arg\min} \Big\{ &\|L^{log}(r) - A^{log}(r) - 1 s^{log}(r) - H^{log}(r)\|_2^2 + \dots \\
&\dots + P_{\text{albedo}}(A^{log}) + P_{\text{shading}}(s^{log}) + P_{\text{spec}}(H^{log}) + J(A^{log}, s^{log}) \Big\}.
\end{aligned}
\tag{5}
$$

The priors P_{albedo} and P_{shading} for albedo and shading essentially apply the key ideas in intrinsic image decomposition to every subaperture image. They are defined in Sect. 4. The specularity prior P_{spec} is specific to light fields, and a main contribution of our work. It is described in detail in Sect. 5. Finally, the smoothing prior J across ray space encourages spatial smoothness and in particular consistency across different subaperture images. It relies on disparity, and is described together with the optimization framework in Sect. 6.

4   Albedo and Shading Priors

We start with describing the priors, which are the key for obtaining an accurate solution for intrisic light field decomposition from the variational model (5). In this section, we introduce the priors P_{albedo} and P_{shading} for albedo and shading, respectively.

Albedo. To model albedo, we combine ideas of Retinex theory, which is widely used to decompose an image into shading and reflectance components [16,17,33], with the idea that pixels with equal chromaticity are likely to have similar albedo [13,26,34]. Thus, the prior for albedo is the sum of two energies, $P_{\text{albedo}}(A^{log}) = E_{\text{retinex}}(A^{log}) + E_{\text{chroma}}(A^{log})$, corresponding to these two models.

Under the simplifying assumption that image derivatives in the log-domain are caused either by shading or reflectance, we classify the derivative at every ray as caused by shading or albedo. The idea is to compute a modified gradient field $\hat{g}$ which assigns a zero value to all derivatives that are caused by shading. The derivative classification is done with approach similar to Color Retinex used in [17,18]. A partial spatial derivative L_x of the light field is classified as albedo if neighbouring RGB vectors into the direction of differentiation are not parallel, or if it is above a certain magnitude. Thus, the modified derivative is

$$
\hat{g}_x = \begin{cases} L_x & \text{if } c_{x+1,y} \cdot c_{x,y} < \tau_{col} \text{ or } |L_x| > \tau_{grad}, \\ 0 & \text{otherwise.} \end{cases}
\tag{6}
$$

Above, $c = (r, g, b)^T$, the constant $\tau_{col} > 0$ is a threshold above which two vectors are assumed to be parallel, and $\tau_{grad} > 0$ is another user-defined constant. In a similar way, we estimate the modified partial derivative $\hat{g}_y$ in the second spatial direction.

The gradient of the albedo should be equal to the gradient field modified by retinex, thus we finally obtain the retinex energy

$$E_{retinex}(A^{log}) = \lambda_{retinex} \int_{\mathcal{R}} \left\| \partial_x A^{log}(r) - \hat{g}_x(r) \right\|^2 + \left\| \partial_y A^{log} r - \hat{g}_y(r) \right\|^2 dr.$$

(7)

The second regularization term is based on chromaticity similarities between adjacent rays. The basic idea is that if two neighboring rays of the same view have close chromaticity values, they have the same albedo. We use the chromaticity measure described by Chen and Koltun [13], which gives a weight $\alpha_{r,q}$ for how likely it is that two rays r and q have the same albedo,

$$\alpha_{r,q} = \left(1 - \frac{\left\| L^{ch}(r) - L^{ch}(q) \right\|}{\max\limits_{r' \in \Omega, \, q' \in N_A(r')} \left\| L^{ch}(r') - L^{ch}(q') \right\|} \right) \sqrt{L^{lum}(r) L^{lum}(q)}, \quad (8)$$

where $N_A(r)$ is a neighborhood of r, and L^{ch} and L^{lum} are chromaticity and luminance. The chromaticity energy

$$E_{chroma}(A^{log}) = \lambda_{chroma} \int_{\mathcal{R}} \sum_{N_A(r)} \alpha_{r,q} \left\| A^{log}(r) - A^{log}(q) \right\|^2 dr$$

(9)

now penalizes dissimilarity of albedos that have chromaticity measure $\alpha_{r,q}$ close to one. Note that we use a mixed continuous/discrete notation for r and q, as our choice of neighbourhood is inherently discrete, while we require a variational rayspace model in the optimization framework, see Sect. 6.

To construct the neighborhoods $N_A(r)$ for every ray $r \in \mathcal{R}$, we impose the assumption that spatially close points in $\mathbb{R}^3$ probably have similar albedo. We select k_A nearest neighbors in $\mathbb{R}^3$ for the point P on the scene surface intersected by r, and choose m_A out of k_A neighbors randomly. We believe that this connectivity strategy has several advantages over fully random connectivity: by defining neighbors we increase the chance to meet points with similar chromaticity, and by random connectivity within neighboring points we avoid disconnected chromaticity clusters.

Shading. The shading prior is also the sum of two components, $P_{shading} = E_{normal} + E_{spatial}$. To model the first component, we adopt the well-known assumption [13,26,35] that scene points which are spatially close to each other and share the same orientation are likely to have similar shading. To facilitate this, we construct the six-dimensional set

$$\Gamma := \left\{ \left(P(r), \, n(P(r)) \right) \, : \, r \in \mathcal{R} \right\},$$

where $P(r)$ is again the point of the scene surface intersected by r, and $n(P(r))$ the corresponding outer normal. The set of neighbours $N_S(r)$ now consists of

Fig. 2. The left image shows the center view of a light field captured with a Lytro Illum camera. The right image shows the specular mask obtained by our method.

the k_N-nearest neighbours of $r \in \mathcal{R}$ in the six-dimensional space Γ. The regularization term

$$E_{normal}(s^{log}) = \lambda_{normal} \int_{\mathcal{R}} \sum_{q \in N_S(r)} (s^{log}(r) - s^{log}(q))^2 \, dr \qquad (10)$$

thus penalizes shading components to be the same if corresponding 3D points are spatially close to each other and their outer normals have similar orientations.

To account for indirect shading, which is caused by inter-reflections between objects in a scene, we also include a purely spatial regularization term

$$E_{spatial}(s^{log}) = \lambda_{space} \int_{\mathcal{R}} \sum_{q \in N_D(r)} (s^{log}(r) - s^{log}(q))^2 \, dr, \qquad (11)$$

where the neighborhood $N_D(r)$ denotes the k_D nearest neighbors of the 3D scene point first intersected by r.

5   Prior for the Specular Component

In this section, we describe the specular prior in the variational energy (5). We first discuss the modeling assumptions, then show how to compute a mask for candidate specular pixels based on these assumptions, and finally construct the prior P_{spec}.

Modeling Assumptions. We combine several approaches to model specularity [2,10,14,28–30,36]. According to the specular motion model [28,29], specularity changes depend on surface geometry. For instance, regions of low curvature on a specular object create color intensity changes within different views. Specular regions of high curvature result in high pixel intensities in all subaperture views. Thus, curvature information can be useful while estimating specularity. In practice, however, it turns out that curvature estimation is very sensitive to inaccuracies of the 3D model of a scene. Imperfect disparity maps lead to a certain amount of noise in the estimated spatial coordinates, thus curvature information becomes highly unreliable. Instead of using curvature information directly, we therefore propose a heuristic approach that estimates candidate regions where specularity or highlights can occur. Our main modeling assumptions are thus:

S1. Specularity is view dependent.
S2. If a projected 3D point has high pixel intensities and its color is constant across all subaperture views, then the point may be part of a specular surface.
S3. If a projected 3D point has high variation in pixel intensities, and the color of the corresponding rays changes across subaperture views, then the point may belong to a specular surface.
S4. If a point is classified as specular, then it is a part of specular surface, and its local neighborhood in $\mathbb{R}^3$ may result in specular pixels from a certain viewing angle.
S5. The distribution of specularity is sparse.

Potential specular objects are identified based on magnitude and variation of pixel values over different views. We compute a specular mask for the center view, and propagate it to the remaining views according to disparity.

Computing the Specular Mask. Our proposed algorithm proceeds in 4 steps:

1. Let Ω_c be the image plane for the center view, and $V = \{(s_1, t_1), ..., (s_N, t_N)\}$ the set of remaining N view points.
   For every $p \in \Omega_c$, we compute the vector $\boldsymbol{\omega_p}$ of color intensity changes with respect to V according to

$$w_p^i = L_i(p + v_i d(p)), \; i = 1, \ldots, N. \tag{12}$$

   where $v_i = (s_c - s_i, t_c - t_i)$ is the view point displacement and $d(p)$ the estimated scalar disparity of p.
2. Identify pixels where color and intensity vary within subaperture views in three steps according to assumptions (S1) and (S3):
   - Filter out a percentage $\%n_{var}$ of pixels that have low luminance variation $\sigma(\boldsymbol{\omega_p})$, where by Ω_c^* we define a set of remaining pixels.
   - Exclude occlusion boundaries from Ω_c^*. To find occlusion boundaries, we compute the k-nearest neighbors in the image domain, and corresponding spatial coordinates in $\mathbb{R}^3$. If neighboring pixels in Ω_c^* are far away in $\mathbb{R}^3$, with distances larger than d_{occ}, then we classify those pixels as occlusion boundaries.
   - From the remaining pixels, finally exclude the percentage $\%n_{conf}$ with the lowest confidence scores similar to the approach proposed by Tao et al. [14].
   To compute confidence, we cluster the corresponding values of $\boldsymbol{\omega_p}$ in two groups using K-means. Let $m(p)$ be the cluster centroid with the larger mean $\mu(m)$. The confidence is computed as

$$c(p) = \exp\left(-\frac{1}{\sigma_{spec}^2}\left(\frac{\beta_0}{\mu(m)} + \frac{\beta_1}{\xi(m)}\right)\right), \tag{13}$$

Fig. 3. Estimated disparity maps for scenes captured with a Lytro Illum camera. From left to right: an outdoor scene with the ceramic owl, a tinfoil swan, an indoor scene with the same owl and a candle. Disparities range between -1.5 and 1.5.

where $\xi(\boldsymbol{m})$ denotes the sum of all distances within the cluster.

The confidence score grows with mean intensity and variation within the brightest cluster. Thus, we obtain pixels with varying values within sub-aperture views. Above, β_0 and β_1 are user-defined parameters that control exponential decay of brightness and distance terms, σ_{spec} scales the confidence function. We fix $\beta_0 = 0.5$, $\beta_1 = 10^{-3}$, $\sigma_{spec} = 2$.

3. Identify pixels where intensity is high and color not changing within all subaperture views according to assumption (S2). According to Tian and Clark [36], regions with high unnormalized Wiener entropy, which is defined as the product of RGB values over all pixels, are likely to be specular. We adopt their approach and also identify those regions.

4. Combine pixels found in steps 2 and 3 into the specular mask

$$h_{mask} = \begin{cases} 1, \text{ specular} \\ 0, \text{ non-specular,} \end{cases} \tag{14}$$

which is then grown according to assumption (S4) to include all k_{spec}-nearest neighbors for each specular pixel in the initial mask.

An example specular mask for a Lytro dataset is shown in Fig. 2.

Final Prior on Specularity. The specular component should be non-zero only within the candidate specular region given by the mask h_{mask} defined above. We therefore strongly penalize non-zero values outside this region by defining the final sparsity prior as

$$P_{\text{spec}}(H^{log}) = \lambda_{spec} \int_{\mathcal{R}} \gamma_w (1 - h_{\text{mask}}) \|H^{log}(\boldsymbol{r})\|^2 \, d\boldsymbol{r} + \lambda_{sparse} \|H^{log}\|_1. \tag{15}$$

where $\gamma_w \gg 0$ is a constant. We include an additional sparsity norm on H^{log} to account for assumption (S5).

Table 1. Main parameters for intrinsic image decomposition used in implementation.

Priors weights	Retinex	Optimization	Neighborhood	Specularity
$\lambda_{chroma} = 0.25$	$\tau_{col} = 0.99$	Global iterations = 10	$k_A = 100$	$\%n_{var} = 75\%$
$\lambda_{retinex} = 0.5$	$\tau_{grad} = 0.2$	Local iterations = 10	$m_A = 10$	$d_{occ} = 0.1$
$\lambda_{normal} = 0.5$		$\mu = 0.01$	$k_N = 20$	$\%n_{conf} = 85\%$
$\lambda_{space} = 0.5$		$\lambda = 0.25$	$k_D = 20$	Wiener entropy = 99 %
$\lambda_{spec} = 0.5$				$k_{spec} = 60$
$\lambda_{spec\_sparse} = 0.1$				$\gamma_w = 10$

6 Ray Space Regularization and Optimization

We summarize the previously defined terms in the variational energy (5) as a functional F, so that to obtain the light field decomposition we have to solve

$$\underset{(A^{log}, s^{log}, H^{log})}{\arg\min} \left\{ F(A^{log}, s^{log}, H^{log}) + J(A^{log}, s^{log}) \right\}. \tag{16}$$

As typical in intrinsic image decomposition, the overall optimization problem is rather complex. However, taking a detailed look at the individual terms, it turns out that we have a convex objective F. Furthermore, our intention is to define the global smoothness term J on ray space in a way that it enforces spatial smoothness within the views, as well as consistency with the disparity-induced structure on the epipolar plane images. Thus, the complete objective function exactly fits the light field optimization framework for inverse problems on ray space proposed by Goldluecke and Wanner [12]. The key advantage of this framework is that it is computationally efficient since it allows to solve subproblems for each epipolar plane image and view independently. Also, it is generic in the sense that we just need to provide a way to compute F and related proximity operators. We thus adopt their method to solve our problem.

In [12], the light field regularizer J in (16) is a sum of several contributions. First, there are individual regularizers J_{xs} and J_{yt} for each epipolar plane image, which depend on the disparity map and employ an anisotropic total variation to enforce consistency of the fields in the arguments with the linear patterns on the epipolar plane images, see Fig. 1. Second, for each view, there is a regularizer J_{st}, and as in the basic framework in [12], we use a simple total variation term for efficiency. In future work, we intend to move to something more sophisticated here.

Albedo and shading are independent of view point, thus their values should not vary between views. We want A^{log} and s^{log} to be constant in the direction of d, except at disparity discontinuities. We also regularize both components within each individual view as noted above. The complete regularizer can thus be written as

$$J(A^{log}, s^{log}) = \mu J_{xs}(A^{log}, s^{log}) + \mu J_{yt}(A^{log}, s^{log}) + \lambda J_{st}(A^{log}, s^{log}), \tag{17}$$

where $\lambda, \mu > 0$ are user-defined constants which correspond to the amount of smoothing on the separate views and EPIs, respectively. The objective is convex,

Fig. 4. Center view images showing the light field decomposition. The first row depicts a decomposition with our approach into: albedo, shading, and specularity. The second row illustrates Chen and Koltun's algorithm [13], where the center view image is decomposed into albedo and shading with the additional input of our generated depth map. The third row illustrates original image, diffuse, and specular images obtained by Tao et al.'s method [10]. Due to EPI constraints and the specularity term, our shading component does not include specular highlights as Chen and Koltun's result. Also, we removed most cast shadows from albedo image, while smoothness priors prevent albedo and shading discontinuities. Tao et al. detects less of the specular regions compared to ours. Their algorithm identifies mostly boundaries of specular regions, and removes those boundaries from the diffuse image. Disparity errors create variation of intensity values in subaperture views, which erroneously are classified as specularity on occlusion boundaries. Our algorithm detects the complete specular regions, since it is more robust to inaccurate disparity estimation.

so that we achieve global optimality. For details and the actual optimization algorithm we refer to [12].

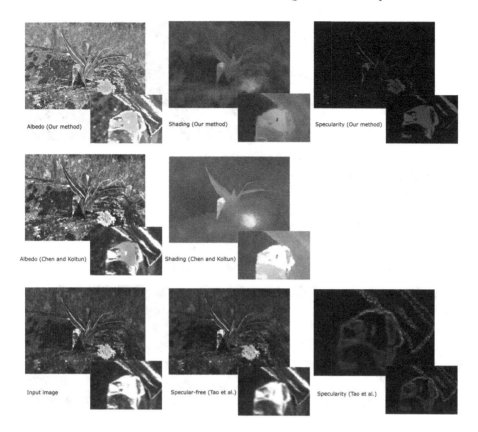

Fig. 5. Center view of the outdoor scene with an origami swan made from aluminum foil. Comparing shading and albedo images, we conclude that our algorithm detects more cast shadows than Chen and Koltun's algorithm. Our specular component contains more correctly classified glossy regions than the one produced by algorithm of Tao et al. We observe that their approach predominantly detects boundaries of specular regions, thus only these are removed in the generated specular free image.

7   Results

We validate our decomposition method on light fields captured with a Lytro Illum plenoptic camera, as well as on synthetic and gantry data sets provided by Wanner et al. [37]. In the paper, we present selected results for real world indoor and outdoor scenes, the rest we show in the supplementary material. While benchmark datasets for evaluating intrinsic image decomposition are presented in [18,25,38], those data sets are designed for algorithm evaluation on single RGB, RGB+D images, or on optic flow; they are not applicable to our light field based method. Since there are thus no ground truth intrinsic light fields available so far, we evaluate our method visually and with qualitive comparisons, deferring rendering of a novel benchmark to future work.

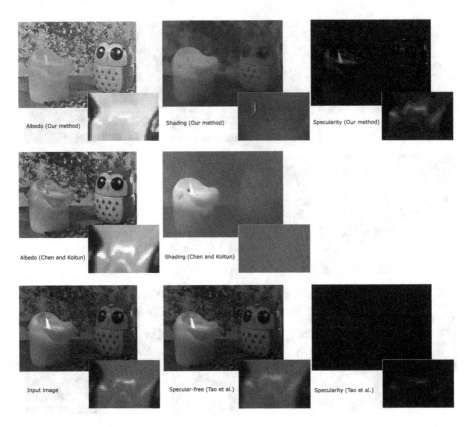

Fig. 6. Center view of an indoor scene with several light sources and non trivial chromaticity. We observe a difference in lighting of albedo images. In our approach, the albedo image is illumination free, compared to albedo images produced with Chen and Koltun's algorithm. A reason can be that we first compute illumination color and exclude it from the optimization model, while in Chen and Koltun's approach illumination color is included in the optimization. Both shading components are specular-free, while the albedo component of Chen and Koltun algorithm contains specularity. There is a near to zero specular component detected with Tao et al.'s algorithm. This can be explained by the bad initial disparity estimation which causes erroneous pixel classification. Our approach outperforms Tao et al., since our specular detection algorithm is occlusion aware and also analyses regions where pixels have high intensities.

To recover a 3D model and estimate normals, we perform disparity estimation with the multi-view stereo method described in [9], with an improved more occlusion-aware data term and refined with further smoothing using a generalized total variation regularizer, see estimated disparity labels in Fig. 3. The main algorithm parameters and their values are presented in Table 1. Our method is implemented in Matlab, version R2015b, with run-times on a PC with Intel(R) Core i7-4790 CPU 3.60 GHz and an NVIDIA GeForce GTX 980.

Fig. 7. Potential locations for specularity that were detected with our algorithm. The left image shows candidate specular regions for the origami swan, the right image depicts the specular mask for the owl and candle scene.

Evaluation Results. Since there are no intrinsic image decomposition algorithms that consider specularity, to compare results of our specular term we select a recent algorithm for depth estimation and specular removal developed for light field cameras by Tao et al. [10]. To compare albedo and shading terms, we investigated recently published algorithms that employ 3D information. There are several papers where depth information is used for intrinsic image decomposition [13,21,26,27]. We selected the algorithm developed by Chen and Koltun [13] to compare against, since it outperforms other algorithms that use 3D information. For both comparisons, we use the authors implementations with default parameter setting. Figures 4, 5 and 6 illustrate original image, our proposed decomposition method, Chen and Koltun [13], and Tao et al. [10]. For all images, contrast was enhanced using the Matlab function `imadjust` for better visualization. Figure 7 illustrates specular masks for origami swan and owl with candle light fields.

We also compared runtime of the algorithms. The Chen and Koltun algorithm converges in 20–30 min for a single image, the method by Tao et al. (including depth estimation) takes 60 min. Our approach evaluated on a light field with a cross-hair shaped subset of 17 views from a light field with 9×9 views in total converges in 30–40 min, which amounts to 1.7–2.4 min per frame.

8 Conclusions

In this work, we propose the first approach towards solving the intrinsic 4D light field decomposition problem while leveraging the disparity-induced structure on the epipolar plane images. In contrast to existing intrinsic image algorithms, the dense collection of views in a light field allows us to define an additional specular term in the decomposition model, so that we can optimize over the specular component as well as albedo and shading by minimizing a single variational functional. As the inverse decomposition problem is embedded in a recent framework for light field labeling [12], we can ensure that albedo and shading estimates are consistent across and use information from all views. Experiments demonstrate that we outperform both a state-of-the-art intrinsic image decomposition method employing additional depth information [13], as well as a light field based method for specular removal [10,14] on challenging non-Lambertian scenes.

Acknowledgements. This work was supported by the ERC Starting Grant "Light Field Imaging and Analysis" (LIA 336978, FP7-2014).

References

1. Land, E.H., McCann, J.J.: Lightness and retinex theory. J. Opt. Soc. Am. **61**, 1–11 (1978)
2. Shroff, N., Taguchi, Y., Tuzel, O., Veeraraghavan, A., Ramalingam, S., Okuda, H.: Finding a needle in a specular haystack. In: 2011 IEEE International Conference on Robotics and Automation (ICRA), pp. 5963–5970 (2011)
3. Beigpour, S., van de Weijer, J.: Object recoloring based on intrinsic image estimation. In: IEEE International Conference on Computer Vision (2016)
4. Wang, T.C., Chandraker, M., Efros, A., Ramamoorthi, R.: SVBRDF-invariant shape and reflectance estimation from light-field cameras. In: Proceedings of International Conference on Computer Vision and Pattern Recognition (2016)
5. Finlayson, G.D., Hordley, S.D., Drew, M.S.: Removing shadows from images using retinex. In: Color Imaging Conference: Color Science and Engineering Systems, Technologies, and Applications (2002)
6. Finlayson, G.D., Hordley, S.D., Lu, C., Drew, M.S.: On the removal of shadows from images. IEEE Trans. Pattern Anal. Mach. Intell. **28**, 59–68 (2006)
7. Shafer, S.: Using color to separate reflection components. Color Res. Appl. **10**, 210–218 (1985)
8. Levoy, M.: Light fields and computational imaging. Computer **39**, 46–55 (2006)
9. Wanner, S., Goldluecke, B.: Variational light field analysis for disparity estimation and super-resolution. IEEE Trans. Pattern Anal. Mach. Intell. **36**, 606–619 (2014)
10. Tao, M., Su, J.C., Wang, T.C., Malik, J., Ramamoorthi, R.: Depth estimation and specular removal for glossy surfaces using point and line consistency with light-field cameras. IEEE Trans. Pattern Anal. Mach. Intell. **38**, 1155–1169 (2015)
11. Tao, M., Srinivasan, P., Hadap, S., Rusinkiewicz, S., Malik, J., Ramamoorthi, R.: Shape estimation from shading, defocus, and correspondence using light-field angular coherence. IEEE Trans. Pattern Anal. Mach. Intell. **39**(3), 546–560 (2016)
12. Goldluecke, B., Wanner, S.: The variational structure of disparity and regularization of 4D light fields. In: Proceedings of International Conference on Computer Vision and Pattern Recognition (2013)
13. Chen, Q., Koltun, V.: A simple model for intrinsic image decomposition with depth cues. In: Proceedings of International Conference on Computer Vision (2013)
14. Tao, M.W., Wang, T.-C., Malik, J., Ramamoorthi, R.: Depth estimation for glossy surfaces with light-field cameras. In: Agapito, L., Bronstein, M.M., Rother, C. (eds.) ECCV 2014. LNCS, vol. 8926, pp. 533–547. Springer, Cham (2015). doi:10. 1007/978-3-319-16181-5_41
15. Barrow, H.G., Tenenbaum, J.M.: Recovering intrinsic scene characteristics from images. Comput. Vis. Syst. **23**, 3–26 (1978)

16. Tappen, M.F., Freeman, W.T., Adelson, E.H.: Recovering intrinsic images from a single image. IEEE Trans. Pattern Anal. Mach. Intell. **27**, 1459–1472 (2005)
17. Chung, Y., Cherng, S., Bailey, R.R., Chen, S.W.: Intrinsic image extraction from a single image. J. Inf. Sci. Eng. **25**, 1939–1953 (2009)
18. Grosse, R., Johnson, M.K., Adelson, E.H., Freeman, W.T.: Ground truth dataset and baseline evaluations for intrinsic image algorithm. In: Proceedings of International Conference on Computer Vision (2009)
19. Barron, J.T., Malik, J.: High-frequency shape and albedo from shading using natural image statistics. In: Proceedings of International Conference on Computer Vision and Pattern Recognition (2011)
20. Barron, J.T., Malik, J.: Color constancy, intrinsic images, and shape estimation. In: Proceedings of European Conference on Computer Vision (2012)
21. Barron, J.T., Malik, J.: Intrinsic scene properties from a single RGB-D image. IEEE Trans. Pattern Anal. Mach. Intell. **38**, 690–703 (2015)
22. Shen, L., Tan, P., Lin, S.: Intrinsic image decomposition with non-local texture cues. In: Proceedings of International Conference on Computer Vision and Pattern Recognition (2008)
23. Finlayson, G.D., Drew, M.S., Lu, C.: Intrinsic images by entropy minimization. In: Pajdla, T., Matas, J. (eds.) ECCV 2004. LNCS, vol. 3023, pp. 582–595. Springer, Heidelberg (2004). doi:10.1007/978-3-540-24672-5_46
24. Gehler, P.V., Rother, C., Kiefel, M., Zhang, L., Schölkopf, B.: Recovering intrinsic images with a global sparsity prior on reflectance. In: NIPS (2011)
25. Bell, S., Bala, K., Snavely, N.: Intrinsic images in the wild. ACM Trans. Graph. (SIGGRAPH) **33**, 159:1–159:12 (2014)
26. Lee, K.J., Zhao, Q., Tong, X., Gong, M., Izadi, S., Lee, S.U., Tan, P., Lin, S.: Estimation of intrinsic image sequences from image+depth video. In: Fitzgibbon, A., Lazebnik, S., Perona, P., Sato, Y., Schmid, C. (eds.) ECCV 2012. LNCS, vol. 7577, pp. 327–340. Springer, Heidelberg (2012). doi:10.1007/978-3-642-33783-3_24
27. Jeon, J., Cho, S., Tong, X., Lee, S.: Intrinsic image decomposition using structure-texture separation and surface normals. In: Fleet, D., Pajdla, T., Schiele, B., Tuytelaars, T. (eds.) ECCV 2014. LNCS, vol. 8695, pp. 218–233. Springer, Cham (2014). doi:10.1007/978-3-319-10584-0_15
28. Blake, A., Bülthoff, H.: Shape from specularities: computation and psychophysics. Phil. Trans. R. Soc. Lond. B **331**, 237–252 (1991)
29. Swaminathan, R., Kang, S.B., Szeliski, R., Criminisi, A., Nayar, S.K.: On the motion and appearance of specularities in image sequences. In: Heyden, A., Sparr, G., Nielsen, M., Johansen, P. (eds.) ECCV 2002. LNCS, vol. 2350, pp. 508–523. Springer, Heidelberg (2002). doi:10.1007/3-540-47969-4_34
30. Adato, Y., Vasilyev, Y., Ben-Shahar, O., Zickler, T.: Toward a theory of shape from specular flow. In: Proceedings of International Conference on Computer Vision (2007)
31. Bolles, R.C., Baker, H.H., Marimont, D.H.: Epipolar-plane image analysis: an approach to determining structure from motion. Int. J. Comput. Vision **1**, 7–55 (1987)
32. Yang, K., Gao, S., Li, Y.: Efficient illuminant estimation for color constancy using grey pixels. In: Proceedings of International Conference on Computer Vision and Pattern Recognition (2015)
33. Weiss, Y.: Deriving intrinsic images from image sequences. In: Proceedings of International Conference on Computer Vision (2001)
34. Finlayson, G.D., Drew, M.S., Lu, C.: Entropy minimization for shadow removal. IJCV **85**(1), 35–57 (2009)

35. Tao, M., Srinivasan, P., Malik, J., Rusinkiewicz, S., Ramamoorthi, R.: Depth from shading, defocus, and correspondence using light-field angular coherence. In: Proceedings of International Conference on Computer Vision and Pattern Recognition (2015)

36. Tian, Q., Clark, J.J.: Real-time specularity detection using unnormalized wiener entropy. In: Computer and Robot Vision (CRV), pp. 356–363 (2013)

37. Wanner, S., Meister, S., Goldluecke, B.: Datasets and benchmarks for densely sampled 4D light fields. In: Vision, Modelling and Visualization (VMV) (2013)

38. Butler, D.J., Wulff, J., Stanley, G.B., Black, M.J.: A naturalistic open source movie for optical flow evaluation. In: Fitzgibbon, A., Lazebnik, S., Perona, P., Sato, Y., Schmid, C. (eds.) ECCV 2012. LNCS, vol. 7577, pp. 611–625. Springer, Heidelberg (2012). doi:10.1007/978-3-642-33783-3_44

Dense Depth-Map Estimation and Geometry Inference from Light Fields via Global Optimization

Lipeng Si and Qing Wang[✉]

School of Computer Science, Northwestern Polytechnical University,
Xi'an 710072, People's Republic of China
qwang@nwpu.edu.cn

Abstract. Light field camera captures abundant and dense angular samplings in a single shot. The surface camera (SCam) model is an image gathering angular sample rays passing through a 3D point. By analyzing the statistics of SCam, a consistency-depth measurement is evaluated for depth estimation. However, local depth estimation still has limitations. A global method with pixel-wise plane label is presented in this paper. Plane model inference at each pixel not only recovers depth but also local geometry of scene, which is suitable for light fields with floating disparities and continuous view variation. The 2nd order surface smoothness is enforced to allow local curvature surfaces. We use a random strategy to generate candidate plane parameters and refine the plane labels to avoid falling in local minima. We cast the selection of defined labels as fusion move with sequential proposals. The proposals are elaborately constructed to satisfy sub-modular condition with 2nd order smoothness regularizer, so that the minimization can be efficiently solved by graph cuts (GC). Our method is evaluated on public light field datasets and achieves the state-of-the-art accuracy.

1 Introduction

Light field camera captures not only the intensity but also the direction of rays in a 3D space, thus light field data contains geometry structure information and reflectance properties of scene. The information of light fields plays an essential role for many applications in image rendering and computational photography. However, recovering the implicit geometry information involving depth estimation from light fields still remains a challenging problem.

Traditional depth recovery approaches, known as stereo matching, extracting disparities by finding correspondence from binocular or multi-view cameras. Light field camera like moving camera on gantry [1] or camera array [2] is no difference from multi-view cameras. But from the perspective of rays, as the view number ascends, the statistical property on angular sampling of light fields is becoming remarkable. The commercial light field camera such as Lytro [3] and Raytrix [4] is capable of capturing a few hundred of views. That dense angular sampling produces special structure of light fields involving depth cues more than

© Springer International Publishing AG 2017
S.-H. Lai et al. (Eds.): ACCV 2016, Part III, LNCS 10113, pp. 83–98, 2017.
DOI: 10.1007/978-3-319-54187-7_6

disparities, such as defocus and Epipolar plane images (EPIs) [5]. These structure and depth cues benefit light field methods outperforming stereo matching on depth estimation.

There has been a lot of work studied on EPI based depth estimation approaches, but most work is local in nature, global method which is widespread in stereo matching, is seldom involved in depth estimation for light field camera for two reasons: First, light field camera has many views so that the computational cost of global method would be too heavy. Second, EPIs separate neighboring pixels into slices with horizontal or vertical direction, so that patchwise smoothness regularizer is hardly enforced. In this paper, we consider the views in one shot by back projecting rays passing through the same 3D point to view plane into an image called surface camera (SCam) [6]. We analyze the statistics of SCam and partition pixels of SCam into on-surface and off-surface parts. Then we build a cost function with on-surface part for every possible SCam, and encode it in a global optimization framework.

Markov random field (MRF) model [7] is widely used in global method, for its applicable with various energy functions. The optimal solution forms an assignment with multi-label. However, light fields have continuous view variation and narrow disparity range, then using discrete disparity label in MRF would cause 'stair-case' effect. For acquiring accurate depth estimation, labels with floating value disparity in continuous variation space should be assigned to each pixel. In this paper, we combine the depth cue of surface camera and scene geometry prior. Under the assumption that natural scenes are consisted of locally planar structures, each pixel is assigned a 3D label represented by a plane parameter, and 2nd order smooth regularizer is enforced in the global optimization. The data cost based on analysis of SCam is presented in Sect. 3. The global optimization procedure is described in Sect. 4. Experimental results are presented in Sect. 5.

2   Related Work

2.1   Depth from Light Fields

Light field is the concept of acquiring the entire information about light emitted by a scene, including every 3D point, ray direction, wavelength and time, so called 7D plenoptic function [8,9]. A 4D light field [1] or Lumigraph [10] reduced the dimensionality of plenoptic function to 4D. Under the two plane parametrization [1], each ray is determined with a 4D coordinates by intersecting with two parallel plane. Thus, a light field is denoted as $L(x, y, s, t)$. (s, t) can be seen as the index of sub views, and (x, y) denotes the coordinates of each view.

Bolles et al. [5] first used EPIs to reconstruct 3D geometry by detecting edges, peaks and troughs with line fitting in the EPI. Criminisi et al. [11] proposed an iterative procedure to segment EPI lines into EPI-tubes representing different depth layers, EPI lines detected by shearing and analyzing photo-consistency in vertical direction. Wanner et al. [12] employed a structure tensor to produce an orientation estimation at each pixel on EPI and coherence of structure tensor to measure the confidence level for every estimation. Their method can give a local depth estimation at every pixel on the EPIs. However, the local estimation suffer from several limitations, such as noises, occlusion, depth discontinuous area, texture-less and specular areas. They proposed a global integration to combine the local estimation on vertical EPI and horizontal EPI of the same point. Thus, by estimating a depth map at a given view (s^*, t^*), only the EPIs on a "cross line" of fixing s^* and t^* are considered. Tosic et al. [13] transformed the EPIs onto a scale-depth space to detect a local minima, but still only the cross-hair views are used. Kim et al. [14] extracted EPI edge to estimate the depth and take a consistency measure of the line on EPI as reliability. It is still a local method but they use a fine to coarse process propagating depth estimation with high reliability to smoothen outliers. All above work does not involve global method with patch-wise smoothness regularization. There are some methods using global optimization with the cost function similar to traditional stereo matching.

Tao et al. [15] proposed an approach combining defocus cue and correspondence as cost function. But the global regularizer they enforced is front-parallel plane assumption, which causes the results with severe 'stair-case' effect. Chen et al. [16] presented a new depth cue on SCam. Every point at a given depth is projected onto all the sub-view image, and the projected pixels consist of an angular sample image called surface camera. They proposed a bilateral metric to estimate the probability of pixel to be occluded. The statistical property is apparently effcctive only for light field with large numbers of densely sampled views. Wang et al. [17] proposed a different occlusion predicting model on SCam. They found that the occlusion is related with edge on spatial image patch. Based on the predictor, un-occluded pixels are selected and consistency and defocus cost is obtained. They used MRF model to enforce neighbor patch smoothness, but still using front-parallel plane prior.

2.2   Global Optimization in Stereo Matching

It seeks a soft label assignment to each pixel in global optimization. A global energy function to minimize is accumulated of all pixels defined on label space. The label costs usually consist of a data term and a smooth term. The smooth term is mostly defined on a 2D MRF model, which enforce the smoothness on nearest neighbor pixels. The energy minimization problem can be solved using various methods, such as graph cut (GC) [18–20], belief propagation (BP) [21].

The computational cost of global optimization depends on the size of label space and resolution of images. The label space generally consist of limited discrete labels. The intuitive way is taking discrete depth value as label, which implicitly assumes that the neighboring pixels should lay on the same depth slice, also called frontal parallel plane assumption. It is common used in global stereo matching methods because it can be efficiently solved by standard graph cuts. However, the parallel plane assumption would cause 'stair-case' effect on depth map when the neighboring pixels lay on a slanted plane or continuous surface. Recently, 3D labels are used by assigning a local 3D plane $d = a_p x + b_p y + c_p$ to each pixel. It can not be directly optimized because the label space $(a, b, c) \in \mathbb{R}^3$

is nearly infinite and hard to discretize. Segmentation based methods [22] generate plane candidates by plane fitting on each segment according to a roughly depth estimation by SFM (structure from motion). Ladicky et al. [23] proposed an iterative manner jointly optimize object-level segmentation and depth estimation. Segmentation based method can only reconstruct piecewise planar geometry structure and segmentation itself is a difficult problem in computer vision.

[24,25] adopted pixel-wise plane labels which could enforce 2nd order surface smoothness. They also need an external method to get a rough depth map for plane fitting. [24] fitted plane on various segments sets with various segmentation algorithm and parameters. [25] generated plane candidates at a random chosen pixel and its neighboring patch. The energy minimization problem is solved by QPBO-GC [26], which is less efficient than standard-GC, since the smooth terms on their plane label no longer satisfies the sub-modular condition. Other work use PatchMatch [27] to infer 3D planes at each pixel, and propagation planes from pixel to pixel in sequential way. In work [27], PMBP [28] is used for MRF optimization. Taniai et al. [29] introduced the spatial propagation in GC optimization, called locally shared labels, also used in this paper. GC is more efficient since GC update the labels of all nodes in MRF in a global way, not sequentially like PM and BP.

In this work, we enforce 2nd order surface smoothness for light field depth estimation. We build the data term from depth cue of SCam based on Chen et al.'s analysis [16] but with two revisions, including that the central pixel value of Scam is substituted by mean value which is robust to noises according to [17] and we revised the probability function to be more suitable for general situation than just occlusion model. The main contribution of this paper is that we impose a pixel-wise geometry inference by assigning random plane labels in global energy minimization framework. The smooth term is similar to Olsson et al. [25], but the optimization procedure is much simplified. The key improvement is the generation of proposals. The proposal scheme of [25] is more related to Woodford et al. [24]. They generated proposals by fitting local planes using RANSAC. So They need an initial depth map which should be obtained by a fast local stereo matching method. We don't fit any plane but generate all the plane parameters randomly, then we need no initialization by external method. And because of the design of proposals, our algorithm could use standard-GC which has faster convergency rate than QPBO-GC used in [25].

3   Surface Camera

3.1   Surface Camera Construction

Under the 2PP light field, we describe how to construct a surface camera. Our analysis is all based on reference view (s_r, t_r), e.g., the central camera of the light field. Suppose p is a pixel in image $I_{(s_r, t_r)}$ with coordinates (x_p, y_p). Along the ray $[x_p, y_p, s_r, t_r]$, suppose there is a 3D space point p_d denoted as (p, d), where d is depth value as distance from camera plane, (p, d) means a space point projected at pixel p on reference image $I_{(s_r, t_r)}$ with depth d. We project a ray from every camera (s,t) passing through p_d to sensor plane at $[x(s,t), y(s,t)]$. Then we collect every angular sample $I_{(s,t)}[x(s,t), y(s,t)]$ of p_d from all cameras and organize them as an st image with resolution equal to the number of cameras. The image is called a surface camera (SCam) or surface light field at p_d, denoted as A_{p_d}.

The pixel $[x(s,t), y(s,t)]$ could be floating value in practice, then we use bilateral interpolation to construct the SCam.

3.2   Consistency Cost of SCam

Actually, when p_d is a scene point, all pixels of SCam would be the correspondence point of pixel p. According to photo-consistency assumption in stereo matching, the variance or other consistency measure of the pixels in SCam should give the lowest value among all possible d. If we construct the cost function by computing consistency measure involving all pixels in SCam, there would be no difference with a traditional stereo matching method. In fact, SCam shows view dependent property due to many limitations, such as noises, occlusion and non-Lambertian surfaces. But in most situation, only part of views are affected by those limitations. It is to say, SCam shows statistically clustering-consistency. We assume at least half of rays can be correctly projected to p_d. Let $\Omega^v_{p_d}$ be the set of rays that could reach p_d, then the consistency measure on SCam is defined as

$$C(p,d) = \frac{1}{|\Omega^v_{p_d}|} \sum_{(s,t)\in\Omega^v_{p_d}} \rho(A_{p_d}(s,t) - \bar{A}_{p_d}), \tag{1}$$

where $|\Omega|$ is the size of set Ω and $\rho(x) = 1 - e^{-\frac{x^2}{2\sigma^2}}$ is a distance function, where σ is a predefined parameter to control the sensitivity of the function to large distances. $\bar{A}_{p_d}$ is the mean of the SCam A_{p_d}, in this paper, we use a MeanShift to obtain $\bar{A}_{p_d}$.

We adopt a consistency metric $P_{A_{p_d}}$ to estimate the probability choosing which pixel in A_{p_d} into $\Omega^v_{p_d}$.

$$P_{A_{p_d}}(s,t) = e^{-\frac{(A_{p_d}(s,t) - \bar{A}_{p_d})^2}{2\sigma_c^2}}, \tag{2}$$

Then we choose at least half of the views with the highest probability in $\Omega^v_{p_d}$ by sorting $P_{A_{p_d}}(s,t)$ of all pixels in A_{p_d}

$$\Omega^v_{p_d} = \{(s,t)|P_{A_{p_d}}(s,t) \geq \min(P_{Thresh}, P^{N_v}_{A_{p_d}})\}, \tag{3}$$

where P_{Thresh} is a predefined threshold and $P^{N_v}_{A_{p_d}}$ is the N_v-th highest probability $P_{A_{p_d}}(s,t)$ among all pixels in A_{p_d}.

$C(p,d)$ gives the consistency cost for pixel p at given depth d. We compute $C(p,d)$ at all possible d to obtain a consistency-depth profile, then a local depth estimation is chosen at the lowest value on the profile for each pixel. But the local estimation is not confident at many pixels especially in textureless region. So we need global method to optimize the result.

4   Global Optimization

4.1   Energy Function

This section describes the denotation of the energy function we are about to minimize. To build an MRF model suitable for solving by GC, we construct a graph with every pixel of reference view image $p \in (P \subset \mathbb{Z}^2)$as the nodes. The energy function is about the label space. Each node is assigned a depth label in label space $f_p \in (S \subset \mathbb{R})$, then a dense depth-map or a configuration of depth label $f: P \rightarrow S$ is obtained. The goal of optimization is to seek an optimal configuration to minimize the energy function as

$$E(f) = \sum_{p \in P} U_p(f_p) + \lambda \sum_{p \in P} \sum_{q \in N(p)} V_{pq}(f_p, f_q), \tag{4}$$

where the first term is called data cost, $U_p(f_p)$ represent the cost of pixel p to be assigned the label f_p. The second term is smooth term. V_{pq} defines a pairwise cost of pixel (p, q) to have labels (f_p, f_q). $q \in N(p)$ is the neighboring pixel of p. λ control the weight between data cost and smooth cost.

Data Term. In this paper, a depth label f_p represents a 3D plane parameters $\{a_p, b_p, c_p\} \in (S \subset \mathbb{R}^3)$, and a pixel q with coordinates (x_q, y_q) belonging to the plane would have depth value $d_q(f_p) = a_p x_q + b_p y_q + c_p$. We already have the consistency cost $C(p, d)$ for each pixel p of reference view image at any floating depth value d discussed in Sect. 3. For scene geometry inference, the data cost U of pixel p should be measured by accumulating the cost of pixels within a large patch W_p at the same label assumption f_p

$$U_p(f_p) = \sum_{q \in W_p(R)} \omega_{pq} C(q, d_q(f_p)), \tag{5}$$

where the patch $W_p(R)$ is centered at p with radius R. ω_{pq} is the adaptive patch weight to measure the coherence of pixel p and its neighboring pixels according to their color intensity, defined as

$$\omega_{pq} = e^{-\|I(p) - I(q)\|_1 / \gamma}, \tag{6}$$

where γ is a predefined parameter.

Smooth Term. The smooth term is a curvature-based, second-order smooth regularization term [25] defined as

$$V_{pq}(f_p, f_q) = max(\omega_{pq}, \varepsilon) min(\bar{V}_{pq}(f_p, f_q), \tau_{dis}), \tag{7}$$

where ω_{pq} weights the coherence of neighboring pixels defined as equation, for discontinuity preserving, neighboring pixels with similar color intensity should probably lay on the same smooth surface. ε is a lower bound to increase the

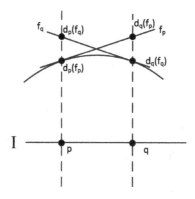

Fig. 1. Geometric interpretation of second order surface smoothness penalty

robustness of algorithm. $\bar{V}_{pq}(f_p, f_q)$ penalizes pixels pair (p, q) when they have different labels (f_p, f_q) defined as

$$\bar{V}_{pq}(f_p, f_q) = |d_p(f_p) - d_p(f_q)| + |d_q(f_q) - d_q(f_p)|, \tag{8}$$

as shown in Fig. 1, f_p, f_q are two plane assumptions at pixel p, q, when p, q are on the same plane, $\bar{V}_{pq}(f_p, f_q) = 0$; the closer of depth value of p, q, and the smaller variation of the normals of f_p, f_q, the lower penalty the term gives. Thus the smooth term encourage a local smooth curvature. The smooth term should satisfy the sub-modular condition that the optimization could be solved by standard GC, which would be discussed in next section.

4.2 Energy Minimization

Graph cuts method solves a binary minimization problem as

$$\min_{x \in L^n} E(x) = \sum_{p \in P} U_p(x_p) + \lambda \sum_{p \in P} \sum_{q \in N(p)} V_{pq}(x_p, x_q), \tag{9}$$

with label set $L = \{0, 1\}$, and the smooth term V_{pq} should satisfy the sub-modular condition

$$V(0, 0) + V(1, 1) \leq V(0, 1) + V(1, 0), \tag{10}$$

α-expansion [19] extends label set to multi-label. While at each expansion move, it is still a binary selection problem by choosing the label changed to label $\alpha \in L$ or not. Lempitsky et al. [30] proposed a method called fusion move to minimize (9) when the label space is in a continuous space $L \in \mathbb{R}$. At each move, given two configurations $f_0\colon x_p = x_0$ and $f_1\colon x_p = x_1$, by solving

$$\min_{z \in \{0,1\}^n} E(z \cdot x_1 + (1 - z) \cdot x_0), \tag{11}$$

the two configurations are fused. Apparently, α expansion is a special case of fusion move. The result in [26] proved

$$E(z \cdot x_1 + (1 - z) \cdot x_0) \leq \min(E(x_0), E(x_1)), \tag{12}$$

then, in the optimization process, f_0 is seen as current configuration, and f_1 as a proposal, therefore the energy can be reduced iteratively by fusing current configuration and different proposals until convergency. The key step is how the proposal is generated from the label space and guarantees the sub-modular condition is satisfied on each proposal.

Random Plane Generation. Firstly, we generate the label space $L \subset \mathbb{R}^3$. Since the parameter space is 3D and continuous, it is impossible to enumerate or even discretize the whole space into limited sampling. We adopt a random strategy to generate plane parameters samplings, and refine them in each iteration step. To get sufficiently large number of samplings but avoid too many combinations of labels for proposal generating, we generate a label set including a few random plane parameters for each pixel respectively, and neighboring pixels could propagate their label set to each other.

Suppose the label set for each pixel p has K labels, as $L_p = \{l_p^{(0)}, l_p^{(1)}, ..., l_p^{(K-1)}\}$, $l_p^{(i)} = (a, b, c)^T \in S$. Every label $l_p^{(i)}$ for pixel p is generated in this way: first give a random depth value z_0 in a reasonable depth range, then generate a random unit vector $n = (n_x, n_y, n_z)^T$ as the normal direction of plane, thus the plane parameter can be obtained according to the coordinates $(x_p, y_p)^T$ of pixel p

$$a_p = -n_x/n_z, b_p = -n_y/n_z, c_p = (n_x x_p + n_y y_p + n_z z_0)/n_z, \tag{13}$$

To propagate pixels' labels in a larger region for enforcing big scale smoothness and speeding up convergency of algorithm, we also adopt a region label set which is shared for all the pixels in the region. We divide the image into regular grids, then each region has a 2D coordinates and neighboring regions just like pixels. Suppose for each region R, the region label set L_R has K_R labels. Region label set is constructed by randomly choosing K_R pixels in region R, and randomly taking one label of each pixel's label set.

Every pixel or region could propagate their label set to neighboring pixels or regions. Thus for a pixel p in region R_p, the pixel could be assigned a candidate label from actually the union of label sets of all the neighboring pixels of p and neighboring regions of R_p

$$C_p = \{ \bigcup_{q \in N_p} L_q \} \cup \{ \bigcup_{R \in N_{R_p}} L_R \}, \tag{14}$$

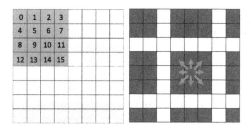

Fig. 2. The image is partitioned into regions with 4×4 pixels, indexed by $0 \sim 15$, and label is propagated to 3×3 neighboring pixels in a proposal, left a pixel-width area unlabeled to ensure sub-modularity.

where N_p represents the 3×3 patch centered at p, and N_{R_p} represents the 3×3 neighboring regions union around region R_p, as shown in Fig. 2. The optimal label for pixel p would be obtained from C_p during the optimization.

Proposals Generation. This section describes how the proposals are generated and the proof of smooth term's sub-modularity on the proposals. The proposal is a configuration that each pixel is assigned a label from candidate label set C_p. We generate a serial proposals $\{f^{(0)}, f^{(1)}, ..., f^{(n-1)}\}$ and sequentially fuse them with the current configuration.

The proposals generation is as follow. To simplify the statement, we only describe the proposals generated from pixel label sets. We divide the image into regions with 4×4 pixels. Pixels in a region are indexed by $0 \sim 15$. Take the label $l_i^{(K_i)}$ from label set of the pixel with index i in every region, and propagate the label to 3×3 neighboring pixels at pixel i. So the proposal $f^{(i \times K + K_i)}$ is constructed. There would be $16 \times K$ proposals in a serial only from pixel label sets. As shown in Fig. 2. At each proposal, all the pixels in a 3×3 area share one label. There are blank areas with one pixel width between red areas. Pixels in blank area are not assigned label in a proposal. Those unlabeled pixels are set to invalid label and given ∞ in cost function, that could guarantee the sub-modular condition of smooth term is satisfied. Let us specify the discussion: Suppose x_p, x_q represent the labels of neighboring pixels p, q in current configuration f, α_p, α_q represent the labels in proposal f_α. If both p, q are in the shared area, then they share the same label α in f_α. $V(\alpha, \alpha) = 0$, it is easy to prove $V(x_p, x_q) \leq V(x_p, \alpha) + V(\alpha, x_q)$; If p is in shared area, and q is in blank area, then q is unlabeled in f_α, smooth term becomes unary function about p vice versa; If both p, q are in blank area, smooth term becomes a constant. From the above discussion can be inferred the usage of the blank area, if there is no blank area, in other word, all shared area is closer next to each other, then when neighboring pixels p, q are on two shared areas with different labels in a proposal, $V(\alpha_p, \alpha_q)$ could be any value, thus the condition (10) might be violated.

The proposals from region label sets are generated using similar method, just seeing each region as a pixel. All pixels in the region share the same region label in a proposal.

Plane Label Refinement. When all proposals have been fused, the label sets is refined by adding random perturbation to the labels in previous iteration. For a label l_p of pixel p, the perturbation is generated by the similar way as pixel label set generation. First, random perturbation on depth value $\Delta d \in [-r_d, r_d]$ and unit vector Δn are generated. Then, we obtain $\bar{d}_p = d_p + \Delta d$ and $\bar{n} = n + r_n \Delta n$. Thus, the perturbated label $\bar{l}_p$ could be calculated by (13). Where, $r_d = (\max depth - \min depth)/2$, $r_n = 1$ in the first iteration, afterwards $r_d \leftarrow r_d/2, r_n \leftarrow r_n/2$ in next iteration. We add perturbation to all labels of the candidate label set C_p to get $\bar{C}_p$. Then we choose the best $K - 1$ labels from $\bar{C}_p$ to have the lowest local energy as follow at pixel p

$$E(s) = U_p(s) + \sum_q \in N(p) V_p q(s, f_q^{(t)}), \tag{15}$$

adding label $f_p^{(t)}$, we construct K labels of refined pixel label set $\bar{L}_p$, where $f_p^{(t)}, f_q^{(t)}$ represent the label of p, q in current configuration. Refined region label set $\bar{L}_R$ is constructed by randomly choosing K_R pixels in region R, and taking the current label of each pixel. Then the new proposals could be generated from the refined label set and starting next cycle of fusion moves.

The whole procedure of the algorithm is as follow.

Algorithm 1. Global optimization procedure with random plane labels

---

Randomly generate pixel label sets$\{L_p\}$ and region label set $\{L_R\}$
Initialize the current configuration $\{f_p^{(t)}\}$
repeat   Generate proposals $\{f^{(0)}, f^{(1)}, ..., f^{(n-1)}\}$
    for each $i \in (1, n - 1)$
        Solve $f^* = \arg\min_{z \in \{0,1\}} E(z \cdot f_p^{(t)} + (1 - z) \cdot f_p^{(i)})$
        Update $f_p^{(t)} = f^*$
    end for
    Refine $\{L_p\}\{L_R\}$
until if convergency

---

5   Experimental Results

We verify our algorithm on the synthetic scenes of HCI datasets [31].

Setting. All the predefined parameters of our algorithm in the experiments are set as follow. In consistency cost of SCam, the parameters are $\{\sigma = 1.0, \sigma_c = 3.0, P_{Thresh} = 0.5\}$. In energy function, data cost is set by accumulating consistency cost of 41×41 pixels in a large patch W_p with radius $R = 20$, weight between data cost and smooth cost is set $\lambda = 100$. The other parameters are set as $\{\gamma = 10, \varepsilon = 0.01, \tau_{dis} = 1.0\}$. For the label sets, all the size K of pixel label set and K_R of region label set are set as 2. We use 3 scales of region label set with 5×5, 25×25 and 100×100 pixels in each region.

Experiments on HCI. HCI contains 7 synthetic scenes. Each scene has 9×9 views with resolution of 768×768 pixels in each view. The depth map is only recovered for the centered view. In our algorithm, we take 10 iterations for each scene.

Figure 3 shows the results of every iteration in the optimization process. It demonstrates the advantage of our method using standard graph cut for two aspects: we start from a random depth initialization without any external constraints on our 3D plane labels, but the convergence is very fast. Only after 3 times iterations, the result is nearly close to the optimal one.

Surface Geometry Inference. Our method estimates a plane model at each pixel. The experimental results contain geometry information with plane parameters and normal direction at each pixel. Figure 4 shows the reconstructed point

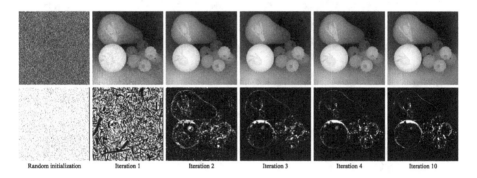

Random initialization   Iteration 1   Iteration 2   Iteration 3   Iteration 4   Iteration 10

Fig. 3. Depth results in the optimization process, showing disparity and 0.1-pixel error maps.

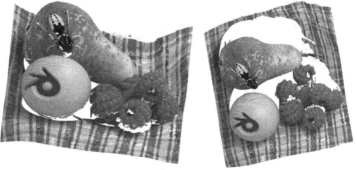

Fig. 4. Reconstructed point cloud of 'StillLife'

cloud of scene 'StillLife'. The normal direction is allowed changing smoothly in our algorithm so that continuous curvature surfaces can be reconstructed correctly.

Depth Results Comparison with State-of-Art Methods. To evaluate the performance of the proposed method, we compare depth estimation results with some state-of-the-art algorithms, including Wanner et al. [12], Yu et al. [32] and Wang et al. [17].

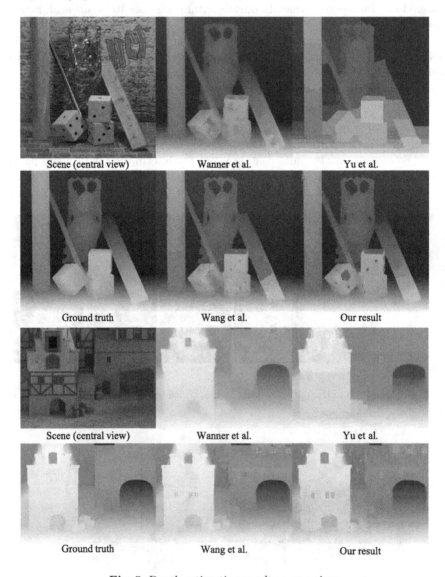

Fig. 5. Depth estimation results comparison

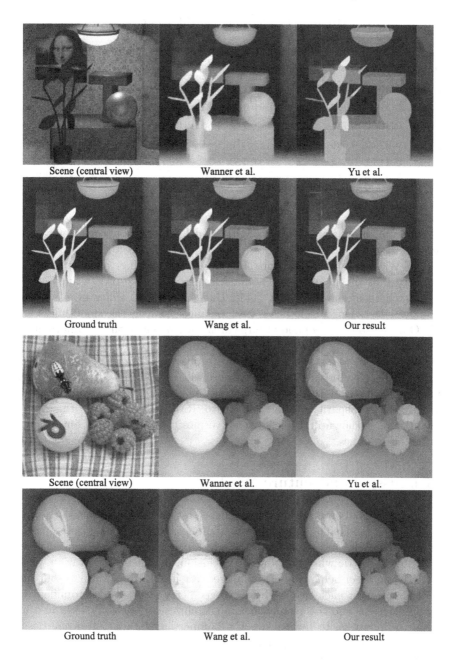

Fig. 6. Depth estimation results comparison

Figures 5 and 6 show results comparison using different methods on four synthetic scenes ('Buddha2', 'Medieval', 'Mona' and 'StillLife' from top to bottom) with ground truth disparity maps. Wanner et al. [12] is robust at local depth

Table 1. Depth RMSE on HCI datasets.

Scene	Ours	Wanner et al. [12]	Yu et al. [32]	Chen et al. [16]	Wang et al. [17]
Buddha	0.1086	0.079	0.134	**0.057**	0.095
Buddha2	**0.0787**	0.094	0.179	0.139	0.107
Horses	0.2054	0.163	0.188	**0.122**	0.140
Medieval	**0.1060**	0.111	0.144	0.129	0.115
Mona	**0.0712**	0.096	0.119	0.077	0.089
StillLife	**0.1111**	0.184	0.150	0.113	0.212
Papillon	**0.0856**	0.158	0.406	0.108	0.125

estimation, but it is hard to support MRF model. Their global consistency label makes the result a little over-smooth, but still well infers a sketchy geometry structure. Wang et al. [17] is the current state-of-art, which give better results in detail, especially on the occlusion boundaries. Yu et al. [32] used a line-assisted stereo matching method for light fields. But their results suffer from a severe 'stair-case' effect. Our result has a best balance between the smoothness and detail. Our method is very good at large area with strong geometry information, e.g., all the floor in the scenes and the curvature surfaces in 'StillLife' are best reconstructed. But we have to say that our results are not good enough at depth discontinuities or boundaries, especially those with tiny local structures. Furthermore, we quantitatively evaluated our algorithm by computing root-mean-square error (RMSE) of disparity with ground truth. RMSE comparison with state-of-art algorithms is shown in Table 1, which verifies that our method outperforms others in most scenes.

6   Discussion and Future Work

We propose a global method to optimize depth estimation on analysis of surface camera of light fields. The data cost is based on consistency measure by partial selection of pixels in surface camera. The selection strategy chooses closest pixels on consistency metric which is robust to occlusion. In global optimization, pixel-wise 3D plane label helps the algorithm recover not only depth-map, but also normal direction at each pixel. The scene geometry structure is prominent for many applications and the geometric compatibilities could also be used in other inference frameworks. In future, the non-Lambertian's statistical properties in surface camera and how those properties benefit the depth recovery for complex surfaces should be considered.

Acknowledgement. The work in the paper is supported by NSFC funds (61272287, 61531014).

References

1. Levoy, M., Hanrahan, P.: Light field rendering. In: ACM SIGGRAPH, pp. 64–71 (1996).
2. Vaish, V., Wilburn, B., Joshi, N., Levoy, M.: Using plane + parallax for calibrating dense camera arrays. In: Proceedings of the 2004 IEEE Computer Society Conference on Computer Vision and Pattern Recognition, CVPR 2004, vol. 1, pp. I-2–I-9 (2004)
3. Lytro: Lytro redefines photography with light field cameras (2011). http://www.lytro.com
4. Raytrix: Raytrix lightfield camera (2012). http://www.raytrix.de
5. Bolles, R.C., Baker, H.H., Marimont, D.H.: Epipolar-plane image analysis: an approach to determining structure from motion. Int. J. Comput. Vis. **1**, 7–55 (1987)
6. Yu, J., McMillan, L., Gortler, S.: Surface camera (scam) light field rendering. Int. J. Image Graph. **4**, 605–625 (2004)
7. Szeliski, R., Zabih, R., Scharstein, D., Veksler, O., Kolmogorov, V., Agarwala, A., Tappen, M., Rother, C.: A comparative study of energy minimization methods for Markov random fields with smoothness-based priors. IEEE Trans. Pattern Anal. Mach. Intell. **30**, 1068–1080 (2007)
8. Adelson, E.H., Bergen, J.R.: The plenoptic function and the elements of early vision. In: Computational Models of Visual Processing, pp. 3–20 (1991)
9. McMillan, L., Bishop G.: Plenoptic modeling: an image-based rendering system. In: ACM SIGGRAPH, pp. 39–46 (1995)
10. Gortler, S.J., Grzeszczuk, R., Szeliski, R., Cohen, M.F.: The lumigraph. In: ACM SIGGRAPH, pp. 43–54 (1996)
11. Criminisi, A., Kang, S.B., Swaminathan, R., Szeliski, R., Anandan, P.: Extracting layers and analyzing their specular properties using epipolar-plane-image analysis. Comput. Vis. Image Underst. **97**, 51–85 (2005)
12. Wanner, S., Goldluecke, B.: Globally consistent depth labeling of 4D light fields. In: 2012 IEEE Conference on Computer Vision and Pattern Recognition (CVPR), pp. 41–48. IEEE (2012)
13. Tosic, I., Berkner, K.: Light field scale-depth space transform for dense depth estimation. In: Proceedings of the IEEE Conference on Computer Vision and Pattern Recognition Workshops, pp. 435–442 (2014)
14. Kim, C., Zimmer, H., Pritch, Y., Sorkine-Hornung, A., Gross, M.: Scene reconstruction from high spatio-angular resolution light fields. ACM Trans. Graph. **32**, 96 (2013)
15. Tao, M., Hadap, S., Malik, J., Ramamoorthi, R.: Depth from combining defocus and correspondence using light-field cameras. In: Proceedings of the IEEE International Conference on Computer Vision, pp. 673–680 (2013)
16. Chen, C., Lin, H., Yu, Z., Kang, S., Yu, J.: Light field stereo matching using bilateral statistics of surface cameras. In: Proceedings of the IEEE Conference on Computer Vision and Pattern Recognition, pp. 1518–1525 (2014)
17. Wang, T.C., Efros, A.A., Ramamoorthi, R.: Occlusion-aware depth estimation using light-field cameras. In: Proceedings of the IEEE International Conference on Computer Vision, pp. 3487–3495 (2015)
18. Boykov, Y., Kolmogorov, V.: An experimental comparison of min-cut/max-flow algorithms for energy minimization in vision. IEEE Trans. Pattern Anal. Mach. Intell. **26**, 1124–1137 (2004)

19. Boykov, Y., Veksler, O., Zabih, R.: Fast approximate energy minimization via graph cuts. IEEE Trans. Pattern Anal. Mach. Intell. **23**, 1222–1239 (2001)
20. Kolmogorov, V., Zabin, R.: What energy functions can be minimized via graph cuts? IEEE Trans. Pattern Anal. Mach. Intell. **26**, 147–159 (2004)
21. Felzenszwalb, P.F., Huttenlocher, D.P.: Efficient belief propagation for early vision. Int. J. Comput. Vis. **70**, 41–54 (2004)
22. Hong, L., Chen, G.: Segment-based stereo matching using graph cuts. In: Proceedings of the 2004 IEEE Computer Society Conference on Computer Vision and Pattern Recognition, CVPR 2004, vol. 1, pp. I-74. IEEE (2004)
23. Ladický, L., Sturgess, P., Russell, C., Sengupta, S., Bastanlar, Y., Clocksin, W., Torr, P.H.S.: Joint optimization for object class segmentation and dense stereo reconstruction. Int. J. Comput. Vis. **100**, 1–12 (2010)
24. Woodford, O., Torr, P., Reid, I., Fitzgibbon, A.: Global stereo reconstruction under second-order smoothness priors. IEEE Trans. Pattern Anal. Mach. Intell. **31**, 2115–2128 (2009)
25. Olsson, C., Ulén, J., Boykov, Y.: In defense of 3d-label stereo. In: Proceedings of the IEEE Conference on Computer Vision and Pattern Recognition, pp. 1730–1737 (2013)
26. Rother, C., Kolmogorov, V., Lempitsky, V., Szummer, M.: Optimizing binary MRFs via extended roof duality. In: IEEE Conference on Computer Vision and Pattern Recognition, CVPR 2007, pp. 1–8. IEEE (2007)
27. Bleyer, M., Rhemann, C., Rother, C.: Patchmatch stereo-stereo matching with slanted support windows. In: BMVC, vol. 11, pp. 1–11 (2011)
28. Besse, F., Rother, C., Fitzgibbon, A., Kautz, J.: PMBP: patchmatch belief propagation for correspondence field estimation. Int. J. Comput. Vis. **110**, 2–13 (2014)
29. Taniai, T., Matsushita, Y., Naemura, T.: Graph cut based continuous stereo matching using locally shared labels. In: Proceedings of the IEEE Conference on Computer Vision and Pattern Recognition, pp. 1613–1620 (2014)
30. Lempitsky, V., Rother, C., Roth, S., Blake, A.: Fusion moves for Markov random field optimization. IEEE Trans. Pattern Anal. Mach. Intell. **32**, 1392–1405 (2010)
31. Wanner, S., Meister, S., Goldluecke, B.: Datasets and benchmarks for densely sampled 4D light fields. In: VMV, pp. 225–226. Citeseer (2013)
32. Yu, Z., Guo, X., Lin, H., Lumsdaine, A., Yu, J.: Line assisted light field triangulation and stereo matching. In: Proceedings of the IEEE International Conference on Computer Vision, pp. 2792–2799 (2013)

Direct and Global Component Separation from a Single Image Using Basis Representation

Art Subpa-asa[1]([⊠]), Ying Fu[2], Yinqiang Zheng[3], Toshiyuki Amano[4], and Imari Sato[1,3]

[1] Tokyo Institute of Technology, Tokyo, Japan
art.s.aa@m.titech.ac.jp
[2] The University of Tokyo, Tokyo, Japan
fuying@ut-vision.org
[3] National Institute of Informatics, Tokyo, Japan
{yqzheng,imarik}@nii.ac.jp
[4] Wakayama University, Wakayama, Japan
amano@sys.wakayama-u.ac.jp

Abstract. Previous research showed that the separation of direct and global components could be done with a single image by assuming neighboring scene points have similar direct and global components, but it normally leads to loss of spatial resolution of the results. To tackle such problem, we present a novel approach for separating direct and global components of a scene in full spatial resolution from a single captured image, which employs linear basis representation to approximate direct and global components. Due to the basis dependency of these two components, high frequency light pattern is utilized to modulate the frequency of direct components, which can effectively improve stability of linear model between direct and global components. The effectiveness of our approach is demonstrated on both simulated and real images captured by a standard off-the-shelf camera and a projector mounted in a coaxial system. Our results show better visual quality and less error compared with those obtained by the conventional single-shot approach on both still and moving objects.

1 Introduction

When a scene is illuminated by a light source, each surface point interacts with light differently and its intensity often includes not only direct reflection but also complex phenomena, such as interreflection, subsurface scattering, and volumetric scattering. Direct component represents the direct reflection of the point caused by the light source. Global component is the brightness of the point caused by the illumination coming from other points in the scene including interreflection and scattering. The goal of this paper is to separate direct and global components from a single image without sacrificing their spatial resolution.

Electronic supplementary material The online version of this chapter (doi:10. 1007/978-3-319-54187-7_7) contains supplementary material, which is available to authorized users.

S.-H. Lai et al. (Eds.): ACCV 2016, Part III, LNCS 10113, pp. 99–114, 2017.
DOI: 10.1007/978-3-319-54187-7_7

(a)   (b)   (c)   (d)   (e)   (f)   (g)

Fig. 1. Separation of direct and global components. (a) Single input image. (b) and (c) Ground truth direct and global components obtained using multiple images. (d) and (e) Direct and global components recovered from a single image by conventional approach. (f) and (g) Direct and global components recovered from a single image by our method.

Nayar *et al.* proposed to separate the observed pixel intensities into direct and global components using high frequency light pattern [1]. By projecting a series of high frequency light patterns into a scene, the direct and global components of the scene are nicely separated by examining intensity differences of lit and unlit pixels under different light patterns. Figure 1(a) illustrates an input image captured under high frequency lighting, (b) and (c) show the separation results of this method from multiple input images.

This separation technique greatly helps understand the relationships among surface points and a light source, and has been applied into various computer vision tasks, such as shape reconstruction [2,3] and image descattering [4,5]. Generally, this approach requires multiple images taken under a series of high frequency light patterns and thus cannot be directly extended to moving scenes without motion compensation [6]. While it has been demonstrated that the separation could be done with a single captured image by assuming neighboring scene points have similar direct and global components [1], this approach sacrifices spatial resolution, and thus direct and global components are recovered at a lower spatial resolution than the captured image (Fig. 1(d) and (e)). Separating a single image into direct and global components without loss of resolution is very challenging and has not been achieved using single-shot measurement.

In this paper, leveraging on the linear basis representation and the modulation effect of high frequency lighting, we present a novel approach for separating direct and global components from a single image, without sacrificing their spatial resolution (see Fig. 1(f) and (g) for an example). Although linear basis representation can effectively describe these two components on low dimensional space and reduce the number of unknown variables, we can not obtain unique separation results because of linear dependency of bases for these two components. We carefully examine this dependency issue and show that high frequency lighting contributes to resolving it, and consequently improving accuracy and robustness of the separation. The effectiveness of our approach is demonstrated on simulated and real images/videos captured by a standard off-the-shelf camera and a projector mounted in a coaxial system.

In summary, our major contributions are that we

- Propose to represent direct and global components based on linear basis representation for full resolution recovery of these components from a single image;
- Carefully examine dependency of direct and global components and show that high frequency light pattern contributes to improving separation accuracy and robustness;
- Set up a coaxial system by using a standard off-the-shelf camera and a projector to capture real images of still and moving scenes, and demonstrate the effectiveness of our separation method.

The rest of this paper is organized as follows. Section 2 provides literature reviews of direct and global component separation. Section 3 discusses the ambiguity of direct and global components observed in a single image and describes our approach for robust separation. Our experiment setup and results are shown in Sect. 4. Finally, conclusions are drawn and future directions of our research are discussed in Sect. 5.

2   Related Works

Conventional method [1] uses multiple images under shifting high frequency light patterns to separate the observed pixel intensities into direct and global components, by examining intensity differences of lit and unlit pixels. As a consequence, it can only be used for static scenes. By assuming neighboring scene points have similar direct and global components, this method can be adapted to separate these two components using a single input image. Unfortunately, the resolution of resulting images is greatly reduced, and error is usually found in the area with sharp depth and/or color variation.

The original multiple image approach has been improved in various ways. The number of images has been reduced to three by using multiplex sinusoid light patterns from multiple projectors [7]. Motion compensation technique [6] has been adopted for scenes with small movement like human body action. Also multiple projector focus points [8,9] have been suggested to improving accuracy of separation.

There are also some researches on how to separate additional components. Global component can be separated further into near range and far range by using multiple binary [10] or sinusoid patterns [11]. A generalized matrix of direct and global components, which indicates relation between light source and image, can be obtained by using iteration of forward and inverse transport matrix [12]. More complex components have been separated in many specific domain such as face [13,14], volumetric fluid [5], translucent object [15,16] and outdoor scene [17].

Special equipments are also introduced for this separation task. For example, a high speed camera is used to capture fast dithering between frames of a DLP projector, which projects complementary patterns for real time separation [18].

A time-of-flight camera is used to separate short pulse of light [19]. Moreover, conventional equipments like LCD panels have been used to create primal-dual coding system to separate the whole transport matrix [20]. Digital Micro-mirror Devices (DMDs) are employed in the stereo projector-camera system to obtain the direct and global components [9]. Homogeneous code between light and sensors can also separate direct, near-range global and far-range global in real time [21].

On the contrary, we separate direct and global components without sacrificing their spatial resolution from a single image, using a standard off-the-shelf camera and a projector.

3   Direct and Global Component Separation

In this section, we describe our approach to separate direct and global components using a single image without sacrificing spatial resolution.

3.1   Direct and Global Component Model

Image i at pixel p under uniform lighting is composed of two components, i.e. direct component i_d and global component i_g, which can be described as

$$i(p) = i_d(p) + i_g(p). \tag{1}$$

When projecting a light pattern into the scene, it affects these two components in completely different ways. Specifically, direct component is modulated by the light pattern in the 2D image space, while global component is virtually independent of the spatial distribution of the light pattern. This is due to the fact that global component is a combination of complex optical phenomena, such as interreflection and scattering, which are caused by the illumination coming from other points in the scene. According to [1], global component under high frequency light pattern is k-percent proportional to that under uniform lighting, where k is the average intensity of light pattern. This effect of high frequency light pattern on direct and global components is illustrated in Fig. 2.

Let us vectorize the image with N pixels into $\boldsymbol{i} = [i(p_1), i(p_2), \cdots, i(p_N)]^T$, whose corresponding light pattern is denoted by $\boldsymbol{l} = [l(p_1), l(p_2), \cdots, l(p_N)]^T$.

Fig. 2. Direct and global components under different lightings. Under high frequency light pattern (second column), direct component is modulated by light pattern spatially, while global component keeps constant up to a scaling factor.

The relationship between the light pattern and the captured image can be described as

$$i = Li_d + \left(\frac{\int l(p)\,dp}{N} \right) i_g = Li_d + ki_g = Li_d + \hat{i}_g, \tag{2}$$

where $L = diag(l)$. $i_d = [i_d(p_1), i_d(p_2), \cdots, i_d(p_N)]^T$ and $i_g = [i_g(p_1), i_g(p_2), \cdots, i_g(p_N)]^T$ represent direct and global components under the uniform lighting in vectorized form, respectively.

According to Eq. (2), it is infeasible to directly solve i_d and $\hat{i}_g$, since the number of unknown variables is twice as many as the number of measurements N in i. To reduce unknown variables, we describe direct and global components by linear basis representation in the following.

3.2   Variable Reduction Using Linear Basis Representation

Assuming that direct and global components are relatively smooth over 2D image space, they can be approximated by linear combination of some predefined bases and unknown coefficients as

$$i_d = D\alpha, \quad \hat{i}_g = G\beta, \tag{3}$$

where $D_{N \times n}$ and $G_{N \times m}$ are the bases for the direct and global components, respectively. $\alpha_{n \times 1}$ and $\beta_{m \times 1}$ are the corresponding coefficient vectors.

Substituting Eq. (3) into (2), we obtain

$$i = Li_d + \hat{i}_g = LD\alpha + G\beta = \begin{bmatrix} LD & G \end{bmatrix} \begin{bmatrix} \alpha^T & \beta^T \end{bmatrix}^T, \tag{4}$$

in which the number of unknown variable has been reduced to $m + n$, which is usually less than the number of image pixels N.

In this paper, we study two types of bases: Fourier and principal component analysis (PCA) basis.

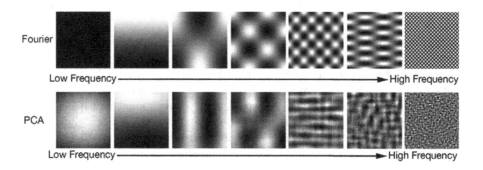

Fig. 3. Samples of Fourier basis and PCA basis

Fourier basis has been widely used for image compression with proven ability to represent general spatially smooth images and is independent of the input data. In this work, we use the 2D Fourier basis to describe direct and global components.

PCA basis can be trained from correlated observations. PCA basis represents in the order of the variance, which means that higher order basis is more significant than the lower order. Since there do not exist large database of direct and global components for general scenes, we use general image dataset [22] as training set to perform PCA[1].

Samples of Fourier and PCA bases used in our experiment are shown in Fig. 3.

3.3   Independence of Linear Equation

To solve Eq. (4) reliably, the matrix $M = \begin{bmatrix} LD & G \end{bmatrix}$ has to be invertible, which means that all columns in M should be independent. The matrix M consists of two parts, LD and G. LD is a product of the light pattern and the bases of direct component, while G is the set of bases for global component only.

In our work, we set bases for direct and global components the same, i.e. $D = G$. In other words, under uniform light ($L = I$), direct and global components could not be directly separated, since $M = \begin{bmatrix} D & G \end{bmatrix}$ is not invertible. The independence between LD and G has to rely on the light pattern L, which modulates D only. We have to set up reasonable L such that columns in M become sufficiently independent.

Selection of Light Pattern. The selection of the light pattern has to meet two conditions. Firstly, the light pattern should be of high frequency to keep global component constant as described in Sect. 3.1. Secondly, the light pattern has to be able to differentiate modulated bases LD of direct component and normal bases G of global component. In this work, we choose to use checker pattern, considering its simplicity in generation and flexibility in changing frequency. In the following, we suggest a frequency domain analysis viewpoint and a linear algebra viewpoint on the effect of the high frequency checker light pattern.

Frequency Analysis. It is well known that high frequency light pattern [1] can be employed to separate direct and global components in spatial domain. However, its effect in the frequency domain has not been analyzed.

As shown in Fig. 4(a), we select image i_d to be direct component, i_g to be global component, which are obtained by the conventional multiple images method [1]. Frequencies of direct and global components are both densely located in low frequency. Frequency of mixing image $i_d + i_g$ with the uniform light is also densely located in low frequency, thus these two components cannot be directly

---

[1] All these images contain both direct and global components, thus it is reasonable to learn the PCA basis on this general image dataset.

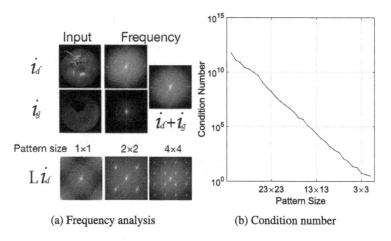

(a) Frequency analysis (b) Condition number

Fig. 4. Analysis on the independence between direct and global components. (a) The intensity of images in frequency domain while illuminating different frequency light pattern; (b) Condition number of the linear equation under different frequency light pattern in logarithm scale. The checker pattern size varies from 32 px × 32 px to 1 px × 1 px.

separated under the uniform lighting. However, after using high frequency light pattern L to modulate i_d, the frequency of i_d has been shifted into higher frequency by L, while the global component is still in the low frequency. Thus, the high frequency light L makes the frequency of direct component i_d different from global component i_g by modulating i_d.

This phenomenon implies that, under high frequency light pattern, direct component i_d can be extracted from the shifted high frequency and global component i_g can be extracted from low frequency. On the other hand, under lower frequency light pattern (i.e. larger pattern size), shifted frequency from the light pattern Li_d is closer to the frequency of global component i_g. Therefore, to use lower frequency light pattern L to separate direct and global components will reduce quality of the separation.

Linear Algebra. To measure the independence of this linear system, we also calculate the condition number of the matrix M, where a smaller condition number indicates higher degree of independence. We conduct experiments by creating matrix M with varying sizes of checker light pattern. The input image is decomposed to 128 px × 128 px and the obtained number of bases is 128 × 128, in which the first 256 bases are selected in order. Figure 4(b) shows that the condition number decreases exponentially as the size of checker light pattern reduces.

In summary, both experiments in Fig. 4 demonstrate that high frequency light pattern can be used to differentiate direct and global components in the presence of a single input image.

3.4   Frequency Relationship Between Basis and Light Pattern

Furthermore, we analyze the relationship between basis and light pattern. From Sect. 3.3, we find that higher frequency of light pattern generates better separation results. Due to physical restrictions, the highest frequency of checker pattern that can be generated by a projector is limited. Here, we discuss the optimal number of bases under this practical limitation. To find optimal number of bases, we measure Root Mean Square Error (RMSE) on direct and global components between results from conventional multiple images method and our proposed method. 1400 patches with the size of 128 px × 128 px are randomly selected from 16 images and used for this simulation.

Firstly, we perform analysis on Fourier basis whose frequency distribution is independent of the input data. We first consider the effect of frequency of light pattern on separation quality. The experimental results in Fig. 5(a) and (b) confirm that higher frequency of light pattern always leads to lower RMSE

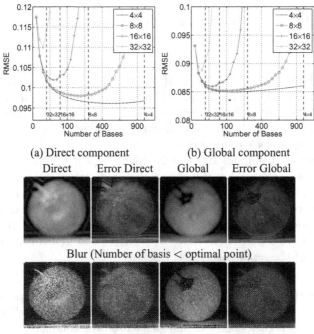

(a) Direct component   (b) Global component

Blur (Number of basis < optimal point)

Vibrant pattern (Number of basis > optimal point)

(c) Error in separation results

Fig. 5. Analysis on the frequency of light pattern and the number of basis. (a) and (b) Comparison of RMSE between different frequency of light pattern when changing the number of basis. *The vertical dotted lines* indicate frequency of basis that equals to frequency of light pattern. (c) Two types of error: (Top) - Blur due to lack of bases, (Bottom) - Vibrant pattern due to overfitting bases.

under the same number of bases. Then, we consider the relationship between the number of bases and separation quality. We find that quality of separation decreases when the number of bases goes beyond the optimal point. Moreover, we notice two properties of the optimal point that (i) the highest frequency of optimal bases is always lower than the frequency of light pattern, and (ii) the number of optimal bases is increasing as the frequency of light pattern increases. Therefore, separation quality of our method is largely limited by the frequency of light pattern.

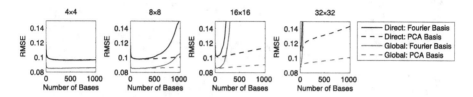

Fig. 6. RMSE comparison on Fourier and PCA bases with respect to the frequency of light pattern and the number of bases.

We also analyze the error from the experiment. The visualization of results with error are shown in Fig. 5(c). These errors are caused by two conflicting effects, blur and vibrant patterns. Blur occurs when frequency of basis is insufficient to represent the source. On the contrary, vibration occurs when frequency of basis exceeds the frequency of light pattern. The vibrant error occurs uniformly in the shape of light pattern which indicates mixing of direct and global components in the high frequency domain.

Lastly, we conduct analysis to compare between Fourier and PCA bases in terms of optimal number of bases and robustness (Fig. 6). Fewer optimal number of basis indicates how well basis can represent the components and reduce computation time. From the experiment, we find similarity on both bases in terms of optimal number. However, on robustness, PCA basis has major advantage over Fourier basis. After surpassing optimal number, RMSE of PCA basis increases in significantly slower rate than Fourier basis. Such capability is extremely useful in real environment where external factors easily change frequency of light pattern and affect its high frequency quality. The robustness of the basis can compensate such error and provide better separation results. Thus, PCA basis delivers more stable results when optimal number is unknown.

4 Experimental Results

We set up the experiment process and hardware as shown in Fig. 7 and evaluate the effectiveness of our method on both simulated and real images. For real scenes, both still and moving scenes are captured by an off-the-shelf camera and a projector mounted in a coaxial system. Such system is used to obtain pixel correspondences between the camera and projected image regardless of the geometry of the scene. The Projection Center Calibration Technique [23] is further employed for precise alignment between the camera and the projector. The coaxial captured system consists of 1920 px × 1080 px resolution projector and 1600 px × 1080 px resolution camera (Fig. 7(a)).

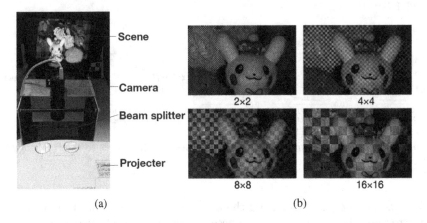

(a)   (b)

Fig. 7. Experimental setup. (a) Coaxial captured system: camera and projector are colocated with a plane beam splitter. (b) Captured images under different frequency of lightings.

4.1   Simulated Images

We first evaluate the accuracy of our separation method on simulated images. The baseline of direct and global components are obtained by using the conventional multiple images method [1], as shown in Fig. 8(a). We synthesize the input images from the obtained baseline under different frequencies of light pattern. Our proposed method and conventional single image method are performed on these synthetic data. Our method uses the simulated data that are illuminated by 4 px × 4 px, 8 px × 8 px, 16 px × 16 px and 32 px × 32 px checker patterns. The number of bases for each pattern is selected according to the optimal point that has the lowest error from the experiment in Sect. 3.4 for both Fourier and PCA bases. Since the conventional single image method performs better under the stripe light pattern, we compute direct and global components using simulated images under 4 px, 8 px, 16 px and 32 px vertical stripe pattern with window size equal to (2 × pattern size + 1 px).

Table 1 shows the RMSE of direct and global components between our proposed and conventional methods. The results clearly show that, given the same

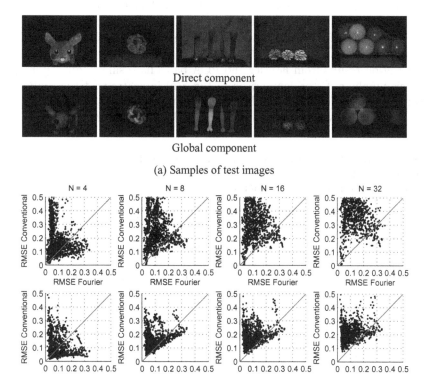

Direct component

Global component

(a) Samples of test images

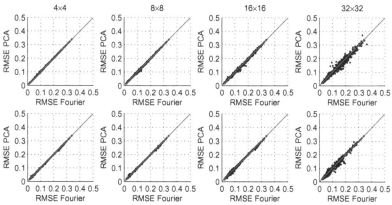

(b) Comparison of RMSE : Conventional single image method and our method under different frequencies of light patterns ($Npx \times Npx$ checker pattern for our Fourier basis method and Npx horizontal stripe pattern for conventional single image method).

(c) Comparison of RMSE: Fourier and PCA bases using different frequencies of light patterns.

Fig. 8. Experimental evaluation on simulated images.

light pattern size, our method has lower RMSE than the conventional single image approach. Moreover, under lower frequency light pattern, our method significantly outperforms the conventional single image method, as shown in Fig. 8(b).

Separation results from Fourier and PCA bases have similar quality in optimal basis numbers. Error comparison in Fig. 8(c) indicates that the results are similar regardless of difference of basis representation, thus the quality of separation results mainly depends on frequency of the light pattern.

Table 1. Average RMSE between baseline and simulated results from each method. Our method simulated under $Npx \times Npx$ checker patterns. Conventional single image method simulated under Npx horizontal stripe patterns.

Method	Direct component				Global component			
	$N = 4$	$N = 8$	$N = 16$	$N = 32$	$N = 4$	$N = 8$	$N = 16$	$N = 32$
Conventional [1]	0.1955	0.3282	0.4061	0.4897	0.0976	0.1703	0.1950	0.2242
Our(Fourier)	**0.0961**	0.0980	0.1019	0.1099	0.0850	0.0852	0.0860	0.0887
Our(PCA)	0.0963	**0.0979**	**0.1010**	**0.1080**	**0.0849**	**0.0851**	**0.0858**	**0.0879**

4.2   Real Images

Visual comparison between our method and conventional method on real images are shown in Fig. 9. In the experiment, we use 8 px × 8 px checker light pattern for our method and conventional multiple images method, and 8 px horizontal stripe light pattern for conventional single image method. We conduct the experiments for both still and moving objects.

As shown in Fig. 9(a) and (b), the separation results from our method are close to the baseline from conventional multiple images method. Our method outperforms the conventional single image method both in accuracy and resolution on the still objects. Taking the mouth and eyes of the doll as examples, we can see that the direct component from conventional single image contains white border around mouth and eyes, which are caused by large max-min window. Besides, our method generates full spatial resolution results, while the resolution of conventional single image method is only 12.5% of the full resolution.

Moreover, as shown in Fig. 9(c), our method has advantages over the conventional method when performing on moving objects. The direct component from multiple images method contains border surrounding edge of the fingers caused by object motion. While our method requires only one frame for separation, it performs better than the conventional multiple images method in this scenario. Lastly, we perform our method and conventional single image method on sequential frames, as shown in Fig. 10, and demonstrate the capability of our method for fully dynamic scenes.

Conventional   Conventional single   Our method: PCA   Our method:
multiple images   image   Fourier

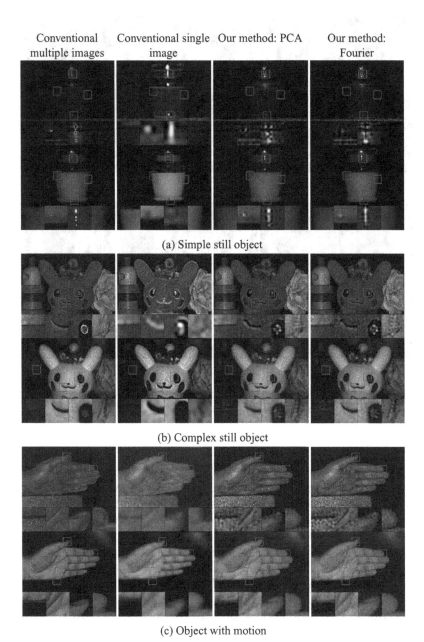

(a) Simple still object

(b) Complex still object

(c) Object with motion

Fig. 9. Separation results of real data. Direct component (Top) and global component (Bottom). From left to right: separation results from conventional multiple images method, conventional single image method, our proposed method using PCA basis, and our proposed method using Fourier basis. (a) and (b) Comparison results of still objects. (c) Comparison results of object with motion.

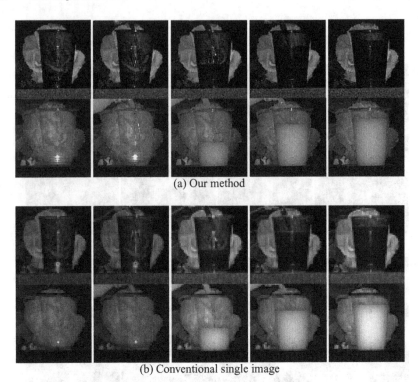

(a) Our method

(b) Conventional single image

Fig. 10. Separation results of a video sequence. From top to bottom: (a) Our method - direct, global. (b) Conventional single image - direct, global.

5   Conclusion

We propose a novel approach for separating direct and global components of a scene in full spatial resolution from a single image by introducing linear basis models to approximate these two components. We have explored the dependency issue of direct and global components, illustrated how high frequency lighting contributes to resolving the ambiguity between them and providing unique separation results. The effectiveness of our approach has been demonstrated on simulated and real images of static and dynamic scenes captured by a standard off-the-shelf camera and a projector. Our results show better visual quality and less error compared with those obtained by the conventional single image approach.

Our method assumes stable condition and strict alignment in a coaxial system, thus factors such as depth difference between calibrated plane and real scene, dithering between frames of projector, and noise from camera, will affect quality of the captured images in real experiments. It is worth investigating how to relax such constraints and ensure stability in the future.

Acknowledgements. This work was supported in part by Grant-in-Aid for Scientific Research on Innovative Areas (No.15H05918) from MEXT, Japan.

References

1. Nayar, S., Krishnan, G., Grossberg, M.D., Raskar, R.: Fast separation of direct and global components of a scene using high frequency illumination. ACM Trans. Graph. Proc. ACM SIGGRAPH **25**(3), 935–944 (2006)
2. Morris, N., Kutulakos, K.: Reconstructing the surface of inhomogeneous transparent scenes by scatter-trace photography. In: IEEE 11th International Conference on Computer Vision (ICCV), pp. 1–8 (2007)
3. Chen, T., Seidel, H.P., Lensch, H.: Modulated phase-shifting for 3D scanning. In: IEEE Conference on Computer Vision and Pattern Recognition (CVPR), pp. 1–8 (2008)
4. Talvala, E.V., Adams, A., Horowitz, M., Levoy, M.: Veiling glare in high dynamic range imaging. In: ACM SIGGRAPH 2007 Papers, SIGGRAPH 2007, New York, NY, USA (2007)
5. Gupta, M., Narasimhan, S., Schechner, Y.: On controlling light transport in poor visibility environments. In: IEEE Conference on Computer Vision and Pattern Recognition (CVPR), pp. 1–8 (2008)
6. Achar, S., Nuske, S., Narasimhan, S.: Compensating for motion during direct-global separation. In: IEEE International Conference on Computer Vision (ICCV), pp. 1481–1488 (2013)
7. Gu, J., Kobayashi, T., Gupta, M., Nayar, S.K.: Multiplexed illumination for scene recovery in the presence of global illumination. In: IEEE International Conference on Computer Vision (ICCV), pp. 1–8 (2011)
8. Gupta, M., Tian, Y., Narasimhan, S., Zhang, L.: A combined theory of defocused illumination and global light transport. Int. J. Comput. Vis. **98**, 146–167 (2012)
9. Achar, S., Narasimhan, S.G.: Multi focus structured light for recovering scene shape and global illumination. In: Fleet, D., Pajdla, T., Schiele, B., Tuytelaars, T. (eds.) ECCV 2014. LNCS, vol. 8689, pp. 205–219. Springer, Heidelberg (2014). doi:10.1007/978-3-319-10590-1_14
10. Gupta, M., Agrawal, A., Veeraraghavan, A., Narasimhan, S.G.: A practical approach to 3D scanning in the presence of interreflections, subsurface scattering and defocus. Int. J. Comput. Vis. **102**, 33–55 (2013)
11. Reddy, D., Ramamoorthi, R., Curless, B.: Frequency-space decomposition and acquisition of light transport under spatially varying illumination. In: Fitzgibbon, A., Lazebnik, S., Perona, P., Sato, Y., Schmid, C. (eds.) ECCV 2012. LNCS, vol. 7577, pp. 596–610. Springer, Heidelberg (2012). doi:10.1007/978-3-642-33783-3_43
12. Bai, J., Chandraker, M., Ng, T.T., Ramamoorthi, R.: A dual theory of inverse and forward light transport. In: European Conference on Computer Vision, pp. 1–8 (2010)
13. Tariq, S., Gardner, A., Llamas, I., Jones, A., Debevec, P., Turk, G.: Efficient estimation of spatially varying subsurface scattering parameters for relighting. ICT Technical Report ICT TR 01 2006, University of Southern California Institute for Creative Technologies (2006)
14. Ghosh, A., Hawkins, T., Peers, P., Frederiksen, S., Debevec, P.: Practical modeling and acquisition of layered facial reflectance. ACM Trans. Graph. **27**, 139:1–139:10 (2008)

15. Mukaigawa, Y., Suzuki, K., Yagi, Y.: Analysis of subsurface scattering based on dipole approximation. Inf. Media Technol. **4**, 951–961 (2009)
16. Munoz, A., Echevarria, J.I., Seron, F.J., Lopez-Moreno, J., Glencross, M., Gutierrez, D.: BSSRDF estimation from single images. Comput. Graph. Forum **30**, 455–464 (2011)
17. Liu, Y., Qin, X., Xu, S., Nakamae, E., Peng, Q.: Light source estimation of outdoor scenes for mixed reality. Vis. Comput. **25**, 637–646 (2009)
18. Narasimhan, S.G., Koppal, S.J., Yamazaki, S.: Temporal dithering of illumination for fast active vision. In: Proceedings of European Conference on Computer Vision, pp. 830–844 (2008)
19. Wu, D., O'Toole, M., Velten, A., Agrawal, A., Raskar, R.: Decomposing global light transport using time of flight imaging. In: IEEE Conference on Computer Vision and Pattern Recognition (CVPR), pp. 366–373 (2012)
20. Owu'Toole, M., Raskar, R., Kutulakos, K.N.: Primal-dual coding to probe light transport. ACM Trans. Graph. **31**, 39:1–39:11 (2012)
21. O'Toole, M., Achar, S., Narasimhan, S.G., Kutulakos, K.N.: Homogeneous codes for energy-efficient illumination and imaging. ACM Trans. Graph. **34**, 35:1–35:13 (2015)
22. Yang, C.Y., Yang, M.H.: Fast direct super-resolution by simple functions. In: Proceedings of IEEE International Conference on Computer Vision (2013)
23. Amano, T.: Projection center calibration for a co-located projector camera system. In: IEEE Conference on Computer Vision and Pattern Recognition Workshops (CVPRW), pp. 449–454 (2014)

Ultra-Shallow DoF Imaging Using Faced Paraboloidal Mirrors

Ryoichiro Nishi$^{(\boxtimes)}$, Takahito Aoto, Norihiko Kawai, Tomokazu Sato,
Yasuhiro Mukaigawa, and Naokazu Yokoya

Graduate School of Information Science,
Nara Institute of Science and Technology, Ikoma, Japan
{nishi.ryoichiro.ne6,takahito-a,norihi-k,tomoka-s,
mukaigawa,yokoya}@is.naist.jp

Abstract. We propose a new imaging method that achieves an ultra-shallow depth of field (DoF) to clearly visualize a particular depth in a 3-D scene. The key optical device consists of a pair of faced paraboloidal mirrors with holes around their vertexes. In the device, a lens-less image sensor is set at one side of their holes and an object is set at the opposite side. The characteristic of the device is that the shape of the point spread function varies depending on both the positions of the target 3-D point and the image sensor. By leveraging this characteristic, we reconstruct a clear image for a particular depth by solving a linear system involving position-dependent point spread functions. In experiments, we demonstrate the effectiveness of the proposed method using both simulation and an actually developed prototype imaging system.

1 Introduction

Shallow DoF (depth-of-field) imaging highlights a target in a photograph by de-focusing undesired objects that exist outside of a certain depth range. As an extreme condition, a microscope achieves ultra-shallow DoF imaging by putting a target object very close to the lens. In this case, the objects except for the tiny target, e.g. like a cell, are extremely blurred and we can see what we want to see by precisely adjusting the focus. Here, the range of DoF depends on the combination of the distance to a target and the aperture size. One problem is that the aperture size cannot be larger than the physical lens size. Although ultra large lenses are required to construct ultra shallow DoF imaging systems for standard size objects, it is almost impossible to produce such large lenses.

To solve this problem, a variety of synthetic aperture methods have been investigated and are classified into two categories: One physically captures images from multiple viewpoints using only standard cameras and the other virtually generates multi-view point images using cameras and some optical components. The former methods use a moving camera [1] or a multiple camera array system [2,3]. Although they can widen the aperture using multi-view images taken from different viewpoints and synthesize full resolution images, they require time consuming camera calibration, or complex multi-camera devices.

© Springer International Publishing AG 2017
S.-H. Lai et al. (Eds.): ACCV 2016, Part III, LNCS 10113, pp. 115–128, 2017.
DOI: 10.1007/978-3-319-54187-7_8

In the latter category, a micro lens array [4–6], mask [7,8], and a micro mirror array [9–11] have been employed to adjust the DoF. Since the aperture size in the systems cannot exceed that of the original camera lens equipped in their systems, it is practically difficult for these methods to achieve shallower DoF. As another method in the latter category, Tagawa et al. [12] proposed a specially designed polyhedral mirror called "turtleback reflector", which can possibly achieve infinite-size aperture by reflecting light rays on mirrors arranged on a hemisphere placed in front of a camera. Although this optical system achieves the ultra-shallow DoF imaging, the resolution of the synthesized image is quite low because all images captured by multiple virtual cameras are recorded as one image in reality.

Another related approach is a confocal imaging one which highlights a specific 3-D point on a target by setting both the optical center of a camera and a point light source at the focus of a lens [13–15]. In the systems based on this approach, a half-mirror enables to set them at the same focus position. Since the systems cannot highlight all the positions on the target at the same time, they physically has to scan the target while changing the highlighted positions to reconstruct the complete image. In addition, since the systems can highlight only the target point within the DoF which are determined by the physical lens size, the size of the target object is still limited by available lenses.

In this paper, we propose a novel imaging device that consists of a pair of faced paraboloidal mirrors for achieving ultra-shallow DoF imaging. Such a device has first been developed for displaying 3D objects and is called "Mirage" [16]. This device is, for example, used for an interactive display [17]. In this study, we leverage this device to capture a cross-sectional image for a specific depth of a 3-D object. To the best of our knowledge, this study is the first one to use the paraboloidal mirrors-based device as an imaging device. The proposed system achieves much larger Numerical Aperture (NA) than those of existing lens-based camera systems, and has capability to handle larger size of objects than conventional microscope systems, while preserving the original image resolution of an image sensor.

2   Device for Ultra Shallow DoF Imaging

This section introduces the proposed ultra shallow DoF imaging device that can capture a specific layer of an object that consists of multiple layers. Figures 1 and 2 show the developed prototype system of the proposed imaging device and its internal structure. The proposed imaging device consists of a pair of same shaped paraboloidal mirrors whose vertex and focal point correspond with each other and has holes at the vertexes of the paraboloidal mirrors for setting an image sensor without a lens at one side and a target object at the other side. It should be noted that paraboloidal mirrors have a feature that light rays from the focal point of a paraboloidal mirror become parallel after reflecting at the mirror and parallel light rays gather at the focal point after reflecting at the mirror as shown in Figs. 3(a) and (b). Therefore, light rays from an object

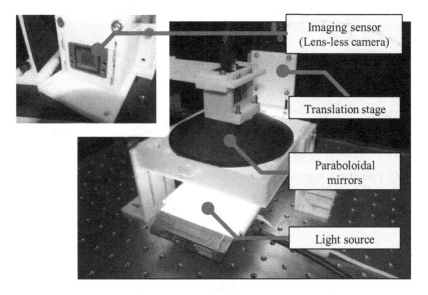

Fig. 1. Our prototype system.

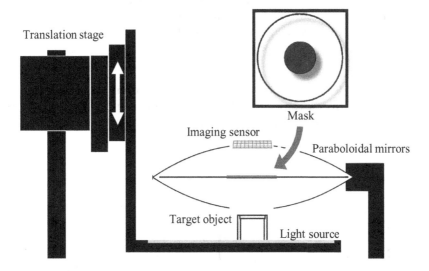

Fig. 2. Internal structure of prototype system.

at the focal point of an upper paraboloidal mirror gather at an image sensor at the focal point of the lower one. Here, since the direct light rays from the object to the image sensor disturb the visualization of an internal layer, a thin mask is placed at the center of the proposed device for light rays from the object not to directly reach the image sensor as shown in Fig. 2. In addition, if an object

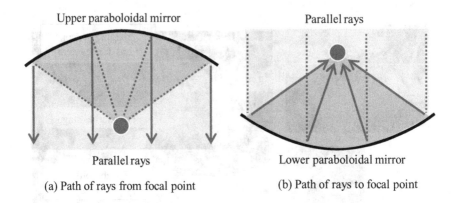

Upper paraboloidal mirror

Parallel rays

Parallel rays

Lower paraboloidal mirror

(a) Path of rays from focal point

(b) Path of rays to focal point

Fig. 3. Path of light rays in paraboloidal mirrors.

moves from the focal point in the direction perpendicular to the image sensor plane (referred to as depth direction), light rays from the object do not gather at the image sensor, resulting in generating a blurred image. Therefore, we can visualize only a specific layer that exists at the focal depth.

Here, we discuss the Numerical Aperture (NA) of the proposed imaging device. The range of the NA is $[0,1]$ and the DoF gets shallower as the NA gets higher. The NA is defined in general as follows:

$$NA = n \sin \theta, \tag{1}$$

where n is the index of refraction of the medium between a target object and an image sensor. In our case, $n = 1.0$ because the medium is the air. θ is the aperture angle which means the maximum angle between the optical axis and

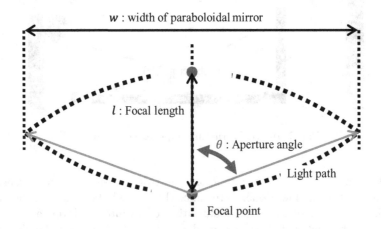

w : width of paraboloidal mirror

l : Focal length

θ : Aperture angle

Light path

Focal point

Fig. 4. Aperture angle in our system.

available light rays as shown in Fig. 4. We can calculate $\sin\theta$ using the width w and focal length l of the paraboloidal mirror as follows.

$$\sin\theta = \frac{\frac{w}{2}}{\sqrt{\left(\frac{w}{2}\right)^2 + \left(\frac{l}{2}\right)^2}}. \tag{2}$$

Here, since the expression of the paraboloidal mirror where $y = 0$ in the coordinate system as shown in Fig. 5 is expressed as $x^2 = 4lz$, the relationship between the width w and the focul length l can be calculated from the following equation.

$$\left(\frac{w}{2}\right)^2 = 4l\left(\frac{l}{2}\right). \tag{3}$$

By replacing w with l, the NA of the proposed device is calculated as follows:

$$NA = \sin\theta = \frac{\sqrt{2}l}{\sqrt{2l^2 + \frac{l^2}{4}}} = \frac{2\sqrt{2}}{3} \simeq 0.94. \tag{4}$$

It should be noted here that the NA of the proposed device is a constant even if the scale of the paraboloidal mirror changes because the NA does not depend on the focal length and width of the paraboloidal mirror as indicated in Eq. (4).

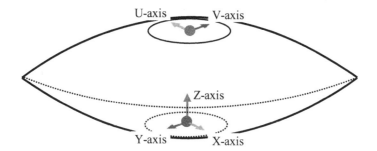

Fig. 5. Definition of coordinate systems.

Table 1 shows the comparison of the NAs of the proposed device and various commercial lenses. In this table, we calculated approximate NAs from F numbers using the following equation

$$NA = \frac{1}{2F} \tag{5}$$

From the table, we can confirm that the NA of the proposed device is much larger than those of various camera lenses and has competitive performance with objective lenses of microscopes. It should be noted that our system can handle relatively larger objects without constructing an ultra-large lens system which is practically almost impossible.

Table 1. Comparison of Numerical Aperture of various lenses.

NA	Lens type	Trade name
0.94	Mirror lens	Our approach
0.85	Objective lens (microscope)	WRAYMER GLF-ACH60X
0.59	Large aperture lens	HandeVision IBELUX 0.85/40MM
0.53	Fixed focal length lens	SCHNEIDER F0.95 Fast C-Mount Lens
0.36	Large aperture lens	Canon EF35mm F1.4L II USM

For the proposed device, depth of field d can be determined geometrically as follows:

$$d = d_{far} - d_{near},\qquad(6)$$

where d_{near} and d_{far} are distances from the image sensor to the nearest and farthest points that are in focus, respectively, and are calculated as follows:

$$d_{near} = \frac{l^2}{l + 2Fc},\qquad(7)$$

$$d_{far} = \frac{l^2}{l - 2Fc},\qquad(8)$$

where c is the size of a circle of confusion, which is the size of a pixel of an image sensor. For example, when c is 0.01 mm, l is 100 mm, and the F number is 0.53 (i.e., NA is 0.94), depth of field d becomes 0.0212 mm.

3   Experiments

In experiments, first, in order to check the raw performance of the proposed imaging device, the effect of geometric aberration is evaluated by measuring PSFs (point spread functions) which vary depending on the 3-D position of a target point, using a simulation environment. Layered real-images are then captured using a prototype imaging device and compared with images captured by a conventional lens-based camera device which has large NA in order to show the feasibility and the advantage of the proposed device. In addition to these two basic experiments for the proposed device, we further show the possibility to remove blurs on captured images using measured PSFs.

3.1   Characteristic of Faced Paraboloidal Mirror-Based Imaging Device

The shape of the PSF, which is a response of an impulse input from a point light source, varies depending on the 3-D position of the point light source placed in the proposed device. In order to analyze the characteristic of the proposed device, shapes of PSFs for different light source positions in the proposed imaging device are measured in a simulation environment. Experimental setting in this simulation is as follows: Focal length p is set to 65 mm. Width w of the mirror device is determined as 184 mm from Eq. (3). An imaging device (20 mm × 20 mm, 201 × 201 pixels) is fixed at one of the two vertex positions. While moving the position of a point light source, we observed shapes of PSFs by this imaging device.

Figures 6 and 7 show the PSFs captured on the image plane (U, V) while moving a light source along the Z axis and X axis shown in Fig. 5, respectively. Since the proposed imaging device is rotationally symmetric, we can know the characteristic of the device using these two axes. From these figures, we can see that the shape of PSF drastically changes when the light source moves along Z axis as shown in Fig. 6, while the change of the PSF along X axis is comparatively moderate as shown in Fig. 7. This indicates that the object moving along Z axis from the vertex position immediately blurs, in contrast to the case for X direction. From this simulation, we can confirm the desirable characteristic of the proposed device for achieving ultra-shallow DoF imaging.

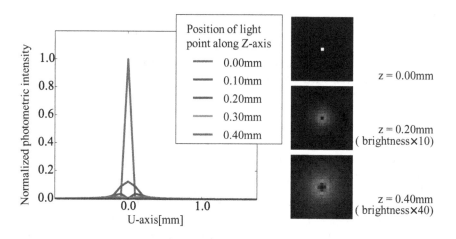

Fig. 6. Observed PSFs for different Z positions of a point light source with $X = 0$. (left: slice of PSF with $V = 0$, right: cropped PSF image)

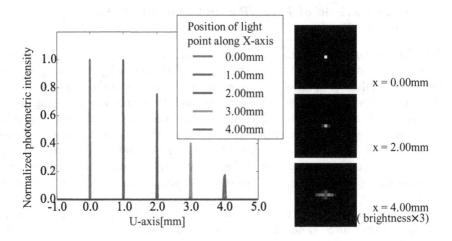

Fig. 7. Observed PSFs for different X positions of a point light source with $Z = 0$. (left: slice of PSF with $V = 0$, right: cropped PSF image)

3.2   Ultra Shallow DoF Imaging Using Prototype

We have constructed the prototype device shown in Fig. 1. In this device, Point Grey Grasshopper2 (1,384 × 1,036 pixels, CCD) without a lens is employed as the image sensor and fixed at the vertex of the upper paraboloidal mirror. A target object is set at the vertex of the lower paraboloidal mirror and its depth can be adjusted by using a translation stage as shown in Fig. 2.

Figure 8 shows target objects 1 to 3 in this experiment which consist of two layered flat surfaces which are transparent films with 0.1 mm thickness where different images are printed. The sizes of surfaces are 20 mm × 20 mm and two layers are separated with 1.2 mm empty gap.

Target 1 and 2 shown in Fig. 8(a), (b) are layered objects where a grid mask texture which contains high frequency component is commonly used as upper layer images and low and high frequency textures are used for lower layer images respectively. Target 3 (Fig. 8(c)) has low frequency texture image for upper layer and high frequency texture image for lower layer where lower layer is almost completely blinded with upper layer in a standard camera image.

Figures 9, 10 and 11(a) show images captured by the camera (Grasshopper2) with a small DoF lens (Schneider Fast C-Mount Lens, 17 mm FL, F = 0.95 (NA = 0.53)) for different height positions Z of the target objects. As we can see in these figures, the lower layer images are partially blinded by the grid patterns for the target 1 and 2, and the lower image is completely blinded for the target 3 even when we have employed relatively shallow DoF lens.

In contrast to this, one of two layer images is largely blurred by the proposed device not depending on the combination of low and high frequency textures, and the other layer image is focused as shown in Figs. 9, 10 and 11(b). Even for the target 3, the characters behind the upper layer image are readable as

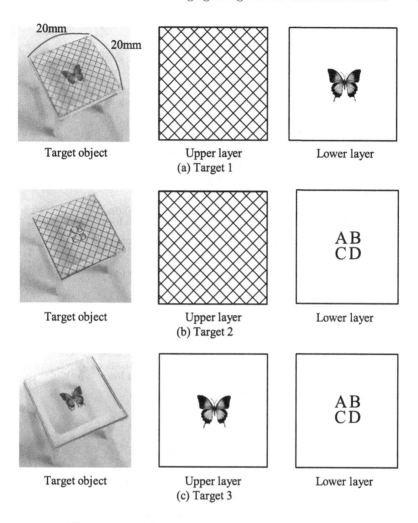

Fig. 8. Target objects and texture images for layers.

shown in Fig. 11(b). By this comparison, we can conclude that our system can achieve much shallower DoF imaging than the conventional lens-based system. However, we also confirmed that the proposed device still has two problems: (1) The images captured by the proposed device blur in the peripheral regions more than the conventional lens, which means that the proposed device has worse geometric aberration than conventional lenses, and (2) textures from the other non-focused layer still remain a little in the captured images.

3.3   Reconstruction of Layer Images Using PSFs

As described in the previous section, the proposed device has the weakness about the blurring effect caused by the aberration and the influences from other layers.

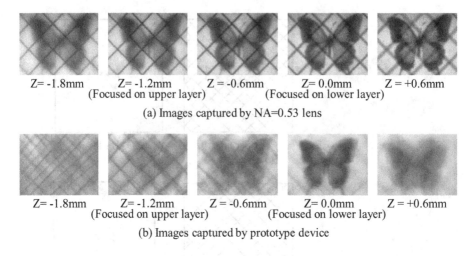

Z= -1.8mm   Z= -1.2mm   Z = -0.6mm   Z= 0.0mm   Z = +0.6mm
           (Focused on upper layer)   (Focused on lower layer)

(a) Images captured by NA=0.53 lens

Z= -1.8mm   Z= -1.2mm   Z = -0.6mm   Z= 0.0mm   Z = +0.6mm
           (Focused on upper layer)   (Focused on lower layer)

(b) Images captured by prototype device

Fig. 9. Experimental results for target 1.

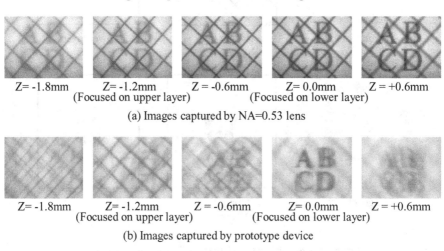

Z= -1.8mm   Z= -1.2mm   Z = -0.6mm   Z= 0.0mm   Z = +0.6mm
           (Focused on upper layer)   (Focused on lower layer)

(a) Images captured by NA=0.53 lens

Z= -1.8mm   Z= -1.2mm   Z = -0.6mm   Z= 0.0mm   Z = +0.6mm
           (Focused on upper layer)   (Focused on lower layer)

(b) Images captured by prototype device

Fig. 10. Experimental results for target 2.

In order to confirm the future possibility to overcome this weakness, here we simply deblur the observed images using measured PSFs by the following manner.

In the proposed system, theoretically, vectorized observed image $\mathbf{o}$ can be represented as the weighted sum of the multiplication of intensity w_k of the k-th point light source and vectorized PSF $\mathbf{p}_k$ for the light-source position k in the device as follows:

$$\mathbf{o} = \sum_k w_k \mathbf{p}_k, \tag{9}$$

where we ignore occlusion effects for simplicity. Here, if $\mathbf{p}_k$ is given by the calibration as shown in the first experiment and we have multiple observed images $\mathbf{o}$

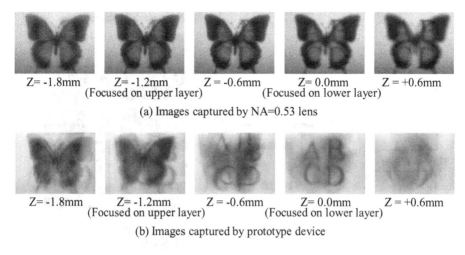

Z= -1.8mm   Z= -1.2mm   Z = -0.6mm   Z= 0.0mm   Z = +0.6mm
                (Focused on upper layer)   (Focused on lower layer)

(a) Images captured by NA=0.53 lens

Z= -1.8mm   Z= -1.2mm   Z = -0.6mm   Z= 0.0mm   Z = +0.6mm
                (Focused on upper layer)   (Focused on lower layer)

(b) Images captured by prototype device

Fig. 11. Experimental results for target 3.

with different depths, we can easily estimate w_k, which means that the aberration is suppressed and the captured images are decomposed into ones for respective layers, by minimizing the sum of error $\|\mathbf{o} - \sum_k w_k \mathbf{p}_k\|^2$ subject to $w_k \geq 0$ with convex optimization [18].

In this experiment, we have tested this method using the same device configuration with the second experiment in the simulated environment. The target object here consists of two layered films with 0.5 mm gap where different images are printed as in Fig. 12, which is more severe situation with narrower gap than that in the previous experiment. For this target, we have captured five images by moving the height of the object (focus point of the device) with 0.25 mm interval as shown in this figure.

Figure 13 shows the effect of decomposition. In this figure, (a) shows original layer images, (b) shows the images captured by focusing on upper and lower layers, (c) shows the decomposed results from the two images of (b), and (d) shows the decomposed results from all the five images. As shown in (b), even if the focus point is precisely adjusted to the position where target layer exists, the effect from the other layer image cannot be avoided, resulting blur effect in this severe situation as similar to the results in the previous experiment. As shown in (c) and (d) in Fig. 13, using this comparatively simple decomposition algorithm, the blurs were successfully reduced even by the two images, and were almost completely removed by the five images.

On the other hand, in the real world, to decompose images captured by the prototype system, we need to know a PSF for each spatial position in a target scene. A straightforward way to measure the PSFs is to prepare a hole whose size is smaller than that of a circle of confusion and align it with each spatial point. However, it is almost infeasible to create such a small hole and align it with each position accurately without special and expensive devices. Therefore,

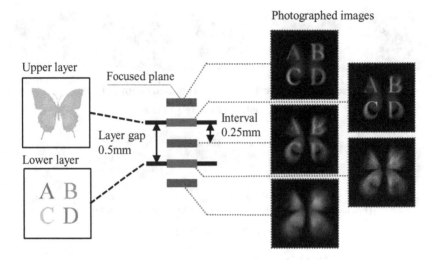

Fig. 12. Texture images for layers and captured input images for different height positions in simulation.

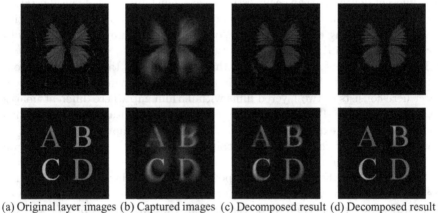

(a) Original layer images (b) Captured images (c) Decomposed result (d) Decomposed result
from two images   from five images

Fig. 13. Effect of decomposition. Top and bottom row show upper and lower layer images, respectively.

we should develop a method to measure PSFs that is an alternative to the way above in the future.

4 Conclusion

This paper has proposed an ultra-shallow DoF imaging method using faced paraboloidal mirrors that can visualize a specific depth. We constructed a prototype system and confirmed the proposed device can capture a specific depth using objects with two layers clearer than a small DoF lens. In the experiment using a simulation environment, we analyzed the characteristic of the proposed device, and we showed that the proposed system also can suppress the aberration and decompose layered images into clear ones using measured PSFs. In future work, we will develop a decomposition method considering occlusions and apply it to images captured by the developed prototype system.

Acknowledgement. This work was supported by JSPS Grant-in-Aid for Research Activity Start-up Grant Number 16H06982.

References

1. Levoy, M., Hanrahan, P.: Light field rendering. In: Proceedings SIGGRAPH, pp. 31–42 (1996)
2. Vaish, V., Wilburn, B., Joshi, N., Levoy, M.: Using plane + parallax for calibrating dense camera arrays. In: Proceedings CVPR, vol. 1, pp. I-2–I-9 (2004)
3. Wilburn, B., Joshi, N., Vaish, V., Talvala, E.V., Antunez, E., Barth, A., Adams, A., Horowitz, M., Levoy, M.: High performance imaging using large camera array. ACM Trans. Graph. **24**(3), 765–776 (2005)
4. Adelson, E.H., Wang, J.Y.A.: Single lens stereo with a plenoptic camera. IEEE Trans. PAMI **14**(2), 99–106 (1992)
5. Ng, R., Levoy, M., Bredif, M., Duval, G., Horowitz, M., Hanrahan, P.: Light field photography with a hand-held plenoptic camera. Proc. CTSR **2**(11), 1–11 (2005)
6. Cossairt, O., Nayar, S., Ramamoorthi, R.: Light field transfer: global illumination between real and synthetic objects. ACM Trans. Graph. **27**(3), 57:1–57:6 (2008)
7. Veeraraghavan, A., Raskar, R., Agrawal, A., Mohan, A., Tumblin, J.: Dappled photography: mask enhanced cameras for heterodyned light fields and coded aperture refocusing. ACM Trans. Graph. **26**(3), 69–76 (2007)
8. Liang, C., Lin, T., Wong, B., Liu, C., Chen, H.H.: Programmable aperture photography: multiplexed light field acquisition. ACM Trans. Graph. **27**(5), 55-1–55-10 (2008)
9. Unger, J., Wenger, A., Hawkins, T., Gardner, A., Debevec, P.: Capturing and rendering with incident light fields. In: Proceedings EGSR, pp. 141–149 (2003)
10. Lanman, D., Crispell, D., Wachs, M., Taubin, G.: Spherical catadioptric arrays: construction, multi-view geometry, and calibration. In: Proceedings 3DPVT, pp. 81–88 (2006)
11. Levoy, M., Chen, B., Vaish, V., Horowitz, M., McDowall, I., Bolas, M.: Synthetic aperture confocal imaging. In: Proceedings SIGGRAPH, pp. 825–834 (2004)
12. Tagawa, S., Mukaigawa, Y., Kim, J., Raskar, R., Matsushita, Y., Yagi, Y.: Hemispherical confocal imaging. IPSJ Trans. CVA **3**, 222–235 (2011)
13. Minsky, M.: Microscopy apparatus. US Patent 3013467 (1961)
14. White, J., Amos, W.B.: An evaluation of confocal versus conventional imaging of biological structures by fluorescence light microscopy. JCB **501**(1), 41–48 (1987)
15. Tanaami, T., Otsuki, S., Tomosada, N., Kosugi, Y., Shimizu, M., Ishida, H.: High-speed 1-frame/ms scanning confocal microscope with a microlens and Nipkow disks. Appl. Opt. **41**(22), 4704–4708 (2002)

16. Adhya, S., Noé, J.: A complete ray-trace analysis of the 'Mirage' toy. In: Proceedinigs SPIE ETOP, pp. 966518-1–966518-7 (2007)
17. Butler, A., Hilliges, O., Izadi, S., Hodges, S., Molyneaux, D., Kim, D., Kong, D.: Vermeer: direct interaction with a 360° viewable 3D display. In: Proceedings UIST, pp 569–576 (2011)
18. Gabay, D., Mercier, B.: A dual algorithm for the solution of nonlinear variational problems via finite-element approximations. Comput. Math. Appl. **2**, 17–40 (1976)

ConvNet-Based Depth Estimation, Reflection Separation and Deblurring of Plenoptic Images

Paramanand Chandramouli[(✉)], Mehdi Noroozi, and Paolo Favaro

Department of Computer Science, University of Bern, Bern, Switzerland
chandra@iam.unibe.ch

Abstract. In this paper, we address the problem of reflection removal and deblurring from a single image captured by a plenoptic camera. We develop a two-stage approach to recover the scene depth and high resolution textures of the reflected and transmitted layers. For depth estimation in the presence of reflections, we train a classifier through convolutional neural networks. For recovering high resolution textures, we assume that the scene is composed of planar regions and perform the reconstruction of each layer by using an explicit form of the plenoptic camera point spread function. The proposed framework also recovers the sharp scene texture with different motion blurs applied to each layer. We demonstrate our method on challenging real and synthetic images.

1 Introduction

When imaging scenes with transparent surfaces, the radiance components present behind and in front of a transparent surface get superimposed. Separating the two layers from a composite image is inherently ill-posed since it involves determining two unknowns from a single equation. Consequently, existing approaches address this problem through additional information obtained by capturing a sequence of images [1–3], or by modifying the data acquisition modality [4–6], or by imposing specific priors on layers [7,8].

A light field camera has the ability to obtain spatial as well as angular samples of the light field of a scene from a single image [9]. With a single light field (LF) image, one can perform depth estimation, digital refocusing or rendering from different view points. This has led to an increased popularity of plenoptic cameras in the recent years [10,11]. While layer separation from a single image is severely ill-posed in conventional imaging, in light field imaging the problem is made feasible as demonstrated in recent works [12–14]. These methods obtain the light field by using a camera array. We propose to use instead microlens array-based plenoptic cameras, because they are more compact and portable. However, both depth estimation and layer separation become quite challenging with a plenoptic camera due to the significantly small baseline [13,14]. Thus,

---

Electronic supplementary material The online version of this chapter (doi:10. 1007/978-3-319-54187-7_9) contains supplementary material, which is available to authorized users.

© Springer International Publishing AG 2017
S.-H. Lai et al. (Eds.): ACCV 2016, Part III, LNCS 10113, pp. 129–144, 2017.
DOI: 10.1007/978-3-319-54187-7_9

we develop a novel technique to estimate depth and separate the reflected and transmitted radiances from a single plenoptic image.

Because of merging of intensities from the two layers, the standard multi-view correspondence approach cannot be used for depth estimation. We develop a neural network-based classifier for estimating depth maps. Our classifier can also separate the scene into reflective and non-reflective regions. The depth estimation process has a runtime of only *a few seconds* when run on current GPUs. For recovering scene radiance, we consider that each of the two layers have a constant depth. We relate the observed light field image to a texture volume which consists of radiances from the reflected and transmitted layers through a point spread function (PSF) by taking into account the scene depth values and optics of the plenoptic camera. We solve the inverse problem of reconstructing the texture volume within a regularization framework. While imaging low-light scenes or scenes with moving objects, it is very common for motion blur to occur. If reflections are present in such scenarios, conventional deblurring algorithms fail because they do not model superposition of intensities. However, if such scenes are imaged by a plenoptic camera, our framework can be used to reverse the effect of motion blur. Note that motion deblurring along with layer separation is quite a challenging task because not only the number of unknowns that have to be simultaneously estimated is high but also blind deconvolution is known to be inherently ill-posed.

Figure 1 shows a real-world example of a scene imaged by a plenoptic camera. One can observe the effect of mixing of intensities of the transmitted and reflected layers in the refocused image generated using Lytro rendering software (Fig. 1(a)). Figures 1(b) and (c) show the result of our texture reconstruction algorithm which was preceded by the depth estimation process.

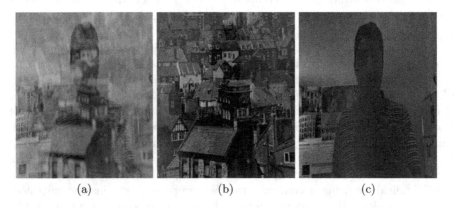

(a)   (b)   (c)

Fig. 1. Layer separation example: (a) Rendered image by Lytro Desktop software. Estimated textures of (b) transmitted and (c) reflected layers.

2   Related Work

We briefly review prior works related to reflection separation, plenoptic cameras and motion blur removal.

Reflection Separation. Many techniques make specific assumptions on the scene texture to achieve layer separation from a single image. Levin et al. assume that the number of edges and corners in the scene texture should be as low as possible [7]. In [15], a small number of manually labeled gradients corresponding to one layer are taken as input. By using the labeled gradients and a prior derived from the statistics of natural scenes, layer separation is accomplished. Li and Brown develop an algorithm that separates two layers from a single image by imposing different prior distributions on each layer [8]. Because of the inherent difficulty in the problem, the performance of single image-based techniques is not satisfactory and therefore, additional cues are incorporated [2].

The use of motion parallax observed in a sequence of images has been one of the popular approaches for layer decomposition [1–3,16,17]. As compared to earlier methods the technique of Li and Brown does not enforce restrictions on the scene geometry and camera motion [3]. Based on SIFT-flow they align the images and label the edges as belonging to either of the two layers. The layers are then reconstructed by minimizing an objective function formulated using suitable priors. The recent framework proposed by Xue et al. is applicable for reflection separation as well as occlusion removal without the assumption of a parametric camera motion [2]. Their method incorporates a robust initialization for the motion field and jointly optimizes for the motion field and the intensities of the layers in a coarse-to-fine manner. In another recent work, Shih et al. propose a layer separation scheme which is based on ghosting effects of the reflected layer [18].

Wanner and Goldlücke address the problem of estimating the geometry of both the transmitted and reflected surfaces from a 4D light field [12]. They identify patterns on the epipolar plane image due to reflections and derive local estimates of disparity for both the layers using a second order structure tensor. Wang et al. build an LF array system for capturing light fields [13]. They estimate an initial disparity map, form an image stack that exhibits low rank property and determine the separation and depth map through an optimization framework based on Robust Principle Component Analysis. The work closest to ours is that of Johannsen et al. [14]. The technique in [12] does not lead to good depth estimates on data from real microlens-array based plenoptic cameras. In [14], the authors propose improvements to the depth estimation technique in [12]. They also propose a variational approach for layer separation given the disparities of each layer. The main advantage of our work over this method is that we estimate the radiances of the layers at a much higher resolution. Our method can also be used in the presence of motion blur. Additionally, for real-word data, the technique in [14] requires user-assisted masks to separate reflecting and Lambertian surfaces. In contrast, our method *automatically* distinguishes between reflective and non-reflective surfaces.

Plenoptic cameras. Although light fields can also be captured from camera arrays, for brevity we focus our discussion on microlens array-based cameras. Ng et al. proposed a portable design for capturing light fields by placing a microlens array between the camera lens and the sensors [9]. This design also enables post capture refocusing, rendering with alternate viewpoint [9] and depth estimation [19]. To overcome the limited spatial resolution in plenoptic images, superresolution techniques have been proposed [19,20]. Georgiev et al. [20] directly combine the information from different angular views to obtain a high resolution image. Bishop and Favaro [19] relate the LF image with the scene radiance and depth map through geometric optics. They initially estimate the depth map and subsequently the high resolution scene radiance through deconvolution. In [21], along with disparity estimation, an input light field is super-resolved not only in spatial domain but also in angular domain.

In recent years, decoding, calibration and depth estimation techniques have been developed for commercially available cameras. Cho et al. [22] develop a method for rectification and decoding of light field data and render the super-resolved texture using a learning-based interpolation method. The calibration techniques in [23,24], develop a model to relate a pixel of the light field image with that of a ray in the scene. A significant number of depth estimation algorithms exist for plenoptic cameras. Most of these methods assume that the scene consists of Lambertian surfaces [25–29]. In a recent work, Tao et al. propose a scheme for depth estimation and specularity removal for both diffuse and specular surfaces [30]. Since view correspondences do not hold for specular surfaces, based on a dichromatic model, they investigate the structure of pixel values from different views in color space. Based on their analysis, they develop schemes to robustly estimate depth, determine light source color and separate specularity. These techniques cannot be used for reflective surfaces since they do not handle superposition of radiances from two layers.

Motion Deblurring. Recovering the sharp image and motion blur kernel from a given blurry image has been widely studied in the literature [31,32]. Although the blind deconvolution problem is inherently ill-posed, remarkable results have been achieved by incorporating suitable priors on the image and blur kernel [33–36]. While the standard blind deconvolution algorithms consider the motion blur to be uniform across the image, various methods have been proposed to handle blur variations due to camera rotational motion [37–39], depth variations [40–42], and dynamic scenes [43]. Although there have been efforts to address the issue of saturated pixels in the blurred observation [44], no deblurring algorithm exists that can handle the merging of image intensities due to transparent surfaces.

Our contributions can be summarized as follows: (i) We model the plenoptic image of a scene that contains reflected and transmitted layers by taking into account the effect of camera optics and scene depth. (ii) Ours is the first method that uses convolutional neural networks for depth estimation on images from plenoptic cameras. Our classifier can efficiently estimate the depth map of both the layers and works even when there are no reflections. (iii) Our PSF-based model inherently constrains the solution space thereby enabling layer separation. (iv) We also extend our framework to address the challenging scenario of joint reflection separation and motion deblurring.

3   Plenoptic Image Formation of Superposed Layers

Initially, let us consider a constant depth scene without any reflections. The LF image of such a scene can be related to the scene radiance through a point spread function (PSF), which characterizes the plenoptic image formation process [19,45,46]. The PSF is dependent on the depth value and the camera parameters, and encapsulates attributes such as disparity across sub-aperture images and microlens defocus. Mathematically, one can express the light field image l formed at the camera sensors in terms of the scene radiance f_d and a PSF H_d as a matrix vector product. Note that both l and f_d are vectorial representations of a 2D image. The subscript d in the PSF H_d indicates the depth label. The entries of the matrix H_d can be explicitly evaluated from the optical model of the camera [19,45]. The PSF H_d is defined by assuming a certain resolution for the scene texture. Defining the scene texture on a finer grid would lead to more columns in the PSF H_d. A column of the matrix denotes an LF image that would be formed from a point light source at a location corresponding to the column index. Note that in practice, the light field image and the texture will be of the order of millions of pixels and generating PSF matrices for such sizes is not feasible. However, the PSF has a repetitive structure because the pattern formed by the intersection of the blur circles of the microlens array and the main lens gets repeated. Based on this fact, the matrix vector product $H_d f_d$ can be efficiently implemented through a set of parallel convolutions between the scene texture and a small subset of the elements of the matrix H_d [45].

While imaging scenes consisting of transparent surfaces, radiances from two different layers get superimposed. Let f_{d_t} and f_{d_r} denote the radiances of the transmitted and reflected layers, and d_t and d_r denote the depth values of the transmitted and reflected layers, respectively. Then the LF image l can be expressed as

$$l = H_{d_t} f_{d_t} + H_{d_r} f_{d_r} = Hf \tag{1}$$

where the variable f which is referred to as texture volume is composed of radiances of the two layers, i.e., $f = [f_{d_t}^T \ f_{d_r}^T]^T$. The matrix H is formed by concatenating the columns of PSF matrices corresponding to depths d_t and d_r, i.e., $H = [H_{d_t} \ H_{d_r}]$.

According to the imaging model, the plenoptic images corresponding to the transmitted layer lie in the subspace spanned by the columns of H_{d_t}, span$\{H_{d_t}\}$, and those of the reflected layer lie in span$\{H_{d_r}\}$. The attributes of an LF image such as extent of blurring and disparity between sub-aperture images vary as the depth changes. We assume that the depth values of the transmitted and the reflected layers are quite different.

4   Proposed Method

In real scenarios, the local intensities of an LF image can consist of components from either the transmitted layer or the reflected layer, or both. Due to superposition of the two layers, the standard approach of establishing correspondences across views would not be applicable. To detect the depth, we train a classifier using a convolutional neural network (ConvNet). Subsequently, we solve the inverse problem of estimating the high-resolution textures of the two layers within a regularization framework. In our texture estimation procedure, we consider that the two layers have constant depth. However, our depth estimation technique is applicable even if there are depth variations within each layer.

4.1   Depth Estimation

In our experiments, we use the Lytro Illum camera. Consequently some of the details of our depth estimation scheme are specific to that camera. However, our method can be easily adapted to other plenoptic cameras as well. We consider that the scene depth ranges from 20 cm to 2.5 m. The depth range is divided into 15 levels denoted by the set $\Delta = \{d_1, d_2, \ldots, d_N\}$ ($N = 15$). The quantization of depth range is finer for smaller depth values and gets coarser as the values increase. For a pair of depths, as their magnitudes increase, their corresponding PSFs become more indistinguishable. Beyond 2.5 m there would be hardly any variations in the PSF.

A patch of an LF image can have intensities from either only one layer (if there is no reflection) or from two layers. We define a set of labels $\Lambda = \{\lambda_1, \lambda_2, \ldots \lambda_L\}$ wherein each label denotes either a combination of a pair of depths or individual depths from the set Δ. We assume that the two layers have significantly different depth values and do not include all possible pairs from Δ in Λ. Instead, we choose 45 different labels in the set Λ.

We perform depth estimation on patches of 2D raw plenoptic images in which the micolenses are arranged on a regular hexagonal grid and avoid the interpolation effects that occur while converting the 2D image to a 4D representation [23]. In an aligned 2D plenoptic image of Lytro Illum camera, the microlens arrangement pattern repeats horizontally after every 16 pixels and vertically every 28 pixels. Consequently, we define one unit to consist of 28 × 16 pixels. For both training and evaluation we use a patch from an LF image consisting of 10 × 10 units. We convert a patch into a set of views wherein, each view is obtained by sampling along horizontal and vertical directions. We discard those views that correspond to the borders of the microlenses. This rearranged set of views is considered as the input for which a label is to be assigned. Including the three color channels, the dimensions of the input corresponding to a patch of a plenoptic image are 10 × 10 × 888. This input data contains disparity information across the 888 views similar to a set of sub-aperture images (illustrated in the supplementary material). Our ConvNet is trained using labeled inputs in this format.

Network Architecture. As depicted in the Fig. 2 our ConvNet contains five layers. The first three are convolutional, and the fourth and fifth layers are fully connected. No pooling layer is used. We use 3×3 filters in the convolutional layers without padding. Our choice of the size of filters leads to a receptive field of 7×7. This in turn corresponds to an area in the LF image that is sufficiently large enough to capture features.

Training. For generating the training data, we used real LF images. Using a projector, we displayed a set of natural images on a planar surface. The LF images were captured by placing the camera at distances $d_1, d_2, \ldots, d_N$. To obtain the data for a label λ, if the label corresponded to a combination of two depths, we superimposed the LF images of the two depths, else we directly used patches from the captured images. Our training dataset consisted of about 16,000 patches per label. We implemented stochastic gradient descent with a batch size of 256 patches. We used the step strategy during the training with 0.01 as the baseline learning rate, and 50,000 as step. The training converged after 170,000 iterations using batch normalization [47] and took about 10 hours with one Titan X.

Efficient Evaluation. Given an LF image as the input, from the ConvNet, we arrive at a label map which contains the label for each unit (28×16 pixels). One could follow a straightforward approach of evaluating the label for every patch through the ConvNet. This would involve cropping patches from the LF image through a sliding window scheme (with a shift of one unit), rearranging every patch in the format of dimensions $10 \times 10 \times 888$ and feeding the cropped patches to the network as input. As there is overlap corresponding to one unit amongst these patches, this approach involves redundant time consuming calculations in the convolutional layers and thereby is inefficient. We address this problem by separating convolutional and fully connected layer operations. Instead of cropping patches, we rearrange the entire LF image into a set of views (following a procedure similar to that of a patch) and drop the views corresponding to the borders of the microlenses. This gives us a set of views that are large in size and having dimensions $W \times H \times 888$. Note that each $10 \times 10 \times 888$ subregion of this array corresponds to one patch in the original LF image and vice versa. We feed this large array as input to the convolutional layers. The last (third) convolutional layer feature map would be of size $(W-6) \times (H-6) \times 384$ (refer to Fig. 2). To determine the label of a patch in the LF image we find its corresponding $4 \times 4 \times 384$ subregion in the third convolutional layer feature map and feed it to the fully connected layers as input. With this alternate approach, we can calculate the depth map of the full LF image of size of $6,048 \times 8,640$ pixels in about 3 s with one Titan X.

We convert the labels to two depth maps by assuming that the depth values of one layer is always greater than the other. The non-reflective regions are also automatically indicated by labels that correspond to individual depth entries.

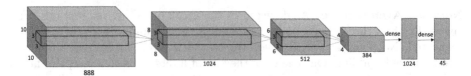

Fig. 2. ConvNet architecture used for depth estimation. Excluding the last layer, all other layers are followed by ReLu (not indicated for the sake of brevity). Since convolutions are performed without padding, and with a stride of one, spatial extent of the data decreases by two after each convolutional layer.

4.2 Texture Reconstruction

Our texture reconstruction algorithm is restricted to scenes wherein the two layers can be approximated by fronto-parallel planes. We evaluate the median values of the two depth maps to arrive at the depth values d_t and d_r. The PSF entries are then evaluated with the knowledge of camera parameters [19]. We define the texture resolution to be one-fourth of the sensor resolution. We formulate a data fidelity cost in terms of the texture volume through the correct PSF. We impose total variation (TV) regularization for each layer separately and arrive at the following objective function

$$\min_f \|l - Hf\|^2 + \nu\|f_{d_t}\|_{TV} + \nu\|f_{d_r}\|_{TV} \tag{2}$$

where $\|\cdot\|_{TV}$ denotes total variation and ν is the regularization parameter. We minimize Eq. (2) by gradient descent to obtain the texture volume, which consists of textures corresponding to each layer.

In our method, the PSF and TV prior impose constraints on the solution space of the layer separation problem. As an example, consider that the true texture $\hat{f}_{d_t}$ has an edge at a particular location. The PSF corresponding to depth d_t enforces that the edge gets repeated across k_{d_t} microlenses in the LF image corresponding to depth d_t. For the other layer, the number of microlenses, k_{d_r}, in which a feature gets repeated would be quite different from k_{d_t}, since we consider significant depth differences across layers. Similarly, the other attributes, such as microlens blur and disparity across views, also vary with depth. In our formulation we look for the texture volume that best explains the observed LF image. In the inverse problem of texture volume estimation, the constraints inherently imposed by the PSF avoids those solutions that generate attributes different from that of the observed LF image.

4.3 Motion Blur Scenario

A relative motion between the scene and the camera would lead to a motion blurred light field image. We model the blurry LF image as

$$l = H_{d_t}f_{d_t} + H_{d_r}f_{d_r} = H_{d_t}M_t u_t + H_{d_r}M_r u_r \tag{3}$$

where u_t and u_r denote the sharp texture, and M_t and M_r denote the blurring matrices of the transmitted and reflected layers respectively. We consider that the blur is uniform and the matrix vector products ($f_{d_t} = M_t u_t$ and $f_{d_r} = M_r u_r$) denote convolutions. In this scenario, we can write the objective function in terms of the sharp texture layers as

$$\min_{u_t,u_r} \|l - H_{d_t} M_t u_t - H_{d_r} M_r u_r\|^2 + \nu \|u_t\|_{TV} + \nu \|u_r\|_{TV}. \tag{4}$$

The objective function in terms of the motion blur kernels is given by

$$\min_{m_t,m_r} \|l - H_{d_t} U_t m_t - H_{d_r} U_r m_r\|^2$$
$$\text{subject to } m_t \succeq 0, m_r \succeq 0, \quad \|m_t\|_1 = 1, \|m_r\|_1 = 1 \tag{5}$$

where U_t and U_r are matrices corresponding to the textures u_t and u_r, respectively, and m_t and m_r denote the vectors corresponding to motion blurs kernels of the transmitted and reflected layers, respectively. We minimize the objective function given in Eqs. (4) and (5) by following an approach similar to the projected alternating minimization algorithm of [36]. In real scenarios, the reflected layer also undergoes the effect of ghosting due to the optical properties of the surface being imaged [18]. The combined effect of ghosting and motion blur can lead to significantly large shifts. Hence deblurring of the reflected layer may not work well in practical situations. However, our model works well even when there is ghosting.

5   Experimental Results

Firstly, we tested our layer separation algorithm on synthetic images while ignoring the effect of motion blur. For a set of depth values ranging from 0.35 m to 2.3 m (in steps of 0.15 m), we simulated LF observations and reconstructed the texture volume. We used four different natural textures. We assumed that the true depth values were known. A representative example is shown in Fig. 3. To quantify the performance we evaluate the normalized cross correlation (NCC) between the true and estimated textures of both the layers and average it over all the four pairs of textures. Figure 3(e) shows plots of mean NCC for cases wherein one of the layers had depth values fixed at 35 cm, 80 cm, and 1.7 m. From the plot, it is clear that the performance of layer separation is good only when the depths of the two layers are far apart. We also note that at 50 cm, there is a dip in the score for all three plots. This is because, this depth value was close to the camera main lens focal plane, wherein it is not possible to recover high resolution texture [19]. For synthetic experiments with motion blur, the average NCC was 0.897 when evaluated on the result of joint motion deblurring and layer separation (see the supplementary material for further details).

We perform real experiments using the Lytro Illum camera. The scene in Fig. 1 consisted of a computer screen and a person holding the camera in a room. The estimated depth maps of the two layers are shown in Fig. 4(b) and (c).

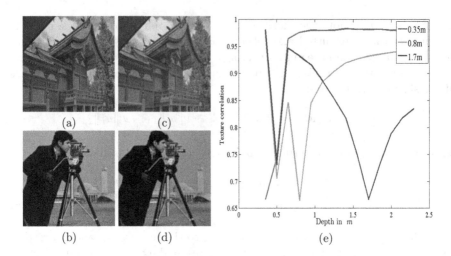

Fig. 3. Synthetic experiment: (a) and (b) true textures. (c) and (d) recovered textures. (e) Mean correlation between true and recovered texture at different depths (legend indicates the depth value of the transmitted layer).

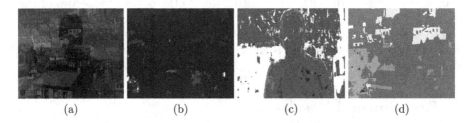

Fig. 4. Depth estimation: (a) Raw LF image. Depth map of (b) transmitted layer and (c) reflected layer (blue indicates presence of components from only one layer). (d) Estimated depth map using the technique of [29] (Color figure online)

The depth map of the transmitted layer is uniform throughout except for some artifacts, thereby correctly depicting the computer screen. In the second depth map, the blue color indicates the locations at which radiance components from only one layer were present. In the reflected layer, the regions corresponding to the white wall, person's face and the camera were completely textureless. Hence these locations were marked as regions without reflections. The depth map of the reflected layer correctly denotes the separation between the person and the background (refer to Fig. 1). Figure 4(d) shows that depth estimation failed when reflections were not accounted for. Despite the assumption of constant depth for each layer, we see that the textures of the two layers have been well separated.

Figure 5 shows the result of depth estimation and layer separation on two other scenes that contained a Lambertian surface as well as a reflective surface. In both these images, our method correctly labels regions corresponding to Lambertian surfaces. In both these images, there are large regions with *limited texture*

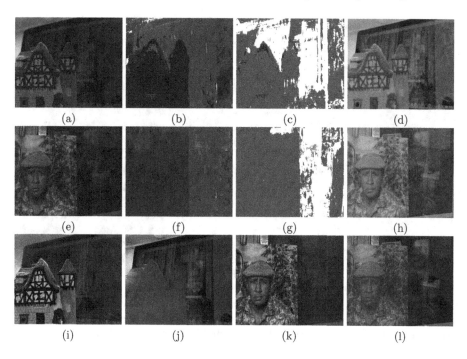

Fig. 5. Depth estimation: (a) and (e) Raw LF images. (b) and (f) Transmitted layer depth map (c) and (g) Reflected layer depth map (blue indicates presence of components from only one layer). (d) and (h) Lytro rendered image. (i)-(l) Recovered textures of the two layers.

in the reflected layer. Consequently, we see that even the textureless regions get marked as region without reflections. Note that in Fig. 5(c), the depth map correctly depicts the white regions corresponding to background which is far from the reflective surface and also shows a small grey region that corresponds to the reflection of the house model seen in the computer screen. In Figs. 5 (i)–(l), we show the recovered textures of the two layers corresponding to the two scenes. We observe artifacts in regions that are close to the camera. In this experiment, objects were as close as about 20 cm from the camera. At this range, depth changes induce large disparity changes and therefore constant depth assumption for the entire layer can lead to artifacts.

In our next experiment, we illuminated a region in front of a computer screen and imaged the screen with a plenoptic camera. The rendering of the LF image from the Lytro Desktop software obtained by refocusing at the screen surface is shown in Fig. 6(b). The result of the proposed depth estimation and layer separation method is shown in Figs. 6(c) and (d). For purpose of comparison, we also captured images of the scene by avoiding the mixing of layers. The Lytro rendering of an image captured without illuminating the reflecting surface and by increasing the brightness of the computer screen is shown in Fig. 6(e).

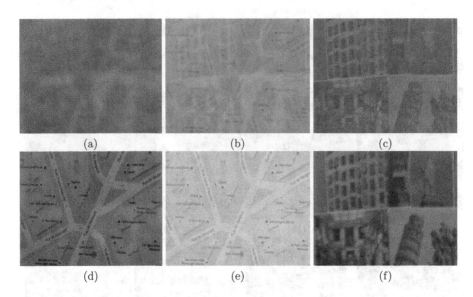

Fig. 6. (a) Raw image. (b) Lytro Desktop rendering of the observation. Recovered (c) reflected and (d) transmitted layer from the proposed scheme. Reference observations corresponding to (e) transmitted and (f) reflected layers *without* superposition.

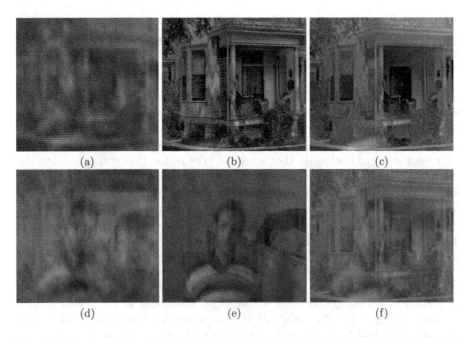

Fig. 7. Layer separation: (a) Raw LF image. (d) Rendered image by Lytro Desktop software. Estimated textures of (b) transmitted and (e) reflected layers from the proposed method. (c) and (f) show results of the algorithm in [8] on the Lytro rendered image.

Fig. 8. Layer separation and motion deblurring: (a) and (g) Raw LF image. (b) and (h) Rendered image from Lytro software. Estimated textures of transmitted (c, i) and reflected (d, j) layers. Motion deblurred transmitted layer (e, k) and reflected layer (f, l) (with motion blur kernel shown as inset).

The reference image for the reflected layer shown in Fig. 6(f) was captured by turning off the screen. Note that in our result, the layers have been well separated even in the presence of textures with high-contrast. Furthermore, when one compares, Figs. 6(d) and (e), our result (from the superimposed image) has a better resolution as against Lytro rendering (of the scene which did not have any layer superposition). Another real example is shown in Fig. 7. In Figs. 7(e) and (f) we see that the result obtained by applying the technique in [8] was not satisfactory.

We next present results on motion blurred scenes captured with a handheld camera. In Fig. 8, we show the raw LF image, rendering by Lytro software (by manually refocusing at the transmitted layer), results of the proposed method with and without motion blur compensation. For the first scene, while the transmitted layer is that of a poster pasted at a window, the reflected layer is that of a person with different objects in the background. The second scene had movie posters in both the layers. In each of the two layers there was a poster with similar content. Through visual inspection of the results on these images we can see that our proposed method produces consistent results.

6   Conclusions

We developed a technique to address the depth estimation and layer separation problem from a single plenoptic image. Not only our ConvNet-based approach enables depth estimation, but also detects the presence of reflections. With the estimated depth values we demonstrated that our PSF-based model enforces constraints that render layer separation feasible. Within our framework we also addressed the challenging problem of motion deblurring. In texture reconstruction we considered that both layers have a constant depth. In the future, our objective is to relax this assumption. Moreover, the performance of the ConvNet-based classifier can be further improved by having more variations in the training data and by including more depth labels.

References

1. Guo, X., Cao, X., Ma, Y.: Robust separation of reflection from multiple images. In: CVPR, pp. 2195–2202 (2014)
2. Xue, T., Rubinstein, M., Liu, C., Freeman, W.T.: A computational approach for obstruction-free photography. ACM Trans. Graph. **34**, 7901–7911 (2015)
3. Li, Y., Brown, M.: Exploiting reflection change for automatic reflection removal. In: ICCV, pp. 2432–2439 (2013)
4. Schechner, Y.Y., Kiryati, N., Basri, R.: Separation of transparent layers using focus. Int. J. Comput. Vis. **39**, 25–39 (2000)
5. Agrawal, A., Raskar, R., Nayar, S.K., Li, Y.: Removing photography artifacts using gradient projection and flash-exposure sampling. ACM Trans. Graph. (TOG) **24**, 828–835 (2005). ACM
6. Kong, N., Tai, Y.W., Shin, J.S.: A physically-based approach to reflection separation: from physical modeling to constrained optimization. IEEE Trans. Patt. Anal. Mach. Intell. **36**, 209–221 (2014)

7. Levin, A., Zomet, A., Weiss, Y.: Separating reflections from a single image using local features. In: CVPR (2004)
8. Li, Y., Brown, M.: Single image layer separation using relative smoothness. In: CVPR, pp. 2752–2759 (2014)
9. Ng, R., Levoy, M., Brédif, M., Duval, G., Horowitz, M., Hanrahan, P.: Light field photography with a hand-held plenoptic camera. Comput. Sci. Tech. Rep. CSTR **2**(11), 1–11 (2005)
10. Lytro. https://www.lytro.com/
11. Raytrix. (http://www.raytrix.de/)
12. Wanner, S., Goldlücke, B.: Reconstructing reflective and transparent surfaces from epipolar plane images. In: GCPR (2013)
13. Wang, Q., Lin, H., Ma, Y., Kang, S.B., Yu, J.: Automatic layer separation using light field imaging. arXiv preprint arXiv:1506.04721 (2015)
14. Johannsen, O., Sulc, A., Goldluecke, B.: Variational separation of light field layers. In: Vision, Modelling and Visualization (VMV) (2015)
15. Levin, A., Weiss, Y.: User assisted separation of reflections from a single image using a sparsity prior. IEEE Trans. Pattern Anal. Mach. Intell. **29**, 1647–1654 (2007)
16. Szeliski, R., Avidan, S., Anandan, P.: Layer extraction from multiple images containing reflections and transparency. In: CVPR, vol. 1, pp. 246–253. IEEE (2000)
17. Tsin, Y., Kang, S.B., Szeliski, R.: Stereo matching with linear superposition of layers. IEEE Trans. Pattern Anal. Mach. Intell. **28**, 290–301 (2006)
18. Shih, Y., Krishnan, D., Durand, F., Freeman, W.T.: Reflection removal using ghosting cues. In: 2015 IEEE Conference on Computer Vision and Pattern Recognition (CVPR), pp. 3193–3201. IEEE (2015)
19. Bishop, T., Favaro, P.: The light field camera: extended depth of field, aliasing and superresolution. IEEE Trans. Pattern Anal. Mach. Intell. **34**, 972–986 (2012)
20. Lumsdaine, A., Georgiev., T.: Full resolution lightfield rendering. Technical report, Adobe Systems (2008)
21. Wanner, S., Goldluecke, B.: Variational light field analysis for disparity estimation and super-resolution. IEEE Trans. Patt. Anal. Mach. Intell. **36**, 606–619 (2014)
22. Cho, D., Lee, M., Kim, S., Tai, Y.W.: Modeling the calibration pipeline of the lytro camera for high quality light-field image reconstruction. In: Proceedings ICCV (2013)
23. Dansereau, D.G., Pizarro, O., Williams, S.B.: Decoding, calibration and rectification for lenselet-based plenoptic cameras. In: Proceedings CVPR (2013)
24. Bok, Y., Jeon, H.G., Kweon, I.S.: Geometric calibration of micro-lens-based light-field cameras using line features. In: Proceedings ECCV (2014)
25. Tao, M., Hadap, S., Malik, J., Ramamoorthi, R.: Depth from combining defocus and correspondence using light-field cameras. In: Proceedings ICCV (2013)
26. Sabater, N., Seifi, M., Drazic, V., Sandri, G., Perez, P.: Accurate disparity estimation for plenoptic images. In: Proceedings ECCV Workshops (2014)
27. Yu, Z., Guo, X., Ling, H., Lumsdaine, A., Yu, J.: Line assisted light field triangulation and stereo matching. In: ICCV. IEEE (2013)
28. Jeon, H.G., Park, J., Choe, G., Park, J., Bok, Y., Tai, Y.W., Kweon, I.S.: Accurate depth map estimation from a lenslet light field camera. In: CVPR (2015)
29. Wang, T.C., Efros, A., Ramamoorthi, R.: Occlusion-aware depth estimation using light-field cameras. In: Proceedings of the IEEE International Conference on Computer Vision (ICCV) (2015)

30. Tao, M., Su, J., Wang, T., Malik, J., Ramamoorthi, R.: Depth estimation and specular removal for glossy surfaces using point and line consistency with light-field cameras. IEEE Trans. Pattern Anal. Mach. Intell. **38**(6), 1155–1169 (2015)
31. Fergus, R., Singh, B., Hertzmann, A., Roweis, S.T., Freeman, W.T.: Removing camera shake from a single photograph. ACM Trans. Graph. **25**, 787–794 (2006)
32. Levin, A., Weiss, Y., Durand, F., Freeman, W.: Efficient marginal likelihood optimization in blind deconvolution. In: CVPR, pp. 2657–2664 (2011)
33. Cho, S., Lee, S.: Fast motion deblurring. ACM Trans. Graph. **28**, 1–8 (2009)
34. Xu, L., Jia, J.: Two-phase kernel estimation for robust motion deblurring. In: ECCV (2010)
35. Xu, L., Zheng, S., Jia, J.: Unnatural L0 sparse representation for natural image deblurring. In: CVPR (2013)
36. Perrone, D., Favaro, P.: Total variation blind deconvolution: the devil is in the details. In: CVPR (2014)
37. Whyte, O., Sivic, J., Zisserman, A., Ponce, J.: Non-uniform deblurring for shaken images. In: Proceedings CVPR (2010)
38. Gupta, A., Joshi, N., Zitnick, L., Cohen, M., Curless, B.: Single image deblurring using motion density functions. In: Proceedings ECCV (2010)
39. Hirsch, M., Schuler, C.J., Harmeling, S., Scholkopf, B.: Fast removal of non-uniform camera shake. In: Proceedings ICCV (2011)
40. Hu, Z., Xu, L., Yang, M.H.: Joint depth estimation and camera shake removal from single blurry image. In: CVPR (2014)
41. Paramanand, C., Rajagopalan, A.N.: Non-uniform motion deblurring for bilayer scenes. In: CVPR (2013)
42. Sorel, M., Flusser, J.: Space-variant restoration of images degraded by camera motion blur. Trans. Img. Proc. **17**, 105–116 (2008)
43. Kim, T.H., Ahn, B., Lee, K.M.: Dynamic scene deblurring. In: The IEEE International Conference on Computer Vision (ICCV) (2013)
44. Whyte, O., Sivic, J., Zisserman, A.: Deblurring shaken and partially saturated images. Int. J. Comput. Vis. **110**, 185–201 (2014)
45. Broxton, M., Grosenick, L., Yang, S., Cohen, N., Andalman, A., Deisseroth, K., Levoy, M.: Wave optics theory and 3-D deconvolution for the light field microscope. Opt. Express **21**, 25418–25439 (2013)
46. Liang, C.K., Ramamoorthi, R.: A light transport framework for lenslet light field cameras. ACM Trans. Graph. **34**, 16:1–16:19 (2015)
47. Ioffe, S., Szegedy, C.: Batch normalization: accelerating deep network training by reducing internal covariate shift. In: Proceedings of the 32nd International Conference on Machine Learning, pp. 448–456. ACM (2015)

Learning a Mixture of Deep Networks for Single Image Super-Resolution

Ding Liu[1]([✉]), Zhaowen Wang[2], Nasser Nasrabadi[3], and Thomas Huang[1]

[1] Beckman Institute, University of Illinois at Urbana-Champaign,
Urbana, IL, USA
dingliu2@illinois.edu
[2] Adobe Research, San Jose, CA, USA
[3] Lane Department of Computer Science and Electrical Engineering,
West Virginia University, Morgantown, WV, USA

Abstract. Single image super-resolution (SR) is an ill-posed problem which aims to recover high-resolution (HR) images from their low-resolution (LR) observations. The crux of this problem lies in learning the complex mapping between low-resolution patches and the corresponding high-resolution patches. Prior arts have used either a mixture of simple regression models or a single non-linear neural network for this propose. This paper proposes the method of learning a mixture of SR inference modules in a unified framework to tackle this problem. Specifically, a number of SR inference modules specialized in different image local patterns are first independently applied on the LR image to obtain various HR estimates, and the resultant HR estimates are adaptively aggregated to form the final HR image. By selecting neural networks as the SR inference module, the whole procedure can be incorporated into a unified network and be optimized jointly. Extensive experiments are conducted to investigate the relation between restoration performance and different network architectures. Compared with other current image SR approaches, our proposed method achieves state-of-the-arts restoration results on a wide range of images consistently while allowing more flexible design choices.

1 Introduction

Single image super-resolution (SR) is usually cast as an inverse problem of recovering the original high-resolution (HR) image from the low-resolution (LR) observation image. This technique can be utilized in the applications where high resolution is of importance, such as photo enhancement, satellite imaging and SDTV to HDTV conversion [1]. The main difficulty resides in the loss of much information in the degradation process. Since the known variables from the LR image is usually greatly outnumbered by that from the HR image, this problem is a highly ill-posed problem.

A large number of single image SR methods have been proposed in the literature, including interpolation based method [2], edge model based method [3] and

© Springer International Publishing AG 2017
S.-H. Lai et al. (Eds.): ACCV 2016, Part III, LNCS 10113, pp. 145–156, 2017.
DOI: 10.1007/978-3-319-54187-7_10

example based method [4–9]. Since the former two methods usually suffer the sharp drop in restoration performance with large upscaling factors, the example based method draws great attention from the community recently. It usually learns the mapping from LR images to HR images in a patch-by-patch manner, with the help of sparse representation [6,10], random forest [11] and so on. The neighbor embedding method [4,7] and neural network based method [8] are two representatives of this category.

Neighbor embedding is proposed in [4,12] which estimates HR patches as a weighted average of local neighbors with the same weights as in LR feature space, based on the assumption that LR/HR patch pairs share similar local geometry in low-dimensional nonlinear manifolds. The coding coefficients are first acquired by representing each LR patch as a weighted average of local neighbors, and then the HR counterpart is estimated by the multiplication of the coding coefficients with the corresponding training HR patches. Anchored neighborhood regression (ANR) is utilized in [7] to improve the neighbor embedding methods, which partitions the feature space into a number of clusters using the learned dictionary atoms as a set of anchor points. A regressor is then learned for each cluster of patches. This approach has demonstrated superiority over the counterpart of global regression in [7]. Other variants of learning a mixture of SR regressors can be found in [13–15].

Recently, neural network based models have demonstrated the strong capability for single image SR [8,16,17], due to its large model capacity and the end-to-end learning strategy to get rid of hand-crafted features. Cui et al. [16] propose using a cascade of stacked collaborative local autoencoders for robust matching of self-similar patches, in order to increase the resolution of inputs gradually. Dong et al. [8] exploit a fully convolutional neural network (CNN) to approximate the complex non-linear mapping between the LR image and the HR counterpart. A neural network that closely mimics the sparse coding approach for image SR is designed by Wang et al. [17,18]. Kim et al. proposes a very deep neural network with residual architecture to exploit contextual information over large image regions [19].

In this paper, we propose a method to combine the merits of the neighborhood embedding methods and the neural network based methods via learning a mixture of neural networks for single image SR. The entire image signal space can be partitioned into several subspaces, and we dedicate one SR module to the image signals in each subspace, the synergy of which allows a better capture of the complex relation between the LR image signal and its HR counterpart than the generic model. In order to take advantage of the end-to-end learning strategy of neural network based methods, we choose neural networks as the SR inference modules and incorporate these modules into one unified network, and design a branch in the network to predict the pixel-level weights for HR estimates from each SR module before they are adaptively aggregated to form the final HR image.

A systematic analysis of different network architectures is conducted with the focus on the relation between SR performance and various network architectures

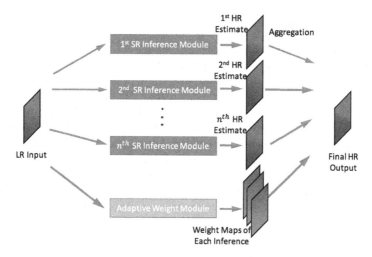

Fig. 1. The overview of our proposed method. It consists of a number of SR inference modules and an adaptive weight module. Each SR inference module is dedicated to inferencing a certain class of image local patterns, and is independently applied on the LR image to predict one HR estimate. These estimates are adaptively combined using pixel-wise aggregation weights from the adaptive weight module in order to form the final HR image.

via extensive experiments, where the benefit of utilizing a mixture of SR models is demonstrated. Our proposed approach is contrasted with other current popular approaches on a large number of test images, and achieves state-of-the-arts performance consistently along with more flexibility of model design choices.

The paper is organized as follows. The proposed method is introduced and explained in detail in Sect. 2. Section 3 describes our experiments, in which we analyze thoroughly different network architectures and compare the performance of our method with other current SR methods both quantitatively and qualitatively. Finally in Sect. 4 we conclude the paper.

2   Proposed Method

2.1   Overview

First we give the overview of our method. The LR image serves as the input to our method. There are a number of **SR inference modules** $\{B_i\}_{i=1}^{N}$ in our method. Each of them, B_i, is dedicated to inferencing a certain class of image patches, and applied on the LR input image to predict a HR estimate. We also devise an **adaptive weight module**, T, to adaptively combine at the pixel-level the HR estimates from SR inference modules. When we select neural networks as the SR inference modules, all the components can be incorporated into a unified neural network and be jointly learned. The final estimated HR image is adaptively aggregated from the estimates of all SR inference modules. The overview of our method is shown in Fig. 1.

2.2 Network Architecture

SR Inference Module: Taking the LR image as input, each SR inference module is designed to better capture the complex relation between a certain class of LR image signals and its HR counterpart, while predicting a HR estimate. For the sake of inference accuracy, we choose as the SR inference module a recent sparse coding based network (SCN) in [17], which implicitly incorporates the sparse prior into neural networks via employing the learned iterative shrinkage and thresholding algorithm (LISTA), and closely mimics the sparse coding based image SR method [20]. The architecture of SCN is shown in Fig. 2. Note that the design of the SR inference module is not limited to SCN, and all other neural network based SR models, e.g. SRCNN [8], can work as the SR inference module as well. The output of B_i serves as an estimate to the final HR frame.

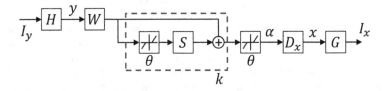

Fig. 2. The network architecture of SCN [17], which serves as the SR inference module in our method.

Adaptive Weight Module: The goal of this module is to model the selectivity of the HR estimates from every SR inference module. We propose assigning pixel-wise aggregation weights of each HR estimate, and again the design of this module is open to any operation in the field of neural networks. Taking into account the computation cost and efficiency, we utilize only three convolutional layers for this module, and ReLU is applied on the filter responses to introduce non-linearity. This module finally outputs the pixel-level weight maps for all the HR estimates.

Aggregation: Each SR inference module's output is pixel-wisely multiplied with its corresponding weight map from the adaptive weight module, and then these products are summed up to form the final estimated HR frame. If we use $\mathbf{y}$ to denote the LR input image, a function $W(\mathbf{y}; \theta_w)$ with parameters θ_w to represent the behavior of the adaptive weight module, and a function $F_{B_i}(\mathbf{y}; \theta_{B_i})$ with parameters θ_{B_i} to represent the output of SR inference module B_i, the final estimated HR image $F(\mathbf{y}; \Theta)$ can be expressed as

$$F(\mathbf{y}; \Theta) = \sum_{i=1}^{N} W_i(\mathbf{y}; \theta_w) \odot F_{B_i}(\mathbf{y}; \theta_{B_i}), \tag{1}$$

where $\odot$ denotes the point-wise multiplication.

2.3   Training Objective

In training, our model tries to minimize the loss between the target HR frame and the predicted output, as

$$\min_{\Theta} \sum_j \| F(\mathbf{y}_j; \Theta) - \mathbf{x}_j \|_2^2, \tag{2}$$

where $F(\mathbf{y}; \Theta)$ represents the output of our model, $\mathbf{x}_j$ is the j-th HR image and $\mathbf{y}_j$ is the corresponding LR image; Θ is the set of all parameters in our model.

If we plug Eq. 1 into Eq. 2, the cost function then can be expanded as:

$$\min_{\theta_w, \{\theta_{B_i}\}_{i=1}^N} \sum_j \| \sum_{i=1}^N W_i(\mathbf{y}_j; \theta_w) \odot F_{B_i}(\mathbf{y}_j; \theta_{B_i}) - \mathbf{x}_j \|_2^2. \tag{3}$$

3   Experiments

3.1   Data Sets and Implementation Details

We conduct experiments following the protocols in [7]. Different learning based methods use different training data in the literature. We choose 91 images proposed in [6] to be consistent with [11,13,17]. These training data are augmented with translation, rotation and scaling, providing approximately 8 million training samples of 56×56 pixels.

Our model is tested on three benchmark data sets, which are Set5 [12], Set14 [21] and BSD100 [22]. The ground truth images are downscaled by bicubic interpolation to generate LR/HR image pairs for both training and testing.

Following the convention in [7,17], we convert each color image into the YCbCr colorspace and only process the luminance channel with our model, and bicubic interpolation is applied to the chrominance channels, because the visual system of human is more sensitive to details in intensity than in color.

Each SR inference module adopts the network architecture of SCN, while the filters of all three convolutional layers in the adaptive weight module have the spatial size of 5×5 and the numbers of filters of three layers are set to be $32, 16$ and N, which is the number of SR inference modules.

Our network is implemented using Caffe [23] and is trained on a machine with 12 Intel Xeon 2.67 GHz CPUs and 1 Nvidia TITAN X GPU. For the adaptive weight module, we employ a constant learning rate of 10^{-5} and initialize the weights from Gaussian distribution, while we stick to the learning rate and the initialization method in [17] for the SR inference modules. The standard gradient descent algorithm is employed to train our network with a batch size of 64 and the momentum of 0.9.

We train our model for the upscaling factor of 2. For larger upscaling factors, we adopt the model cascade technique in [17] to apply $\times 2$ models multiple times until the resulting image reaches at least as large as the desired size. The resulting image is downsized via bicubic interpolation to the target resolution if necessary.

3.2   SR Performance vs. Network Architecture

In this section we investigate the relation between various numbers of SR inference modules and SR performance. For the sake of our analysis, we increase the number of inference modules as we decrease the module capacity of each of them, so that the total model capacity is approximately consistent and thus the comparison is fair. Since the chosen SR inference module, SCN [17], closely mimics the sparse coding based SR method, we can reduce the module capacity of each inference module by decreasing the embedded dictionary size n (i.e. the number of filters in SCN), for sparse representation. We compare the following cases:

- one inference module with $n = 128$, which is equivalent to the structure of SCN in [17], denoted as *SCN (n = 128)*. Note that there is no need to include the adaptive weight module in this case.
- two inference modules with $n = 64$, denoted as *MSCN-2 (n = 64)*.
- four inference modules with $n = 32$, denoted as *MSCN-4 (n = 32)*.

The average Peak Signal-to-Noise Ratio (PSNR) and structural similarity (SSIM) [24] are measured to quantitatively compare the SR performance of these models over Set5, Set14 and BSD100 for various upscaling factors ($\times 2, \times 3, \times 4$), and the results are displayed in Table 1.

Table 1. PSNR (in dB) and SSIM comparisons on Set5, Set14 and BSD100 for $\times 2$, $\times 3$ and $\times 4$ upscaling factors among various network architectures. Red indicates the best and blue indicates the second best performance.

Benchmark		SCN ($n = 128$)	MSCN-2 ($n = 64$)	MSCN-4 ($n = 32$)
Set5	$\times 2$	36.93/0.9552	37.00/0.9558	36.99/0.9559
	$\times 3$	33.10/0.9136	33.15/0.9133	33.13/0.9130
	$\times 4$	30.86/0.8710	30.92/0.8709	30.93/0.8712
Set14	$\times 2$	32.56/0.9069	32.70/0.9074	32.72/0.9076
	$\times 3$	29.41/0.8235	29.53/0.8253	29.56/0.8256
	$\times 4$	27.64/0.7578	27.76/0.7601	27.79/0.7607
BSD100	$\times 2$	31.40/0.8884	31.54/0.8913	31.56/0.8914
	$\times 3$	28.50/0.7885	28.56/0.7920	28.59/0.7926
	$\times 4$	27.03/0.7161	27.10/0.7207	27.13/0.7216

It can be observed that *MSCN-2 (n = 64)* usually outperforms the original SCN network, i.e. *SCN (n = 128)*, and *MSCN-4 (n = 32)* can achieve the best SR performance by improving the performance marginally over *MSCN-2 (n = 64)*. This demonstrates the effectiveness of our approach that each SR inference model is able to super-resolve its own class of image signals better than one single generic inference model.

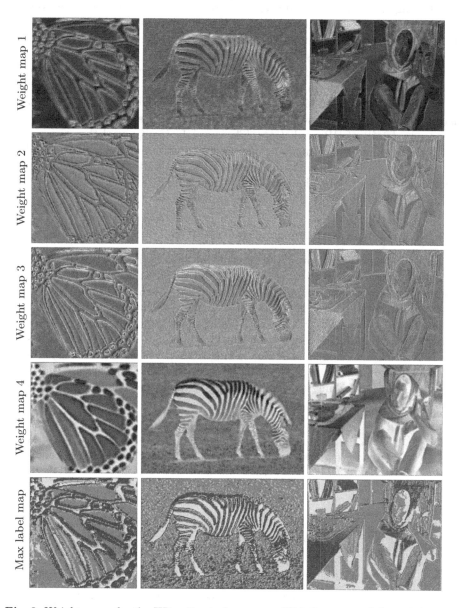

Fig. 3. Weight maps for the HR estimate from every SR inference module in *MSCN-4* are given in the first four rows. The map (*max label map*) which records the index of the maximum weight across all weight maps at every pixel is shown in the last row. Images from left to right: the *butterfly* image upscaled by ×2; the *zebra* image upscaled by ×2; the *barbara* image upscaled by ×2.

In order to further analyze the adaptive weight module, we select several input images, namely, *butterfly*, *zebra*, *barbara*, and visualize the four weight maps for every SR inference module in the network. Moreover, we record the index of the maximum weight across all weight maps at every pixel and generate a *max label map*. These results are displayed in Fig. 3.

From these visualizations it can be seen that weight map 4 shows high responses in many uniform regions, and thus mainly contributes to the low frequency regions of HR predictions. On the contrary, weight map 1, 2 and 3 have large responses in regions with various edges and textures, and restore the high frequency details of HR predictions. These weight maps reveal that these sub-networks work in a supplementary manner for constructing the final HR predictions. In the *max label map*, similar structures and patterns of images usually share with the same label, indicating that such similar textures and patterns are favored to be super-resolved by the same inference model.

3.3   Comparison with State-of-the-Arts

We conduct experiments on all the images in Set5, Set14 and BSD100 for different upscaling factors ($\times 2$, $\times 3$, and $\times 4$), to quantitatively and qualitatively compare our own approach with a number of state-of-the-arts image SR methods. Table 2 shows the PSNR and SSIM for adjusted anchored neighborhood regression (A+) [13], SRCNN [8], RFL [11], SelfEx [9] and our proposed model, *MSCN-4 (n = 128)*, that consists of four SCN modules with $n = 128$. The single generic SCN without multi-view testing in [17], i.e. *SCN (n = 128)* is also included for comparison as the baseline. Note that all the methods use the same 91 images [6] for training except SRCNN [8], which uses 395,909 images from ImageNet as training data.

Table 2. PSNR (SSIM) comparison on three test data sets for various upscaling factors among different methods. The best performance is indicated in red and the second best performance is shown in blue. The performance gain of our best model over all the other models' best is shown in the last row.

Data set	Set5			Set14			BSD100		
Upscaling	$\times 2$	$\times 3$	$\times 4$	$\times 2$	$\times 3$	$\times 4$	$\times 2$	$\times 3$	$\times 4$
A+ [13]	36.55	32.59	30.29	32.28	29.13	27.33	31.21	28.29	26.82
	(0.9544)	(0.9088)	(0.8603)	(0.9056)	(0.8188)	(0.7491)	(0.8863)	(0.7835)	(0.7087)
SRCNN [8]	36.66	32.75	30.49	32.45	29.30	27.50	31.36	28.41	26.90
	(0.9542)	(0.9090)	(0.8628)	(0.9067)	(0.8215)	(0.7513)	(0.8879)	(0.7863)	(0.7103)
RFL [11]	36.54	32.43	30.14	32.26	29.05	27.24	31.16	28.22	26.75
	(0.9537)	(0.9057)	(0.8548)	(0.9040)	(0.8164)	(0.7451)	(0.8840)	(0.7806)	(0.7054)
SelfEx [9]	36.49	32.58	30.31	32.22	29.16	27.40	31.18	28.29	26.84
	(0.9537)	(0.9093)	(0.8619)	(0.9034)	(0.8196)	(0.7518)	(0.8855)	(0.7840)	(0.7106)
SCN [17]	36.93	33.10	30.86	32.56	29.41	27.64	31.40	28.50	27.03
	(0.9552)	(0.9144)	(0.8732)	(0.9074)	(0.8238)	(0.7578)	(0.8884)	(0.7885)	(0.7161)
MSCN-4	37.16	33.33	31.08	32.85	29.65	27.87	31.65	28.66	27.19
	(0.9565)	(0.9155)	(0.8740)	(0.9084)	(0.8272)	(0.7624)	(0.8928)	(0.7941)	(0.7229)
Our improvement	0.23	0.23	0.22	0.29	0.24	0.23	0.25	0.16	0.16
	(0.0013)	(0.0011)	(0.0008)	(0.0010)	(0.0034)	(0.0046)	(0.0044)	(0.0056)	(0.0068)

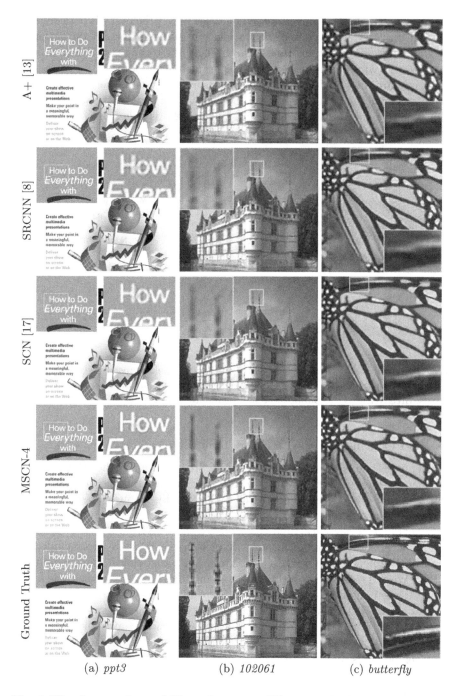

(a) *ppt3*   (b) *102061*   (c) *butterfly*

Fig. 4. Visual comparisons of SR results among different methods. From left to right: the *ppt3* image upscaled by ×3; the *102061* image upscaled by ×3; the *butterfly* image upscaled by ×4.

It can be observed that our proposed model achieves the best SR performance consistently over three data sets for various upscaling factors. It outperforms *SCN (n = 128)* which obtains the second best results by about 0.2 dB across all the data sets, owing to the power of multiple inference modules.

We compare the visual quality of SR results among various methods in Fig. 4. The region inside the bounding box is zoomed in and shown for the sake of visual comparison. Our proposed model *MSCN-4 (n = 128)* is able to recover sharper edges and generate less artifacts in the SR inferences.

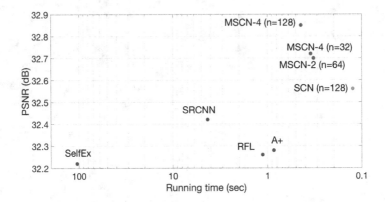

Fig. 5. The average PSNR and the average inference time for upscaling factor ×2 on Set14 are compared among different network structures of our method and other SR methods. SRCNN uses the public slower implementation of CPU.

3.4   SR Performance vs. Inference Time

The inference time is an important factor of SR algorithms other than the SR performance. The relation between the SR performance and the inference time of our approach is analyzed in this section. Specifically, we measure the average inference time of different network structures in our method for upscaling factor ×2 on Set14. The inference time costs versus the PSNR values are displayed in Fig. 5, where several other current SR methods [8,9,11,13] are included as reference (the inference time of SRCNN is from the public slower implementation of CPU). We can see that generally, the more modules our network has, the more inference time is needed and the better SR results are achieved. By adjusting the number of SR inference modules in our network structure, we can achieve the tradeoff between SR performance and computation complexity. However, our slowest network still has the superiority in term of inference time, compared with other previous SR methods.

4   Conclusions

In this paper, we propose to jointly learn a mixture of deep networks for single image super-resolution, each of which serves as a SR inference module to handle a certain class of image signals. An adaptive weight module is designed to predict pixel-level aggregation weights of the HR estimates. Various network architectures are analyzed in terms of the SR performance and the inference time, which validates the effectiveness of our proposed model design. Extensive experiments manifest that our proposed model is able to achieve outstanding SR performance along with more flexibility of design. In the future, this approach of image super-resolution will be explored to facilitate other high-level vision tasks [25].

References

1. Park, S.C., Park, M.K., Kang, M.G.: Super-resolution image reconstruction: a technical overview. IEEE Sig. Process. Mag. **20**, 21–36 (2003)
2. Morse, B.S., Schwartzwald, D.: Image magnification using level-set reconstruction. In: CVPR 2001, vol. 1, 1–333. IEEE (2001)
3. Fattal, R.: Image upsampling via imposed edge statistics. In: ACM Transactions on Graphics (TOG), vol. 26, p. 95. ACM (2007)
4. Chang, H., Yeung, D.Y., Xiong, Y.: Super-resolution through neighbor embedding. In: Proceedings of the 2004 IEEE Computer Society Conference on Computer Vision and Pattern Recognition, CVPR 2004, vol. 1, p. 1. IEEE (2004)
5. Glasner, D., Bagon, S., Irani, M.: Super-resolution from a single image. In: ICCV 2009, pp. 349–356. IEEE (2009)
6. Yang, J., Wright, J., Huang, T.S., Ma, Y.: Image super-resolution via sparse representation. IEEE Trans. Image Process. **19**, 2861–2873 (2010)
7. Timofte, R., De Smet, V., Van Gool, L.: Anchored neighborhood regression for fast example-based super-resolution. In: 2013 IEEE International Conference on Computer Vision (ICCV), pp. 1920–1927. IEEE (2013)
8. Dong, C., Loy, C.C., He, K., Tang, X.: Image super-resolution using deep convolutional networks. TPAMI **38**(2), 295–307 (2015)
9. Huang, J.B., Singh, A., Ahuja, N.: Single image super-resolution from transformed self-exemplars. In: 2015 IEEE Conference on Computer Vision and Pattern Recognition (CVPR), pp. 5197–5206. IEEE (2015)
10. Wang, Z., Yang, Y., Wang, Z., Chang, S., Yang, J., Huang, T.S.: Learning super-resolution jointly from external and internal examples. IEEE Trans. Image Process. **24**, 4359–4371 (2015)
11. Schulter, S., Leistner, C., Bischof, H.: Fast and accurate image upscaling with super-resolution forests. In: Proceedings of the IEEE Conference on Computer Vision and Pattern Recognition, pp. 3791–3799 (2015)
12. Bevilacqua, M., Roumy, A., Guillemot, C., Alberi-Morel, M.L.: Low-complexity single-image super-resolution based on nonnegative neighbor embedding (2012)
13. Timofte, R., De Smet, V., Van Gool, L.: A+: adjusted anchored neighborhood regression for fast super-resolution. In: Cremers, D., Reid, I., Saito, H., Yang, M.-H. (eds.) ACCV 2014. LNCS, vol. 9006, pp. 111–126. Springer, Heidelberg (2015). doi:10.1007/978-3-319-16817-3_8
14. Dai, D., Timofte, R., Van Gool, L.: Jointly optimized regressors for image super-resolution. In: Eurographics, vol. 7, p. 8 (2015)
15. Timofte, R., Rasmus, R., Van Gool, L.: Seven ways to improve example-based single image super resolution. In: 2016 IEEE Conference on Computer Vision and Pattern Recognition (CVPR). IEEE (2016)

16. Cui, Z., Chang, H., Shan, S., Zhong, B., Chen, X.: Deep network cascade for image super-resolution. In: Fleet, D., Pajdla, T., Schiele, B., Tuytelaars, T. (eds.) ECCV 2014. LNCS, vol. 8693, pp. 49–64. Springer, Heidelberg (2014). doi:10.1007/978-3-319-10602-1_4

17. Wang, Z., Liu, D., Yang, J., Han, W., Huang, T.: Deep networks for image super-resolution with sparse prior. In: Proceedings of the IEEE International Conference on Computer Vision, pp. 370–378 (2015)

18. Liu, D., Wang, Z., Wen, B., Yang, J., Han, W., Huang, T.S.: Robust single image super-resolution via deep networks with sparse prior. IEEE Trans. Image Process. **25**, 3194–3207 (2016)

19. Kim, J., Lee, J.K., Lee, K.M.: Accurate image super-resolution using very deep convolutional networks. In: 2016 IEEE Conference on Computer Vision and Pattern Recognition (CVPR). IEEE (2016)

20. Yang, J., Wang, Z., Lin, Z., Cohen, S., Huang, T.: Coupled dictionary training for image super-resolution. IEEE Trans. Image Process. **21**, 3467–3478 (2012)

21. Zeyde, R., Elad, M., Protter, M.: On single image scale-up using sparse-representations. In: Boissonnat, J.-D., Chenin, P., Cohen, A., Gout, C., Lyche, T., Mazure, M.-L., Schumaker, L. (eds.) Curves and Surfaces 2010. LNCS, vol. 6920, pp. 711–730. Springer, Heidelberg (2012). doi:10.1007/978-3-642-27413-8_47

22. Martin, D., Fowlkes, C., Tal, D., Malik, J.: A database of human segmented natural images and its application to evaluating segmentation algorithms and measuring ecological statistics. In: Proceedings of the Eighth IEEE International Conference on Computer Vision, ICCV 2001, vol. 2, pp. 416–423. IEEE (2001)

23. Jia, Y., Shelhamer, E., Donahue, J., Karayev, S., Long, J., Girshick, R., Guadarrama, S., Darrell, T.: Caffe: convolutional architecture for fast feature embedding. In: Proceedings of the ACM International Conference on Multimedia, pp. 675–678. ACM (2014)

24. Wang, Z., Bovik, A.C., Sheikh, H.R., Simoncelli, E.P.: Image quality assessment: from error visibility to structural similarity. IEEE Trans. Image Process. **13**, 600–612 (2004)

25. Wang, Z., Chang, S., Yang, Y., Liu, D., Huang, T.: Studying very low resolution recognition using deep networks. In: 2016 IEEE Conference on Computer Vision and Pattern Recognition (CVPR). IEEE (2016)

A Fast Blind Spatially-Varying Motion Deblurring Algorithm with Camera Poses Estimation

Yuquan Xu[1(✉)], Seiichi Mita[1], and Silong Peng[2]

[1] Research Center for Smart Vehicles, Toyota Technological Institute,
Nagoya, Japan
`yuquan.xu86@toyota-ti.ac.jp`
[2] Institute of Automation, Chinese Academy of Sciences,
Beijing, China

Abstract. Most existing non-uniform deblurring algorithms model the blurry image as a weighted summation of several sharp images which are warped by one latent image with different homographies. These algorithms usually suffer from high computational cost due to the huge number of homographies to be considered. In order to solve this problem, we introduce a novel single image deblurring algorithm to remove the spatially-varying blur. Since the real motion blur kernel is very sparse, in this paper we first estimate a feasible active set of homographies which may hold large weights in the blur kernel and then compute the corresponding weights on these homographies to reconstruct the blur kernel. Since the size of the active set is quite small, the deblurring algorithm will become much faster. Experiment results show that the proposed algorithm can effectively and efficiently remove the non-uniform blur caused by camera shake.

1 Introduction

When taking photographs, the image blur arisen from camera shake is one of the prime undesirable phenomenon to degrade our photo especially in the dim environment. The aim of blind image deblurring is to tackle this problem by recovering the clear latent image from only a blurry, compromised observation. To model the blurry images, a simple and common approach is to model the blurred image as the convolution of a sharp image with a uniform blur kernel or point spread function (PSF) which describes the related motion trajectory between the camera and the object. In this case, the image deblurring issue is a blind deconvolution process which is a well-known ill-posed inverse problem. For the past few years, this uniform image deblurring problem has been studied extensively and gained dramatically improvement [1–7]. However, in the space-invariant blur all the pixels in the blurry image are assumed to share a same PSF and this assumption is sufficient only if the blur is mainly caused by the in-plane translation of the camera. This assumption does not hold in general

© Springer International Publishing AG 2017
S.-H. Lai et al. (Eds.): ACCV 2016, Part III, LNCS 10113, pp. 157–172, 2017.
DOI: 10.1007/978-3-319-54187-7_11

[8] since the real camera shake not only translates the sensor but also tilts and rotates it and results in a space-variant blur.

For the spatially-varying blur, the PSF at each pixel location is different and estimating these diverse kernels is an over intricate mission. Hence, Tai et al. [9] proposed a projective motion path blur (PMPB) model to depict the spatially variant blur caused by the camera shake. In this model, the non-uniform blurred image is formulated as a weighted summation of a series of clear images which are "seen" by the camera during the exposure time. These clear images can be regarded as the transformed version with different homographies of one latent image. By this way, the PSF estimation problem changes into computing the weights of different camera poses or homographies rather than directly estimating enormous blur kernels. However even in this model the dense sampling of the high dimension variable space make the computational cost still too much heavy. The full camera motion space is six-dimensional, i.e. three for rotation and three for translation. Considering only ten values are sampled in each dimension, there are one million unknown weights need to be determined. Therefore to decrease the complexity of this problem, most existing algorithms approximate the six dimensional real camera trajectory into three dimensions with two main approaches, three rotations [8] or in-plane translation and rotation around the visual axis [10,11]. However, there are still thousands of weights need to be computed and the resulting algorithms are time-consuming.

The main cause of the heavy computational cost comes from the tremendous quantity of camera poses (homographies) needed to be considered, and during the optimization many intermediate warped images should be computed. The most direct idea to resolve this problem is to reduce the number of homographies, fortunately, by reviewing the real motion kernel, the camera motion trajectory is always very sparse, so the number of active camera poses which hold large weights is much smaller than the number of whole motion space which means that the majority of weights are zero or close to zero. Hence, to accelerate the non-uniform deblurring, we introduce a novel method that estimates a small set of active camera poses during the optimization instead of estimating thousands of weights on the dense sampling variable space. Hu et al. [12] introduced a fast algorithm which also utilized the small camera pose subspace to leverage the computational issues of non-uniform deblurring. However the camera pose subspace used in their algorithm highly depended on an initialization step and they obtained the new poses only by sampling based on the previous set using a Gaussian distribution. Thus, if the initialization poses did not contain the real poses, their algorithm can seldom find the correct active camera poses. Cho et al. [13] presented a registration based non-uniform deblurring algorithm which estimated a set of homographies by registering the latent image with the residual image. However in their method, they estimated the homographies one by one without considering the relation of these homographies and their method need multiple input images.

In this paper, we propose a fast single image deblurring algorithm to remove the non-uniform blur. In our method, we directly estimate the active subspace with a multi-scale framework incorporated with the modified L_0 regularized method [7] by alternating direction method (ADM). We update the set of homographies together in each iteration and then compute the weights on these poses. Besides in our algorithm, the camera motion are not parameterized by six-dimensional motion parameters but a full eight-dimensional homography matrix by which our method do not need the information of the camera intrinsic parameters, such as the focal length of the camera which is needed in [8,12], and the three motion dimensions assumption [8,10,11]. Extensive experiments compared with state-of-the-art methods demonstrate the effectiveness and efficiency of the proposed method.

2   Related Work

Image deblurring is one of the basic problems in low level vision and has gained major advancements in blind image deblurring problem, where both the latent image and PSF are unknown. For the uniform blur, since the naive maximum a posterior (MAP) algorithm will fail on natural images and leads to the trivial no-blur solution, there are three major classes of successful blind deblurring algorithms used in the literature, (1) Variational Bayesian (VB) framework or called "MAP_k" algorithm, (2) MAP with special regularization terms and we called it "MAP_1" algorithm, (3) MAP with an ad-hoc prediction step based on shock filter and we called it "MAP_2" algorithm. For the VB framework, Fergus et al. [1] proposed a VB approach with natural image prior and was the first successful algorithm to estimate the complex motion kernel. Levin et al. [6] analyzed why using VB instead of simple MAP framework in the image deblurring issue and introduced a marginal likelihood optimization "MAP_k" algorithm to estimate the PSF. Babacan et al. [14] presented a general method for blind image deconvolution using Bayesian inference with general sparse image priors. For "MAP_1" methods, Krishnan et al. [5] introduced an "L_1/L_2" normalized regularization term while Xu et al. [7] used a unnatural sparse "L_0" term to estimate the blur kernel. For "MAP_2" algorithms, Cho and Lee [3] introduced the shock filter into the naive MAP framework to force the sparse gradient in latent image during the optimization and raised a fast uniform deblurring framework. Xu and Jia [4] proposed an "r-map" to filter out the texture part of the image and selected the useful image edges after the shock filter. There are also some other image deblurring techniques which need additional information except the single input blurry observation, like additional equipment [15], multiple images [16,17] and alpha matting [18].

Recently many researchers have make great efforts to deal with the spatially variant blur caused by the camera shake. In [19,20], they segmented the input degraded image into multiple overlapped or non-overlapped parts, each of which was assumed to be uniform blur and deblurred individually. Tai et al. [9] introduced the PMPB model and most existing non-uniform deblurring algorithms are based on this model with some modification. The majority of techniques adopted in the non-uniform deblurring problem are just like the uniform deblurring issue.

Several algorithms utilized additional information to help the spatially variant motion deblurring, like additional equipment [21], multiple images [8,13] and alpha matting [22]. It is worth noting that Whyte et al. [8] introduced a three rotations model and extended three uniform deblurring algorithms, "MAP_k" method [1] and "MAP_2" method [3] and noisy/blurry pair method [16], to the non-uniform case. Hirsch et al. [11] proposed a "MAP_2" method with an effective forward blur based on the efficient filter flow (EFF) framework in which the image is partitioned into regions and then estimate a uniform blur kernel for each region with a weighting function for smoothing the boundaries between two adjacent regions. Zhang and Wipf [23] presented a "MAP_k" method with a spatially-adaptive sparse prior. In [7], Xu et al. presented a "MAP_1" method to remove the spatially-varying blur.

3   Problem Formulation

Follow the PMPB model, the blurry image is formulated as a weighted summation of a series clear images.

$$B = \sum_{i}^{N} \omega_i I_i \tag{1}$$

where B is the blurry observation, N is the discrete sampling rate of the shutter period as well as the total number of the intermediate images, I_i is the ith clear image and ω_i is the weight of I_i which represents the ratio of time on that camera pose, each intermediate image I_i is a transformed vision of the latent image I with homography $W(\mathbf{x}; \mathbf{p}_i)$

$$I_i(\mathbf{x}) = I(W(\mathbf{x}; \mathbf{p}_i)) = I\left(\frac{(p_{i1} + 1)x + p_{i2}y + p_{i3}}{p_{i7}x + p_{i8}y + 1}, \frac{p_{i4}x + (p_{i5} + 1)y + p_{i6}}{p_{i7}x + p_{i8}y + 1}\right). \tag{2}$$

where $\mathbf{p} = (p_1, p_2, ..., p_8)^T$ is the eight parameters in homography, $\mathbf{x} = (x, y)^T$ denotes the coordinate of the image, respectively. So plus the inevitable noise ε, the blurry image is finally modeled as:

$$B(\mathbf{x}) = \sum_{i}^{N} \omega_i I(W(\mathbf{x}; \mathbf{p}_i)) + \varepsilon \tag{3}$$

In most non-uniform motion deblurring algorithms, N is very huge because of the dense sampling on the high-dimensional parameter space, which make the non-uniform motion deblurring algorithm time-consuming. However, the real camera motion trajectory is very sparse, so most of ω_i is zero or very close to zero and only a few weights occupy the majority of the energy in the blur kernel. As a result, if we can estimate the active set of homographies which

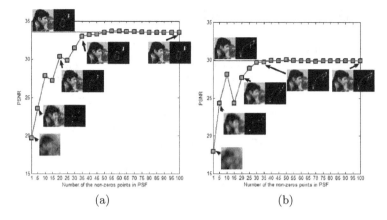

(a)   (b)

Fig. 1. Sampled kernel result. We use 2 PSFs to blur the image "Cameraman" and then use different sampled kernels to deblur the images. Blue lines in the chart show the PSNRs of deblurred results with the groundtruth kernels, the close ups are the deblurred results and the groundtruth kernels. X axis 1 means the sampled PSFs just have one non-zero pixel which means the deblurred results are exactly the blurred inputs and the close-ups in this point show the blurred images. The other close-ups show the sampled kernels with different number of non-zero points and the corresponding deblurred results. (Color figure online)

take up larger weights, during the optimization we do not need to compute the enormous number of intermediate warped images, which will lead to the fast non-uniform image deblurring algorithm perform much faster.

Actually we find that in the motion blur case 40–50 intermediate images are enough for a moderate 27×27 blur kernel, we show this example in Fig. 1. We use 2 PSFs to blur the "Cameraman" image, which are real camera shake blur kernels from the dataset of Levin et al. [6] with size 21×21 and 27×27. We then sample these kernels to different number of non-zero pixels and use these sampled kernels to restore the latent image with the deconvolution algorithm proposed in [24] with the same parameter setting. In the figure, we show the Peak Signal to Noise Ratio (PSNR) value of the deblurred results and in some data points we show the closeups of cropped deblurred images and the corresponding sampled kernels. This figure shows that the PSNR of the deblurred result generally increases when more non-zeros pixels in PSF are used for both four cases. However in the motion blur case, when the number reaches 40, the PSNR of the sampled kernel is closed to the real kernel. The homography number is related to the kernel size, if the blur is serious, a larger number should be used whereas with a slight blur we can employ a small number to accelerate the performance of the algorithm. Typically, in our experiment we set $N = 40$ for all the examples.

4   Blind Deblurring Algorithm

In (3), there are three variables to be determined for our algorithm, which are the active homography parameters $\mathbf{p}_i$, the corresponding weights ω_i and the latent image I. The $\mathbf{p}_i$ and ω_i in our model can be seen as the PSF of the blur image, and following previous deblurring work, we use a two step method to remove the non-uniform motion blur: (1) kernel estimation and (2) sharp image restoration. Firstly we use the derivative domain of images which will obtain better performance on the PSF estimation [6] and propose a "MAP_1" algorithm with L_0 regularized term to estimate the $\mathbf{p}_i$ and ω_i. After that we utilized a restoration step to recover the latent image with the PSF estimation in step 1. Figure 2 shows the flowchart of the proposed algorithm.

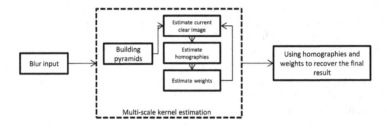

Fig. 2. Flowchart of the proposed method.

4.1   PSF Estimation

In this subsection, we describe our PSF estimation method, which estimates the PSF from a given blurred image. Our PSF is parameterized by the homography parameters $\mathbf{p}_i$ and the corresponding weights ω_i. With the L_0 regularization, we derive following cost function

$$\min_{\tilde{I},\mathbf{P},\Omega} \sum_{\mathbf{x}} ||\tilde{B}(\mathbf{x}) - \sum_{i}^{N} \omega_i \tilde{I}(W(\mathbf{x};\mathbf{p}_i))||^2 + \lambda_I ||\tilde{I}||_0 + \lambda_p ||\mathbf{P}||^2 + \lambda_\omega ||\Omega||^2 \quad (4)$$

where $\tilde{I}$ and $\tilde{B}$ denote the gradients of the latent image I and blur image B, $||\cdot||_0$ represents the L_0 norm of the image gradients and $\mathbf{P}$ and Ω denote the set of homography parameters and weights, i.e. $\mathbf{P} = (\mathbf{p}_1^T, \mathbf{p}_2^T, ..., \mathbf{p}_N^T)^T$, $\Omega = (\omega_1, \omega_2, ..., \omega_N)^T$, respectively. λ_I, λ_p and λ_ω are regularization parameters of the latent image $\tilde{I}$, parameter $\mathbf{P}$ and weight Ω, respectively. We propose the alternately iterative method to solve this optimization problem and derive our method in a multi-scale setting to avoid poor local minima and increase the stability of the algorithm. Specifically we alternately estimate the $\mathbf{P}$, Ω and $\tilde{I}$

from coarse to fine levels, where the estimation at a coarse level are upsampled and used as initial values in the next finer level as an incremental multi-resolution technique.

Update $\tilde{I}$. To obtain the $\tilde{I}$, from (4) we have

$$
\mathbf{l} = \arg\min_{\mathbf{l}} ||\mathbf{b} - \sum_i^N \omega_i \mathbf{H}_i \mathbf{l}||^2 + \lambda_I ||\mathbf{l}||_0 \tag{5}
$$

where $\mathbf{b}$ and $\mathbf{l}$ denote the vector representations of the $\tilde{B}$ and $\tilde{I}$, and $\mathbf{H}_i$ is the transformation matrix that produces a projective transform of the image $\mathbf{l}$ with homography parameter $\mathbf{p}_i$, respectively. To solve (5), we first introduce a intermediate variables $\mathbf{w}$ and convert (5) to the following equivalent problem:

$$
\begin{aligned}
&\underset{\mathbf{l}, \mathbf{w}}{\text{minimize}} \quad ||\mathbf{b} - \sum_i^N \omega_i \mathbf{H}_i \mathbf{l}||^2 + \lambda_I ||\mathbf{w}||_0 \\
&\text{subject to } \mathbf{w} = \mathbf{l}.
\end{aligned} \tag{6}
$$

Then, we have the following Lagrangian function:

$$
L(\mathbf{l}, \mathbf{w}, \mathbf{u}) = ||\mathbf{b} - \sum_i^N \omega_i \mathbf{H}_i \mathbf{l}||^2 + \lambda_I ||\mathbf{w}||_0 + \mathbf{u}^T(\mathbf{l} - \mathbf{w}) + \frac{\mu}{2}||\mathbf{l} - \mathbf{w}||^2 \tag{7}
$$

where $\mathbf{u}$ is the Lagrange multiplier and μ is the penalty parameter. The ADM algorithm is applied to the Lagrangian function in (7) by alternatively solving following subproblems,

$$
\mathbf{l}^{n+1} = \arg\min_{\mathbf{l}} ||\mathbf{b} - \sum_i^N \omega_i \mathbf{H}_i \mathbf{l}||^2 + \frac{\mu}{2}||\mathbf{l} - \mathbf{w}^n + \frac{1}{\mu}\mathbf{u}^n||^2 \tag{8}
$$

$$
\mathbf{w}^{n+1} = \arg\min_{\mathbf{w}} \lambda_I ||\mathbf{w}||_0 + \frac{\mu}{2}||\mathbf{l}^n - \mathbf{w} + \frac{1}{\mu}\mathbf{u}^n||^2 \tag{9}
$$

$$
\mathbf{u}^{n+1} = \mathbf{u}^n + \mu(\mathbf{l}^n - \mathbf{w}^n) \tag{10}
$$

It is easy to see that (8) is a quadratic function which can be effectively solved by conjugate gradient (CG) method. To solve (9), we can obtain its explicit solution as,

$$
\mathbf{w}^{n+1} = \begin{cases} \mathbf{l}^n + \frac{1}{\mu}\mathbf{u}^n, & (\mathbf{l}^n + \frac{1}{\mu}\mathbf{u}^n)^2 \geq \frac{2\lambda_I}{\mu} \\ 0, & \text{otherwise} \end{cases} \tag{11}
$$

Hence, the ADM algorithm for solving (7) is outlined in Algorithm 1.

Algorithm 1. ADM for solving (5)

1: **Input: b, P, Ω**
2: **Initialization:** $l = b$, $w = u = 0$, $\mu = 0.01$, $n = 0$
3: **repeat**
4:      Compute l^n by solving (8)
5:      Compute w^n by (11)
6:      Compute u^n by (10)
7:      Update $\mu = 3\mu$, $n = n + 1$
8: **until** $n = 15$
9: **return** Latent image l

Update P. With the gradient image $\tilde{I}$, we compute the homography parameters **P** by solving

$$\min_{\mathbf{P}} \sum_{\mathbf{x}} ||\tilde{B}(\mathbf{x}) - \sum_{i}^{N} \omega_i \tilde{I}(W(\mathbf{x}; \mathbf{p}_i))||^2 + \lambda_p ||\mathbf{P}||^2 \tag{12}$$

To optimize the expression in (12), we assume a current estimate of $\mathbf{P} = (\mathbf{p}_1^T, \mathbf{p}_2^T, ..., \mathbf{p}_N^T)^T$ is known and then iteratively solves for increments to the parameters $\Delta\mathbf{P} = (\Delta\mathbf{p}_1^T, \Delta\mathbf{p}_2^T, ..., \Delta\mathbf{p}_N^T)^T$. Hence, we approximately minimizes:

$$\min_{\Delta\mathbf{P}} \sum_{\mathbf{x}} ||\tilde{B}(\mathbf{x}) - \sum_{i}^{N} \omega_i \tilde{I}(W(W(\mathbf{x}; \Delta\mathbf{p}_i); \mathbf{p}_i))||^2 + \lambda_p ||\Delta\mathbf{P}||^2 \tag{13}$$

To solve (13), we apply the first Taylor expansion

$$\min_{\Delta\mathbf{P}} \sum_{\mathbf{x}} ||\tilde{B}(\mathbf{x}) - \sum_{i}^{N} \omega_i (\tilde{I}_i + \nabla\tilde{I}_i \frac{\partial W}{\partial \mathbf{p}_i} \Delta\mathbf{p}_i)||^2 + \lambda_p ||\Delta\mathbf{P}||^2 \tag{14}$$

where $\tilde{I}_i = \tilde{I}(W(\mathbf{x}; \mathbf{p}_i))$ is the warped image, $\nabla\tilde{I}_i$ is the gradient of $\tilde{I}_i$ and $\partial W / \partial \mathbf{p}$ is the Jacobian of the warp function (2),

$$\frac{\partial W}{\partial \mathbf{p}} = \begin{bmatrix} \frac{\partial W_x}{\partial p_1} & \frac{\partial W_x}{\partial p_2} & \cdots & \frac{\partial W_x}{\partial p_8} \\ \frac{\partial W_y}{\partial p_1} & \frac{\partial W_y}{\partial p_2} & \cdots & \frac{\partial W_y}{\partial p_8} \end{bmatrix} = \frac{1}{D} \begin{bmatrix} x & y & 1 & 0 & 0 & 0 & -x'x & -y'x \\ 0 & 0 & 0 & x & y & 1 & -y'x & -y'y \end{bmatrix} \tag{15}$$

where

$$D = p_7 x + p_8 y + 1;$$

$$x' = \frac{(p_1 + 1)x + p_2 y + p_3}{D}$$

$$y' = \frac{p_4 x + (p_5 + 1)y + p_6}{D}$$

Since $\partial W / \partial \mathbf{p}_i$ is evaluated at $(\mathbf{x}; \mathbf{0})$, all the $\partial W / \partial \mathbf{p}_i$ with different i share the same value with $D = 1$, $x' = x$ and $y' = y$, then

$$\frac{\partial W}{\partial \mathbf{p}_i} = \begin{bmatrix} x\ y\ 1\ 0\ 0\ 0 -x^2 -xy \\ 0\ 0\ 0\ x\ y\ 1 -xy -y^2 \end{bmatrix} \tag{16}$$

Let $e = \tilde{B} - \sum_{i=1}^{N} \omega_i \tilde{I}_i$ is reconstruction residual image by the previous estimation, $\mathbf{J} = (\omega_1 \nabla \tilde{I}_1 \frac{\partial W}{\partial \mathbf{p}}, \omega_2 \nabla \tilde{I}_2 \frac{\partial W}{\partial \mathbf{p}}, ..., \omega_N \nabla \tilde{I}_N \frac{\partial W}{\partial \mathbf{p}})$, (14) is simplified to

$$\min_{\Delta \mathbf{P}} \sum_{\mathbf{x}} ||e - \mathbf{J} \Delta \mathbf{P}||^2 + \lambda_p ||\Delta \mathbf{P}||^2 \tag{17}$$

Equation (18) can be directly solved by

$$\Delta \mathbf{P} = \left(\sum_{\mathbf{x}} \mathbf{J}^T \mathbf{J} + \lambda_p \mathbf{E} \right)^{-1} \sum_{\mathbf{x}} \mathbf{J}^T e \tag{18}$$

where $\mathbf{E}$ is the identity matrix. After $\Delta \mathbf{P}$ is computed, we update each $\mathbf{p}_i$ with

$$\begin{bmatrix} p_{i1}\ p_{i2}\ p_{i3} \\ p_{i4}\ p_{i5}\ p_{i6} \\ p_{i7}\ p_{i8}\ 1 \end{bmatrix} \propto \begin{bmatrix} p_{i1}\ p_{i2}\ p_{i3} \\ p_{i4}\ p_{i5}\ p_{i6} \\ p_{i7}\ p_{i8}\ 1 \end{bmatrix} \cdot \begin{bmatrix} \Delta p_{i1}\ \Delta p_{i2}\ \Delta p_{i3} \\ \Delta p_{i4}\ \Delta p_{i5}\ \Delta p_{i6} \\ \Delta p_{i7}\ \Delta p_{i8}\ 1 \end{bmatrix} \tag{19}$$

In our experiment, we iteratively applying (18) and (19) for 20 iterations to obtain the final $\mathbf{P}$ as shown in Algorithm 2.

Algorithm 2. Update the active camera poses

1: **Input:** $\tilde{I}$, Ω, initial poses $\mathbf{P}$
2: **Pre-compute:** Jacobian $\frac{\partial W}{\partial \mathbf{p}}$
3: **repeat**
4:     Warp $\tilde{I}$ to a set of warped images $\tilde{I}_i$ with $\mathbf{P}$
5:     Compute $e = \tilde{B} - \sum_{i=1}^{N} \omega_i \tilde{I}_i$ and $\mathbf{J} = (\omega_1 \nabla \tilde{I}_1 \frac{\partial W}{\partial \mathbf{p}}, \omega_2 \nabla \tilde{I}_2 \frac{\partial W}{\partial \mathbf{p}}, ..., \omega_N \nabla \tilde{I}_N \frac{\partial W}{\partial \mathbf{p}})$
6:     Compute $\Delta \mathbf{P}$ by (18)
7:     Update $\mathbf{P}$ by (19)
8: **until** Meets maximum iterations
9: **return** The new poses $\mathbf{P}$

Update Ω. When $\mathbf{P}$ and $\tilde{I}$ is estimated, from (4) we can reformulate it to:

$$\Omega = \arg\min_{\Omega} ||\mathbf{b} - \mathbf{S}\Omega||^2 + \lambda_\omega ||\Omega||^2 \tag{20}$$

where $\mathbf{S} = [\mathbf{s}_1, \mathbf{s}_2, ..., \mathbf{s}_N]$, $\mathbf{s}_i$ is the $M \times 1$ vector representations of $\tilde{I}_i$, M is the total number of pixels in the image, respectively. So $\mathbf{S}$ is a $M \times N$ matrix. Since the motion blur is very sparse, the intermediated image number N is small, e.g. 40 in our experiment, we can directly construct the matrix $\mathbf{S}$, then together with some common constraint on the PSF, we can easily solve the constraint optimization problem to obtain $\mathbf{\Omega}$:

$$\arg\min_{\mathbf{\Omega}} ||\mathbf{b} - \mathbf{S}\mathbf{\Omega}||^2 + \lambda_\omega ||\mathbf{\Omega}||^2$$
$$\text{subject to } \sum_{i=1} \omega_i = 1, \omega_i \geq 0 \tag{21}$$

By alternatively updating $\mathbf{P}$, $\mathbf{\Omega}$ and $\mathbf{l}$, the final PSF estimation algorithm is summarized in Algorithm 3. The iteration number for each scale is set to be 10.

Algorithm 3. PSF estimation

1: **Input**: The blurred image B
2: Construct pyramids for the input image $B_1, ..., B_n$, n is number of pyramid levels.
3: Set all the weights $\omega = 1/N$ and initial $\mathbf{p}_i$ to the identity warp.
4: $l \leftarrow 1$
5: **repeat**
6:     **repeat**
7:         Estimate $\tilde{I}$ by Algorithm 1
8:         Update the $\mathbf{P}$ by Algorithm 2
9:         Update the $\mathbf{\Omega}$ by solving (21)
10:    **until** Meets maximum iterations
11:    Upsample the image $I_{l+1} \leftarrow I_l \uparrow$, flows $\mathbf{P}_{l+1} \leftarrow \mathbf{P}_l \uparrow$ to the next level
12: **until** $l = n$,
13: **return** Homography parameters $\mathbf{P}$ and weights $\mathbf{\Omega}$

4.2   Final Image Restoration

After having estimated and fixed the homography parameters $\mathbf{P}$ and weights $\mathbf{\Omega}$, we restore the latent image by using the similar method as updating $\mathbf{l}$. We replace the L_0 norm in (5) by the natural image prior as in [24,25], i.e. we minimize

$$I = \arg\min_I ||B - \sum_i^N \omega_i \mathbf{H}_i I||^2 + \gamma ||\nabla I||_{0.8} \tag{22}$$

We also use ADM method to solve (22) by alternatingly minimizing following subproblems

$$I^{n+1} = \arg\min_I ||B - \sum_i^N \omega_i \mathbf{H}_i I||^2 + \frac{\mu}{2} ||\nabla I - \mathbf{w}^n + \frac{1}{\mu} \mathbf{u}^n||^2 \tag{23}$$

$$\mathbf{w}^{n+1} = \arg\min_{\mathbf{w}} \gamma ||\mathbf{w}||_{0.8} + \frac{\mu}{2} ||\nabla I^n - \mathbf{w} + \frac{1}{\mu} \mathbf{u}^n||^2 \tag{24}$$

$$\mathbf{u}^{n+1} = \mathbf{u}^n + \mu(\nabla I^n - \mathbf{w}^n) \tag{25}$$

Equation (23) also can be minimized by CG method and for the point-wise updating function (24), we can efficiently solve it by a lookup table algorithms as in [25].

4.3 Parameter Setting

To determine the iteration number in Algorithms 1, 2 and 3, we plot in Fig. 3 the energy value at different iteration numbers of these three steps at the finest scale on a synthetic example, the "Cameraman" image is blurred by a rotation kernel. We can see that the convergence of Algorithms 1 and 3 is very fast, so we set the iteration number of these two steps to be 10, while the convergence of Algorithm 2 is a little bit slower so the iteration number is set to be 20. Besides in Algorithm 1, we set $\mu = 0.01$ $\lambda_I = 0.01$, $\lambda_p = \lambda_\omega = 5$ in (12) and (20), and we set $\gamma = 0.008$ in the final image restoration step. For multi-scale settings, The upsampling factor in our algorithm from one level to the next is 2, and after the algorithm finishes ar one scale, we remove the smallest homographies with their weights smaller than the 5% of the maximum weight and we add new homographies by sampling based on the left estimation homographies using a Gaussian distribution as in [12].

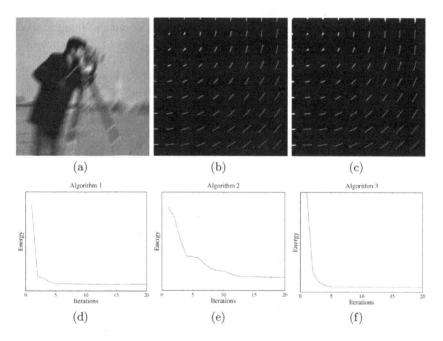

Fig. 3. Convergence analysis. (a) is the synthetic blurry image, (b) is the groundtruth rotation kernel generated with 80 homographies, (c) is the estimated kernel with 40 homographies, (d)–(f) are the Energy value of Algorithms 1, 2 and 3 with different iterations, respectively

5   Experimental Results

In this section, we show the experiment on various blurry examples including both the quality and quantity results. We implemented our method in Matlab 2011b on Window 7 system with a single core of Intel Core i3-3220 processor and 8 GB RAM. The pixel values of all the images are scaled to $[0, 1]$. When estimating the blur kernel, the color image is firstly changed into gray image by the command "rgb2gray" in Matlab.

For the quantity results, we test our algorithm on the dataset provided in [26], which contains 48 images of size 800×800. In this dataset, there are 4 groundtruth sharp images each of which is blurred by 12 non-uniform blur kernels. We compare the performance of the proposed algorithm with two non-uniform methods [7,8,11]. Since in [13], they estimate the blur kernel by the estimating the homographies one by one using a image pair, here we also change our homographies updating method "Algorithm 2" to the similar routine and compare the results, we call this algorithm "Cho Single". The average PSNR value of each image by three algorithms are illustrated in Fig. 4, from which we can see that the proposed algorithm perform better than [8,11] and one by one updating method as [13] and have similar performance with [7]. An example of these results are shown in Fig. 5.

In Table 1, we present the running time of different algorithms on several deblurring examples. The running time of [7,12] is carried out by their open code and the running time of [11] is from their published paper with a single core of an Intel Core i5. It is a remarkable fact that the running time of [12] is much bigger than the time mentioned in their paper which is 600 s on 441×611 image. The reason may be that the algorithm [12] need a very large RAM over 8 GB RAM in our computer, therefore their code may perform much slower in this case. From the Table 1, we can see that the proposed method is faster than other non-uniform

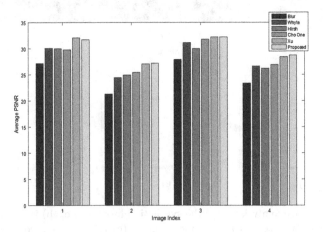

Fig. 4. Quantitative comparison on the dataset [26].

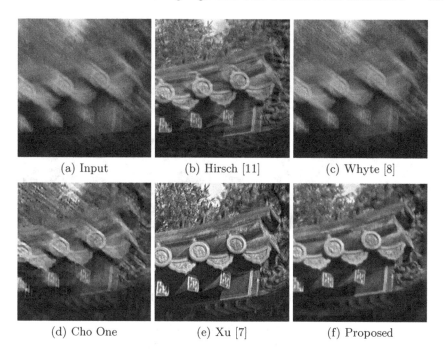

(a) Input   (b) Hirsch [11]   (c) Whyte [8]

(d) Cho One   (e) Xu [7]   (f) Proposed

Fig. 5. Visual quality comparison on the dataset [26]

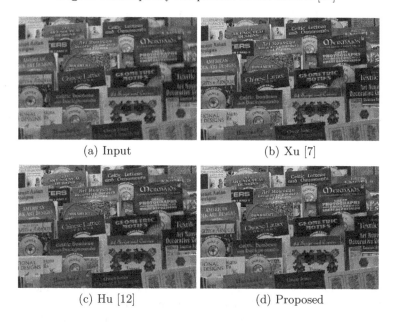

(a) Input   (b) Xu [7]

(c) Hu [12]   (d) Proposed

Fig. 6. One deblurring result of Table 1

(a) Input   (b) Cho [13]   (c) Proposed

(d) Input   (e) Joshi [21]   (f) Proposed

Fig. 7. Non-uniform deblurring results.

Table 1. Running time of different algorithms

Image size	400×400	768×512	441×611
Proposed	**531 s**	**1254 s**	**758 s**
Xu [7]	763 s	1731 s	1164 s
Hu [12]	1230 s	3497 s	1957 s
Hirsch [11]	N/A	N/A	1567s

algorithms. The results on 768×512 image are shown in Fig. 6, in which our result is comparable with other two algorithms. In Fig. 7, we compare with other spatially variant deblurring algorithms [13, 21]. Comparing with other algorithms, our method can produce the comparable or better results.

6 Conclusion and Discussion

In this paper, we have presented a fast algorithm to deal with the spatially-varying blur caused by camera shake. Differing from previous efforts on both uniform and non-uniform PSF estimation, in which they first constraint the range of camera motion to a parameter space by users, e.g. demanding the user to input a kernel size of the blurred image, or have a default range, then the weights on these parameters are estimated. In our algorithm, since the real motion blur is very sparse we first directly estimate the probable motion poses during the exposure time and then estimate the correspond weights on these poses. A L_0 sparse image prior term is utilized and effectively solved by ADM method. Then we show that our algorithm can effectively deal with the non-uniform blurry images. However, the execution time of the proposed algorithm highly depends on the homography number N. As a result, when the blur size is large, more homographies are needed to represent the blur and our algorithm will become slower. In this case, the EFF forward framework can be used and we do not need to pre-compute thousands of point spread function basis [11] but only a small set at estimated homographies.

References

1. Fergus, R., Singh, B., Hertzmann, A., Roweis, S.T., Freeman, W.T.: Removing camera shake from a single photograph. ACM Trans. Graph. (TOG) **25**, 787–794 (2006). ACM
2. Shan, Q., Jia, J., Agarwala, A.: High-quality motion deblurring from a single image. ACM Trans. Graph. (TOG) **27** (2008). Article no. 73. ACM
3. Cho, S., Lee, S.: Fast motion deblurring. ACM Trans. Graph. (TOG) **28** (2009). Article no. 145. ACM
4. Xu, L., Jia, J.: Two-phase kernel estimation for robust motion deblurring. In: Daniilidis, K., Maragos, P., Paragios, N. (eds.) ECCV 2010. LNCS, vol. 6311, pp. 157–170. Springer, Heidelberg (2010). doi:10.1007/978-3-642-15549-9_12
5. Krishnan, D., Tay, T., Fergus, R.: Blind deconvolution using a normalized sparsity measure. In: 2011 IEEE Conference on Computer Vision and Pattern Recognition (CVPR), pp. 233–240. IEEE (2011)
6. Levin, A., Weiss, Y., Durand, F., Freeman, W.T.: Efficient marginal likelihood optimization in blind deconvolution. In: 2011 IEEE Conference on Computer Vision and Pattern Recognition (CVPR), pp. 2657–2664. IEEE (2011)
7. Xu, L., Zheng, S., Jia, J.: Unnatural LO sparse representation for natural image deblurring. In: 2013 IEEE Conference on Computer Vision and Pattern Recognition (CVPR). IEEE (2013)
8. Whyte, O., Sivic, J., Zisserman, A., Ponce, J.: Non-uniform deblurring for shaken images. Int. J. Comput. Vis. **98**, 168–186 (2012)
9. Tai, Y.W., Tan, P., Brown, M.S.: Richardson-Lucy deblurring for scenes under a projective motion path. IEEE Trans. Pattern Anal. Mach. Intell. **33**, 1603–1618 (2011)
10. Gupta, A., Joshi, N., Lawrence Zitnick, C., Cohen, M., Curless, B.: Single image deblurring using motion density functions. In: Daniilidis, K., Maragos, P., Paragios, N. (eds.) ECCV 2010. LNCS, vol. 6311, pp. 171–184. Springer, Heidelberg (2010). doi:10.1007/978-3-642-15549-9_13
11. Hirsch, M., Schuler, C.J., Harmeling, S., Scholkopf, B.: Fast removal of non-uniform camera shake. In: 2011 IEEE International Conference on Computer Vision (ICCV), pp. 463–470. IEEE (2011)
12. Hu, Z., Yang, M.H.: Fast non-uniform deblurring using constrained camera pose subspace. In: BMVC, pp. 1–11 (2012)
13. Cho, S., Cho, H., Tai, Y.W., Lee, S.: Registration based non-uniform motion deblurring. Comput. Graph. Forum **31**, 2183–2192 (2012). Wiley Online Library
14. Babacan, S.D., Molina, R., Do, M.N., Katsaggelos, A.K.: Bayesian blind deconvolution with general sparse image priors. In: Fitzgibbon, A., Lazebnik, S., Perona, P., Sato, Y., Schmid, C. (eds.) ECCV 2012. LNCS, vol. 7577, pp. 341–355. Springer, Heidelberg (2012). doi:10.1007/978-3-642-33783-3_25
15. Ben-Ezra, M., Nayar, S.K.: Motion deblurring using hybrid imaging. In: Proceedings of 2003 IEEE Computer Society Conference on Computer Vision and Pattern Recognition, vol. 1, p. I-657. IEEE (2003)

16. Yuan, L., Sun, J., Quan, L., Shum, H.Y.: Image deblurring with blurred/noisy image pairs. ACM Trans. Graph. (TOG) **26**(3) (2007). Article no. 1. ACM

17. Lasang, P., Ong, C.P., Shen, S.M.: CFA-based motion blur removal using long/short exposure pairs. IEEE Trans. Consum. Electron. **56**, 332–338 (2010)

18. Jia, J.: Single image motion deblurring using transparency. In: IEEE Conference on Computer Vision and Pattern Recognition, CVPR 2007, pp. 1–8. IEEE (2007)

19. Šorel, M., Šroubek, F., Flusser, J.: Recent advances in space-variant deblurring and image stabilization. In: Byrnes J. (ed.) Unexploded Ordnance Detection and Mitigation. NATO Science for Peace and Security Series B: Physics and Biophysics, pp. 259–272. Springer, Dordrecht (2009)

20. Harmeling, S., Hirsch, M., Schölkopf, B.: Space-variant single-image blind deconvolution for removing camera shake. In: NIPS, pp. 829–837 (2010)

21. Joshi, N., Kang, S.B., Zitnick, C.L., Szeliski, R.: Image deblurring using inertial measurement sensors. ACM Trans. Graph. (TOG) **29** (2010). Article no. 30

22. Dai, S., Wu, Y.: Motion from blur. In: IEEE Conference on Computer Vision and Pattern Recognition, CVPR 2008, pp. 1–8. IEEE (2008)

23. Zhang, H., Wipf, D.: Non-uniform camera shake removal using a spatially-adaptive sparse penalty. In: Advances in Neural Information Processing Systems, pp. 1556–1564 (2013)

24. Levin, A., Fergus, R., Durand, F., Freeman, W.T.: Image and depth from a conventional camera with a coded aperture. ACM Trans. Graph. (TOG) **26** (2007). Article no. 70

25. Krishnan, D., Fergus, R.: Fast image deconvolution using hyper-laplacian priors. In: Advances in Neural Information Processing Systems, pp. 1033–1041 (2009)

26. Köhler, R., Hirsch, M., Mohler, B., Schölkopf, B., Harmeling, S.: Recording and playback of camera shake: benchmarking blind deconvolution with a real-world database. In: Fitzgibbon, A., Lazebnik, S., Perona, P., Sato, Y., Schmid, C. (eds.) ECCV 2012. LNCS, vol. 7578, pp. 27–40. Springer, Heidelberg (2012). doi:10.1007/978-3-642-33786-4_3

Removing Shadows from Images of Documents

Steve Bako[1]([✉]), Soheil Darabi[2], Eli Shechtman[2], Jue Wang[2],
Kalyan Sunkavalli[2], and Pradeep Sen[1]

[1] University of California, Santa Barbara, Santa Barbara, CA, USA
stevebako@umail.ucsb.edu
[2] Adobe Research, Seattle, WA, USA

Abstract. In this work, we automatically detect and remove distracting shadows from photographs of documents and other text-based items. Documents typically have a constant colored background; based on this observation, we propose a technique to estimate background and text color in local image blocks. We match these local background color estimates to a global reference to generate a *shadow map*. Correcting the image with this shadow map produces the final unshadowed output. We demonstrate that our algorithm is robust and produces high-quality results, qualitatively and quantitatively, in both controlled and real-world settings containing large regions of significant shadow.

1 Introduction

Images of documents, receipts, menus, books, newspapers, flyers, signs, and other text are frequently captured. Whether we are sending a page from a textbook highlighting important information, taking a picture of an ancient engraving while on vacation, or just saving an illustration from a loved one, such images show up in our everyday lives. However, these images are highly susceptible to shadows due to occlusions of ambient light by the photographer or other objects in the environment. These shadows cause distracting artifacts that can make an image difficult to interpret or use.

In this paper, we propose an automatic technique to detect and remove shadows from images of documents. Our main observation is that documents typically have a constant colored background; for example, the actual color of the paper typically does not change throughout a document. However, illumination effects like shadows and shading cause changes in observed images intensities. Our technique detects these changes and enforces a consistent background color to produce the unshadowed output. In particular, we estimate text and background colors in local blocks of the image and generate a *shadow map* that uses a per-pixel gain to match these local background estimates to a global reference. We evaluate the robustness of our approach, both quantitatively and qualitatively, on a variety of controlled and real-world examples.

Electronic supplementary material The online version of this chapter (doi:10. 1007/978-3-319-54187-7_12) contains supplementary material, which is available to authorized users.

© Springer International Publishing AG 2017
S.-H. Lai et al. (Eds.): ACCV 2016, Part III, LNCS 10113, pp. 173–183, 2017.
DOI: 10.1007/978-3-319-54187-7_12

2   Previous Work

There are two main categories of work in shadow removal. The first category focuses on removing shadows from general images, such as typical outdoor pictures, that have strong distracting shadows. For example, Guo et al. [1] remove shadows from natural images by finding corresponding shadow and non-shadow regions and performing a per-pixel relighting. Gong et al. [2] demonstrate even more robust results by using manually specified well-lit and shadow regions. A recent method from Gryka et al. [3] uses a learning approach with user-provided brush strokes to relight the shadow regions appropriately. Moreover, there are intrinsic imaging approaches that separate an image into its reflectance and shading components [4–8]. Lastly, there are algorithms that use shadow estimation for a specific application such as video relighting [9] or shape recovery [10]. All these methods tend to have artifacts when applied to document images.

A second category of techniques has been specifically developed to remove shadows from document images. Some such methods [5,11,12], inspired by general intrinsic approaches, correct geometric distortions and can estimate illumination within their framework to address shading artifacts. Our approach is more similar to the state-of-the-art method of Oliveira et al. [13], where a constant color for the document background is assumed to generate a gain map. However, they detect background-only regions and interpolate the remaining areas of the gain map. Therefore, their method can fail to remove shadows when excessive interpolation is required to fill in the holes of the gain map. Our approach, on the other hand, can effectively estimate the gain map in text and background regions, so it does not suffer from interpolation inaccuracies. Finally, Adobe Acrobat uses an "Enhance" feature [14] on images of documents that is typically used to brighten dark images. However, since this applies a global correction, it fails to remove local shadow regions and leaves residual shadows. Our approach performs analysis on small overlapping blocks throughout the document and is thus able to correct localized shadows.

Finally, there are image binarization methods that segment an image into black and white and discard all color information [15–22]. We note that, although related to our work, these approaches have the fundamentally different goal of creating a binary image that is more effective for optical character recognition (OCR) applications. On the other hand, we aim to improve documents by removing their shadows while still keeping the same color and tone as the original. Thus, we view this field of work as orthogonal to ours. In fact, in Sect. 5, we demonstrate how our method can be used as a pre-process to improve binarization techniques and thus the OCR applications that use them.

3   Algorithm

Our main observation is that documents tend to have a constant colored background throughout, so the unshadowed output should have this property as well. We propose to apply a factor, α_i, as determined by our computed shadow map at each of the i pixels in the input image in order to match the local background color intensities with the global reference color. Specifically, we calculate:

$$\tilde{\mathbf{c}}_i = \frac{\mathbf{c}_i}{\boldsymbol{\alpha}_i}, \tag{1}$$

where $\mathbf{c}_i$ and $\tilde{\mathbf{c}}_i$ are the RGB color intensities of the shadowed input and the unshadowed output at pixel i, respectively.

We find the local and global background intensities in the document and normalize the local background intensities by the global reference color to generate the per-pixel RGB shadow map, $\boldsymbol{\alpha}$. Applying this shadow map (Eq. 1) to the input image produces the final result.

Local and Global Background Colors: To find the local background colors, we begin by dividing the input image into small overlapping blocks. In each of these blocks, we cluster the pixel intensities into two clusters that we label as either the paper background or text. For clustering both the local and global data, we used Gaussian mixture models (GMM) fit with Expectation-Maximization (EM) and initialized with k-means clustering. In general, documents typically have dark colored text on a bright background. Based on this, we assign the cluster center with the higher mean as the local background RGB color, $\boldsymbol{\ell}_i$, where i denotes the pixel at the center of the current block. In the case of a constant colored block (i.e., all background), the clusters have almost identical averages so selecting the higher one is still valid.

Next, we find the global reference background color. We take the pixel intensities from the entire input, rather than a local region, and cluster them into two categories of paper and text, as before. Again, the cluster mean that has a higher value is labeled as the background intensity. Finally, we search all of the intensities in the original input and assign the one closest to the background cluster mean as the final global background RGB reference, $\mathbf{g}$. Note that although the cluster mean could be used in this step instead of the closest intensity, we empirically found that this approach slightly improves results.

Computing Shadow Map: We expect the global background color to be the true background color. The local background color deviates from this because of illumination effects such as shadows and shading. In order to remove the influence of illumination, we compute the ratio of the local and global background colors to generate the shadow map as:

$$\boldsymbol{\alpha}_i = \frac{\boldsymbol{\ell}_i}{\mathbf{g}}, \tag{2}$$

where $\boldsymbol{\ell}_i$ is the local background intensity at pixel i and $\mathbf{g}$ is the global background reference for all the pixels. Moreover, $\boldsymbol{\alpha}_i$ maps each input pixel to the reference background color and, when applied to the input image (Eq. 1), produces the final unshadowed result.

Figure 1 shows the general framework of our proposed approach along with an example of a shadow map, which accurately detects regions of shadow and their relative intensities.

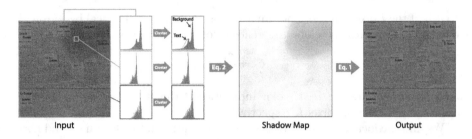

Fig. 1. Our approach takes an input document image that contains shadow and produces a high-quality shadow-free result. We analyze image intensities in local blocks (light blue and purple) and partition them into two clusters – background and text. We also do this for the whole image (green) to estimate a global background reference. Applying Eq. 2 to the background mean clusters results in a shadow map that is used in Eq. 1 to produce the final unshadowed result (Color figure online).

Implementation Details: We implemented our algorithm in C++ and it takes roughly 2 s to process a 1024×1024 image, with clustering being the most costly sub-process. Thus, for acceleration, we randomly sample 150 pixels in each block (21×21) for local clustering and 1000 pixels throughout the entire image for global clustering. For further speed-up, we do not perform local clustering at each pixel. Instead, we only consider pixels at specific strides (i.e., 20 in our implementation) for calculating the local background intensities. Thus, our shadow map is stride times smaller than the input image. To get the full resolution shadow map, we upsample using an 8×8 Lanczos filter. Note that since we use a stride, the low resolution shadow map can have some slight noise due to differences in cluster means, so we first apply a 3×3 median filter followed by a Gaussian filter ($\sigma = 2.5$) to smooth it out and avoid small blotchy artifacts in the final result. Since illumination tends to vary smoothly and generates soft shadows in practice, computing a downsampled shadow map does not adversely affect our results. Finally, text can have a lot of variation in intensity values and two clusters is not always sufficient to capture accurate statistics. Thus, we found that using three means for both the local and global clustering worked better than two.

4   Results

We begin by validating the correctness of our method by creating a dataset in a controlled environment with a set illumination and obtain a ground truth image for comparison. Specifically, we captured 81 shadow images of 11 documents each with 5–9 variations of shadow intensity and shape using a Canon 5D Mark II DSLR camera, a tripod, a photography lamp, and light-blocking objects. Furthermore, for each document, we captured the ground truth without shadow, where the light-blocking object was removed while illumination was

Fig. 2. We show an example from a controlled environment used to validate the correctness of our approach relative to the ground truth. We also provide the results and MSE scores of related approaches for comparison. In general, our output has minimal artifacts and is closest to ground truth in terms of average and median MSE (Table 1). See the supplemental materials for additional results from this controlled setting.

Table 1. A comparison with previous approaches of average and median MSE on our controlled dataset consisting of 81 images (11 documents and 5–9 variations of shadow).

Method	Avg. MSE	Median MSE
Bell et al. [6]	125.44	119.94
Gong et al. [2]	390.98	172.57
Pilu et al. [18]	67.38	53.54
Wagdy et al. [21]	74.06	43.73
Oliveira et al. [13]	23.08	19.01
Ours	22.26	18.45

kept constant. In Fig. 2, we demonstrate that we can match ground truth closely with minimal artifacts. Note that to avoid brightness differences we match the average color of our output to that of the ground truth.

Since we have the ground truth image, we can also provide quantitative comparisons. In Fig. 2, we show visual and MSE comparisons for the intrinsic approach of Bell et al. [6], the general shadow removal method of Gong et al. [2], the Adobe Acrobat "Enhance" feature [14], and the state-of-the-art algorithm in document de-shadowing by Oliveira et al. [13] on one specific example. In addition, Table 1 reports the overall performance of each method[1] over our entire dataset. Table 1 also includes statistics for two additional Retinex-based approaches [18,21] that are designed for document binarization rather than our application. In general, we found that these two approaches can leave residual shadow and blur out detail; this can be observed in the results in the supplementary material. As shown, our method performs the best relative to ground truth in terms of MSE. It is worth noting that, since we captured this dataset in a controlled environment with simple conditions (e.g., a single light source, flat documents, etc.), our dataset is not as difficult as the typical scenarios found in

---

[1] We exclude Acrobat "Enhance" from the table as it requires manual interaction for each image.

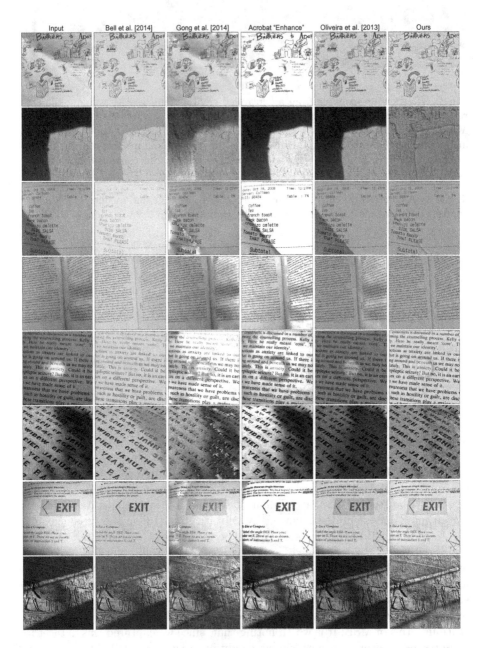

Fig. 3. We provide comparisons on challenging real-world examples from Flickr. Our method consistently produces high-quality results where other approaches fail. See the supplemental materials for full resolution comparisons on our complete real-world dataset.

practice. Thus, although we significantly outperform other approaches, we have only slightly better MSE relative to the method of Oliveira et al. [13].

In Fig. 3, we show comparisons for a subset of our more challenging real-world dataset obtained from Flickr. Note the shadows appear as is in the original image and were not altered in any way. As shown, all of the previous methods produce significant artifacts for these examples. The reflectance image from Bell et al. [6] loses some of the text to the shading layer and still has residual shadows. The general method from Gong et al. [2] cannot accurately estimate the statistics of the shadow region, despite the user interaction, and introduces saturation artifacts. The Adobe Acrobat "Enhance" feature [14] applies a global transformation on the image, so it fails to remove the local shadows. Finally, Oliveira et al.'s method [13] relies on interpolation so it cannot always remove the shadows. On the other hand, our approach works robustly across a wide range of difficult examples and generates high-quality results. Note that comparisons for all 16 images of our real-world dataset can be found in the supplemental materials along with additional comparisons to Retinex methods [18,21].

5   Discussion

In Fig. 4, we demonstrate that our algorithm can also be used as a pre-process for image binarization, which is typically used for OCR applications. Here, we compare the binarization method of Su et al. [22] with and without our shadow removal on an example from the Document Image Binarization Contest 2013 (DIBCO 2013) [23] dataset. When using our approach as a pre-process, there is a clear improvement in both visual quality and PSNR[2]. Note this binarization

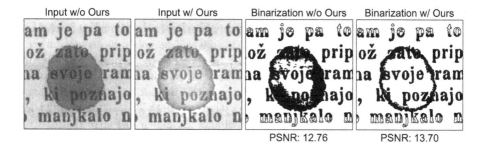

Fig. 4. Our approach can be used as a pre-process to improve the performance of OCR applications. We take a single example from the DIBCO 2013 [23] dataset containing a shadow (leftmost image) and run our shadow removal algorithm on it (center left image). We see a clear improvement both perceptually and in PSNR when applying the image binarization method of Su et al. [22] on these inputs (two rightmost images).

---

[2] The DIBCO 2013 dataset provides ground truth images to evaluate the performance of binarization algorithms. Furthermore, almost all of the images in this dataset are already shadow-free, so we report PSNR on just a single example containing shadow.

Fig. 5. Our algorithm is designed for document images and can introduce artifacts when non-text elements (e.g., figures, charts, pictures, etc.) are present. However, we are able to handle such cases by taking as input a user-defined mask that specifies such regions. We ignore these areas when clustering and perform interpolation to fill the holes and obtain our final shadow map. We show how this mask helps reduce artifacts in this real-world example.

method is already robust against illumination changes like shadow, yet there is still a benefit from running our shadow removal algorithm. This improvement will help the performance of OCR applications that take binary images as input.

One limitation of our method is that we assume that the input documents contain only intensities that correspond to either paper or text (i.e., no figures, pictures, graphs, etc.) so that we can cluster local and global pixel intensities into two categories: paper background and text. If the pixel intensities correspond to additional regions (e.g., a picture in the document), then the cluster means will be biased, cause incorrect background estimates, and generate overly bright or overly dark regions in the final result.

However, figures and pictures typically only account for a small region of the document. Thus, we can take a user-defined mask specifying these regions and ignore them when performing our shadow map calculation. After we have our shadow map, we can use interpolation (e.g., natural neighbor interpolation) to fill in the holes left by the mask and use it to generate the result. As shown in Fig. 5, using a mask allows us to improve results for such cases.

If the document background color changes throughout the document then the global reference background intensity that we find is unreliable and the final result will have brightness or color artifacts, as shown in Fig. 6(a). Finally, with intense hard shadows, there can be artifacts at the shadow boundaries since local blocks straddle regions with both well-illuminated and shadowed text and background. In such areas our cluster means are inaccurate creating a visible boundary between the well-illuminated and unshadowed regions (Fig. 6(b)). See the supplemental materials for full resolution examples of the various limitations.

Input   Ours

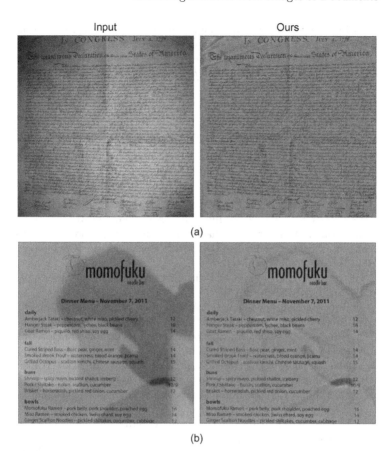

(a)

(b)

Fig. 6. Our approach has some limitations. For example, in (a), the background color changes throughout the document, so we produce slight color and brightness artifacts. In (b), we show that in cases with intense hard shadows, our method leaves small residual artifacts at the shadow boundaries. These artifacts result from unreliable clustering at either the global (a) and local (b) levels.

6   Conclusion

We have presented an approach for removing shadows from images of text (e.g., documents, menus, receipts) by generating a shadow map, or per-pixel scaling, that matches local background colors to a global reference. Our approach works robustly, as demonstrated qualitatively and quantitatively, on a wide range of examples containing large amounts of strong shadow in both controlled and real-world settings.

Acknowledgement. This work was supported in part by NSF awards IIS-1321168 and RI-1619376 and by a gift from Adobe. We thank Daniel Oliveira for providing code to run comparisons. The images in Figs. 3, 5 and 6 were used through the Creative

Commons 2.0 License without modification. The title and photographers from Flickr (unless otherwise noted) in order of appearance are: *Open Textbook Summit 2014* by BCcampus_News, *Army in the Shadows, Army in the Light* by Cuzco84, *That Please* by Kimli, *Medieval text in the Christ Church Archive* by -JvL-, *Untitled* by Jacek.NL, *Cartmel Priory* by Rosscophoto, *Transfer Damaged Textbook* by Enokson, *Untitled* by Colin Manuel, *find* by PHIL, Declaration of Independence photo by taliesin at Morguefile.com (Morguefile License), and *Momofuku - Menu w/ Shadow puppets* by Lawrence. Please see supplementary materials for links to the images and license.

References

1. Guo, R., Dai, Q., Hoiem, D.: Paired regions for shadow detection and removal. IEEE Trans. Pattern Anal. Mach. Intell. **35**, 2956–2967 (2013)
2. Gong, H., Cosker, D.: Interactive shadow removal and ground truth for variable scene categories. In: Proceedings of the British Machine Vision Conference. BMVA Press (2014)
3. Gryka, M., Terry, M., Brostow, G.J.: Learning to remove soft shadows. ACM Trans. Graph. **34**(5), 153:1–153:15 (2015)
4. Zhao, Q., Tan, P., Dai, Q., Shen, L., Wu, E., Lin, S.: A closed-form solution to retinex with non-local texture constraints. PAMI **34**(7), 1437–1444 (2012)
5. Yang, Q., Tan, K.H., Ahuja, N.: Shadow removal using bilateral filtering. IEEE Trans. Image Process. **21**, 4361–4368 (2012)
6. Bell, S., Bala, K., Snavely, N.: Intrinsic images in the wild. ACM Trans. Graph. (SIGGRAPH) **33**(4), 159:1–159:12 (2014)
7. Barron, J.T., Malik, J.: Shape, illumination, and reflectance from shading. TPAMI **37**(8), 1670–1687 (2015)
8. Zhou, T., Krahenbuhl, P., Efros, A.A.: Learning data-driven reflectance priors for intrinsic image decomposition, pp. 3469–3477 (2015)
9. Sunkavalli, K., Matusik, W., Pfister, H., Rusinkiewicz, S.: Factored time-lapse video. ACM Trans. Graph. **26** (2007). Proceedings of the SIGGRAPH
10. Abrams, A., Hawley, C., Pless, R.: Heiometric stereo: shape from sun position. In: European Conference on Computer Vision (ECCV) (2012)
11. Brown, M.S., Tsoi, Y.C.: Geometric and shading correction for images of printed materials using boundary. Trans. Img. Proc. **15**, 1544–1554 (2006)
12. Zhang, L., Yip, A.M., Tan, C.L.: Removing shading distortions in camera-based document images using inpainting and surface fitting with radial basis functions. In: 9th International Conference on Document Analysis and Recognition, ICDAR 2007, 23–26 September, Curitiba, Paraná, Brazil, pp. 984–988 (2007)
13. Oliveira, D.M., Lins, R.D., França Pereira e Silva, G.: Shading removal of illustrated documents. In: Kamel, M., Campilho, A. (eds.) ICIAR 2013. LNCS, vol. 7950, pp. 308–317. Springer, Heidelberg (2013). doi:10.1007/978-3-642-39094-4_35
14. Acrobat, A.: Enhance document photos captured using a mobile camera (2016). https://helpx.adobe.com/acrobat/using/enhance-camera-images.html
15. Otsu, N.: A threshold selection method from gray-level histograms. Automatica **11**, 23–27 (1975)
16. Pal, N.R., Pal, S.K.: A review on image segmentation techniques. Patt. Recogn. **26**, 1277–1294 (1993)
17. Sauvola, J., Pietikäinen, M.: Adaptive document image binarization. Patt. Recogn. **33**, 225–236 (2000)

18. Pilu, M., Pollard, S.: A light-weight text image processing method for handheld embedded cameras (2002)
19. Shi, Z., Govindaraju, V.: Historical document image enhancement using background light intensity normalization. In: 2004 Proceedings of the 17th International Conference on Pattern Recognition, ICPR 2004, vol. 1, pp. 473–476. IEEE (2004)
20. Gatos, B., Pratikakis, I., Perantonis, S.J.: Adaptive degraded document image binarization. Patt. Recogn. **39**, 317–327 (2006)
21. Wagdy, M., Faye, I., Rohaya, D.: Fast and efficient document image clean up and binarization based on retinex theory. In: 2013 IEEE 9th International Colloquium on Signal Processing and its Applications (CSPA), pp. 58–62. IEEE (2013)
22. Su, B., Lu, S., Tan, C.L.: Robust document image binarization technique for degraded document images. IEEE Trans. Image Process. **22**, 1408–1417 (2013)
23. Pratikakis, I., Gatos, B., Ntirogiannis, K.: ICDAR 2013 document image binarization contest (DIBCO 2013). In: 2013 12th International Conference on Document Analysis and Recognition (ICDAR), pp. 1471–1476. IEEE (2013)

Video Enhancement via Super-Resolution Using Deep Quality Transfer Network

Pai-Heng Hsiao$^{(\boxtimes)}$ and Ping-Lin Chang

Umbo CV Inc., Taipei City, Taiwan
{paiheng.hsiao,ping-lin.chang}@umbocv.com

Abstract. Streaming low bitrate while preserving high-quality video content is a crucial topic in multimedia and video surveillance. In this work, we explore the problem of spatially and temporally reconstructing high-resolution (HR) frames from a high frame-rate low-resolution (LR) sequence and a few temporally subsampled HR frames. The targeted problem is essentially different from the problems handled by typical super-resolution (SR) methods such as single-image SR and video SR, which attempt to reconstruct HR images using only LR images. To tackle the targeted problem, we propose a deep quality transfer network, based on the convolutional neural network (CNN), which consists of modules including generation and selection of HR pixel candidates, fusion with LR input, residual learning and bidirectional architecture. The proposed CNN model has real-time performance at inference stage. The empirical studies have verified the generality of the proposed CNN model showing significant quality gains for video enhancement.

1 Introduction

Super-resolution (SR) has been an important technique in many applications such as video enhancement, multimedia and video surveillance. Due to the limited bandwidth of the Internet, reducing streaming bitrate for exploiting the bandwidth while preserving high frame-rate and high-resolution video is highly desirable. In addition to using existing hard-coded compression encoder/decoder such as H.264 [1], another plausible approach is by streaming only a high frame-rate LR sequence together with a few corresponding low frame-rate HR sequence, so that they can be fused together to interpolate a high frame-rate HR sequence on demand. The interpolation, as the result of the fusion, requires an algorithm that can transfer high-quality video content from both spatial and temporal domain.

SR methods are typically used for enhancing image and video quality. They can be categorized into various types solving different problems. Single-image SR (a.k.a. Spatial SR) aims to recover fine image details given a LR image [2–4]. Video SR (a.k.a. multi-frame SR) attempts to estimate a HR image from a sequence of LR observations [5–8]. Temporal SR on the other hand is a process for interpolating temporal frames to recover rapid motions that occur faster than the recording frame-rate causing invisibility in the video sequence [8]. Yet another SR type uses

© Springer International Publishing AG 2017
S.-H. Lai et al. (Eds.): ACCV 2016, Part III, LNCS 10113, pp. 184–200, 2017.
DOI: 10.1007/978-3-319-54187-7_13

both LR and subsampled HR images jointly to reconstruct a high frame-rate HR sequences [9], which aligns with our target problem the most.

Video enhancement with hybrid LR and HR input sequence transferring high-quality video content from both spatial and temporal domain of HR to the reconstructed LR frame requires a special treatment. Despite many SR methods have been proposed for various scenarios, they cannot be simply applied to our target problem. Single-image SR concerns only single LR frame at once and ignores neighbor LR frames, failing in transferring temporal information. Also, given auxiliary HR frames, single-image SR cannot help transfer the high-quality spatial content into the reconstruction process, nor can conventional video SR methods do, since they are originally designed for dealing with only LR input frames. One previous work on video enhancement tackles fusion with both low frame-rate HR and high frame-rate LR using explicit optical flow estimation for finding correspondences for motion compensation [9]. Estimating optical flow however is computationally very expensive, which makes the method not feasible for real-time applications.

Based on the recent success of deep convolutional neural network (CNN) [10,11] used by many SR methods [2,3,5,6], we propose an end-to-end trainable CNN architecture, *deep quality transfer network* (DQTN). The proposed CNN does not explicitly estimate optical flow [12] but instead takes into account an important observation - Video content is usually with continuous motions and hence explicit optical flow estimation can be in fact unnecessary for video SR [13,14]. Specifically, we rely on CNN's superior data-driven learning ability to train a HR pixel candidate layer followed by a pixel selection layer and a fusion layer using a set of ground truth LR and HR frames as input. This greatly resolves the difficulty of collecting large amount of ground truth optical flow data for training [6]. In addition, the designed network adopts residual learning and has an bidirectional architecture which sequentially bundles input and output, i.e., the last reconstruction output HR frame is used as input to the current reconstruction.

The proposed CNN model is termed deep quality transfer network (DQTN) for its superior ability of transferring high-quality video content to the reconstruction process of video enhancement. Empirical studies have verified its generality, efficiency and effectiveness that it can reconstruct a 96×96 image in 40 ms during inference and has shown a significant improvement on SR quality compared with previous SR methods in both intra- and inter-dataset studies.

2   Related Work

The proposed method is inspired by the methods of single-image and video SR, optical flow and CNN. In this section, we review existing techniques for these subjects and highlight the distinct features of the proposed method.

2.1   Single Image Super-Resolution

Depending on used priors, single-image SR can be classified into four types including prediction models, edge-based methods, image statistical methods and patch-based methods [15]. Prediction models reconstruct a HR image by a LR image using a linear relationship of pixel neighborhood such as bilinear and bicubic interpolation. The edge-based methods consider edge as a prior and preserve edges when processing SR reconstruction. Image statistical methods take various image properties into account for the SR process. Patch-based methods attempt to learn the nonlinear mapping function between a pair of corresponding LR and HR patch. Among these four types, the patch-based methods generally perform better [2].

Deep CNN has been used for single-image SR because of its outstanding ability at learning non-linearity from data for regression and feature extraction. Super-resolution convolutional neural network (SRCNN) constructs a three-layer network and is able to be trained end-to-end in which it achieves competitive SR results [2]. Kim et al. proposed a deep CNN using adjustable gradient clipping and residual learning to enable fast convergence [16], and later on improved the architecture by a recursive network which significantly reduces the number of parameters [3].

The fusion component in the proposed CNN architecture is inspired by the previous CNN-based single-image SR methods which are able to transfer high-quality spatial image content from ground truth training HR images to general LR images. This however does not exploit high-quality temporal information provided by the ground truth HR sequence. We solve the issue by adding video SR component into the proposed CNN model.

2.2   Video Super-Resolution

Video or multi-frame SR was first proposed by Baker and Kanade where they used optical flow to reconstruct and fuse sequential HR images from LR images [17]. Shahar et al. proposed a spatial-temporal patch-based method to spatially and temporally reconstruct HR images from LR ones [8].

Huang et al. devised a bidirectional recurrent convolutional network (BRCN) which leverages CNN and recurrent neural network (RNN) to exploit forward and backward LR feed to reconstruct HR frames [5]. Liao et al. proposed deep draft-ensemble learning which uses handcrafted optical flow models to generate a number of SR drafts, i.e., HR image patch candidates, to fuse later with LR images by using a CNN [6].

Another setup of video SR fuses high frame-rate LR sequence and low frame-rate subsampled HR frames to reconstruct a high frame-rate HR sequence. The setup is particularly pragmatic for nowadays camcorders and surveillance cameras where the on-board chip can output video and capture low frame-rate HR still images simultaneously. Such reconstruction setup is however very challenging since it requires jointly optimizing single-image and video SR together with the consideration of motion compensation (i.e. the optical flow of the LR sequence).

To tackle this problem, Gupta et al. proposed a two-stage reconstruction method, including two-step optical flow process and rendering algorithm [9]. Other related works either relies on optical flow estimation [18] or handcrafted feature [19].

Inspired by the previous CNN-based video SR methods, the proposed CNN model reconstructs a HR image by (1) forwarding temporally subsampled HR images through a HR pixel candidates generation layer followed by a pixel selection layer to extract most likely HR pixels and (2) fusing the selected HR pixels with pixels in an input LR image. The reconstructed HR image is then used immediately as input with a next LR image for reconstructing the next HR image. The network sequentially bundles the inputs and outputs and employs bidirectional feed for boosting the reconstruction accuracy.

2.3   Optical Flow

The accuracy of optical flow has a significant effect on video SR [6]. Optical flow can be estimated by conventional handcrafted models such as [20] or by more sophisticated models such as TV-l_1 [6]. Many CNN-based optical flow methods are studied to enhance flow accuracy or solve complex handcrafted designed models. For example, DeepFlow deals with large displacement flow by using a matching algorithm [21]. EpicFlow extracts optical flow with both image content and contour information [22]. FlowNet performs end-to-end training and uses a CNN employing multi-resolution network to reconstruct final optical flow [12].

The proposed method has advantages over the previous approaches mainly by two perspectives. First, compared to traditional handcrafted optical flow models, the HR pixel candidate generation layer and the pixel selection layer implicitly learn the accurate flow from training data. Second, instead of performing flow estimation and motion compensation separately such as in [6,9], the proposed CNN model learns the optimal HR pixels by considering the motion compensation with the input LR images during training. Consequently, the model does not require high-quality ground truth optical flow during training. This can be considered a significant improvement in model training since ground truth optical flow is in practice extremely difficult to collect.

3   The Proposed Approach

The proposed method contains several key modules. Each module is described in this section.

3.1   Problem Statement

Given two synchronous LR and HR sequences with a certain frame-rate, let T be the sampling frequency for the HR sequence, i.e., $T = 5$ means sampling one HR frame for every five LR frames. An input dataset is defined as $D = \{\mathcal{Y}_t, \mathcal{L}_t, \mathcal{H}_t\}_{t=1}^N$

where t is the frame index, N the number of total frames, $\mathcal{Y}_t$ the ground truth HR frames, $\mathcal{L}_t$ the LR input frames, $\mathcal{H}_t$ the HR input frames, and

$$\mathcal{H}_t = \begin{cases} \mathcal{Y}_t, & \text{if } t = kT + 1, k \geq 0, \\ 0, & \text{otherwise.} \end{cases} \tag{1}$$

The goal is to reconstruct unsampled HR frames $\mathcal{H}_t$ at $t \neq kT + 1$ by using both $\mathcal{L}_t$ and $\mathcal{H}_t$ at $t = kT + 1$ with the proposed CNN architecture. Note that the input LR frames are first up-scaled to be with the same resolution of HR frames by bicubic interpolation.

3.2   Network Design

The proposed method is devised based on an observation that video usually contains continuous motion. One frame can be reconstructed by its previous frame using motion estimation/motion compensation (MEMC) with errors $\mathcal{E}$. The MEMC works as first to find pixel correspondences in two consecutive frames, and second to compensate the displacement of two correspondent pixels from the previous to the current frame. This can be formulated as

$$\hat{\mathcal{Y}}_{t+1} = \text{MEMC}\left(\mathcal{Y}_{t+1}, \hat{\mathcal{Y}}_t\right) + \mathcal{E}_{t+1}, \tag{2}$$

where $\hat{\mathcal{Y}}_{t+1}$ and $\hat{\mathcal{Y}}_t$ are the reconstructed frames by MEMC at time $t + 1$ and t respectively, $\mathcal{Y}_{t+1}$ is the input frame at time $t + 1$, and $\mathcal{E}_{t+1}$ is the error at time $t + 1$.

Generation of HR Pixel Candidates and Selection. To reconstruct HR frames $\mathcal{Y}$ from LR frames $\mathcal{L}$ and HR frames $\mathcal{H}$, based on the concept of MEMC but instead of finding pixel correspondences followed by pixel compensation, we flip the operation order by generating HR pixel candidates followed by finding pixel correspondence. This is an important step toward data-driven MEMC learning without ground truth optical flow training dataset.

Based on the assumption of continuous motion in video content, the displacement of each pixel of two continuous frames is generally within a local neighborhood. In other words, the correspondent pixels can be located in the neighborhood region. We can use convolution to model such observation. A convolution operation is in fact the summation over element-by-element product between the input data and the flipped convolution kernel. In 2D convolution, a pixel can be moved from one location to another using a specific kernel. This is a key idea for the proposed CNN model illustrated in Eq. (3) - A 5×5 matrix is convolved by a 3×3 kernel with only one non-zero entry 1 at the top-left corner. The result of this convolution is equivalent to shifting each element in the 5×5 matrix toward the top-left direction.

$$\begin{bmatrix} x_{11} & x_{12} & x_{13} & x_{14} & x_{15} \\ x_{21} & x_{22} & x_{23} & x_{24} & x_{25} \\ x_{31} & x_{32} & x_{33} & x_{34} & x_{35} \\ x_{41} & x_{42} & x_{43} & x_{44} & x_{45} \\ x_{51} & x_{52} & x_{53} & x_{54} & x_{55} \end{bmatrix} * \begin{bmatrix} 1 & 0 & 0 \\ 0 & 0 & 0 \\ 0 & 0 & 0 \end{bmatrix} = \begin{bmatrix} x_{22} & x_{23} & x_{24} & x_{25} & 0 \\ x_{32} & x_{33} & x_{34} & x_{35} & 0 \\ x_{42} & x_{43} & x_{44} & x_{45} & 0 \\ x_{52} & x_{53} & x_{54} & x_{55} & 0 \\ 0 & 0 & 0 & 0 & 0 \end{bmatrix}. \tag{3}$$

Based on this observation, pixels can be shifted around various locations, if the image is convolved by different kernels with different sizes. Therefore we can apply convolution using multiple kernels to generate a number of HR pixel candidates which can be seen as HR motion compensation candidates. A special case emerges when the convolution is with a 1×1 kernel acting like a pixel selector. A convolution layer with a 1×1 kernel is attached right after the HR pixel candidates generation layer for pixel selection in order to select the optimal HR pixel candidate for later fusion with the input LR frames.

Fusion of HR Pixel Candidates with Input LR Frames and Post-refinement. In practice, some pixels may violate the assumption of MEMC, such as pixels around image border or being occluded in which they cannot be found for a correspondence. These problematic pixels can be handled by fusion of both HR pixel candidates and a current input LR image. The basic idea is that if the network cannot figure out an optimal HR candidate for a pixel, it simply fuses the pixel with its co-located LR pixel. The fusion consists of multiple convolution layers similar to single-image SR to refine and super resolve the output image [2].

Residual Learning. Natural image can be generally decomposed into low-frequency (i.e., homogeneous areas) and high-frequency (i.e. detailed textures) parts. In order to preserve image details during SR reconstruction, we incorporate CNN-based residual learning into the model design. It has been shown that focusing on learning high-frequency data can improve overall SR accuracy and speed training convergence [16].

Bidirectional Architecture. The longer the temporal distance from a LR frame $\mathcal{L}_{t+j}$ to its nearest input HR frame $\mathcal{H}_t$, the more difficult to accurately reconstruct the HR frame $\mathcal{R}_{t+j}$. This is because the reconstruction error will propagate through temporal domain. Bidirectional architecture is thus adopted for the proposed CNN architecture to alleviate the error propagation problem since it can shorten the temporal distance between each LR frame to its nearest HR frame. Specifically, the temporal distance between $\mathcal{L}$ to $\mathcal{H}$ can be shortened by a half of sample rate T.

Deep Quality Transfer Networks. The proposed CNN model has four main components, i.e., generation of HR pixel candidates and selection, fusion of HR candidates with LR input and post-refinement, residual learning, and bidirectional architecture. Since the model has a deep structure for inferring temporal information from the HR input image and fusing the spatial HR compensation results with the input LR frames, it is therefore termed as deep quality transfer networks (DQTN).

3.3 Object Function

We can formulate the devised network as Eq. (4) where $f(\cdot)$ denotes the generation of HR pixel candidates and the selection of optimal candidates, $g(\cdot)$ the fusion of HR candidates with LR input and post-refinement, together with the consideration of the error term $\mathcal{E}_{t+1}$. The residual learning takes place at the addition operation between $\mathcal{L}_{t+1}$ and the fusion result.

$$\mathcal{Y}_{t+1} = \mathcal{L}_{t+1} + \left(g\Big(\mathcal{L}_{t+1}, f(\mathcal{Y}_t)\Big) + \mathcal{E}_{t+1} \right). \tag{4}$$

Given a sample rate T, considering the input HR frames $\mathcal{H}_t$, the predicted HR frames $\mathcal{P}_{t+1,\{f,b\}}$ with subscript forward (f) or backward (b), LR frame $\mathcal{L}_{t+1}$ at $t \neq kT + 1$ and the bidirectional fusion $\mathcal{B}(\cdot)$, we define the objective function by Eq. (5), where $\mathcal{R}_{t+1}$ is the final reconstructed HR frame. We aim to train the proposed CNN to fit $\mathcal{R}_{t+1}$ to ground truth $\mathcal{Y}_{t+1}$.

$$\mathcal{P}_{t+1,\{f,b\}} = \begin{cases} \mathcal{L}_{t+1} + \Big(g(\mathcal{L}_{t+1}, f(\mathcal{H}_t)) + \mathcal{E}_{t+1}\Big), & \text{if } t = kT + 1, k \geqslant 0, \\ \mathcal{L}_{t+1} + \Big(g(\mathcal{L}_{t+1}, f(\mathcal{P}_t)) + \mathcal{E}_{t+1}\Big), & \text{otherwise,} \end{cases}$$

$$\mathcal{R}_{t+1} = \mathcal{B}(\mathcal{P}_{t+1,f}, \mathcal{P}_{t+1,b}). \tag{5}$$

3.4 Network Architecture

For the first stage, generation of HR pixel candidates and selection, to enrich the HR pixel candidates of various motion conditions (i.e., still, small and large), we implement the proposed DQTN by using 5 different kernel sizes including 1×1, 3×3, 7×7, 15×15 and 31×31. Note that one may add more kernels with different sizes based on the motion characteristic of applied dataset. For the selection layer, we simply apply 1×1 convolution kernels. At the second stage, fusion with HR pixel candidates with LR input and post-refinement, we apply 3×3 convolution kernels for local fusion. Next, to refine the fusion results containing both HR and LR pixels, we apply two 3×3 convolution layers for post-refinement followed by one 3×3 output layer with three convolution kernels for transforming the result to RGB domain.

For the residual learning, we directly add the LR input $\mathcal{L}_{t+1}$ and the output of the second stage. Therefore, the first stage $f(\cdot)$ and the second stage $g(\cdot)$ are trained to learn the high frequency part for the reconstruction. As for the final bidirectional architecture, we first concatenate two output RGB images of

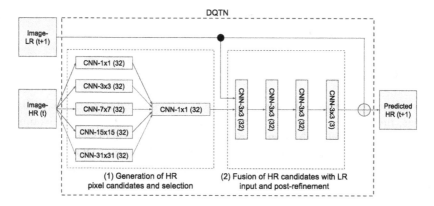

Fig. 1. Network architecture of proposed deep quality transfer network (DQTN). The LR frame is up-scaled to the resolution of HR frame by bicubic method. The number within the parenthesis indicates the number of convolution kernel.

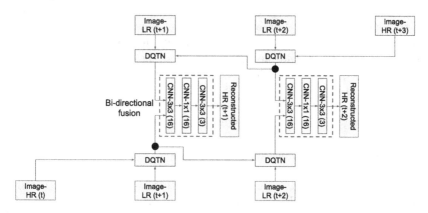

Fig. 2. Bidirectional architecture with proposed deep quality transfer network (DQTN) for $T = 3$ case. The number within the parenthesis indicates the number of convolution kernel.

DQTN, and then merge them by simple three 3×3 convolution layers. For each convolution kernel size, we empirically set the number of feature maps to be 32. Figure 1 illustrates the proposed deep quality transfer networks (DQTN) and Fig. 2 shows the bidirectional architecture for a $T = 3$ case. All operations are purely convolution. There are no pooling layers applied since we aim to keep the resolution of input and output images to be the same. We train the CNN with all three RGB channels jointly.

3.5 Optimization

We apply Mean Square Error (MSE) between the bidirectional prediction $\mathcal{R}_{t+1}$ and the ground truth HR frame $\mathcal{Y}_{t+1}$, i.e., $||\mathcal{R}_{t+1} - \mathcal{Y}_{t+1}||^2$, to optimize Eq. (5)

using stochastic gradient descent (SGD). Practically we set momentum to be 0.9 and the initial learning rate to be $10e^{-5}$.

4   Experiments

We evaluated the proposed bidirectional DQTN on two public datasets with two settings, intra-dataset and inter-dataset experiments. We thoroughly examined the proposed method with different parameters for validating the effectiveness of each proposed component. We then applied the bidirectional DQTN to unseen test datasets to show the generality of the model. We also quantitatively and qualitatively compare the results with previous single-image SR and video SR methods.

4.1   Datasets and Experiment Setup

For intra-dataset setting, we took 25 standard YUV[1] video sequences for generating training and testing datasets. YUV dataset is widely used for evaluating video coding performance and it contains various motion patterns. We transformed YUV into RGB color space and cropped the central 96 × 96 volumes for each video sequence. For each 96 × 96 volume, the first 60% frames are split into training and the rest 40% into testing set. In total there are 6,678 training frames and 4,452 testing frames out of 25 video sequences. For this setting, both the training and testing set might share common data statistics. During training, two consecutive training batches of T frames are closely overlapped by temporal gap one for data augmentation.

For inter-dataset setting, we randomly split 25 96 × 96 sequences into 15 training sequences and 10 testing sequences. In this case, both sets do not share common data statistics. We evaluate our method on the testing data of different resolution. In addition, We directly apply the CNN model trained with the training set of intra-dataset to the testing data provided by [6]. This experiment can help us investigate the generality of the proposed CNN model with respect to totally unseen data.

Our networks are trained from scratch with random initialized filters. More data and using pretrained networks may benefit training networks. However they are beyond the scope of this work.

To obtain the LR frames, we downsample each 96 × 96 frame by bicubic interpolation and upsample them again by bicubic interpolation to make the resolution to be the same as the HR frame. We evaluate the performance with scaling factor 4 since this is considered a very difficult case for SR problems [5]. For the evaluation metrics, we adopt widely used peak signal-to-noise ratio (PSNR) and structural similarity (SSIM). The higher PSNR and SSIM value, the better reconstruction quality. Note that the input HR frames are not included in

---

[1] YUV video dataset: http://www.codersvoice.com/a/webbase/video/08/152014/130.html.

all PSNR/SSIM evaluation for emphasizing the quality transfer ability of our model.

The average bitrate reduction of HR frames temporally subsampled at T compared with a full HR input sequence can be calculated by Eq. (6), where M is assumed to be the average bitrate of HR frame. To evaluate the effectiveness of temporal reconstruction of our method, we conduct experiments under two values of period T, including $T = 2$ (group of 1-HR-1-LR) and $T = 5$ (group of 1-HR-4-LR). When $T = 2$ and $T = 5$, the average video bitrate reduction are around 46% and 75% respectively, which can be considered a very large amount.

$$\text{average bit-rate reduction}(\%) = 1 - \Big(M + M * (T - 1)/\text{scaling factor}^2\Big)/(M * T)$$

$$= 1 - \Big(1 + 1 * (T - 1)/\text{scaling factor}^2\Big)/T. \qquad (6)$$

We compare the proposed bidirectional DQTN with two recent published state-of-the-art CNN-based methods. One is a single-image SR method, SRCNN [2], and the other a video SR method, BRCN [5]. We feed HR frames into BRCN with HR frame sample rate $T = 2$ (group of 1-HR-1-LR) as additional baseline to compare the performance. Bicubic interpolation method is also included as one of baselines.

Implementation and Running Time. The proposed bidirectional DQTN is implemented by Torch7 on a workstation with Intel 2.4 GHz E5-2620V3 CPU, 64 GB memory and one NVIDIA K80 graphic card. The proposed CNN model consists of only convolution layers and does not require pre-processing optical flow, which results in an average inference time for one single 96×96 image for about 40 ms. By computing a batch of patches in parallel, a larger image can be processed in real-time.

4.2   Intra-dataset Studies

The quantitative results of 25 testing sequences of different baselines methods and the proposed bidirectional DQTN are summarized in Table 1. On average the proposed method significantly outperforms the others by 2.05 dB to 5 dB. This is mainly because the proposed CNN model can generate and select optimal HR pixel candidates from neighboring HR frames to help reconstruct LR frames.

For HR frame period $T = 2$, one can observe that the performance of BRCN does not improve. The reason can be that the architecture of BRCN is not designed for transferring HR information at all. In contrast, the proposed bidirectional DQTN with HR input achieves much better results since our model is designed to transfer image quality. Different sample rate at $T = 2$ and $T = 5$ are also self-compared for the proposed model. It can be observed that the performance of $T = 2$ is better than of $T = 5$. This is reasonable since small value of T means a shorter temporal distance from LR frame to its nearest input HR frames $\mathcal{H}$.

Table 1. The intra-dataset testing results of 25 96 × 96 cropped YUV sequences with scaling factor 4. The proposed bidirectional DQTN is abbreviated as bi-DQTN.

PSNR(dB)/SSIM	Bicubic	SRCNN [2]	BRCN [5]	BRCN w/HR	bi-DQTN	bi-DQTN
HR frame period				$T=2$	$T=2$	$T=5$
Average	27.38/0.82	28.00/0.84	27.83/0.84	27.80/0.84	**32.34/0.92**	**30.05/0.88**
01.akiyo	25.49/0.85	26.23/0.86	26.25/0.86	25.87/0.86	32.53/0.96	28.32/0.91
02.bridgeclose	28.64/0.72	29.13/0.75	28.97/0.76	29.03/0.74	36.62/0.93	33.32/0.87
03.bus	19.29/0.51	20.08/0.57	19.96/0.56	19.90/0.56	20.54/0.67	20.39/0.64
04.carphone	28.31/0.88	28.94/0.89	28.90/0.89	28.69/0.89	29.68/0.90	28.93/0.89
05.claire	27.68/0.93	29.60/0.95	29.47/0.94	29.11/0.94	35.63/0.98	31.76/0.96
06.coastguard	19.15/0.42	20.01/0.50	20.04/0.51	19.72/0.47	26.46/0.84	23.85/0.73
07.container	30.18/0.85	31.58/0.87	31.36/0.86	31.47/0.87	41.17/0.97	37.93/0.95
08.football	25.01/0.84	25.64/0.86	25.56/0.86	25.43/0.85	24.48/0.83	24.55/0.82
09.foreman	29.35/0.85	28.74/0.86	28.13/0.85	28.63/0.86	31.82/0.90	29.78/0.87
10.grandma	26.88/0.91	29.32/0.93	29.27/0.93	28.43/0.92	37.42/0.98	33.12/0.96
11.highway	39.73/0.84	36.35/0.84	35.61/0.84	37.40/0.84	39.04/0.87	39.15/0.85
12.ice	28.16/0.86	29.39/0.85	29.43/0.84	29.18/0.87	28.75/0.85	28.74/0.84
13.miss-america	31.18/0.94	31.90/0.95	31.52/0.95	31.56/0.95	36.84/0.97	33.59/0.95
14.mobile	21.55/0.74	22.16/0.77	22.07/0.77	21.85/0.76	28.52/0.93	24.93/0.86
15.mother-daughter	36.14/0.98	35.94/0.98	36.21/0.98	36.25/0.98	41.13/0.99	38.82/0.98
16.news	26.41/0.86	27.85/0.88	27.77/0.88	27.19/0.87	29.94/0.94	28.09/0.92
17.sample	24.87/0.74	25.66/0.77	25.48/0.77	25.38/0.77	31.75/0.92	28.74/0.86
18.silent	30.74/0.85	31.22/0.85	30.91/0.85	30.94/0.85	34.63/0.94	32.58/0.90
19.soccer	29.37/0.93	29.81/0.93	29.67/0.93	29.73/0.93	29.29/0.92	29.09/0.91
20.stefan	22.82/0.70	23.21/0.74	23.20/0.73	23.12/0.72	23.37/0.73	23.10/0.71
21.suzie	30.26/0.91	31.19/0.92	30.85/0.91	30.90/0.92	37.40/0.97	33.91/0.95
22.tempete	21.83/0.72	22.25/0.75	22.24/0.75	22.07/0.74	30.39/0.95	26.46/0.88
23.tennis	25.00/0.87	25.14/0.86	25.06/0.85	25.15/0.87	29.70/0.94	28.55/0.93
24.trevor	29.74/0.87	31.57/0.90	30.81/0.89	30.84/0.89	36.63/0.96	32.60/0.92
25.waterfall	26.80/0.84	27.19/0.85	27.10/0.85	27.09/0.85	34.70/0.96	30.83/0.91

As a result, the error propagation problem is less serious. We visualize and compare reconstructed image of different methods. Qualitative results are shown in Fig. 5 for 3 selected testing videos. One can observe that our method not only has significant quantitative gains over the baselines, but can also transfer HR details to LR frames for reconstruction.

We also highlight the failure cases where the proposed bidirectional DQTN has results similar to, or slightly worse than, SRCNN in 08.football, 12.ice, 19.soccer and 20.stefan sequences. This is mainly due to either large displacement or complex motion happening in these sequences. In such cases, the used 5 various kernel sizes may not be able to capture all different kinds of motions during training, so they fail in inference stage to generate good HR pixel candidates. Figure 6 shows 2 sets of the failure cases. We observe that all methods fail in different presentations. For example, bicubic interpolation, SRCNN and BRCN cannot reconstruct image details while bidirectional DQTN strives to recover the details but fails in blending them in the end. Adding more kernels with larger sizes might improve the results straightforwardly, but how to elegantly deal with radical motion patterns is of great interest to our future work.

Table 2. Comparisons of performance according to different training parameters, including HR frame input $\mathcal{H}$, residual learning and bidirection architecture. Entry with ✓ indicates the component is enabled in the network. The results are in PSNR(dB).

HR frame input $\mathcal{H}$	Residual learning	Bidirection	T = 2	T = 5
✗	✓	✓	27.70	27.83
✓	✗	✗	31.09	29.10
✓	✓	✗	31.24	29.39
✓	✓	✓	32.34	30.05

Table 3. Comparisons of performance with respect to the combination of different kernels used in the stage of generation of HR pixel candidates.

PSNR(dB)	T = 2	T = 5
w/o HR frame $\mathcal{H}$	27.69	27.83
{1}	32.25	29.41
{1,3}	32.23	29.58
{1,3,7}	32.36	29.62
{1,3,7,15}	32.49	29.62
{1,3,7,15,31}	32.34	30.05

Network Configuration Studies. We evaluated various network configurations to verify the contributions of components such as the network for generating and selecting HR pixel candidates from HR frame input $\mathcal{H}$, residual learning and bidirectional architecture. During the comparison, the number of feature maps is fixed to 32 and the kernels used in the stage of generation of HR pixel candidates are fixed with various sizes {1,3,7,15,31} (abbreviation for $\{1 \times 1, 3 \times 3, 7 \times 7, 15 \times 15, 31 \times 31\}$). Table 2 summarizes the verification results excluding HR frames. One can see that significant gains (4.64 dB and 2.22 dB) can be obtained when HR frames are used for both $T = 2$ and $T = 5$, and the gains of $T = 2$ is generally more than that of $T = 5$ if with HR frame input $\mathcal{H}$. This is a strong evidence to validate that the proposed quality transfer network can transfer high-quality content for the reconstruction. It can also be observed

Table 4. Comparisons of performance with respect to different feature maps. Different kernel sizes include {1,3,7,15,31}.

PSNR(dB)	T = 2	T = 5
8 feature maps	32.24	29.42
16 feature maps	32.36	29.52
32 feature maps	32.34	30.05

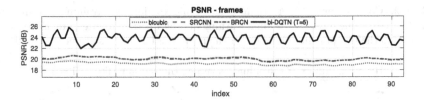

Fig. 3. Testing video: 06.coastguard. PSNR vs. frames with HR frame sampling rate $T = 5$ and scaling factor 4

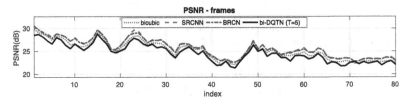

Fig. 4. Testing video: 08.football. PSNR vs. frames with HR frame sampling rate $T = 5$ and scaling factor 4

that both residual learning and bidirectional architecture improve the overall performance.

To validate the effectiveness of using multiple various kernels sizes, we trained and tested with increasing sets of kernel sizes with the number of feature maps fixed to 32. Table 3 summarizes the results showing that when using more different kernels with various sizes, the performance can be improved especially for the case $T = 5$. This indicates that capturing HR pixel candidates for various motion displacement can enhance the overall performance. For analyzing the effect of the number of feature maps, We fixed the kernels and use 8, 16 and 32 feature maps for the model. Table 4 summarizes the results showing that the increasing number of feature maps can also improve overall performance. This shows that more kernels can indeed extract more different types of HR pixel candidates so that the network can better learn to generate and select optimal HR pixel candidates.

Quality Transfer in Temporal Domain. The reconstruction PSNR with respect to each frame are shown in Figs. 3 and 4. Note that HR frames are skipped and not plotted. The peaks indicate the reconstructed HR frames next to the input HR frames. One can observe that the PSNR descends when the temporal distance from LR frame to its nearest HR frame $\mathcal{H}$ increases. This is expected as longer temporal distance causes more errors that are propagated. For sequence with large and complex motion displacement, the proposed method works similar to the baselines as shown in Fig. 4 for the sequence 08.football. To reduce the fluctuation, one may reduce the sampling rate T.

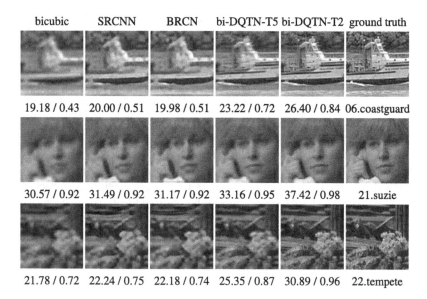

Fig. 5. Reconstruction results of different methods. The numbers below each images are PSNR(dB)/SSIM computed according to ground truth.

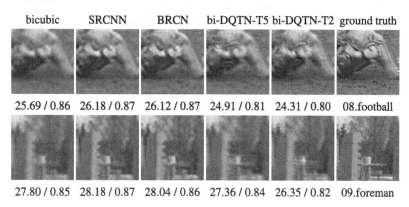

Fig. 6. The failure cases. The numbers below each images are PSNR(dB)/SSIM computed according to ground truth.

4.3 Inter-dataset Studies

We applied the proposed method to unseen testing dataset to evaluate the generality of the devised model. Specifically we trained the proposed bidirectional DQTN with 15 cropped YUV sequences and evaluate the model using the rest 10 sequences. Table 5 shows the quantitative results in which the proposed method outperforms the baselines by 2.19 dB and 0.94 dB for $T = 2$ and $T = 5$ respectively. This verifies the generality of the trained model that it can work with unseen data statistics. In addition, we applied the trained model directly on the

Table 5. Comparisons of performance of inter-dataset YUV results. Feature map number of each kernel size is set to be 32. The types of kernel sizes include {1,3,7,15,31}.

Average PSNR (dB)	Bicubic	SRCNN [2]	BRCN [5]	bi-DQTN $T=2$	bi-DQTN $T=5$
	27.53	28.02	28.05	30.24	28.99

testing data provided by [6], in which each test sequence has 31 frames. Note that the resolution of this testing dataset is different from that of our training data (i.e., YUV dataset). Table 6 summarizes the results of baselines and deep draft-ensemble learning [6] at 16th frame. It can be observed from the results that the proposed method in general outperforms the other. Two reconstructed images are shown in Fig. 7 where one can find that the proposed method preserves more details compared with other baselines SR approaches. The inter-dataset studies have validated the generality of the proposed method to unseen dataset.

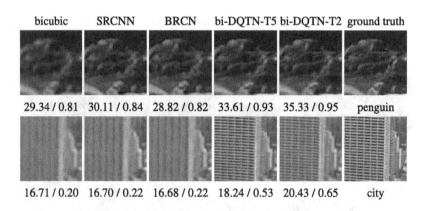

Fig. 7. Reconstruction results of different methods on the testing dataset of [6]. The numbers below each images are PSNR(dB)/SSIM computed according to ground truth.

Table 6. Comparisons of performance with deep ensemble learning. For bi-DQTN, the number of feature maps of each kernel size is set to be 32, and different kernel sizes include {1,3,7,15,31}.

Average PSNR (dB)	Bicubic	SRCNN [2]	BRCN [5]	deepSR [6]	bi-DQTN $T=2$	bi-DQTN $T=5$
Temple	29.79	28.81	28.50	**30.23**	**33.09**	30.11
Penguin	40.01	35.50	33.61	31.87	**41.92**	**40.92**
City	25.63	25.86	25.73	24.89	**29.48**	**26.26**

5 Conclusions

In this paper, we have presented a deep quality transfer network (DQTN) for video enhancement given a high frame-rate LR sequence and a few temporally subsampled HR frames. The proposed CNN model consists of modules including generation and selection of HR pixel candidates, fusion with LR input, residual learning and bidirectional architecture. The devised network can learn motion compensation from scratch from the training data. It therefore does not require pre-processing optical flow and can be trained end-to-end. Empirical intra- and inter-dataset studies on public datasets have validated the generality of the proposed CNN model and have shown its superior performance on video enhancement compared with previous methods.

References

1. Wiegand, T., Sullivan, G.J., Bjøntegaard, G., Luthra, A.: Overview of the h. 264/avc video coding standard. IEEE Trans. Circuits Syst. Video Technol. **13**, 560–576 (2003)
2. Dong, C., Loy, C.C., He, K., Tang, X.: Image super-resolution using deep convolutional networks. IEEE Trans. Pattern Anal. Mach. Intel. (PAMI) **32**, 295–307 (2016)
3. Kim, J., Lee, J.K., Lee, K.M.: Deeply-recursive convolutional network for image super-resolution. In: Proceedings of the IEEE Conference on Computer Vision and Pattern Recognition (CVPR) (2016)
4. Huang, J.B., Singh, A., Ahuja, N.: Single image super-resolution from transformed self-exemplars. In: Proceedings of the IEEE Conference on Computer Vision and Pattern Recognition (CVPR), pp. 5197–5206 (2015)
5. Huang, Y., Wang, W., Wang, L.: Bidirectional recurrent convolutional networks for multi-frame super-resolution. In: Neural Information Processing Systems (NIPS), pp. 235–243 (2015)
6. Liao, R., Tao, X., Li, R., Ma, Z., Jia, J.: Video super-resolution via deep draft-ensemble learning. In: Proceedings of the International Conference on Computer Vision (ICCV), pp. 531–539 (2015)
7. Ma, Z., Liao, R., Tao, X., Xu, L., Jia, J., Wu, E.: Handling motion blur in multi-frame super-resolution. In: Proceedings of the IEEE Conference on Computer Vision and Pattern Recognition (CVPR), pp. 5224–5232 (2015)
8. Shahar, O., Faktor, A., Irani, M.: Space-time super-resolution from a single video. In: Proceedings of the IEEE Conference on Computer Vision and Pattern Recognition (CVPR), pp. 3353–3360 (2011)
9. Gupta, A., Bhat, P., Dontcheva, M., Deussen, O., Curless, B., Cohen, M.: Enhancing and experiencing spacetime resolution with videos and stills. In: Proceedings of the IEEE International Conference on Computational Photography (ICCP), pp. 1–9 (2009)
10. LeCun, Y., Bottou, L., Bengio, Y., Haffner, P.: Gradient-based learning applied to document recognition. Proc. IEEE **86**, 2278–2324 (1998)
11. Krizhevsky, A., Sutskever, I., Hinton, G.E.: ImageNet classification with deep convolutional neural networks. In: Neural Information Processing Systems (NIPS), pp. 1097–1105 (2012)

12. Fischer, P., Dosovitskiy, A., Ilg, E., Häusser, P., Hazırbaş, C., Golkov, V., van der Smagt, P., Cremers, D., Brox, T.: FlowNet: learning optical flow with convolutional networks. In: Proceedings of the International Conference on Computer Vision (ICCV), pp. 2758–2766 (2015)
13. Protter, M., Elad, M., Takeda, H., Milanfar, P.: Generalizing the nonlocal-means to super-resolution reconstruction. IEEE Trans. Image Process. (TIP) **18**, 36–51 (2009)
14. Takeda, H., Milanfar, P., Protter, M., Elad, M.: Super-resolution without explicit subpixel motion estimation. IEEE Trans. Image Process. (TIP) **18**, 1958–1975 (2009)
15. Yang, C.Y., Ma, C., Yang, M.H.: Single-image super-resolution: a benchmark. In: Fleet, D., Pajdla, T., Schiele, B., Tuytelaars, T. (eds.) ECCV 2014. LNCS, vol. 8692, pp. 372–386. Springer, Heidelberg (2014). doi:10.1007/978-3-319-10593-2_25
16. Kim, J., Lee, J.K., Lee, K.M.: Accurate image super-resolution using very deep convolutional networks. In: Proceedings of the IEEE Conference on Computer Vision and Pattern Recognition (CVPR) (2016)
17. Baker, S., Kanade, T.: Super-resolution optical flow. Technical report (1999)
18. Tarvainen, J., Mikko, N., Pirkko, O.: Spatial and temporal information as camera parameters for super-resolution video. In: Proceedings of the International Symposium on Multimedia (ISM), pp. 302–305 (2012)
19. Ancuti, C., Ancuti, C.O., Bekaert, P.: A patch-based approach to restore videos using additional stills. In: Proceedings of the International Symposium ELMAR, pp. 143–146 (2010)
20. Horn, B.K., Schunck, B.G.: Determining optical flow. In: Technical Symposium East. International Society for Optics and Photonics, pp. 319–331 (1981)
21. Weinzaepfel, P., Revaud, J., Harchaoui, Z., Schmid, C.: Deepflow: large displacement optical flow with deep matching. In: Proceedings of the International Conference on Computer Vision (ICCV), pp. 1385–1392 (2013)
22. Revaud, J., Weinzaepfel, P., Harchaoui, Z., Schmid, C.: Epicflow: edge-preserving interpolation of correspondences for optical flow. In: Proceedings of the IEEE Conference on Computer Vision and Pattern Recognition (CVPR), pp. 1164–1172 (2015)

Face and Gestures

Age Estimation Based on a Single Network with Soft Softmax of Aging Modeling

Zichang Tan[1,2], Shuai Zhou[1,3], Jun Wan[1,2(✉)], Zhen Lei[1,2], and Stan Z. Li[1,2]

[1] National Laboratory of Pattern Recognition, Center for Biometrics and Security Research, Institute of Automation, Chinese Academy of Sciences, Beijing, China
jun.wan@ia.ac.cn
[2] University of Chinese Academy of Sciences, Beijing, China
[3] Faculty of Information Technology, Macau University of Science and Technology, Taipa, Macau

Abstract. In this paper, we propose a novel approach based on a single convolutional neural network (CNN) for age estimation. In our proposed network architecture, we first model the randomness of aging with the Gaussian distribution which is used to calculate the Gaussian integral of an age interval. Then, we present a soft softmax regression function used in the network. The new function applies the aging modeling to compute the function loss. Compared with the traditional softmax function, the new function considers not only the chronological age but also the interval nearby true age. Moreover, owing to the complex of Gaussian integral in soft softmax function, a look up table is built to accelerate this process. All the integrals of age values are calculated offline in advance. We evaluate our method on two public datasets: MORPH II and Cross-Age Celebrity Dataset (CACD), and experimental results have shown that the proposed method has gained superior performances compared to the state of the art.

1 Introduction

Until today, age estimation is still a very complex pattern classification problem. We judge a person's age mainly through the shin sheen, smooth degree, wrinkles and others, and such appearances are closely linked with various genes, diets, living and working environment and so on, which make age prediction further complicated. Initial work on age estimation goes back to 1990s [1], which simply classifies face images into several age groups based on facial shape features and skin wrinkle analysis. After this work, age estimation attracts more and more scholars' attention. Moreover, in the initial stage of age estimation research, the available age dataset is extremely limited. Fortunately, with the effort of the scholars all over the world, many large datasets are available for age estimation, like FG-NET [2], MORPH II [3], CACD [4], which increases by hundreds of times compared with datasets in the initial stage of age estimation research and significantly promote the development of age estimation.

Aging is a continuous process and the boundaries between adjacent ages are not obvious. Firstly, each person has different aging speeds and people of the

© Springer International Publishing AG 2017
S.-H. Lai et al. (Eds.): ACCV 2016, Part III, LNCS 10113, pp. 203–216, 2017.
DOI: 10.1007/978-3-319-54187-7_14

same age may appear to be slightly older or younger comparing with each other. For example, two faces come from different people of the same age may look like in different ages. Secondly, aging is a very slow process and faces at close ages would look similar. We believe that faces labeled with particular age are also close those with neighboring ages and even could be labeled with multiple labels in some way. For instance, Geng et al. [5] treated each face image with a age label distribution rather than a single label and thus during learning, it judged each face by considering its real age and adjacent ages. To be more correct, the label of age is more like to be a set of soft labels rather than a specific evidence, taking various factors related to aging into account.

Therefore, we propose an age estimation framework for exploring the aging information based on the CNN framework. The aging model is embedded into the network to explore more efficient features for age estimation. The main contributions of our work are summarized below:

- We model the randomness of aging with the Gaussian distribution for the chronological age. It is used to calculate the Gaussian integral of an age interval around the true age.
- We propose a new loss function: soft softmax regression function. The new function applies the aging modeling to compute the loss in the training phase. Compared with the classic softmax function, the new function considers the age interval instead of the specific age value.
- Compared with the softmax function, the proposed CNN frameworks with soft softmax function can alleviate the overfitting problems according to our experiments.
- Because of the complexity of Gaussian integral, a look up table is built to accelerate this process.

The rest of the paper is organized as follows. Related works are reviewed in Sect. 2. The proposed method is presented in Sect. 3. Then, experiments are provided in Sect. 4 to evaluate our method and compare with the state-of-the-art methods. Section 5 gives some discussions about the proposed method. Finally, a conclusion is drawn in Sect. 6.

2   Related Works

Early methods for age estimation just classify facial images into several age groups according to some hand-crafted features based on facial geometry features and skin wrinkle analysis [1]. The facial geometry features mainly consist of geometric relationships that computed by the sizes and distances between some primary features (e.g. eyes, mouth, nose, etc.). Facial geometry features are used to distinguish babies, and skin wrinkle feature can distinguish young adults from senior adults. Few years later, on the basis of the former works, Horng et al. [6] locate eyes, mouths and noses in face images via sobel edge operator and region labeling, then extract geometric and wrinkle features for age estimation.

In recent years, fortunately, automatic human age estimation receives increasing attention and more and more new methods have been proposed along with the development of the facial analysis technology. Geng et al. [7,8] proposed AGing pattErn Subspace (AGES) approach to model the aging pattern, which achieves the mean absolute error (MAE) on FG-NET database to 6.22 years. Moreover, many methods were proposed for age estimation based on manifold learning [9–11]. Those methods firstly learned facial age features in low-dimensional representation with manifold learning, then defined a regression function to fit manifold data for further age prediction. For example, Guo et al. [9] introduced the age manifold learning scheme to extract facial age features, then proposed locally adjusted robust regressor (LARR) method to predict age for face images, which improved performance significantly and reduces MAE on FG-NET to 5.07 years. More recently, local features become very popular for age estimation, such as Gabor [12], Local Binary Patterns (LBP) [13], Biologically-Inspired Features (BIF) [14]. After features extracted by those local image descriptors, classification or regression methods would be used for predicting the age, such as BIF+SVM [14], BIF+SVR [14], BIF+CCA [15].

In the last few years, the CNN has made a lot of progress in age estimation [16–19]. Comparing with traditional methods, CNN learns useful features autonomously instead of hand-crafted ways. Yi et al. [17] designed 46 parallel CNNs with multi-scale facial image patches as input for age estimation, which reduced the MAE to 3.63 years in MORPH II database and achieved the state-of-the-art performance. The parallel CNNs need pre-partion of facial images into different parts according to facial landmarks and predefined scales, and each part would be processed by a separated CNN. Rothe et al. [16] proposed a new method called Deep EXpectation (DEX) model based on VGG-16 network, which won the 1st place at the ChaLearn LAP challenge 2015. However, such deep CNN needs to be pretrained with a large age database and Rothe et al. collected 0.5 million images to do that.

Multi-label learning (MLL) [20] is also a hot topic especially in age estimation research in recent years. MLL assigns training instances with a set of labels rather than a single label to solve the problem of label ambiguity. Geng et al. [5,21] labeled each facial image with multiple age labels followed a label distribution, showing the advantages dramatically over the methods with single-label.

In this paper, our work is also inspired by the MLL and CNN. We combine the advantages of both MLL and CNN, and propose a novel method with aging information for age estimation.

3   The Proposed Method

In the proposed method, we first model the randomness of aging with Gaussian distribution. Then, we present a soft softmax function using aging modeling. Compared with the softmax function, the new function is more efficient for age estimation. Later, we introduce a simple CNN framework which is similar to AlexNet [22]. Finally, we present a fast calculation method based on a look up table to accelerate the processing time for Gaussian integral.

3.1   Modeling the Randomness of Aging

For MLL problem, each image would be allowed to be labeled by multiple labels. Following the works [5,21], each face image is labeled by an age label distribution that is a set of possibilities which represent the description degrees corresponding to each label. One face image x includes some possible discrete age labels $L = \{l_1, ..., l_k\}$ in our aging model. Let $P(l_c, l_i)$ denotes the probability of the possible label l_i corresponding to the chronological age l_c, where $P(l_c, l_i) \in [0 \; 1]$, $\sum_{i=1}^{k} P(l_c, l_i) = 1$. $P(l_c, l_i)$ is the maximum value when l_i is equal to l_c.

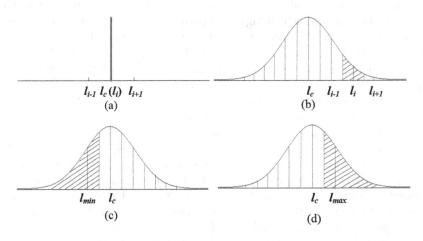

Fig. 1. (a) The Kronecker function $\delta(l_c, l_i)$, $l_c = l_i$; (b) the integral $P(l_c, l_i)$ of gaussian distribution ($l_{min} < l_i < l_{max}$); (c) the integral $P(l_c, l_i)$ of gaussian distribution ($l_i = l_{min}$); (d) the integral $P(l_c, l_i)$ of gaussian distribution ($l_i = l_{max}$).

There are some works [5,21] that allow each face image to be labeled by a Gaussian distribution for age estimation. Following those works, We also model the aging problem with the Gaussian distribution which is used to calculate the integral of an age interval. The formal formulation of aging model is given in Eq. 1.

$$P(l_c, l_i) = \begin{cases} \int_{-\infty}^{l_i+0.5} \frac{1}{\sqrt{2\pi}\sigma} e^{-\frac{(x-l_c)^2}{2\sigma^2}} \, dx & \text{if } l_i = l_{min} \\ \int_{l_i-0.5}^{l_i+0.5} \frac{1}{\sqrt{2\pi}\sigma} e^{-\frac{(x-l_c)^2}{2\sigma^2}} \, dx & \text{if } l_{min} < l_i < l_{max} \\ \int_{l_i-0.5}^{\infty} \frac{1}{\sqrt{2\pi}\sigma} e^{-\frac{(x-l_c)^2}{2\sigma^2}} \, dx & \text{if } l_i = l_{max} \end{cases} \quad (1)$$

where $l_i \in [l_{min} \; l_{max}]$ is the discrete age label, l_c denotes chronological age, l_{min} and l_{max} is the minimum and maximum age, respectively.

In Fig. 1(b–d), it shows the calculation of integrals via Eq. 1. Also, we note that if one face image with a single age label only, it also has a special distribution with $P(l_c, l_i) = \delta(l_c, l_i)$, where δ is the Kronecker function shown in Fig. 1(a).

3.2 Soft Softmax Regression Function

The traditional softmax regression function usually applied remarkably in the field of classification with neural networks [22]. There are also some methods [16,18] for age estimation with the softmax function, which have achieved promising results. However, age estimation is not a simple classification problem because aging is a continuous process. Also, face images at close age would look very similar. On the other hand, each person has different aging speeds because of many intrinsic and extrinsic factors, in which people with the same age would have various aging features. Hence, different age classes are related rather than independent.

From those observations, we believe that the problem of age estimation can be considered with the aging information as described in Sect. 3.1. Rothe et al. [16] calculated the expected value among softmax output probabilities and their corresponding age as final predicted age, which achieved better results comparing with the predict age having the maximum probability of the softmax output. This refinement fuses not only the real age's information but also other ages' in the prediction phase. In this section, we design a soft softmax regression function to push the network learn from both truth ages and their age intervals in Eq. 1 in the training phase.

For the age estimation, $x_c \in R^d$ denotes the CNN output features for the cth sample, and $l_c \in \{l_{min}, ..., l_{max}\}$ is its corresponding age labels. Here, we set $l_{min} = 0$, $l_{max} = k$ for the convenient description. Given a training set includes m samples $S = \{(x_0, l_0), ..., (x_c, l_c), ..., (x_{m-1}, l_{m-1})\}$, $c \in [0 \ m - 1]$, the soft softmax loss function with the aging model is defined as:

$$J(\theta) = -\frac{1}{m} \left[\sum_{c=0}^{m-1} \sum_{i=0}^{k} P(l_c, i) log \frac{e^{\theta_i^T x_c}}{\sum_{j=0}^{k} e^{\theta_j^T x_c}} \right] \tag{2}$$

where θ is the parameter matrix of the soft softmax function; $P(l_c, l_i)$ considers the probabilities of the specific label l_c and its adjacent labels. And when $P(l_c, l_i) = \delta(l_c, l_i)$, the soft softmax function becomes the traditional softmax function.

Then we can calculate the gradient formula of Eq. 2 as:

$$\nabla_{\theta_v} J(\theta) = -\frac{1}{m} \left[\sum_{c=0}^{m-1} \sum_{i=0}^{k} P(l_c, l_i) x_c \left(1\{v = i\} - \frac{e^{\theta_v^T x_c}}{\sum_{j=0}^{k} e^{\theta_j^T x_c}} \right) \right] \tag{3}$$

where ∇_{θ_v} is the gradient vector of the soft softmax parameters for age v, $1\{\bullet\}$ is indicator function which means $1\{v = j\} = 1$ if and only if $v = j$.

Therefore, the optimized parameters θ can be obtained via the SGD algorithm in the proposed method.

3.3 The Network Architecture

Our convolutional neural network is shown in Fig. 2, which is similar to the AlexNet [22]. The input of the network is RGB facial image with size 224×224.

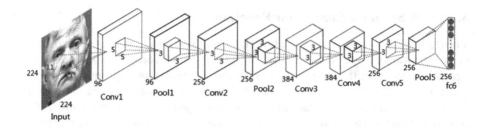

Fig. 2. The network architecture is used in the proposed method.

The network includes 5 convolutional layers, 3 max pooling layers and a fully connected layer. The filter size of each layer is also shown in Fig. 2. All the convolutional layers are followed by Rectified Linear Units (ReLU). The network are optimized by Stochastic Gradient Descent (SGD).

We train the network with proposed soft softmax loss function with 101 output neurons corresponding to age numbers from 0 to 100. And it is ok when the dataset lacks of samples corresponding to the output neuron. There are two ways to predict the age value. For the first way, the predicted value can be obtained via the maximum probability of the softmax output. When we used this way to calculate MAE, we call this way as MAE with maximum probability (MP). For the second way, we can conduct a softmax expected value refinement [16] to improve the accuracy, and the final predicting age is $\sum_{i=0}^{k} p_i y_i$, where p_i is the predicting probability of the corresponding age y_i. We call the second way as MAE with expected value (EV).

3.4   Look up Table for Fast Calculation of Integrals

The integral of Gaussian distribution function among the interval $[a,\ b]$ is:

$$\int_a^b f(x;\mu,\sigma)\mathrm{d}x = \frac{1}{\sqrt{2\pi}\sigma} \int_a^b e^{-\frac{(x-\mu)^2}{2\sigma^2}} \mathrm{d}x \qquad (4)$$

where μ is the expected value and σ^2 is the variance of the distribution.

The Gaussian integral in Eq. 4 is usually efficiently calculated by error function $erf(x) = \frac{2}{\sqrt{\pi}} \int_0^x e^{-t^2} \mathrm{d}t$ [23]. More specifically, the gaussian integral can be calculated with erf as following:

$$\int_a^b f(x;\mu,\sigma)\mathrm{d}x = \frac{1}{\sqrt{2\pi}\sigma} \int_{-\infty}^b e^{-\frac{(x-\mu)^2}{2\sigma^2}} \mathrm{d}x - \frac{1}{\sqrt{2\pi}\sigma} \int_{-\infty}^a e^{-\frac{(x-\mu)^2}{2\sigma^2}} \mathrm{d}x$$

$$= \frac{1}{2}\left[1 + erf\left(\frac{b-\mu}{\sqrt{2}\sigma}\right)\right] - \frac{1}{2}\left[1 + erf\left(\frac{a-\mu}{\sqrt{2}\sigma}\right)\right] \qquad (5)$$

$$= \frac{1}{2}\left[erf\left(\frac{b-\mu}{\sqrt{2}\sigma}\right) - erf\left(\frac{a-\mu}{\sqrt{2}\sigma}\right)\right]$$

From Eq. 5, we can see that each integral calculation need to call erf twice. However, the proposed method needs to calculate integrals both in forward and backward propagation phases for each age label and each face image, which would cost a lot of time accumulatively. Therefore, we prepare an integral table that stores multi-part integrals for each age, thus tedious calculation can be avoided through this table to find corresponding integrals. The consume time of two methods will be discussed in Sect. 5.

4 Experiments

4.1 Datasets

We evaluated the proposed method on MORPH II [3] and CACD [4] datasets, which are available standard datasets for facial age estimation. Some samples from both datasets are shown in Fig. 3.

Fig. 3. Face samples from the MORTH II (see the first row) and CACD (see the second row) datasets.

MORPH II includes about 55,000 face images and age ranges from 16 to 77 years. It provides the personal information, such as age, gender, and ethnicity. This dataset is more abundant in the age information, but the faces are recorded under uneven illumination.

CACD is collected from Internet Movie DataBase (IMDB), and it is the largest public cross-age database. This database includes more than 160 thousands images of 2000 celebrities taken from 2004 to 2013 (10 years in total). The age ranges from 16 to 62. Compared with MORPH II, CACD has the biggest total quantity and average number of each subject.

4.2 Experimental Setting

In our experiments, all images are resized into 224 × 224. We use SGD and mini-batch size of 64. The learning rate starts from 0.001, and the models are trained for up to 300000 iterations. We use a weight decay of 0.0005 and a momentum of 0.9.

Fig. 4. It shows face samples from the MORPH II dataset in the first row, and their corresponding results of face alignment in the second row.

We follow the work [17,24] to split MORPH II into three non-overlapped subsets S_1, S_2, S_3 randomly. These three subsets are constructed by two rules: (1) Male-Female ratio is equal to three; (2) White-Black ratio is equal to one. In our experiments, we totally use the same test protocols[1] provided by Yi et al. [17]. That is all experiments are repeated two times: (1) Training set: S1, testing sets: S2 + S3; (2) Training set: S2, testing sets: S1 + S3.

Only a few works [25] conducted evaluation on the CACD database owing to its noise. Note that only 200 celebrities of the database are checked and their noisy images are removed, and images of other celebrities contain much noises. Thus, these 200 celebrities are used for testing and the others for training in our experiments.

In our experiments, all images would be processed by a face detector [26] and non-face images would be removed. After processing, there are 55244 images in MORPH II and 162941 images in CACD. Then, we use active shape models (ASM) [27] to detect the facial landmarks and all facial images would be aligned and cropped via the locations of the eyes center and the upper lip (see Fig. 4). When evaluating on MORPH II, images for training is about a quarter of testing images, which is extremely insufficient. Therefore, we augment training images with flipping, rotating by $\pm 5^0$ and $\pm 10^0$, and adding Gaussian white noises with variance of 0.001, 0.005, 0.01, 0.015 and 0.02.

To further improve the performance on MORPH II, our network (with the softmax function) are pretrained on IMDB-WIKI [16]. Note that we don't conduct such operation for CACD evaluation because some images from IMDB-WIKI and CACD are duplicated.

4.3 Parameters Discussion

For aging modeling, the parameter σ controls the shape of Gaussian distribution at each age. σ is smaller, the Gaussian distribution is sharper and neighboring

---

[1] http://www.cbsr.ia.ac.cn/users/dyi/agr.html.

ages contribute less to the learning of chronological age in the training stage. Likewise, the contribution of neighboring ages would increase as σ rises.

To find an appropriate value for σ, we conducted experiments with a variety of σ on MORPH II and the results are shown in the Table 1. As we can see, the results is not sensitive to the parameter $\sigma \in [0.5, 1.2]$ and the well-done performance can be achieved when σ is about to 1. Thus, we set σ to 1 in our following experiments.

Moreover, we can see that the performance of MAE with EV is better than MAE with MP used the same value σ in Table 1. The same conclusion will also be shown in the next section.

Table 1. Results with different σ on the MORPH II dataset (the lower the better). The top 2 performances are shown in boldface, which are from the average MAE with MP under training set $S1$ and $S2$ respectively. From the best performances, we set $\sigma = 1$ in our experiments.

σ	0.5	0.6	0.7	0.8	0.9	1.0	1.1	1.2
MAE with MP (Train:S_1, Test:S_2+S_3)	3.37	3.37	3.33	3.31	3.30	3.27	3.30	3.30
MAE with MP (Train:S_2, Test:S_1+S_3)	3.06	3.05	3.09	3.04	3.06	3.05	3.07	3.04
Average MAE with MP	3.215	3.21	3.21	3.175	3.18	**3.16**	3.185	**3.17**
MAE with EV (Train:S_1, Test:S_2+S_3)	3.28	3.29	3.26	3.25	3.25	3.24	3.26	3.25
MAE with EV (Train:S_2, Test:S_1+S_3)	3.01	3.01	3.05	3.01	3.04	3.03	3.05	3.04
Average MAE with EV	3.145	3.15	3.155	**3.130**	3.145	**3.135**	3.155	3.145

4.4   Comparisons

Results on the MORPH II dataset. We conduct our experiments with the softmax and soft softmax regression function respectively and the results are shown in Table 2. We can see that the results of the soft softmax function are superior than the softmax function under MAE with MP or EV. Without pretrained model, the average MAE is 3.16 with MP and 3.14 with EV.

To further improve the performance, we also pretrain the model on the IMDB-WIKI database, and it achieves the best performance with the soft softmax function and EV, which the average MAE is 3.03. Compared with the softmax method, the pretrained model with soft softmax reduce the average MAE dramatically from 3.20 to 3.06 and from 3.08 to 3.03 with MP and EV respectively.

As shown in Table 2, the results show that the soft softmax regression function is superior than the softmax function for age estimation whenever MP, EV or the pretrained model is used.

Table 2. Results based on the CNN mentioned in Sect. 3.3 with different objective functions on the MORPH II dataset (the lower the better). It shows that the best performances are from the soft softmax function with pretrained model whenever MAE with MP or EV is used.

Methods	Train set	Test set	MAE with MP	Avg. MAE with MP	MAE with EV	Avg. MAE with EV
Softmax	$S1$	$S2 + S3$	3.45	3.28	3.28	3.16
	$S2$	$S1 + S3$	3.10		3.03	
Soft softmax	$S1$	$S2 + S3$	3.27	3.16	3.24	3.14
	$S2$	$S1 + S3$	3.05		3.03	
Pretrained model, softmax	$S1$	$S2 + S3$	3.34	3.20	3.19	3.08
	$S2$	$S1 + S3$	3.06		2.97	
Pretrained model, soft softmax	$S1$	$S2 + S3$	3.19	3.06	3.14	**3.03**
	$S2$	$S1 + S3$	2.93		2.92	

Table 3. Comparisons with the state-of-the-art methods on MORPH II under the same testing protocol (the lower the better).

Methods	Train set	Test set	MAE	Avg. MAE
Our method	$S1$	$S2 + S3$	3.14	3.03
	$S2$	$S1 + S3$	2.92	
Multi-scale CNN [17]	$S1$	$S2 + S3$	3.72	3.63
	$S2$	$S1 + S3$	3.54	
BIF+KCCA [15]	$S1$	$S2 + S3$	4.00	3.98
	$S2$	$S1 + S3$	3.95	
BIF+KPLS [28]	$S1$	$S2 + S3$	4.07	4.04
	$S2$	$S1 + S3$	4.01	
BIF+rCCA [15]	$S1$	$S2 + S3$	4.43	4.42
	$S2$	$S1 + S3$	4.40	
BIF+PLS [28]	$S1$	$S2 + S3$	4.58	4.56
	$S2$	$S1 + S3$	4.54	
CNN [29]	$S1$	$S2 + S3$	4.64	4.60
	$S2$	$S1 + S3$	4.55	
BIF+KSVM [15]	$S1$	$S2 + S3$	4.89	4.91
	$S2$	$S1 + S3$	4.92	
BIF+LSVM [15]	$S1$	$S2 + S3$	5.06	5.09
	$S2$	$S1 + S3$	5.12	
BIF+CCA [15]	$S1$	$S2 + S3$	5.39	5.37
	$S2$	$S1 + S3$	5.35	

Table 4. Results based on the CNN mentioned in Sect. 3.3 with different objective functions on the CACD dataset (the lower the better).

Methods	Train set	Test set	MAE with MP	MAE with EV
Softmax	1800 celebrities	200 celebrities	5.43	5.28
Soft softmax (ours)	1800 celebrities	200 celebrities	5.22	**5.19**

Then, we compared our method with the state-of-the-art methods. The comparisons are shown in Table 3 under the same testing protocol. We can see that our method can achieve the best performance with the average MAE of 3.03, which is reduced 0.6 comparing with the previous best method [17] with the average MAE of 3.63. Specially, whenever training with S1 and testing with S2 + S3 or training with S2 and testing with S1 + S3, our method is the best with MAE of 3.14 and 2.92, respectively.

Results on the CACD dataset. The images in this database are taken in the unconstrained environment, which are more close to the real life. We conduct the experiments on this database with the softmax and soft softmax regression function. The experimental results are shown in Table 4. The soft softmax method can achieve better result than the softmax method whenever using MP or EV. Our method reduces the MAE to 5.22 with MP and 5.19 with EV. And the proposed method is also better than the DFDNet method [25] which has achieved 5.57 in MAE.

5   Discussion

5.1   Anti-overfitting Analysis

Fig. 5(a) shows the loss trend when training on S1 and testing on S2 + S3 on the MORPH II dataset. Compared with the soft softmax regression function, the training loss decreases sharper used the softmax function after 15000 iterations. It indicates that the overfitting problem may be more serious when the softmax function is used.

Moreover, both two loss functions are designed to reducing testing error for age estimation. Thus, we draw the MAE trends as well, which are shown in Fig. 5(b). With the softmax function, training MAE are lower while testing MAE are higher than the soft softmax function. Obviously, the model with the softmax function is more likely to be overfitted. That is because it has poor performances compared with the soft softmax function.

Form the above discussions, the soft softmax regression function has the anti-overfitting characteristic in some way.

5.2   Computational Time Analysis

In this section, we mainly compare the consuming time among two methods for integrals calculation in the training process, looking up table and Gaussian

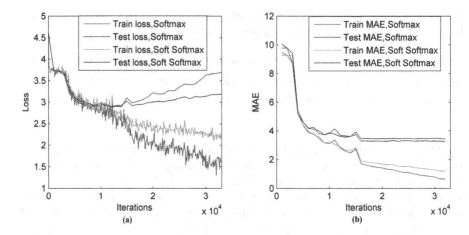

Fig. 5. (a) It shows the training and testing loss by iterations; (b) It shows the MAE trend by iterations.

Table 5. Comparison of the execution time between look up table method and direct Guassian integral.

Methods	Look up table (Our)	Online gaussian integral calculation
Time/10,000 iterations	1702 s	2097 s

integral calculation with the error function [23]. Our comparative experiments conducted under the same conditions with GTX TITAN X GPU and the results are shown in Table 5. Compared with online Gaussian integral calculation, the way of look up table reduce the consuming time by about 18.84%. In the training process with Gaussian integral calculation with the error function, for each instance, we need to calculate the interval for each age in the forward and backward pass phase, and each interval calculation need to call error function twice, which is very time consuming.

6 Conclusion

In this paper, we proposed a novel approach based on a single CNN for age estimation. First, the randomness of aging is modeled by the Gaussian integral that not only considers the chronological age but also includes the age intervals nearby the truth age. Second, we present a soft softmax regression function instead of classic softmax function, which is combined with aging model. Moreover, to further speed up the computation of gaussian integral, we build a look up table to store pre-compute gaussian integrals. So this way only requires one memory access from the look up table. Evaluations on two age datasets show that the proposed method achieves state-of-the-art performances.

Acknowledgement. This work was supported by the National Key Research and Development Plan (Grant No. 2016YFC0801002), the Chinese National Natural Science Foundation Projects ♯61473291, ♯61572501, ♯61502491, ♯61572536, Science and Technology Development Fund of Macau (No. 019/2014/A1), NVIDIA GPU donation program and AuthenMetric *R&D* Funds.

References

1. Kwon, Y.H., da Vitoria Lobo, N.: Age classification from facial images. Comput. Vis. Image Underst. **74**, 1–21 (1999)
2. Lanitis, A., Draganova, C., Christodoulou, C.: Comparing different classifiers for automatic age estimation. IEEE Trans. Syst. Man Cybern. B Cybern. **34**, 621–628 (2004)
3. Rawls, A.W., Ricanek, K.: MORPH: development and optimization of a longitudinal age progression database. In: Fierrez, J., Ortega-Garcia, J., Esposito, A., Drygajlo, A., Faundez-Zanuy, M. (eds.) BioID 2009. LNCS, vol. 5707, pp. 17–24. Springer, Heidelberg (2009). doi:10.1007/978-3-642-04391-8_3
4. Chen, B.-C., Chen, C.-S., Hsu, W.H.: Cross-age reference coding for age-invariant face recognition and retrieval. In: Fleet, D., Pajdla, T., Schiele, B., Tuytelaars, T. (eds.) ECCV 2014. LNCS, vol. 8694, pp. 768–783. Springer, Heidelberg (2014). doi:10.1007/978-3-319-10599-4_49
5. Geng, X., Yin, C., Zhou, Z.H.: Facial age estimation by learning from label distributions. IEEE Trans. Pattern Anal. Mach. Intell. **35**, 2401–2412 (2013)
6. Horng, W.B., Lee, C.P., Chen, C.W.: Classification of age groups based on facial features. Tamkang J. Sci. Eng. **4**, 183–192 (2001)
7. Geng, X., Zhou, Z., Smith-Miles, K.: Automatic age estimation based on facial aging patterns. IEEE Trans. Pattern Anal. Mach. Intell. **29**, 2234–2240 (2007)
8. Geng, X., Zhou, Z.H., Zhang, Y., Li, G., Dai, H.: Learning from facial aging patterns for automatic age estimation. In: Proceedings of the 14th Annual ACM International Conference on Multimedia, pp. 307–316. ACM (2006)
9. Guo, G., Yun, F., Dyer, C.R., Huang, T.S.: Image-based human age estimation by manifold learning and locally adjusted robust regression. IEEE Trans. Image Process. **17**, 1178–1188 (2008)
10. Fu, Y., Xu, Y., Huang, T.S.: Estimating human age by manifold analysis of face pictures and regression on aging features. In: 2007 IEEE International Conference on Multimedia and Expo, pp. 1383–1386. IEEE (2007)
11. Yun, F., Huang, T.S.: Human age estimation with regression on discriminative aging manifold. IEEE Trans. Multimedia **10**, 578–584 (2008)
12. Gao, F., Ai, H.: Face age classification on consumer images with Gabor feature and fuzzy LDA method. In: Tistarelli, M., Nixon, M.S. (eds.) ICB 2009. LNCS, vol. 5558, pp. 132–141. Springer, Heidelberg (2009). doi:10.1007/978-3-642-01793-3_14
13. Günay, A., Nabiyev, V.V.: Automatic age classification with LBP. In: 23rd International Symposium on Computer and Information Sciences, ISCIS 2008, pp. 1–4. IEEE (2008)
14. Guo, G., Mu, G., Fu, Y., Huang, T.S.: Human age estimation using bio-inspired features. In: IEEE Conference on Computer Vision and Pattern Recognition, CVPR 2009, pp. 112–119. IEEE (2009)
15. Guo, G., Mu, G.: Joint estimation of age, gender and ethnicity: CCA vs. PLS. In: 10th IEEE International Conference and Workshops on Automatic Face and Gesture Recognition (FG 2013), pp. 1–6. IEEE (2013)

16. Rothe, R., Timofte, R., Gool, L.: DEX: deep expectation of apparent age from a single image. In: Proceedings of the IEEE International Conference on Computer Vision Workshops, pp. 10–15 (2015)

17. Yi, D., Lei, Z., Li, S.Z.: Age estimation by multi-scale convolutional network. In: Cremers, D., Reid, I., Saito, H., Yang, M.-H. (eds.) ACCV 2014. LNCS, vol. 9005, pp. 144–158. Springer, Heidelberg (2015). doi:10.1007/978-3-319-16811-1_10

18. Levi, G., Hassner, T.: Age and gender classification using convolutional neural networks. In: Proceedings of the IEEE Conference on Computer Vision and Pattern Recognition Workshops, pp. 34–42 (2015)

19. Kuang, Z., Huang, C., Zhang, W.: Deeply learned rich coding for cross-dataset facial age estimation. In: Proceedings of the IEEE International Conference on Computer Vision Workshops, pp. 96–101 (2015)

20. Tsoumakas, G., Katakis, I.: Multi-label classification: An overview. Department of Informatics, Aristotle University of Thessaloniki, Greece (2006)

21. Geng, X., Wang, Q., Xia, Y.: Facial age estimation by adaptive label distribution learning. In: 2014 22nd International Conference on Pattern Recognition (ICPR), pp. 4465–4470. IEEE (2014)

22. Krizhevsky, A., Sutskever, I., Hinton, G.E.: Imagenet classification with deep convolutional neural networks. In: Advances in Neural Information Processing Systems, pp. 1097–1105 (2012)

23. Andrews, L.: Special functions of mathematics for engineers. In: SPIE (1998)

24. Guo, G., Mu, G.: Human age estimation: what is the in fluence across race and gender? In: Proceedings of the IEEE International Conference on Computer Vision Workshops, pp. 71–78 (2010)

25. Liu, T., Lei, Z., Wan, J., Li, S.Z.: DFDnet: discriminant face descriptor network for facial age estimation. In: Yang J., Yang J., Sun Z., Shan S., Zheng W., Feng J. (eds.) Biometric Recognition. LNCS, vol. 9428, pp. 649–658. Springer, Cham (2015). doi:10.1007/978-3-319-25417-3_76

26. Viola, P., Jones, M.: Rapid object detection using a boosted cascade of simple features. In: 2001 IEEE Conference on Computer Vision and Pattern Recognition, vol. 1, p. I-511. IEEE (2001)

27. Cootes, T.F., Taylor, C.J., Cooper, D.M.L., Graham, J.: Active shape models their training and application. Comput. Vis. Image Underst. **61**, 38–59 (1995)

28. Guo, G., Mu, G.: Simultaneous dimensionality reduction and human age estimation via kernel partial least squares regression. In: 2011 IEEE Conference on Computer Vision and Pattern Recognition (CVPR), pp. 657–664. IEEE (2011)

29. Yang, M., Zhu, S., Lv, F., Yu, K.: Correspondence driven adaptation for human profile recognition. In: 2011 IEEE Conference on Computer Vision and Pattern Recognition (CVPR), pp. 505–512. IEEE (2011)

Illumination-Recovered Pose Normalization for Unconstrained Face Recognition

Zhongjun Wu$^{(\boxtimes)}$, Weihong Deng, and Zhanfu An

Beijing University of Posts and Telecommunications, Beijing, China
wuzhongjun1992@126.com, {whdeng,anzhanfu}@bupt.edu.cn

Abstract. Identifying subjects with pose variations is still considered as one of the most challenging problems in face recognition, despite the great progress achieved in unconstrained face recognition in recent years. Pose problem is essentially a misalignment problem together with self-occlusion (information loss). In this paper, we propose a continuous identity-preserving face pose normalization method and produce natural results in terms of preserving the illumination condition of the query face, based on only five fiducial landmarks. "Raw" frontalization is performed by aligning a generic 3D face model into the query face and rendering it at frontal pose, with an accurate self-occlusion part estimation based on face borderline detection. Then we apply Quotient Image as a face symmetrical feature which is robust to illumination to fill the self-occlusion part. Natural normalization result is obtained where the self-occlusion part keeps the illumination conditions of the query face. Large scale face recognition experiments on LFW and MultiPIE achieve comparative results with state-of-the-art methods, verifying effectiveness of proposed method, with advantage of being database-independent and suitable both for face identification and face verification.

1 Introduction

Face recognition has been an active research area for its huge potential in real world applications, such as access control or video surveillance. The focus of face recognition study has shifted from constrained settings to unconstrained settings, as evidenced by the development of face databases, from lab databases, such as FERET [1], MultiPIE [2], to databases in the wild, such as LFW [3]. In unconstrained environment, the irregular conditions of pose, illumination, expression and resolution significantly affects the performance of face recognition system. Among these factors, pose is considered the most challenging one. An excellent solution towards pose variations brings benefit to other tasks such as feature extraction or facial attributes analysis.

Pose problem is essentially a misalignment problem caused by the rigid motion of 3D face structure, resulting in self-occlusion (loss of information) and loss of semantic correspondence [4]. Directly comparing two faces in different poses is difficult. The basic idea is to match pixels in 2D face images to the same semantic 3D facial points by face synthesis.

© Springer International Publishing AG 2017
S.-H. Lai et al. (Eds.): ACCV 2016, Part III, LNCS 10113, pp. 217–233, 2017.
DOI: 10.1007/978-3-319-54187-7_15

Existing pose-invariant face recognition methods can be broadly categorized into two families: 2D-based and 3D-based. In the first class, Li *et al.* in [5] represents a test image using some bases or exemplars and the coefficients can be regarded as one kind of pose-invariant features. Local linear regression (LLR) [6] learned appearance transformation between different poses. However, the performance is limited for the incapability of capturing 3D rotations as well as solving self-occlusion problem with using 2D warping. Recent years, deep learning models, such as FIP [7], SPAE [8], MVP [9], CPF [10] have been designed to learn non-linear transformation to convert a non-frontal face to a canonical (frontal) face or several target poses faces and get high recognition rates. But large and well-arranged data has to be prepared and the distribution of test data is usually different from the training data in real world application.

3D-based methods are usually based on a reference 3D face model or a deformable model with shape and illumination parameters, to handle 3D pose variations intuitively. 3D methods are divided into several categories as followed [11].

Recognition by fitting: 3DMM [12] is a powerful 3D representation for human face which fits parameters of 3D shape, pose and illumination and use them for recognition. But it is hard to implement it in practical system for high computational burden.

Pose synthesis: virtual face images under arbitrary poses can be generated using 3D models constructed from gallery images. Probe face is matched to the virtual images with similar pose to the probe. GEM [13] is an efficient 3D face modeling method, which estimates 3D shape by assigning generic face depth information directly to probe 2D images. However, GEM only deals with frontal face, requiring frontal faces for each identity, which is not always satisfied in unconstrained setting.

Pose normalization: 2D probe image is normalized to a canonical (frontal) view based on a 3D model to simplify unconstrained setting to constrained one in terms of pose variations. Asthana *et al.* [14] synthesized a frontal view of the input face by aligning an averaged 3D face model to it, using view-based AAM. But the self-occlusion part is unfilled. HPEN [15] fit the shape parameters of 3DMM and get a complete identity-preserving normalization results by filling the invisible region naturally. But it is based on 68 landmarks detection, where performance may drop due to unprecise localization. LFW3D [16] employed a generic 3D face model to "frontalize" non-frontal images and synthesized the occlusion part based on face symmetry with occlusion degree estimation. But the lighting conditions of output is not consistent with input face when lighting on both sides of face are different and unnatural results will be produced.

In this paper, we propose a continuous face pose normalization method which is identity-preserving and produces natural results in terms of illumination condition, based on only five fiducial landmarks. First, a generic 3D face model is aligned to the input face image based on the detected five landmarks. Then the face contour is detected for purpose of accurately estimating the self-occlusion

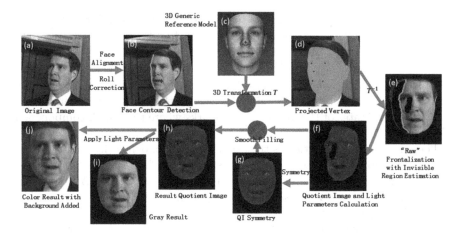

Fig. 1. Visual illustration of proposed pose normalization method. (Color figure online)

part. We can get a "raw" frontalization result by rendering the appearance-assigned 3D mesh at frontal pose with self-occlusion part unfilled (Sect. 2). In order to fill the invisible part naturally, we apply Quotient Image [17] as a face symmetrical feature which is robust to illumination. After estimating lighting parameters and making use of Quotient Image, natural normalization result is obtained where the self-occlusion part is filled with keeping the illumination conditions of input face (Sect. 3). Large scale face recognition experiments on LFW [3] and MultiPIE [2] achieve comparative results with state-of-the-art methods, verifying the effectiveness of proposed method (Sect. 4). The overall procedure of proposed method is shown in Fig. 1.

The advantage of proposed method is that the whole procedure does not depend on any specific training data and can be generalized well in unconstrained setting. Based on only five fiducial landmarks, proposed method is very suitable for practical applications.

2   "Raw" Frontalization

In this part, we will describe the "raw" frontalization process in detail. Inspired by the previous work [14,16] of using single 3D reference model to make pose normalization, we emphasize on keeping the appearance of input face rather than keeping its shape because the shapes produced from different pose images of the same identity are not guaranteed to be similar. Our target is to obtain highly aligned normalization results for better comparison between different face images.

Given a query image, five stable facial landmarks are located automatically or manually (see the blue '+' in Fig. 1(b)). The five fiducial landmarks in the 3D generic reference model (see Fig. 1(c)) have full correspondence with the landmarks of the query image. A 3D-to-2D projection matrix T is fitted using generalized least squares solution to the linear system for least square residual:

$$V_{Q-2d} \sim V_{R-3d}\boldsymbol{T} \tag{1}$$

where V_{Q-2d} is a 5×2 matrix with each row representing the (x, y) coordinates of Query-2d landmarks. V_{R-3d} is a 5×4 matrix with each row representing the $(x, y, z, 1)$ coordinates of Reference-3d landmarks where the fourth component 1 is for translation.

The underlying assumption is that sparse correspondence (five points correspondence) is able to represent dense correspondence of face vertices for the reason that human face can be roughly considered as a rigid structure. Although this assumption can not be strictly satisfied, highly aligned results can be obtained by this way. With projection matrix T, all vertices of reference model are projected onto the query image (see Fig. 1(d)) and the intensities of projected positions are assigned to the corresponding vertices by bi-linear interpolation. By rendering the appearance-assigned reference model at frontal pose, we can obtain an initial frontalization result.

When the landmarks do not include face contour, *e.g.*, the five landmarks we used, the problem that the semantic positions of face contour landmarks changes from pose to pose can be avoided. In addition, some previous works, *e.g.*, [14,15], are based on dozens of landmarks and detecting them accurately for profile faces will be difficult because of severe self-occlusion. With using that five stable landmarks, the ranges that face recognition system can handle with will be extended largely.

2.1   Face Borderline Detection for Self-occlusion Region Estimation

As face deviates from frontal to profile, some regions become invisible due to self-occlusion and it is considered as kind of information loss. Inaccurately estimation of invisible region position will lead to unnatural results because unwanted texture, such as background texture may be introduced to face region. In [14,18], Z-Buffer [19] is applied to estimate the visibility of each vertex. The idea is that the visibility condition of aligned 3D model approximates the visibility condition of the query face. But when the facial shape of query face differs from the generic face shape largely, the estimation will be inaccurate and introduce unwanted texture.

Face borderline is the boundary that separates visible texture and invisible texture. So face borderline detection facilitate accurate estimation of invisible region. Example comparison of results from Z-Buffer and borderline detection is shown in Fig. 2.

In [14], face borderline detection is formulated as finding a curve running from the top row to the bottom row of a certain rectangle with target of maximize edge strength and smoothness. The smoothness is constrained that the difference between adjacent row is within one pixel. But the objection function with only edge strength is too simple and it would fail with complex background texture in unconstrained environment.

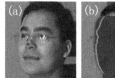

Fig. 2. Comparison of visibility detection from Z-Buffer and borderline detection. (a) Example input image from MultiPIE. (b) Aligned 3D model on input image. The texture on the left side of true borderline (green line) in Z-Buffer method is considered to be visible while actually not. (c) Results of visibility estimation of Z-Buffer. Black pixels indicates invisible. Red ellipse marks the unwanted background texture. (d) Results of visibility estimation using borderline detection which is more accurate. (Color figure online)

We use the information of borderline of projected 3D model and extend the object function in [14] to conduct a robust detection. After aligning the generic reference model into the query image, we can easily detect the borderline of aligned 3D model, which can constrain a certain borderline search region (see the red box in Fig. 3). The gradient magnitude is defined as

$$g(\mathbf{I}) = |\frac{\partial}{\partial x}\mathbf{I}| + |\frac{\partial}{\partial y}\mathbf{I}| \tag{2}$$

$\mathbf{I}$ is the search region of the query image. This magnitude is subtracted and divided by the mean and variance of itself for normalization. Since the direction of borderline are close to vertical, in order to reduce imposters, those pixels with large ratio of vertical gradient to horizontal gradient will not be saved.

Fig. 3. Similarity of found curve and projected borderline. (Color figure online)

With the borderline of projected 3D model, we introduce the term of similarity between found curve and projected borderline. For each pixel in search region, we calculate its tangential direction through its vertical and horizontal gradient (the blue arrow in Fig. 3), represented as $\mathbf{T}_i(x, y)$. For the projected 3D model borderline, the tangential direction of row y can also be calculated (the purple arrow in Fig. 3), represented as $\mathbf{T}_r(y)$. The similarity of direction

$T_i(x, y)$ at pixel (x, y) to projected 3D model borderline is calculated as cosine similarity,

$$s(x, y) = \frac{T_i(x, y) \cdot T_r(y)}{\| T_i(x, y) \| \| T_r(y) \|} \tag{3}$$

The basic idea is that the found curve should share a similar curve shape to the projected 3D borderline. The total optimization problem can be defined as

$$\max_{\{x_i\}} \sum_i g(x_i, y_i) + \lambda \sum_i s(x_i, y_i) \tag{4}$$

with constraint that x_{i-1} and x_i has to be within one pixel. λ is a parameter that balances the importance of gradient magnitude and the importance of curve shape similarity, which is set to 5 in our implementation. This optimization can be solved by dynamic programming and examples of found curves are shown in Fig. 4. The found face contour is back-transformed to the frontal 3D reference model through matrix T^{-1} and we can get a rather accurate visible region mask as our "raw" frontalization result (see Figs. 1(e) and 2(d)). It is noted that the visibility of nose region is estimated using Z-Buffer method [19].

Fig. 4. Examples of face borderline detection from LFW database.

3   Self-occlusion Region Filling

If the yaw angle of face is too large, some face regions become invisible due to self-occlusion. In order to obtain consistent frontalization result for completely texture comparison, the self-occlusion region should be filled naturally. Asthana *et al.* [14] leaves the invisible region unfilled and can not produce a consistent result. Ding *et al.* [18] use mirrored pixels which would produce incoherent face texture especially when the illumination conditions on both sides of face are largely different. The recent work, LFW3D [16] combines mirrored pixels with occlusion degree estimation but still suffers the illumination inconsistence problem. The self-occlusion problem is kind of information loss and the basic idea is to use face symmetry. In order to keep the illumination condition of input image, we are driven to find a feature that is not sensitive to illumination and satisfy face symmetry condition.

Quotient Image [17] is essentially the ratio of surface reflectance (gray-level) of an object against another object. For example, Caucasian face commonly has higher surface reflectance than Black people face and so has higher value of Quotient Image. Quotient Image feature is only relative to surface reflectance and is insensitive to illumination. It also satisfy face symmetry condition which is suitable for filling the self-occlusion region. We first briefly review the Quotient Image.

3.1   Quotient Image

Face, as a class of object, can be considered as Lambertian Surface with a reflection function: $\rho(u,v)n(u,v)^T s$, where $0 \leq \rho(u,v) \leq 1$ is the surface reflectance (gray-level) associated with point u,v in the image, $n(u,v)$ is the surface normal direction associated with point u,v in the image, and s is the (white) light source direction (point light source) and whose magnitude is the light source intensity.

In [17], the concept *Ideal Class of Object*, i.e., objects that have same shape but differ in surface albedo is defined. Under this assumption, the *Quotient Image* $Q_y(u,v)$ of face y against face a is defined:

$$Q_y(u,v) = \frac{\rho_y(u,v)}{\rho_a(u,v)} \qquad (5)$$

where u,v range over the image. Thus, Q_y depends only on the relative surface texture information and is independent of illumination.

A bootstrap set containing N (N is small) identities under M unknown independent illumination (totally $M \times N$ images) is adopted. Q_y of a input image $Y(u,v)$ can be calculated as

$$Q_y(u,v) = \frac{Y(u,v)}{\sum_{j=1}^{M} \bar{A}_j(u,v)x_j} \qquad (6)$$

where $\bar{A}_j(u,v)$ is the average of images under illumination j in the bootstrap set and x_j is linear combination coefficient which can be determined by the bootstrap set images and the input image $Y(u,v)$.

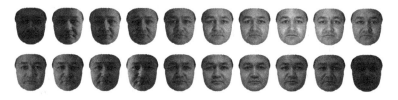

Fig. 5. Example bootstrap images from one identity. The illumination ids are marked as 00−09 in the first row, 10−19 in the second row.

3.2   Illumination Consistence Filling for Self-occlusion Region

We use YUV color space instead of RGB because Y is gray-level and it is independent of the other two channels. We fill the invisible region in Y channels using Quotient Image and combine directly symmetrical UV channels texture to get the final RGB result.

Our bootstrap set is formed by the frontal images from 12 identities under 20 lighting conditions from session *one* in MultiPIE [2] database. The selection of identities hardly affects final result [17]. Example bootstrap set images (gray level) from one identity is shown in Fig. 5.

For the "raw" frontalization result, we have the visible region mask, representing the valid texture, which can be used to estimate Quotient Image and lighting condition. We mask all the images in the bootstrap set using the visible mask of the query image and estimate Quotient Image on valid texture as well as lighting coefficient x_j, which can be represented as:

$$Q_{y-mask}(u,v) = \frac{Y_{mask}(u,v)}{\sum_{j=1}^{M} \bar{A}_{j-mask}(u,v)x_j} \tag{7}$$

Q_{y-mask} denotes Quotient Image of incomplete frontalization result. Y_{mask} denotes "raw" frontalization result. We make symmetry of the visible side and get Q_{y-sym}, which is blended with Q_{y-mask} smoothly using poisson editing [20] mentioned in [15] and finally we get Q_{y-full}. Since we have estimating lighting coefficient x_j to represent lighting conditions, we combine $\bar{A}_{j-full}$ and x_j to get Y_{full}, represented as:

$$Y_{full}(u,v) = Q_{y-full}(u,v) \cdot \sum_{j=1}^{M} \bar{A}_{j-full}(u,v)x_j \tag{8}$$

The basic idea of our filling is that we estimate lighting conditions from incomplete valid texture and use it as global representation. As we mentioned before, UV channels of invisible region is filled by directly mirrored pixels and we can get colored frontalization result by back transforming YUV space into RGB space. The ability of keeping illumination consistence is better viewed in

---

Algorithm 1. Invisible Region Filling

Input: "Raw" frontalization result, bootstrap set images
Output: Full frontalization result
  1: Mask bootstrap set images with same mask of "Raw" frontalization result.
  2: Solve Q_{y-mask} and light coefficient $x_j (1 \le j \le 20)$ according to Eq. 7.
  3: Mirror Q_{y-mask} and get Q_{y-sym}. Blend Q_{y-sym} into Q_{y-mask} smoothly using poisson editing and get Q_{y-full}.
  4: Compute Y_{full} by Q_{y-full}, x_j and full bootstrap set images according to Eq. 8.
  5: Mirror UV channels and back transform to RGB space. Adding background texture using affine transformation and get full frontalization result.

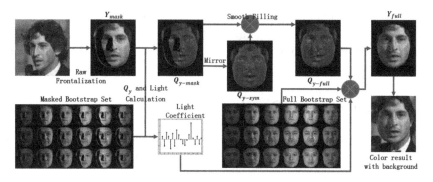

Fig. 6. Process of self-occlusion part filling. The images from 3 identities of 6 lightings in the bootstrap set are shown for convenience. There are actually 12 identities and 20 lightings.

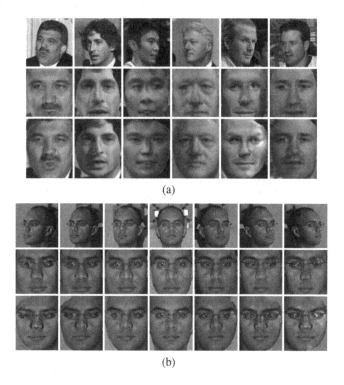

Fig. 7. Example frontalization results from (a) LFW and (b) MultiPIE (pose variation from $-45°$ to $+45°$ in step of $15°$). First Row: Input images. Second Row: Results of LFW3D [16]. Third Row: Results of Proposed Method. Our results keep illumination consistence and produce less artifacts for accurate borderline detection and smooth filling.

gray result. With adding background texture using affine transformation in [15], a complete frontalization is generated. Figure 6 demonstrates process of self-occlusion part filling, which is also summarized in the following algorithm block. Example frontalization results from LFW and MultiPIE are shown in Fig. 7.

4   Experiments and Results

In this section, we evaluate the performance of proposed method on LFW and MultiPIE databases for face verification and face identification settings respectively.

4.1   Face Verification on LFW

Labeled Faces in the Wild (LFW) [3] is the most commonly used database for unconstrained face recognition this years. LFW contains 13233 face images of 5749 persons collected from Internet with large variations including pose, age, illumination, expression, resolution, etc. We report our results following the "View 2" setting which defines 10 disjoint subsets of image pairs for cross validation. Each subset contains 300 matched pairs and 300 mismatched pairs. We follow the "Image-Restricted, Label-Free Outside Data" protocol and outside data includes BFM [21] as 3D reference model and frontal, multiple illumination facial images from MultiPIE [2] as bootstrap set images in Quotient Image.

For an input image, we frontalized it with invisible region filling when the estimated yaw angle is larger than 13°. We compare our method with two state-of-the-art method in terms of 3D face frontalization, HPEN [15] and LFW3D [16]. High dimensional LBP (HD-LBP) [22] is extracted on HPEN and proposed method for comparison. The images released by LFW3D are $90 * 90$ pixels only containing face region, which are not suitable for HD-LBP extraction. So we just extract LBP features on LFW3D and proposed method for comparison. Similarity metric learning (Sub-SML) [23] is adopted to boost face verification performance.

In order to discover how much the face verification performance could be improved by using proposed face frontalization method, we extract LBP on LFW-a [24] and HD-LBP on original LFW images and apply Sub-SML for comparison.

Results and Analysis. Table 1 shows the verification performance on LFW of different methods and Fig. 8(a) shows corresponding ROC curves. We first show the result of directly extracting LBP or HD-LBP with using Sub-SML, which achieves 83.92% and 88.78% respectively. With adding proposed method, we boost the performance to 88.82% and 91.50%, an improvement of 4.90% and 2.72% respectively. The improvement on feature HD-LBP is smaller because HD-LBP is already an excellent and expressive feature. These improvements come from our explicitly frontalization with natural, consistent results generated for directly texture comparison.

Table 1. Verification performance on LFW give by mean accuracy and standard error under image restricted, label-free outside data protocol.

Methods	Accuracy ($\bar{\mu} \pm S_E$)
LBP+Sub-SML [23]	0.8392 ± 0.0065
HD-LBP+Sub-SML	0.8878 ± 0.0046
LFW3D [16]+LBP+Sub-SML	0.8818 ± 0.0047
OURS+LBP+Sub-SML	0.8882 ± 0.0041
HPEN [15]+HD-LBP+Sub-SML	0.9152 ± 0.0037
OURS+HD-LBP+Sub-SML	0.9150 ± 0.0058

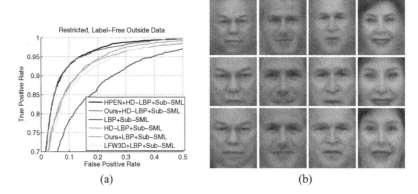

(a)   (b)

Fig. 8. (a) ROC curves on LFW under image restricted, label-free outside data protocol. (b) Mean faces by averaging corresponding multiple images of four subjects from LFW. First Row: Deep-Funneled [25], Second Row: LFW3D [16], Third Row: Proposed Method.

With same condition of using LBP with Sub-SML, proposed method outperforms LFW3D by 0.64% since we accurately estimate the self-occlusion region and fill it smoothly with keeping illumination consistence, resulting in less artifacts than LFW3D. Under the setting of using HD-LBP with Sub-SML, we achieve 91.50%, nearly the same performance as HPEN. It is noticed that HPEN utilizes 68 facial landmarks for shape fitting along with expression normalization. Bad face normalization result may occur due to un-precise 68 landmarks localization under large pose. Our method adopts five stable landmarks which are easier to detect even under large pose, indicating the simpleness and superiority of proposed method.

Qualitative Result. LFW3D [16] shows how well the frontalization method have preserved the texture of input identity by showing mean faces of several subjects. We follow this qualitative experiment and results are shown in Fig. 8(b).

It can be seen that the details around the eyes and mouth are better preserved and more consistent in our method, compared with the other two methods.

4.2 Face Identification Across Pose on MultiPIE

MultiPIE contains 754,204 images of 337 identities, where each identity has images captured under controlled environment with 15 poses and 20 illumination in four sessions during different periods, supporting development of algorithms for face recognition across pose, illumination and expression. A common setting for face recognition across pose, proposed in [5,14], is used for evaluation. This setting adopts images with different poses with neutral illumination marked as ID 07. The first 200 identities in all the *four* sessions are used for training and remaining 137 identities for test. During test, one frontal image of each identity from the earliest session in the test set is selected as gallery. The remaining images from $-45°$ to $+45°$ except $0°$ are selected as probes. This setting evaluates recognition robustness affected by pose, as well as other real-world factors, such as appearance changes by glasses or mustache.

For all gallery images and each probe image, we make frontalization and invisible region filling is used when estimated yaw angle is larger than $13°$, the same as operation on LFW. HD-LBP is adopted as feature extractor. For better comparison, we apply several classifiers, including PCA, LDA and LRA [26]. PCA and LDA are trained on frontalized images from the first 200 identities. LRA is directly trained on frontalized gallery images by mapping gallery faces to equidistant space targets which could enhance the discrimination between similar faces.

We compare our method with several pose normalization methods, including two 3D methods, Asthana11 [14] and HPEN [15], and three 2D methods, MDF [27], FIP [7] and MVP [9], the latter two are representative deep learning methods. Rank-1 identification rates are reported as results.

Table 2. Rank-1 identification rates (Percentage) on **MultiPIE** across pose. The first and the second highest performance are in **Bold**.

Methods	$-45°$	$-30°$	$-15°$	$+15°$	$+30°$	$+45°$	**Avg.**
Asthana11 [14]	74.1	91	95.7	95.7	89.5	74.8	86.9
HPEN+PCA [15]	88.5	95.4	97.2	98.0	95.7	89.0	94.0
HPEN+LDA [15]	97.4	99.5	99.5	**99.7**	99.0	96.7	**98.6**
MDF [27]	93	98.7	**99.7**	**99.7**	98.3	93.6	97.2
FIP [7]+LDA	95.6	98.5	**100**	99.3	98.5	**97.8**	98.3
MVP [9]	93.4	**100**	**100**	**100**	**99.3**	95.6	98.1
OURS+PCA	89.3	97.3	97.3	97.7	94.7	86.0	93.7
OURS+LDA	**98.0**	99.3	99.3	99.0	**99.3**	96.3	98.5
OURS+LRA [26]	**98.7**	**99.7**	**100**	**99.7**	**100**	**98.7**	**99.5**

Results and Analysis. Table 2 presents recognition results on MultiPIE across pose. Asthana11 exploited 3D information learnt from 200 subjects of training set and achieved mean accuracy of about 87%. The shortage is that self-occlusion part and background were not filled. MDF [27] transformed a non-frontal face to a frontal one by face pixels rearrangement, using morphable displacement field learnt from 3D face models and achieved competitive result. Proposed method with LDA outperform above two methods possibly for the natural filling of self-occlusion region and background. Similar to the experiment on LFW, proposed method with PCA as well as LDA achieve very close results to HPEN, indicating the effectiveness of our accurate invisible region estimation and natural filling, by using just five stable landmarks. With applying LRA classifier, we further boost our performance to 99.5% and outperform other methods, especially in large angles ($\pm 45°$). As we can see, FIP and MVP are two representative deep learning methods. They achieve competitive results with taking advantage of same pose distribution of training data and test data. In contrast, proposed method does not utilize any database-dependent information and would generalize well across continuous pose.

4.3 Further Discussion on Illumination Normalization

We apply light coefficients estimated from incomplete texture as global light representation to preserve lighting condition of input face, which inspires an idea of illumination normalization, by applying light coefficients of canonical lighting condition. Concretely, 20 lighting conditions exists in the bootstrap set marked as id $00-19$ (shown in Fig. 5), among which id 07 represents canonical lighting condition. We set $x_j = 1$ $(j = 8)$ and $x_j = 0$ $(j = 1 : 7, 9 : 19)$ in Eq. 8 and get illumination normalization result. Previous illumination normalization methods, such as WA [28] and DCT [29] mainly focus on frontal face while our idea provide a simple, unify framework for illumination normalization after pose normalization.

We perform face identification experiments across pose and illumination variations on MultiPIE. Images of 249 ids from session *one*, covering 7 poses ($-45°$ to $+45°$) and 20 illumination are used. The first 100 ids are for training, and the remaining 149 ids for test. The frontal image under illumination marked as ID 07 of each identity in the test set is chosen as the gallery. The remaining images from $-45°$ to $+45°$ except $0°$ (illumination ID 07 excluded) are selected as probes. We examine the performance of just Pose Normalization (briefly denoted as **PN**) and both Pose and Illumination Normalization (briefly denoted as **PIN**). Feature extractor is LBP and classifiers PCA, LDA and LRA [26] are tested as previous experiments in Sect. 4.2.

Results and Analysis. From Table 3 we can see that, explicitly illumination normalization largely improve total performance towards all classifiers, from 45.4% to 67.8% in PCA, from 63.1% to 75.5% in LDA, from 77.0% to 86.5% in LRA, verifying the effectiveness of proposed idea in solving both pose and

Table 3. Rank-1 Identification Rates (Percentage) on **MultiPIE** Across Illumination. Recognition rate under one illumination condition is the averaged result of 6 possible poses. Pose Normalization is briefly denoted as **PN**. Pose and Illumination Normalization is briefly denoted as **PIN**.

Methods	00	01	02	03	04	05	06	08	09	10
PN+PCA	30.2	20.1	24.6	32.3	48.8	67.8	79.4	79.2	71.0	50.8
PIN+PCA	48.1	35.1	45.3	55.8	68.7	81.3	89.0	88.5	84.3	71.3
PN+LDA	58.7	37.9	46.9	57.3	71.0	84.9	91.7	93.5	87.4	78.3
PIN+LDA	65.1	47.4	58.7	71.4	82.0	92.4	96.9	96.4	92.6	82.0
PN+LRA	64.9	44.1	51.9	67.3	87.3	96.3	99.5	99.7	97.1	88.8
PIN+LRA	79.6	59.3	73.2	85.0	94.9	97.7	99.3	99.0	98.3	94.5
	11	12	13	14	15	16	17	18	19	**Avg.**
PN+PCA	34.1	26.5	20.3	43.4	51.1	57.5	51.2	43.4	31.4	45.4
PIN+PCA	57.7	44.0	32.0	64.3	70.9	80.8	71.0	63.1	46.9	67.8
PN+LDA	63.5	47.6	39.0	70.0	74.6	81.3	74.7	71.6	58.4	63.1
PIN+LDA	69.0	56.0	47.5	80.3	84.9	89.5	82.1	76.0	64.4	75.5
PN+LRA	67.8	51.7	46.4	81.1	88.3	94.2	89.8	82.9	64.4	77.0
PIN+LRA	83.3	70.4	59.5	91.2	95.1	97.8	95.0	90.8	80.5	86.5

illumination problem in such a simple way. We can observe that performance of some lighting condition are relatively low, *e.g.*, 01, 02, 12, 13 and also the improvement from **PN** to **PIN** under these conditions are relatively large. We have selected −45° for demonstration. The 5 lowest performance light conditions and 5 highest performance ones in **PN+LRA** are shown in Fig. 9. In pose normalization results (third row), large illumination variance (strong specular light or dark ambient light) exists in the former group and leads to uneven, unsmooth face texture, resulting in low performance. From another perspective, under former group, our illumination normalization can largely reduce the lighting difference between probes and galleries and thus boost the performance with large proportion. The illumination conditions of latter group are close to gallery and thus achieve higher performance.

4.4   Discussion and Limitations

The normalization process takes about 1.5 s, running on a 2.8 Ghz CPU with matlab code. The bottleneck part is face and background rendering, which takes about 0.8 s and can be accelerated by C++.

In the process of invisible region filling, we use Quotient Image as a feature insensitive to illumination, satisfying face symmetry. The assumption of Quotient Image is Lambertian reflectance surface. When strong specular light occur, it can not model very well and would generate unnatural results. Also, it is hard to

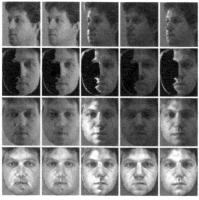 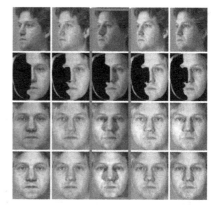

(a) 5 lowest performance light conditions   (b) 5 highest performance light conditions

Fig. 9. Example results from various lighting of $-45°$ from MultiPIE. First Row: Input images. Second Row: "Raw" frontalization results. Third Row: Pose normalization result. Fourth Row: Pose and Illumination Normalization results. (Color figure online)

eliminate the presence of cast shadow which leads to obvious artifacts when applying face symmetry (see the group in red box in Fig. 9(b)).

5   Conclusion

In this paper, considering the pose factor in unconstrained face recognition, we propose a continuous identity-preserving face normalization method which produces natural results in terms of illumination condition. With face borderline detection, the self-occlusion part is accurately detected and natural result is obtained by applying Quotient Image as a face symmetrical feature which is robust to illumination. We also provide a simple idea for illumination normalization in our framework. Our method achieve very competitive performance on LFW and MultiPIE datasets. With using only five stable landmarks and advantage of being database independent, our work is suitable for practical applications. In the future, we will focus on more sophisticated illumination modeling method to handle with strong specular light and cast shadow problem.

Acknowledgments. This work was partially sponsored by supported by the NSFC (National Natural Science Foundation of China) under Grant No. 61375031, No. 61573068, No. 61471048, and No.61273217, the Fundamental Research Funds for the Central Universities under Grant No. 2014ZD03-01, This work was also supported by Beijing Nova Program, CCF-Tencent Open Research Fund, and the Program for New Century Excellent Talents in University.

References

1. Phillips, P.J., Moon, H., Rizvi, S.A., Rauss, P.J.: The feret evaluation methodology for face-recognition algorithms. IEEE Trans. Pattern Anal. Mach. Intell. (TPAMI) **22**, 1090–1104 (2000)
2. Gross, R., Matthews, I., Cohn, J., Kanade, T., Baker, S.: Multi-pie. Image Vis. Comput. **28**, 807–813 (2010)
3. Huang, G.B., Ramesh, M., Berg, T., Learned-Miller, E.: Labeled faces in the wild: a database for studying face recognition in unconstrained environments. Technical Report 07–49, University of Massachusetts, Amherst (2007)
4. Ding, C., Tao, D.: A comprehensive survey on pose-invariant face recognition. arXiv preprint arXiv:1502.04383 (2015)
5. Li, A., Shan, S., Gao, W.: Coupled bias-variance tradeoff for cross-pose face recognition. IEEE Trans. Image Process. **21**, 305–315 (2012)
6. Chai, X., Shan, S., Chen, X., Gao, W.: Locally linear regression for pose-invariant face recognition. IEEE Trans. Image Process. **16**, 1716–1725 (2007)
7. Zhu, Z., Luo, P., Wang, X., Tang, X.: Deep learning identity-preserving face space. In: 2013 IEEE International Conference on Computer Vision (ICCV), pp. 113–120. IEEE (2013)
8. Kan, M., Shan, S., Chang, H., Chen, X.: Stacked progressive auto-encoders (spae) for face recognition across poses. In: 2014 IEEE Conference on Computer Vision and Pattern Recognition (CVPR), pp. 1883–1890. IEEE (2014)
9. Zhu, Z., Luo, P., Wang, X., Tang, X.: Multi-view perceptron: a deep model for learning face identity and view representations. In: Advances in Neural Information Processing Systems, pp. 217–225 (2014)
10. Yim, J., Jung, H., Yoo, B., Choi, C., Park, D., Kim, J.: Rotating your face using multi-task deep neural network. In: Proceedings of the IEEE Conference on Computer Vision and Pattern Recognition, pp. 676–684 (2015)
11. Yi, D., Lei, Z., Li, S.: Towards pose robust face recognition. In: Proceedings of the IEEE Conference on Computer Vision and Pattern Recognition, pp. 3539–3545 (2013)
12. Blanz, V., Vetter, T.: Face recognition based on fitting a 3D morphable model. IEEE Trans. Pattern Anal. Mach. Intell. **25**, 1063–1074 (2003)
13. Prabhu, U., Heo, J., Savvides, M.: Unconstrained pose-invariant face recognition using 3D generic elastic models. IEEE Trans. Pattern Anal. Mach. Intell. **33**, 1952–1961 (2011)
14. Asthana, A., Marks, T.K., Jones, M.J., Tieu, K.H., Rohith, M.: Fully automatic pose-invariant face recognition via 3D pose normalization. In: 2011 IEEE International Conference on Computer Vision (ICCV), pp. 937–944. IEEE (2011)
15. Zhu, X., Lei, Z., Yan, J., Yi, D., Li, S.Z.: High-fidelity pose and expression normalization for face recognition in the wild. In: Proceedings of the IEEE Conference on Computer Vision and Pattern Recognition, pp. 787–796 (2015)
16. Hassner, T., Harel, S., Paz, E., Enbar, R.: Effective face frontalization in unconstrained images. In: Proceedings of the IEEE Conference on Computer Vision and Pattern Recognition, pp. 4295–4304 (2015)
17. Shashua, A., Riklin-Raviv, T.: The quotient image: class-based re-rendering and recognition with varying illuminations. IEEE Trans. Pattern Anal. Mach. Intell. **23**, 129–139 (2001)
18. Ding, L., Ding, X., Fang, C.: Continuous pose normalization for pose-robust face recognition. Sig. Process. Lett. **19**, 721–724 (2012). IEEE

19. Foley, J.D.: Computer Graphics: Principles and Practice, vol. 12110. Addison-Wesley Professional, Reading (1996)
20. Pérez, P., Gangnet, M., Blake, A.: Poisson image editing. ACM Trans. Graph. (TOG) **22**, 313–318. ACM (2003)
21. Paysan, P., Knothe, R., Amberg, B., Romdhani, S., Vetter, T.: A 3D face model for pose and illumination invariant face recognition. In: Sixth IEEE International Conference on Advanced Video and Signal Based Surveillance, AVSS 2009, pp. 296–301. IEEE (2009)
22. Chen, D., Cao, X., Wen, F., Sun, J.: Blessing of dimensionality: high-dimensional feature and its efficient compression for face verification. In: Proceedings of the IEEE Conference on Computer Vision and Pattern Recognition, pp. 3025–3032 (2013)
23. Cao, Q., Ying, Y., Li, P.: Similarity metric learning for face recognition. In: Proceedings of the IEEE International Conference on Computer Vision, pp. 2408–2415 (2013)
24. Wolf, L., Hassner, T., Taigman, Y.: Similarity scores based on background samples. In: Zha, H., Taniguchi, R., Maybank, S. (eds.) ACCV 2009. LNCS, vol. 5995, pp. 88–97. Springer, Heidelberg (2010). doi:10.1007/978-3-642-12304-7_9
25. Huang, G., Mattar, M., Lee, H., Learned-Miller, E.G.: Learning to align from scratch. In: Advances in Neural Information Processing Systems, pp. 764–772 (2012)
26. Deng, W., Hu, J., Zhou, X., Guo, J.: Equidistant prototypes embedding for single sample based face recognition with generic learning and incremental learning. Pattern Recogn. **47**, 3738–3749 (2014)
27. Li, S., Liu, X., Chai, X., Zhang, H., Lao, S., Shan, S.: Morphable displacement field based image matching for face recognition across pose. In: Fitzgibbon, A., Lazebnik, S., Perona, P., Sato, Y., Schmid, C. (eds.) ECCV 2012. LNCS, vol. 7572, pp. 102–115. Springer, Heidelberg (2012). doi:10.1007/978-3-642-33718-5_8
28. Du, S., Ward, R.: Wavelet-based illumination normalization for face recognition. In: IEEE International Conference on Image Processing, ICIP 2005, vol. 2, p. II-954. IEEE (2005)
29. Chen, W., Er, M.J., Wu, S.: Illumination compensation and normalization for robust face recognition using discrete cosine transform in logarithm domain. IEEE Trans. Syst. Man Cybern. Part B Cybern. **36**, 458–466 (2006)

Local Fractional Order Derivative Vector Quantization Pattern for Face Recognition

Jing Li, Nong Sang[✉], and Changxin Gao

National Key Laboratory of Science and Technology
on Multi-Spectral Information Processing, School of Automation,
Huazhong University of Science and Technology, Wuhan 430074, China
verylijing@163.com, {nsang,cgao}@hust.edu.cn

Abstract. Previous works have shown that fractional order derivative can give a better image description compared with conventional integral one in applications of edge detection, image segmentation, image restoration, and so on. Motivated by this conclusion, in this paper, we propose a novel local image descriptor, local fractional order derivative vector quantization pattern (fVQP), based on image local directional fractional order derivative feature vector and vector quantization method for face recognition. Compared with image integral order derivative information based local binary pattern (LBP), local derivative pattern (LDP) and local directional derivative pattern (LDDP), our fVQP image descriptor has the advantages of better image recognition performance and robust to noise. Extensive experimental results conducted on four benchmark face databases demonstrate the superior performance of our fVQP compared with existing state-of-the-art descriptors for face recognition in terms of recognition rate.

1 Introduction

Being a fundamental issue in image processing, pattern recognition and computer vision, face recognition plays a significant role in a wide range of applications. An extensive and systematic face recognition methods summarization can be found in [1]. Among various face recognition methods, local image descriptor based methods have been proven successful over the past few years due to its advantages of discriminability and robustness to various variations.

Local binary pattern (LBP) [2] is one of the most representative local descriptors, which characterizes the spatial structure of a local image texture by thresholding a 3×3 square neighborhood with the value of the center pixel intensity and utilizing the sign information to form its LBP coding value. Due to its impressive computational efficiency and good texture discriminative property, LBP has gained considerable attention.

To further improve its performance, many variants of LBP were explored recently. Heikkilä et al. [3] proposed centre symmetric LBP (CS-LBP), which is different to LBP in that it compares centre symmetric pairs of pixels against a centre pixel, rather than comparing each pixel with the centre. This halves

© Springer International Publishing AG 2017
S.-H. Lai et al. (Eds.): ACCV 2016, Part III, LNCS 10113, pp. 234–247, 2017.
DOI: 10.1007/978-3-319-54187-7_16

the number of bits of binary patterns for the same number of neighbours. Guo et al. [4] proposed the completed LBP (CLBP), which include the information contained in the center pixel and magnitudes of local difference between center and neighboring pixels as complementary to the signs used by the original LBP. Considering that high frequency texture regions have high higher variances and do more contribution to discriminate image, Guo et al. [5] proposed variance weighted LBP (LBPV) histogram image representation method, rather than quantized the variance into N bins. Tan and Triggs [6] proposed local ternary pattern (LTP), which quantizes the image pixel gray-level differences between center and neighbor pixels into a ternary value instead of a binary one. Pan et al. [7] proposed local vector quantization patterns (LVQP), which improves the LBP by replacing the local difference vector binary scalar quantization with vector quantization method, resulting in higher discriminative power as well as better robustness to noise.

Note that LBP and its above-mentioned variants are in essence the encoding of image first-order derivative, which are just forward differences of pixel intensities around a center pixel. Considering that the high order derivative can capture more detailed discriminative information, some works have explored high order derivative based LBP. Zhang et al. [8] proposed local derivative pattern (LDP) for image representation. The m-th order code of LDP is generated by encoding the variations of the image $(m-1)$-th order derivatives. Because there are a lot of pre-specific directions that can be used to estimate the variations of the $(m-1)$-th order local derivative, LDP may result in too many codes for each pixel. Furthermore, it is impossible for LDP to address the rotation invariance issue. Guo et al. [9] proposed the local directional derivative pattern (LDDP), which encode high order directional derivative feature along multi-direction around center pixel, for texture classification. Yuan [10] theoretically analyzed the original LBP from the Taylor expansion of a signal along a direction and proposed high order derivative local binary pattern (DLBP), which is similar to the LDDP. And then by applying the circular shift sub-uniform and scale space operation let the DLBP descriptor has rotation and scale invariance capability for texture classification.

As a generalization of the integral order derivative, fractional order derivative has played a very important role in various physical sciences fields in the last decades, and image process based on image fractional order derivative operation has been succeed used for edge detection [11,12], image segmentation [13,14], image restoration [15,16], fingerprint identification [17] and face recognition [18,19]. Compared with integral order derivative, fractional order derivative has the following attributes or advantages [20]: (1) The integral order derivative of direct current or low-frequency signal is usually zero, while that of fractional one is nonzero, which may carries discriminant information for image representation; (2) Fractional order derivative of image has special Mach effect, and its antagonism characteristics lead it to have a special bionic vision receptive field model; (3) Fractional order derivative is an effective method for nonlinearly enhancing the complex texture detail feature.

Image fractional order derivative can furthest preserve the low-frequency contour feature, while nonlinearly strengthen high-frequency marginal information that gray-level changes greatly, as well as enhance texture details that gray-level does not change evidently.

In this paper, we propose a new image descriptor, local fractional order derivative vector quantization pattern (fVQP), for face image recognition. In summary, the novelty of this paper is twofold: (1) By introducing the advantages of image directional fractional order derivative information, our fVQP is a improvement of conventional LBP and its variants and thus it can give better image representation compared with all of them. (2) Compared with scalar quantization, our applied vector quantization method not only preserves sign and the magnitude information of the local fractional order derivative feature vector simultaneously but also robust to noise. Experimental results on the AR, Extended Yale B, CMU PIE and FERET face image databases demonstrate the effectiveness of our proposed image descriptor compared with existing state-of-the-art methods.

The rest of this paper is organized as follows. In Sect. 2, we briefly review the original LBP and its two improvements based on high order derivative information, i.e. LDP and LDDP. In Sect. 3, we describe our proposed fVQP descriptor based image representation and classification method in detail. Section 4 presents the experimental results on four databases. Finally, we conclude the paper in Sect. 5.

2 Related Work

2.1 Local Binary Pattern

LBP is a gray-scale image descriptor which characterizes the spatial structure of the local image texture. Given a central pixel I_c in the image, the LBP coding number is computed by comparing its value with those of its neighborhoods:

$$LBP_{P,R}(I_c) = \sum_{p=0}^{P-1} S_1(i_p - i_c) \times 2^p \tag{1}$$

$$S_1(x) = \begin{cases} 1, & x \geq 0 \\ 0, & x < 0 \end{cases} \tag{2}$$

where i_c and i_p are the gray value of the central pixel I_c and its neighbors $I_p, p = 0, \ldots, P - 1$ respectively, P is the number of neighbors and R is the radius of the neighborhood. Because only the signs of the differences between the center pixel and its neighbors rather than their exact values are used to define the encoded pattern number, LBP can achieve gray-scale invariance.

2.2   Local Derivative Pattern

Local derivative pattern (LDP) [8] descriptor is a high-order texture descriptor. In a general formulation, the LDP expands the LBP with the high-order derivative direction variations for extracting more detailed discriminative features. For encoding the m-th order LDP of pixel I_c, the $(m-1)$-th order derivatives along $0°$, $45°$, $90°$ and $135°$ directions are denoted as $i_{c,\theta}^{(m-1)}$, where $\theta = 0°$, $45°$, $90°$ and $135°$, which are pre-calculated separately based on

$$i_{c,0°}^{(m-1)} = i_{1,0°}^{(m-2)} - i_{c,0°}^{(m-2)} \tag{3}$$

$$i_{c,45°}^{(m-1)} = i_{1,45°}^{(m-2)} - i_{c,45°}^{(m-2)} \tag{4}$$

$$i_{c,90°}^{(m-1)} = i_{1,90°}^{(m-2)} - i_{c,90°}^{(m-2)} \tag{5}$$

$$i_{c,135°}^{(m-1)} = i_{1,135°}^{(m-2)} - i_{c,135°}^{(m-2)} \tag{6}$$

Then, the m-th order LDP in θ derivative direction at I_c, $LDP_\theta^m(I_c)$, is a binary string describing gradient trend changes in a local region of directional $(m-1)$-th order derivative $I^{(m-1)}$

$$LDP_\theta^{(m)}(I_c) = \sum_{p=0}^{P-1} S_2\left(i_{p,\theta}^{(m-1)}, i_{c,\theta}^{(m-1)}\right) \times 2^p \tag{7}$$

where the Function of $S_2(\cdot, \cdot)$ represents the spatial relationship between the center pixel and its neighborhoods in a given derivative direction. In other words, $S_2(\cdot, \cdot)$ determinates the types of 1D spatial relationship transitions and also labels the neighborhood pixels of the center pixel into a binary code as

$$S_2(x_1, x_2) = \begin{cases} 0, & x_1 \cdot x_2 \geq 0 \\ 1, & \text{otherwise} \end{cases} \tag{8}$$

Finally, the m-th order LDP of image pixel I_c, $LDP^{(m)}(I_c)$, is defined as the concatenation of the four directional LDPs

$$LDP^{(m)}(I_c) = \left\{ LDP_\theta^{(m)}(I_c) | \theta = 0°, 45°, 90°, 135° \right\} \tag{9}$$

2.3   Local Directional Derivative Pattern

Given a central pixel I_c and its P circularly and evenly spaced neighbors $I_p, p = 0, \ldots, P-1$ with radius R, $R+1$, $R+2$ and $R+3$. Its 2nd, 3rd, and 4th order derivatives along p direction could be computed as follows:

$$i_{p,R}^{(2)} = \left(i_p^{R+1} - i_p^R\right) - \left(i_p^R - i_c\right) = i_p^{R+1} - 2i_p^R + i_c \tag{10}$$

$$\begin{aligned} i_{p,R}^{(3)} &= [(i_p^{R+2} - i_p^{R+1}) - (i_p^{R+1} - i_p^R)] - [(i_p^{R+1} - i_p^R) - (i_p^R - i_c)] \\ &= i_p^{R+2} - 3i_p^{R+1} + 3i_p^R - i_c \end{aligned} \tag{11}$$

$$
\begin{aligned}
i_{p,R}^{(4)} &= \{[(i_p^{R+3} - i_p^{R+2}) - (i_p^{R+2} - i_p^{R+1})] - [(i_p^{R+2} - i_p^{R+1}) - (i_p^{R+1} - i_p^{R})]\} \\
&\quad - \{[(i_p^{R+2} - i_p^{R+1}) - (i_p^{R+1} - i_p^{R})] - [(i_p^{R+1} - i_p^{R}) - (i_p^{R} - i_c)]\} \\
&= i_p^{R+3} - 4i_p^{R+2} + 6i_p^{R+1} - 4i_p^{R} + i_c
\end{aligned}
$$

$$(12)$$

Based on the local binary coding strategy of LBP and high order derivative definition Eqs.(10)–(12), the local directional derivative pattern (LDDP) descriptor is defined as [9]:

$$
LDDP_{P,R}^{(m)}(I_c) = \sum_{p=0}^{P-1} S_1 \left(i_{p,R}^{(m)} \right) \times 2^p
$$

$$(13)$$

where $m = 1, 2, \ldots$, is order of derivative pattern and $S_1(x)$ is the binary function as defined in Eq.(2). It is clear that traditional $LBP_{P,R}$ is a special case of $LDDP_{P,R}^{(m)}$, when $m = 1$.

3 Local Fractional Order Derivative Vector Quantization Pattern

3.1 LBG Algorithm

Presented by Linde, Buzo and Gray in 1980, LBG (Linde-Buzo-Gray) algorithm is a widely used vector quantization (VQ) method in pattern recognition and image processing [21]. In general, the problem of VQ design can be stated as follows. Given a training feature vectors $S = \{x_i \in R^d | i = 1, \ldots, N\}$, and generates a representative codebook $C = \{c_j \in R^d | j = 1, \ldots, K\}$ with a specified codebook size $K \ll N$ as output such that the average distortion is minimized. The detailed procedure of the LBG algorithm for a codebook generation is given as follows:

Input: The training vector set $S = \{x_i \in R^d | i = 1, \ldots, N\}$

Initialization: Generate the initial codebook $C = \{c_j \in R^d | j = 1, \ldots, K\}$. Set the iteration counter $t = 0$ and the initial average distortion $D_1 = \infty$. Set the maximum iteration counter T and the distortion threshold ϵ.

- Step 1: For each training vector x_i, find its closest codeword in the current codebook. $x_i \in S_q$ if $\|x_i - c_q\|_2 \leq \|x_i - c_j\|_2$ for $j \neq q$.
- Step 2: Update the codebook $c_j, j = 1, \ldots, K$ by $c_j = \frac{1}{|S_j|} \sum_{x_i \in S_j} x_i$.
- Step 3: Calculate the current average distortion $D_t = \sum_{j=1}^{K} \sum_{x_i \in S_j} \|x_i - c_j\|_2$.
- Step 4: Compute the average distortion ratio ADR_t, where $ADR_t = \frac{D_{t-1} - D_t}{D_t}$
- Step 5: If $ADR_t \leq \varepsilon$ or $t == T$, stop the iterations. Otherwise, $t = t + 1$, return to step 1.

Output: The ultimate codebook $C = \{c_j \in R^d | j = 1, \ldots, K\}$

From the detailed steps of the LBG algorithm, we can observe that the performance of the final codebook depends on the initial codebook. In this work, the initial codebook is composed by randomly selected vectors from the training set.

3.2   Fractional Order Derivative

In the literature, there are several definitions of fractional differential and the Grunwald-Letnikov (G-L), Riemann-Liouville (R-L), and Caputo definitions are the most commonly used ones for the Euclidean measure. The R-L and Caputo definitions use the Cauchy equation, and they are computationally complex. The G-L definition, which expresses a function using a weighted sum around the function, is appropriate for image processing application.

Let $f(t)$ be an analogue signal which will be processed, and $D^k f(t)$ be its fractional differential of order k, where k is any real positive number. The G-L definition [22] is given by

$$D^k f(t) = \lim_{n \to \infty} \left(\frac{n}{t-c}\right)^k \sum_{m=0}^{n} (-1)^m \frac{\Gamma(k+1)}{m!\Gamma(k-m+1)} \times f\left(t - m\frac{t-c}{n}\right) \quad (14)$$

where $\Gamma(\cdot)$ denotes the Gamma Function defined as $\Gamma(\nu) = \int_0^\infty e^{-t}t^{\nu-1}dt$, and $n = t - c/h$. Let $h = 1$, then n should be $t - c$. By differentiating the Function $f(t)$ with respect to t, the difference expression of the fractional derivative of signal $f(t)$ can be expressed as

$$\frac{\partial^k f(t)}{\partial t^k} = f(t) + (-k)f(t-1) + \frac{(-k)(-k+1)}{2!}f(t-2)$$
$$+ \frac{(-k)(-k+1)(-k+2)}{3!}f(t-3) \quad (15)$$
$$\cdots + \frac{\Gamma(-k+1)}{n!\Gamma(-k+n+1)}f(t-n)$$

We can observe that only the first coefficient is constant "1", the other is nonzero. The non-zero coefficients are $1, -k, \frac{(-k)(-k+1)}{2!}, \frac{(-k)(-k+1)(-k+2)}{3!}, \cdots,$ $\frac{\Gamma(-k+1)}{n!\Gamma(-k+n+1)}$. It can be proven that the summation of the nonzero coefficients is not zero, which means that its response value in flat areas of the image is not zero and therefore it can be used to characterize different image feature.

3.3   Local Fractional Order Derivative Vector Quantization Pattern

In this subsection, we will describe our propose local fractional order derivative vector quantization pattern (fVQP) image descriptor. The straightforward idea is that we encode the completed local fractional order derivative feature vector, which is composed of center pixel gray-scale value and directional fractional derivative feature with different order parameter settings, by applying the LBG vector quantization method.

Completed Local Fractional Order Derivative Feature Vector. By adopting the first five items of the fractional derivative Function (15) as approximations for simple calculation, the fractional order derivative of image pixel I_c along the direction d is

$$i_d^{(k)} \approx i_c + (-k)i_c^R + \frac{(-k)(-k+1)}{2!}i_c^{R+1} + \frac{(-k)(-k+1)(-k+2)}{3!}i_c^{R+2}$$
$$+ \frac{(-k)(-k+1)(-k+2)(-k+3)}{4!}i_c^{R+3} \tag{16}$$

where i_c^R, i_c^{R+1}, i_c^{R+2} and i_c^{R+3} are the grayscale values of image pixel I_c^R, I_c^{R+1}, I_c^{R+2} and I_c^{R+3} respectively, which are circularly and evenly spaced neighbors around I_c with radius R, $R+1$, $R+2$, $R+3$ as shown in Fig. 1.

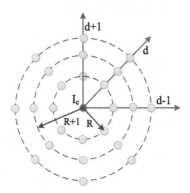

Fig. 1. Central pixel and its P circularly and evenly spaced neighbors pixel with radius R, $R+1$, $R+2$.

For a given image, 20-dimensional fractional order derivative feature vector is extracted with 20 equally spaced fractional order parameters between 0 and 2 (*i.e.*, $\Delta k = 0.1$) for each direction. And for each of the image pixel, D directions fractional order derivative are computed from its neighborhood. In addition, the grayscale value i_c of central pixel also contains effective discriminative information for local pattern encoding. Thus, the completed local fractional order derivative feature vector with the dimensionality of $20 \times D + 1$ for each image pixel is formed as

$$x = \begin{bmatrix} i_c & i_1^{(k1)} & i_2^{(k1)} & \cdots & i_D^{(k1)} & \cdots & i_1^{(k20)} & i_2^{(k20)} & \cdots & i_D^{(k20)} \end{bmatrix} \tag{17}$$

Feature Vector Quantization. The sign and the magnitude information of the local fractional order derivative vector are preserved simultaneously during vector quantization procedure. In the proposed method, we define a local pattern codebook by performing LBG algorithm to the completed local fractional order derivative feature vectors from different kinds of face images. For each completed local fractional order derivative feature vector, finds its best match codeword in the codebook and the best match codeword index is defined as the fVQP code of the central pixel in a local region. Specifically, we first calculates the Euclidean distance between each codeword and the feature vector $x = (x_1, \ldots, x_d)$ as follows

$$D_e(x, c_j) = \sum_l (x_l - c_{j,l})^2 \tag{18}$$

The feature vector x can be quantized into the codeword index j of the best match codeword c_j in the codebook $C = \{c_j \in R^d | j = 1, \ldots, K\}$.

$$fVQP_{R,D}^K(x) = \min_j \{D_e(x, c_j), j = 1, \ldots, K\} \tag{19}$$

Our fVQP is different from the LBP, LDDP and LVQP descriptors on the following aspects: (1) As a generalization of the integral-order derivative, fractional order derivative has better image representation ability. This is the main difference between fVQP and LVQP. (2) By setting multiple values for the order parameter of fractional derivative, our image representation method fused multi-view feature derived from derivative domain, while LBP, LDDP, LVQP and others are all one view image feature of derivative domain. (3) The sign and the magnitude information of the fractional order derivative vector are utilized simultaneously in fVQP. While in LBP and LDDP, the magnitude of difference vector is either ignored or processed separately from the sign information. (4) The scalar quantization process of LBP and LDDP is coarse and sensitive to noise, and the space complexity is 2^n when the dimension of feature vector is n. Whereas our applied Vector Quantization (VQ) method is more robust against noise or flat image areas and the space complexity of corresponding image representation is setable.

3.4 Image Representation and Classification

In this work, the face image is represented by the fVQP histogram, which describes the statistical distribution of the local micro-structures defined by fVQP descriptor. For such consideration, the fVQP coded image I_{fVQP} is first obtained by applying the fVQP operator to each pixel. Then, we divide the encoded image into w non-overlapping rectangular regions $R_0, R_1, \ldots, R_w$ to aggregate the spatial information to the descriptor, and then each of the region is used to compute a histogram of fVQP independently. Finally, all image regional fVQP histograms $H_0, H_1, \ldots, H_w$ are concatenated into one feature vector as the face image representation. Such representation contains information on pixel, regional and global levels.

In this work, we adopt support vector machine (SVM), which has been proved to be a powerful classifier for various classification tasks, as classifier to evaluate face recognition performance of different image descriptors. SVM first maps feature data into a higher dimensional space, and then finds the optimal separating hyperplane with maximal margin to split different classes.

Given N training samples $T = \{(x_i, y_i) | i = 1, \ldots, N\}$, where $x_i \in R^n$ and $y_i \in \{-1, 1\}$, the test feature data x is classified by function:

$$f(x) = sign(\sum_{i=1}^n \alpha_i y_i K(x_i, x) + b) \tag{20}$$

where n is the number of support vector, α_i are Lagrange multipliers of the dual optimization problem, b is the threshold of the hyperplane, and $K(\cdot, \cdot)$ is a kernel function, in this work, we chose the Radial Basis kernel Function.

SVM is binary classifier intrinsically, and the multi-class face recognition classification is achieved by utilizing one-against-rest technique, which constructs k classifiers. Grid-search during 10-fold cross-validation is carried out to select optimal SVM kernel parameters, the parameter setting producing best cross-validation performance is picked. We use the SVM implementation in the publicly available machine learning library LIBSVM [23].

4 Experiments

In this section, we will evaluate the effectiveness of fVQP and compare it with some state-of-the-art algorithms by experimenting on four large publicly available face databases.

4.1 Databases and Experimental Setup

AR face database [24] contains more than 4000 face images of 126 subjects (70 men and 56 women) with different facial expressions, illumination conditions, and occlusions. For each subject, 26 images were taken in two separate sessions (two weeks interval between the two sessions). A subset that contains 100 subjects (50 male and 50 female) is chosen in our experiments and the original images are normalized to 121×100 pixels.

Extended Yale B face database [25] consists of 2414 front-view face images of 38 individuals. There are about 64 images under different laboratory-controlled lighting conditions for each individual, where the horizontal light source angle varies from $-130°$ to $+130°$, while the vertical angle changes from $-40°$ to $+90°$. Original image size is 192×168, in our experiments, the cropped images of size 100×100 are used.

CMU PIE face database [26] contains 68 subjects with 41368 face images as a whole. Images of each person were taken across 13 different poses, under 43 different illumination conditions, and with 4 different expressions. We choose a subset of images from the five near frontal poses (C05, C07, C09, C27, C29) of each person. There are 170 images per subject with all kinds of illuminations and expressions. All images have been cropped and resized to be 64×64 pixels.

FERET face database [27] is a result of the FERET program, which was sponsored by the US Department of Defense through the DARPA Program. It has become a standard database for testing and evaluating state-of-the-art face recognition algorithms. The proposed method was tested on a subset of the FERET database. This subset includes 1,400 images of 200 individuals (each individual has seven images) and involves variations in facial expression, illumination, and pose. In the experiment, the facial portion of each original image was automatically cropped based on the location of the eyes, and the cropped

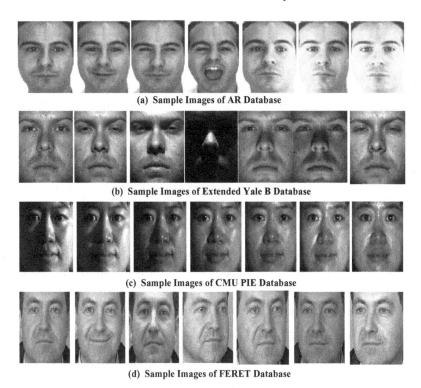

(a) Sample Images of AR Database

(b) Sample Images of Extended Yale B Database

(c) Sample Images of CMU PIE Database

(d) Sample Images of FERET Database

Fig. 2. Sample images of (a) AR face database, (b) the extended Yale B face database, (c) CMU PIE face database and (c) FERET face database.

images was resized to 80×80 pixels. The sample images of these four databases are shown in Fig. 2.

In the experiments, for all databases, we randomly choose half of the images per class for training and the rest ones for testing. In order to obtain reliable results, we repeat the experiment process by 10 times with different random selected training and testing images.

4.2   Parameter Evaluation

As discussed before, there are three parameters that should be selected to optimize the performance of fVQP based face representation. Of them, two parameters (R, D) are of directional fractional order derivative vector formatting, K determines the number of patterns that a face image pixel will be quantized into. In this section, we will study the influence of each parameter by varying their values one at a time while fixing the others. Experimental results on four database are shown in Table 1.

In addition to the optimal parameter for each database, we also observed from the Table that our fVQP descriptor has higher recognition rate than LVQP

Table 1. Recognition rate (%) of fVQP and LVQP versus different parameters setting

(R, D)	AR			Yale B			CMU PIE			FERET		
	(1, 8)	(2, 16)	(3, 24)	(1, 8)	(2, 16)	(3, 24)	(1, 8)	(2, 16)	(3, 24)	(1, 8)	(2, 16)	(3, 24)
$LVQP^{32}$	79.16	82.09	83.52	83.69	82.37	80.18	83.64	82.29	79.83	84.36	84.69	85.13
$fVQP^{32}$	83.23	82.53	83.61	84.10	85.29	81.23	85.44	83.17	81.91	86.54	87.46	86.63
$LVQP^{64}$	88.59	88.30	86.63	84.82	85.38	81.57	86.51	86.63	84.71	87.45	88.54	86.66
$fVQP^{64}$	91.23	91.16	90.57	85.79	87.23	84.69	88.44	89.17	86.23	89.54	90.49	89.57
$LVQP^{128}$	88.92	89.21	87.06	88.66	88.02	86.91	91.52	90.16	89.64	90.68	91.27	81.83
$fVQP^{128}$	92.38	90.55	91.18	93.66	90.43	89.16	93.03	92.59	91.92	92.71	94.03	93.69
$LVQP^{256}$	90.23	93.29	88.79	92.37	88.61	86.25	93.59	93.39	90.53	93.99	94.39	93.27
$fVQP^{256}$	93.19	96.46	92.33	96.86	92.77	89.98	95.86	94.73	93.69	96.25	97.25	95.66

under the same K, R and D. This is justified by the fact that fractional order derivative can better characterize image local patterns feature than integer one.

4.3   Comparisons with Other Descriptors

In this section, we will compare the face recognition performance of our fVQP method with those representative and state of the art methods, including LBP, LDP[3], LDDP[2], LDDP[3], LBPV, CLBP, Bag-of-Word (BoW) [28], histograms of fractional differential gradients (HFDG) [18] and principal patterns of fractional order differential gradients (PPFDG) [19]. For the LBP, LBPV, CLBP and LDP descriptors, we used the parameters $P = 8$ and $R = 1$. For the LDDP descriptor, we use a number of 8 sampling directions for radius $R = 1$. If a point is not in the image grid, its value is estimated by interpolation. For HFDG and PPFDG, unsigned orientation is evenly divided into 8 intervals. Experimental results of these methods in the four databases are illustrated in the Table 2.

Table 2. Recognition rate (%) of different descriptors on four databases

	AR	Yale B	CMUPIE	FERET
LBP	89.53	89.93	89.49	88.48
LDP[3]	93.02	94.64	92.11	90.12
LDDP[2]	94.29	93.85	93.02	92.31
LDDP[3]	91.66	93.93	92.49	92.51
LBPV	93.23	92.67	93.01	91.35
CLBP	92.51	93.49	91.18	87.71
BoW	94.35	93.78	92.26	92.46
HFDG	91.16	89.36	89.74	90.46
PPFDG	91.67	90.12	90.54	90.36
fVQP	**96.46**	**96.87**	**95.86**	**97.25**

From Table 2, we can observe that our proposed fVQP descriptor achieves the performance of 96.46%, 96.87%, 95.86% and 97.25% on the AR, extended Yale B, CMU PIE and FERET databases respectively, which are better than LBP and others.

4.4   Performance Comparison Under Noise

For face recognition tasks, noise is inevitable under uncontrolled circumstances. Therefore, it is essential for an image descriptor to be insensitive to noise. In this section, we will verify the tolerance of the proposed method to two common types of image noise, e.g., additive Gaussian white noise and multiplicative speckle

Table 3. Effects of the additive Gaussian white noise ($\sigma = 0.002$) on the recognition rate (%).

	AR	YaleB	CMU PIE	FERET
LBP	78.83	79.61	78.53	78.44
LDP$^{(3)}$	84.58	86.25	83.34	81.32
LDDP$^{(2)}$	87.10	86.36	85.57	84.66
LDDP$^{(3)}$	83.95	86.17	85.21	84.83
LBPV	85.57	85.50	85.89	83.85
CLBP	84.55	86.14	83.59	80.48
BoW	85.63	85.92	84.76	84.56
HFDG	81.40	80.10	80.23	80.76
PPFDG	81.77	80.16	80.99	81.22
fVQP	**90.31**	**90.61**	**89.01**	**90.99**

Table 4. Effects of the multiplicative speckle noise ($\sigma = 0.002$) on the recognition rate (%).

	AR	YaleB	CMU PIE	FERET
LBP	78.82	79.64	78.70	78.43
LDP$^{(3)}$	84.71	86.83	83.11	81.65
LDDP$^{(2)}$	86.97	86.46	85.66	84.91
LDDP$^{(3)}$	83.94	86.59	84.80	84.81
LBPV	85.88	85.35	85.57	84.32
CLBP	84.69	85.82	83.66	80.62
BoW	85.95	85.78	84.33	84.68
HFDG	82.12	79.71	80.58	81.58
PPFDG	82.56	80.36	81.24	80.36
fVQP	**90.06**	**90.18**	**89.28**	**90.18**

noise. In the experiment, face images without noise serve as training set, and noise polluted images are used for testing set. The resultant values are listed in Tables 3 and 4.

As shown in the Tables, different methods have considerable declines in terms of the recognition rate in the presence of noise compared to the case without it. Among them, we can observe that the LBP is more sensitive to noise than the other approaches. On the contrary, our proposed approach yield the highest tolerance to the both types of noise.

5 Conclusion

In this paper, we propose a robust and discriminative image descriptor, fVQP, to build image feature representation for face recognition. The merits of the proposed new image descriptor fVQP are: (1) It is simple to understand and easy to implement; (2) It introduces the local image directional fractional order derivative information which is the generalization of conventional integral order derivative and thus can extract more discriminant information for image representation; (3) It significantly reduces the dimension of face feature representation, with resulting improved computational efficiency without sacrificing recognition performance; (4) It applies the vector quantization based local feature encoding method which let it robust to noise and its space complexity setable. In our future work, we will apply our descriptor to large-scale image recognition to further validate its generalizability.

References

1. Tolba, A., El-baz, A., El-harby, A.: Face recognition: a literature review. Int. J. Sig. Process. **2**, 88–103 (2006)
2. Ojala, T., Pietikäinen, M., Mäenpää, T.: Multiresolution gray-scale and rotation invariant texture classification with local binary patterns. IEEE Trans. Pattern Anal. Mach. Intell. **24**, 971–987 (2002)
3. Heikkilä, M., Pietikäinen, M., Schmid, C.: Description of interest regions with local binary patterns. Pattern Recogn. **42**, 425–436 (2009)
4. Guo, Z., Zhang, L., Zhang, D.: A completed modeling of local binary pattern operator for texture classification. IEEE Trans. Image Process. **19**, 1657–1663 (2010)
5. Guo, Z., Zhang, L., Zhang, D.: Rotation invariant texture classification using LBP variance (LBPV) with global matching. Pattern Recogn. **43**, 706–719 (2010)
6. Tan, X., Triggs, B.: Enhanced local texture feature sets for face recognition under difficult lighting conditions. IEEE Trans. Image Process. **19**, 1635–1650 (2010)
7. Pan, Z., Fan, H., Zhang, L.: Texture classification using local pattern based on vector quantization. IEEE Trans. Image Process. **24**, 5379–5388 (2015)
8. Zhang, B., Gao, Y., Zhao, S., Liu, J.: Local derivative pattern versus local binary pattern: face recognition with high-order local pattern descriptor. IEEE Trans. Image Process. **19**, 533–544 (2010)
9. Guo, Z., Li, Q., You, J., Zhang, D., Liu, W.: Local directional derivative pattern for rotation invariant texture classification. Neural Comput. Appl. **21**, 1893–1904 (2012)

10. Yuan, F.: Rotation and scale invariant local binary pattern based on high order directional derivatives for texture classification. Digit. Sig. Proc. **26**, 142–152 (2014)
11. Mathieu, B., Melchior, P., Oustaloup, A., Ceyral, C.: Fractional differentiation for edge detection. Sig. Process. **83**, 2421–2432 (2003)
12. Yang, H., Ye, Y., Wang, D., Jiang, B.: A novel fractional-order signal processing based edge detection method. In: Proceedings of International Conference on Control Automation Robotics and Vision, pp. 1122–1127 (2010)
13. Nakib, A., Oulhadj, H., Siarry, P.: Fractional differentiation and non-pareto multi-objective optimization for image thresholding. Eng. Appl. Artif. Intell. **22**, 236–249 (2009)
14. Nakib, A., Oulhadj, H., Siarry, P.: A thresholding method based on two-dimensional fractional differentiation. Image Vis. Comput. **27**, 1343–1357 (2009)
15. Ye, Y., Pan, X., Wang, J.: Identification of blur parameters of motion blurred image using fractional order derivative. In: Proceedings of International Conference on Information Science, Signal Processing and their Applications, pp. 539–544 (2012)
16. Pan, X., Ye, Y., Wang, J.: Fractional directional derivative and identification of blur parameters of motion-blurred image. SIViP **8**, 565–576 (2014)
17. Chen, J., Huang, C., Du, Y., Lin, C.: Combining fractional-order edge detection and chaos synchronisation classifier for fingerprint identification. IET Image Proc. **8**, 354–362 (2014)
18. Yu, L., Ma, Y., Cao, Q.: Face recognition with histograms of fractional differential gradients. J. Electron. Imaging **23**, 033012–033012 (2014)
19. Yu, L., Cao, Q., Zhao, A.: Principal patterns of fractional-order differential gradients for face recognition. J. Electron. Imaging **24**, 013021–013021 (2015)
20. Pu, Y., Zhou, J., Yuan, X.: Fractional differential mask: a fractional differential-based approach for multiscale texture enhancement. IEEE Trans. Image Process. **19**, 491–511 (2010)
21. Linde, Y., Buzo, A., Gray, R.: An algorithm for vector quantizer design. IEEE Trans. Commun. **28**, 84–95 (1980)
22. Pu, Y., Wang, W., Zhou, J., Wang, Y., Jia, H.: Fractional differential approach to detecting textural features of digital image and its fractional differential filter implementation. Sci. China Ser. F Inf. Sci. **51**, 1319–1339 (2008)
23. Chang, C., Lin, C.J.: LIBSVM: a library for support vector machines. ACM Trans. Intell. Syst. Technol. **2**, 8016–8026 (2011)
24. Martinez, A., Benavente, R.: The AR face database. Technical report, Computer Vision Centre, Autonomous University of Barcelona (1998)
25. Lee, K., Ho, J., Kriegman, D.: Acquiring linear subspaces for face recognition under variable lighting. IEEE Trans. Pattern Anal. Mach. Intell. **27**, 684–698 (2005)
26. Sim, T., Baker, S., Bsat, M.: The CMU pose, illumination, and expression database. IEEE Trans. Pattern Anal. Mach. Intell. **25**, 1615–1618 (2003)
27. Phillips, P., Hyeonjoon, M., Rizvi, S., Rauss, P.: FERET evaluation methodology for face-recognition algorithms. IEEE Trans. Pattern Anal. Mach. Intell. **22**, 1090–1104 (2000)
28. Kotani, K., Qiu, C., Ohmi, T.: Face recognition using vector quantization histogram method. In: Proceedings of International Conference on Image Processing, pp. II-105-II-108 (2002)

Learning Facial Point Response for Alignment by Purely Convolutional Network

Zhenqi Xu$^{(\boxtimes)}$, Weihong Deng, and Jiani Hu

Beijing University of Posts and Telecommunications, Beijing, China
{xuzhenqi,whdeng,jnhu}@bupt.edu.cn

Abstract. Face alignment is important for most facial analysis system. Regression based methods directly map the input face to shape space, make them sensitive to the face bounding boxes. In this work, we aim at developing a model that can deal with complex non-linear variations and be invariant to face bounding box distributions, while preserving high alignment accuracy. We define response map for each facial point, which is a 2D probability map indicating the presence likelihood of facial point at the corresponding locations. We solve the face alignment problem by two-stage processes. The first stage is response mapping stage, we use deep Purely Convolutional Network (a specialised Convolutional Neural Network designed for face alignment problem) to reconstruct the response maps. The second stage is shape mapping stage, which processes the response maps to get locations of facial key points. We explored four functions for this stage: *max* function, *max + PCA*, *mean* function and *mean + PCA* function. Experiments done on 300 W dataset show that our algorithm outperforms state-of-the-art methods.

1 Introduction

Face alignment or facial landmarks detection is to automatically find salient facial points such as the center of the pupils, the tip of the nose, the corners of the mouth and so on. It has attracted much interest due to its importance for many facial analysis applications, *e.g.* feature based face recognition [1], head pose estimation [2], 3D face modelling [3] and facial expression analysis [4].

Even though many methods have been developed for face alignment recently, this problem still remains an open issue due to large pose variations and complex expression conditions. Besides, occlusion caused by eyeglasses or hair makes it even harder. It is difficult for a model to predict a location if the edges or corners are occluded.

To solve this problem, Cootes *et al.* [5] proposed Active Appearance Model (AAM), which models the statistics of shape, appearance and their correlation using Principle Component Analysis (PCA). The process of localizing facial points for a face image can be seen as searching a best shape in the projected subspace. AAM is linear model and thus may have limitations on faces with complex non-linear variations.

© Springer International Publishing AG 2017
S.-H. Lai et al. (Eds.): ACCV 2016, Part III, LNCS 10113, pp. 248–263, 2017.
DOI: 10.1007/978-3-319-54187-7_17

Another kind of methods involves several-stage cascaded regression. Supervised Descend Method (SDM) [6] starts from an initialized shape, and gradually regress the residual of the current shape and the target shape. One drawback of SDM is that it need a good initialization. To alleviate this drawback, Cao *et al.* [7] run SDM on multiple initialization and average their outputs. Zhu *et al.* [8] model the shape probability and find the initialization in the shape space. Recently, coarse-to-fine technique and deep learning are applied on the cascaded regression method [8–10]. Deep learning methods are more powerful than linear regression, Sun *et al.* [9] use deep Convolution Neural Networks (CNN) and Zhang *et al.* [10] use deep auto-encoders to solve this problem. Most regression based methods are directly mapping the face image into shape space, which makes them sensitive to the face bounding boxes, since the shape is highly relevant to the face bounding boxes. Yang *et al.* [11] have discussed this sensitivity in details.

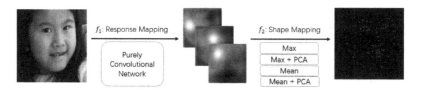

Fig. 1. Overall architecture of our methods. In response mapping stage, we use purely convolutional network to learn the response maps. In shape mapping stage, we explored three different functions: *max, max + PCA, mean* and *mean + PCA*.

In this paper, we aim at developing a model that can deal with complex nonlinear variations and be invariant to face bounding box distributions, while preserving high alignment precision. We define response map for each facial points, the response map is a 2D probability map indicating the presence likelihood of facial points at the corresponding locations. Instead of directly mapping the input face to shape, we divide the process into two stages, the first stage is response mapping stage, in which we get response maps given input face, the second stage is shape mapping stage, in which we process the response maps to get the locations of facial landmarks. The overall architecture can be seen in Fig. 1.

In response mapping stage, we deal with a task of mapping image (input face) to image (response maps) and can be solved by recent Fully Convolutional Network (FCN) [12]. The original FCN architecture is proposed for image segmentation and may not suit face alignment. Thus we make the following modification, firstly, we replace the max pooling by convolutional pooling, secondly, we do not use deconvolution to up-sample and other tricks used in [12] to boost the resolution, thirdly, we train our model from scratch with a KL divergence loss. Despite the non-linear neuron layers, the whole architecture is composed of only convolutional layers, thus we call our architecture Purely Convolutional Network (PCN).

Obtaining the response map from the PCN output, we explore four methods of shape mapping. The first is simply getting a predicted shape by taking the location of the maximum response for each point. This shape looks noisy, we train a PCA shape model from training labels and use this model to refine the shape. The others are to take the expectations of the response maps and also applying PCA on it.

We have done experiments on 300-W dataset [13], and achieve 5.43% mean error, outperform the state-of-the-art performance (5.76%), showing the effectiveness of our methods. Besides, our algorithm is robust to face bounding boxes while most regression based methods do not.

2   Face Alignment Formulation

Firstly, we define shape $s = [s_x, s_y] \in \mathbb{R}^{2K}$ as a vector of concatenating the horizontal and vertical coordinates of K points in a 2D space. The face alignment problem can be formulated as: giving an image $I \in R^{m*n}$ containing a human face, we need a transformation f to find the shape s (the locations of k salient facial points). The transformation:

$$f : I \mapsto s \tag{1}$$

The input image must contain a face, this can be achieved by face detection algorithms [14].

Cascaded Regression Methods: Many methods are based on cascaded regression. This kind of methods starts from a initialization shape s_0, and learn multiple regressors and new shapes iteratively:

$$f_i : \Phi(s_i, I) \mapsto \Delta s_i \tag{2}$$

$$s_i = s_{i\text{-}1} + \Delta s_i \tag{3}$$

where $\Phi(s_i, I)$ is a shape-index feature extractor, which extracts features from face image I based on the current shape s_i. Each stage, a regressor learns a model to predict the residual between the current shape and the label shape s_*. And the regressors are learned by solving:

$$\min_{f_i} ||s_* - s_{i\text{-}1} - \Delta s_i|| \tag{4}$$

SDM [6], ESR [7], LBF [15], Sun et al. [9] and CFAN [10] are all cascaded regression based methods. They differs in the choices of three components: regressors $\{f_i\}$, shape-index feature extractor $\Phi(s_i, I)$ and initialization shape s_0. Table 1 summarizes the five methods. Regression methods are usually sensitive to face bounding boxes, since the shape distribution changes according to face bounding boxes, this may cause two problems, the first is performance degradation due to the variation of face bounding boxes, the second is the difficulty of reimplementing existing algorithms since we need to get the same face bounding boxes distribution.

Response Mapping Methods: The cascaded regression methods map face image to shape directly. Different from them, response mapping methods first

Table 1. Summarises of different regression based methods. They differ from each other by three components: regressors, feature extractor and initialization shape.

Methods	Regressors ($\{f_i\}$)	Feature extractor (Φ)	Initialization shape (s_0)
SDM [6]	Linear regression	SIFT	Mean shape
ESR [7]	Random ferns	Pixel difference	Random
LBF [15]	Linear regression	Local binary features	Mean shape
Sun *et al.* [9]	CNN	Raw pixels	Initialized to the stage output
CFAN [10]	Deep auto-encoder	Raw pixels	Initialized to the stage output

learn the response maps for each facial point and then get a shape based on the response maps. The response mapping, denoted as $C_i \in R^{m*n}$, is the same size as the input face image, and is positive related to the presence probability of salient point at corresponding location. We formulate this process as:

$$f_1 : I \mapsto \{C_{i=1}^K\} \tag{5}$$

$$f_2 : \{C_{i=1}^K\} \mapsto s \tag{6}$$

where, K is the key point numbers.

In this work, we devise a new convolutional network for f_1, which we call it Purely Convolutional Network (PCN), since the architecture only contains convolutional layers except for neural layers. f_2 have multiple choices, a simple method is to get the location of maximum response, or get the expectation location of the response map. Here, we also train a PCA model to reduce the noise in the predicted shape.

Belhumeur *et al.* [16] use SVM to learn the facial point response, and then model the global shape using a consensus of exemplars.

3   Methods

In this section, we introduce our methods step by step, firstly we define the response mapping and the loss function for our deep learning architecture. Secondly, we go to the details of our Purely Convolutional Network, and explain our architecture design patterns. Finally, we show how we get the shape from response maps.

3.1   Reconstructing Response Maps

Our aim is to learn response maps for given input face image. To define a response map, we make the following assumptions:

(1) The response of the exact interested location (x^*, y^*) is highest.
(2) The farther a point (x, y) is away from (x^*, y^*), the lower the response is.
(3) The sum of response equals 1, so it forms a probability distribution.

It's reasonable for us to make these assumptions. Specifically, we choose the probability distribution to be two dimensional Gaussian (and will be normalized to ensure the third assumption).

$$L_i \sim \mathcal{N}((x_i^*, y_i^*), \sigma) \tag{7}$$

Our PCN network reconstructs L_i, we denote the output of PCN as $\tilde{C}_i$, then normalize it to form a probability C_i, this can be achieved by applying softmax function on $\tilde{C}_i$. Since L_i and C_i are both probability distributions, we choose the loss function to be KL divergence loss:

$$\arg\min \frac{1}{N} \sum_{i=1}^{N} \sum_{k=1}^{K} \sum_{j,c} L_k^i(j,c) \log \frac{L_k^i(j,c)}{C_k^i(j,c)} \tag{8}$$

The KL divergence loss is actually the same as cross entropy loss, the residual of these two losses is constant. Another choice of loss function is Euclidean distance between L_i and C_i. We prefer the former because the plateaus are less present in KL divergence loss than in Euclidean loss [17]. Besides, most elements of L_i and C_i are close to zero, makes the network using Euclidean loss easily stuck in bad local minima.

3.2   Purely Convolutional Network

Convolutional neural network (CNN) has been applied broadly on vision tasks, and shows success on image classification [18], object detection [19], face recognition [20] and so on. The vanilla CNN is composed of convolutional layers, pooling layers and fully connected layers. The convolutional layers are to extract dense and local feature maps, and pooling layers especially max pooling layers have two main contributions: firstly, it can reduce dimension, secondly, it is invariant small deformation which is important for classification problem. The fully connected layers accumulate all feature maps.

The vanilla CNN is suitable for classification problem. For face alignment problem, we aim at reconstructing response map for input face, thus we need to revise vanilla CNN. Following FCN [12], we replace the fully connected layers by convolutional layers to make this architecture having the ability of transforming image to image.

Another concern is to replace the common used max pooling with convolutional pooling. The max pooling is to return the maximum value of an area, while convolutional pooling is to assign a weight to each of this area and then accumulate them (see Fig. 2), the weight is shared in the same feature map. The difference of convolutional pooling layer and convolutional layer is that pooling are done on a single input channel (so the number of output channels is the same as the input). The convolutional pooling is better than max pooling for two reasons. One is that max pooling is invariant to small translations like shift and rotation, which is suitable for classification problems. While face alignment problem is translation specific, that is if we shift, scale or rotate the input face,

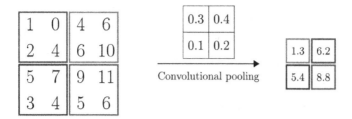

Fig. 2. Illustration of convolutional pooling operation. Convolutional pooling assign a weight to each node and then accumulate them. The weights are shared in the same channel.

the outcome should be change accordingly. Thus, max pooling is not suitable for face alignment. The other factor is that convolutional pooling is parametrized, the kernel is learned from data, thus will increase the representation power of our model.

Since our architecture comprises only convolutional layers (not considering neuron layers), we call it Purely Convolutional Network.

One way to boost the performance of CNN is increasing depth. Inspired by works of [21,22], we designed multiple networks, their depths are $\{11, 16, 26, 36, 41\}$. We find that the deeper, the better. Figure 3 illustrates the architectures of our Purely Convolutional Networks. For face alignment problem, increasing the depth has special meanings. To make discussion clear, we first define the receptive field sizes (rfs) for a node is that the size of area from the input which would influence its activation. The rfs in the last layer of our model increases as the depth increases. For deeper models, the neurons in the last layer see more input area then make a decision. That's also a reason why deep models are better for face alignment.

It is hard to train a deep model due to the notorious problem of vanishing and exploding gradients [17]. Several methods are developed to solve this problem, such as robust initialization [17,23], batch normalization [24], and deep residual learning [22]. We use two methods to train our deep model. Due to the reason that we increase the depth of our model gradually, we initialize the deep models using the last learned model. And the first one was initialized randomly. The size of parameters are not coincident in some layers, thus we use a similar way of Net2Net operation [25] to boost the convergence speed and performance of deep models. We also use the gradient clipping strategy [26] which can be seen as an adaptive learning rate methods to solve the gradient exploding.

3.3   Shape Mapping

In shape mapping stage, we get the predicted shape given response mappings. There are two simple operations to complete this task, one is to get the location of the max response for each point, the other is the get the expectation locations from the response maps.

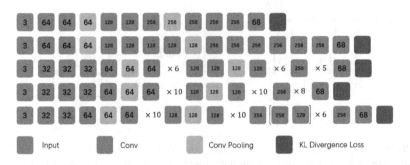

Fig. 3. Architecture of our PCN models. The depths from the top line to bottom line are $\{11, 16, 26, 36, 41\}$. Different colors of blocks mean different layers, and the number in the block means the output channels of the layer. All the filter sizes are set to be 3 with padding 1 to ensure output spatial size not changed. All convolutional (pooling) layers except the last one are followed by ReLU neuron layer.

Through the above operations, we can get a good facial point prediction for input face. But the shapes generated are noisy, see Fig. 4. This noisy comes two factors, the first is that we don't incorporate the global shape constraints into our PCN model explicitly, thus some points without clear edge or corner like the contour of the face is hard to localize. The second is the quantization noise due to pooling operation of our model, the size of response mapping is $1/4$ of the original input image.

3.4   PCA Noise Reduction

Inspired by AAM [5], we learn a PCA model from training shapes ($S = \{l_i \in \mathbb{R}^{2*N}\}$) to reduce the noise of predicted shape. Then, we process shapes from PCN output by projecting them into the PCA space and then reconstructing them. We discuss this process in details.

Training Set Alignment. Several factors can influence a shape, such as shift, rotation, scale, expression and pose. To make our global model to be invariant to scale, rotation, and shift, we align the training and testing set by firstly computing the mean shape $\bar{s}$ of training set, and then find a similarity transform for each training shape by solving:

$$\arg\min_{a,b,\theta} ||\bar{s} - T_{a,b,\theta}(s)|| \tag{9}$$

where $T_{a,b,\theta}(\cdot)$ is similarity transform parametrized by shift parameter $a \in \mathbb{R}^2$, scale parameter $b \in \mathbb{R}$ and rotation angle θ:

$$T_{a,b,\theta}(s) = b \times \begin{bmatrix} \cos\theta & \sin\theta \\ -\sin\theta & \cos\theta \end{bmatrix} \times \begin{bmatrix} s_x^T \\ s_y^T \end{bmatrix} + \begin{bmatrix} a_x \\ a_y \end{bmatrix} \tag{10}$$

We iteratively compute these two steps until the mean shape $\bar{s}$ unchanged.

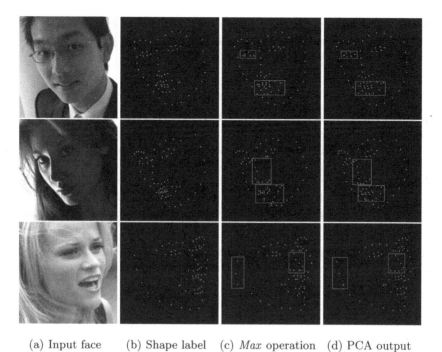

(a) Input face   (b) Shape label   (c) *Max* operation   (d) PCA output

Fig. 4. Effects of PCA shape model. The shape from *max* operation looks noisy (like the noise and mouth part). The shape is refined by PCA noise reduction. (Zoom in to see it clearly.)

PCA Model Learning. We run a PCA model on the aligned training set, ignore the eigenvectors with small eigenvalues and obtain mean shape $\bar{s}$ and a matrix $W \in \mathbb{R}^{2N*t}$, where t is a hyper-parameter representing the number of eigenvectors preserved and can be chosen using a validation set.

PCA Model Inference. Now we have the mean shape $\bar{s}$ of the aligned training set and projection matrix W, we can process the shape from PCN output s_{PCN} by aligning it using formula 9 and obtaining aligned shape s_a and similarity transform $T_{a,b,\theta}$. Then we reduce the noise by computing:

$$s_r = W * W^T * (s_a - \bar{s}) + \bar{s} \tag{11}$$

where s_r is the shape with less noise.

Finally, we re-transform s_r into the original shape space:

$$s_{PCA} = T^{-1}_{a,b,\theta}(s_r) \tag{12}$$

4   Experiments

In this section, we firstly describe the experimental settings and evaluation criteria. Then we discuss about the effectiveness of our architecture including the depth and convolution pooling. Thirdly, we discuss different shape mapping operations and give their performances. We also compare our methods with state-of-the-art face alignment algorithms.

4.1   Datasets

We evaluate our proposed methods on the challenging 300-W dataset [13]. The 300-W dataset consists of 3148 training faces collected from HELEN dataset [27], LFPW dataset [16] and AFW dataset [28]. The testing dataset consists of totally 689 faces coming from LFPW, HELEN and the challenging IBUG subset. Evaluations are done on three parts: common subset contains 554 images from LFPW and HELEN datasets, challenging subset contains 135 images from IBUG and full set. We mainly evaluate the 68 points, but our algorithm can extend to extract other numbers of points easily.

Evaluation Criteria. We use two commonly used criteria to evaluate our algorithm. One is the average error rate, computed as:

$$e = \frac{1}{N} \sum_{i=1}^{N} \frac{||s_i - s_i^*||_2}{68 * D_i} \tag{13}$$

where D_i is the distance between the centres of two eyes computed from label shape s^*. The other is cumulative error distribution (CED), which will treat samples with larger mean error as a failure:

$$CED(\epsilon) = \frac{1}{N} \sum_{i=1}^{N} \mathbb{1} \left(\frac{||s_i - s_i^*||_2}{68 * D_i} \le \epsilon \right) \tag{14}$$

where ϵ is the threshold variable and $\mathbb{1}(\cdot)$ is the indicator function.

Dataset Split. To choose hyper-parameters for our methods, we split the dataset into two part, one is used for training and the other is for validation. The validation set contains about 1/6 of the total images and are randomly chosen. We firstly choose the hyper-parameters for our model, then we fix the hyper-parameters and use the whole training set to retrain our model to fully use the training data.

Data Augmentation. We train our model from scratch without external data. To alleviate over-fitting, we expand the dataset by doing scale and rotation transformation. We first crop the face with bounding box enlarged by $\{1.2\times, 1.3\times, 1.4\times\}$, for each scale, we rotate it with angle $\{\pm30°, \pm25°, \pm20°, \pm15°, \pm10°, \pm5°, 0°\}$.

4.2   Training Purely Convolutional Network

Implementation Details. We implement our Purely Convolutional Network using deep learning toolbox Caffe [29]. We train our PCN model by stochastic gradient descent (SGD) with momentum 0.9 and learning rate 0.02. All the

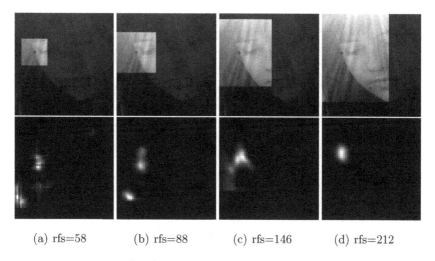

(a) rfs=58   (b) rfs=88   (c) rfs=146   (d) rfs=212

Fig. 5. Illustration of receptive field size (first row) and response map for the first key point (second row). As rfs increases, the model can see more area then make a decision, and thus will get more accurate response map.)

faces are resized to 224×224. The mini-batch size is set to be 32 and we use L2 regularization with decay parameter is 0.0001. We increase the depth from 11 to 41 gradually. The first PCN is trained from scratch, other PCNs are initialized from the model of last PCN to boost the training speed and to get a better result. Deep neural networks are very hard to learn, we use the gradient clipping technique described in [26], and we clip the length of gradient to 10. We don't tune these parameters much, and stop tuning when we get a reasonable convergence. We believe that carefully tune these hyper-parameter will get more precise predictions, but it's not our focus. Table 3 lists the basic results done on 300-W dataset.

Convolutional Pooling vs. Max Pooling. Table 2 shows performance comparison between convolutional pooling and max pooling. We can see that replacing max pooling with convolutional pooling will decrease the error rate by a large margin, which testifies the ability of convolutional pooling for alignment problem.

Depth, Receptive Fields and Performance. The second to fifth rows are our $\{11, 16, 26, 36, 41\}$ layers PCN models. As we increase the depth of our model, the mean error are decreased rapidly from 8.66 to 5.43 mean error, which shows deep models are far more powerful than shallow ones. Figure 5 shows the rfs and response maps, as rfs increases, the model can see more area then make a decision, which results in a better response map. The performance improvement is saturated when the rfs reaches the image size, continuing increasing the depth and rfs no longer gets lower error rate, as can be seen on Fig. 6.

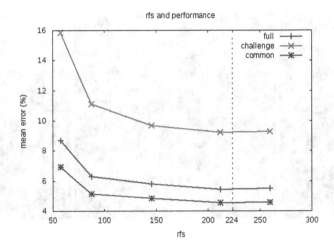

Fig. 6. The figure of rfs and performance. The gray dashed line shows the image size, the error rate decreases as the rfs increases, and saturate at the image size.

Table 2. Comparison between convolutional pooling and max pooling.

Depth	Mean error (full/challenge/common)	
	Max pooling	Convolutional pooling
11	9.77/17.96/7.78	8.66/15.87/6.91
16	6.49/11.39/5.30	6.28/11.08/5.12
26	5.88/9.69/4.95	5.78/9.67/4.83
36	5.57/9.15/4.69	5.43/9.19/4.52

Table 3. Results on 300-W dataset with different architectures.

Depth	rfs	Mean error(full/challenge/common)			
		Max	Max + PCA	Mean	Mean + PCA
11	58	10.57/18.16/8.71	9.30/16.48/7.55	8.91/16.26/7.12	8.66/15.87/6.91
16	88	7.57/12.84/6.28	6.88/11.95/5.64	6.39/11.29/5.19	6.28/11.08/5.12
26	146	6.77/10.91/5.76	6.11/10.03/5.15	5.86/9.85/4.89	5.78/9.67/4.83
36	212	6.21/10.16/5.25	5.71/9.58/4.77	5.46/9.27/4.53	5.43/9.19/4.52
41	260	6.23/10.24/5.25	5.77/9.79/4.79	5.50/9.33/4.57	5.49/9.27/4.57

4.3 Shape Mapping

Max vs. Mean. The mean shape mapping method is always better than max shape mapping method. Our thirty six layer model gives 6.21 mean error for max shape mapping, while 5.46 for mean shape mapping. The reasons come from two factors, the first is that max pooling only predict integral indexes. This will add quantization error on its performance. The other is that mean pooling uses more information in the response map to give a prediction. Since we initially reconstruct a gaussian response mapping, the mean shape mapping can be seen as approximating a gaussian distribution on predicted response map and then returning its mean.

PCA Noise Reduction. We treat our PCA model as noise reduction function. The noise comes from the implicit encoding of global shape from our PCN model and the quantization noise from max operation. From Table 3, PCA takes effects on max and mean shape mapping. And it takes more effect on max operation than on mean operation, it's reasonable since max operation generates more noise due to quantization error. From the depth perspective, the performance gain from PCA reduces as depth increases, this is because deep model has large receptive fields and contains more global information.

4.4   Compared with Other Methods

To illustrate the performance of our model, we make a comparison with other methods. The overall outcome can be seen in Table 4, we report our results obtained from our thirty six layer model and *max + PCA* shape mapping. We achieve 5.43 mean error, outperforms the state-of-the-art 5.76 from CFSS [8]. Figure 7 shows examples of SDM [6], CFAN [10], CFSS [8] and our method. The CED curves are given in Fig. 7.

Table 4. Mean error on 300-W datasets

Methods	Full	Common	Challenging
SDM [6]	7.52	5.60	15.4
RCPR [30]	8.35	6.18	17.26
CFAN [10]	7.69	5.50	16.78
ESR [7]	7.58	5.28	17.00
LBF [15]	6.32	4.95	11.98
CFSS [8]	5.76	4.73	9.98
PCN(ours)	**5.43**	**4.52**	**9.19**

4.5   Comparison with Regression Based Methods

The regression based methods are predicting the facial point locations directly, while our methods firstly learn the response map then apply shape mapping. One good reason for our methods is the response map will shift according to the faces, thus lead to theoretically shift-invariant (ignore the padding effect). To make discuss clear, we define shift-invariant face alignment problem below.

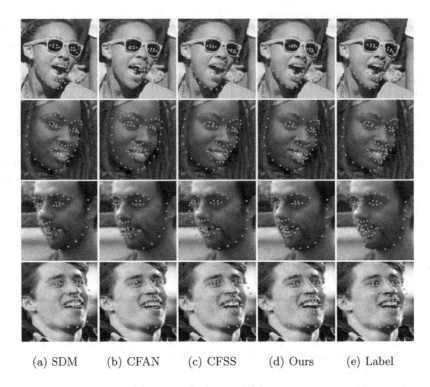

(a) SDM   (b) CFAN   (c) CFSS   (d) Ours   (e) Label

Fig. 7. Examples of SDM [6], CFAN [10], CFSS [8]. The images are from challenging ibug subset from 300 W dataset.

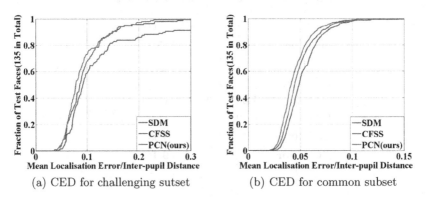

(a) CED for challenging sutset   (b) CED for common subset

Fig. 8. Comparisons of cumulative errors distributions(CED) curves.

Definition of Shift Invariant Face Alignment Problem. Suppose a function to predict a key point on given image, e.g. $f : I \in \mathbb{R}^{m*n} \mapsto (p_x, p_y) \in \mathbb{R}^2$, where I represents an image, and (p_x, p_y) is the location of key point. If $f(I(x, y)) = (p_x, p_y)$, then $f(I(x - k, y + j)) = (p_x + k, p_y - j)$. Thus, we can see that f is shift-invariant: the predicted key point is not changed when image shifts.

The cascaded regression methods deal with the shift translation by cascaded regression. The gap between predicted shape and the label shape is becoming smaller during this progress. But this methods can not ensure shift invariant, thus is not robust to face bounding box. Besides, the initialization shape is important for cascaded regression methods, if the initialization is far away from the label shape, it's hard to correct it later.

By learning the response mappings, our method is more robust to face bounding boxes. The responses can be seen as priori knowledge, and is more specific than shape priori. Since we can get shapes from responses, but we cannot get responses from shapes. Thus, learning the responses is more efficient than learning shapes directly. Besides, if we input a random generated image (not a face), the shape regression will still output a good shape, while our method will output a noise, we can easily tell this noise shape from good shape. So we can build a model to give a confidence for the shape prediction (which can not be done by shape regression method).

5   Conclusions and Future Works

In this work, we solve the face alignment problem by two stages. In the response mapping stage, we reconstruct the response maps for each key point for given face image. In the shape mapping stage, we predict the shape based on response maps. With $mean + PCA$ operation, our model achieve 4.53 mean error on 300 W dataset. Besides, our model is theoretically shift-invariant, which is important in real application and can process images at the speed of 250 fps on PC for 68 points prediction.

In the future, we will incorporate the global shape constraints into the PCN architecture to fully utilize the power of deep models. Besides, the PCN can also be applied to other location related tasks, like face detection.

Acknowledgments. This work was partially sponsored by supported by the NSFC (National Natural Science Foundation of China) under Grant No. 61375031, No. 61573068, No. 61471048, and No. 61273217, the Fundamental Research Funds for the Central Universities under Grant No. 2014ZD03-01, This work was also supported by Beijing Nova Program, CCF-Tencent Open Research Fund, and the Program for New Century Excellent Talents in University.

References

1. Zhao, W., Chellappa, R., Phillips, P.J., Rosenfeld, A.: Face recognition: a literature survey. ACM Comput. Surv. (CSUR) **35**, 399–458 (2003)
2. Murphy-Chutorian, E., Trivedi, M.M.: Head pose estimation in computer vision: a survey. IEEE Trans. Pattern Anal. Mach. Intell. **31**, 607–626 (2009)
3. Blanz, V., Vetter, T.: Face recognition based on fitting a 3D morphable model. IEEE Trans. Pattern Anal. Mach. Intell. **25**, 1063–1074 (2003)

4. Zafeiriou, L., Antonakos, E., Zafeiriou, S., Pantic, M.: Joint unsupervised face alignment and behaviour analysis. In: Fleet, D., Pajdla, T., Schiele, B., Tuytelaars, T. (eds.) ECCV 2014. LNCS, vol. 8692, pp. 167–183. Springer, Heidelberg (2014). doi:10.1007/978-3-319-10593-2_12

5. Cootes, T.F., Edwards, G.J., Taylor, C.J.: Active appearance models. IEEE Trans. Pattern Anal. Mach. Intell. **23**(6), 681–685 (2001)

6. Xiong, X., Torre, F.: Supervised descent method and its applications to face alignment. In: Proceedings of the IEEE Conference on Computer Vision and Pattern Recognition, pp. 532–539 (2013)

7. Cao, X., Wei, Y., Wen, F., Sun, J.: Face alignment by explicit shape regression. Int. J. Comput. Vision **107**, 177–190 (2014)

8. Zhu, S., Li, C., Change Loy, C., Tang, X.: Face alignment by coarse-to-fine shape searching. In: Proceedings of the IEEE Conference on Computer Vision and Pattern Recognition, pp. 4998–5006 (2015)

9. Sun, Y., Wang, X., Tang, X.: Deep convolutional network cascade for facial point detection. In: Proceedings of the IEEE Conference on Computer Vision and Pattern Recognition, pp. 3476–3483 (2013)

10. Zhang, J., Shan, S., Kan, M., Chen, X.: Coarse-to-fine auto-encoder networks (CFAN) for real-time face alignment. In: Fleet, D., Pajdla, T., Schiele, B., Tuytelaars, T. (eds.) ECCV 2014. LNCS, vol. 8690, pp. 1–16. Springer, Heidelberg (2014). doi:10.1007/978-3-319-10605-2_1

11. Yang, H., Jia, X., Loy, C.C., Robinson, P.: An empirical study of recent face alignment methods. arXiv preprint arXiv:1511.05049 (2015)

12. Long, J., Shelhamer, E., Darrell, T.: Fully convolutional networks for semantic segmentation. In: Proceedings of the IEEE Conference on Computer Vision and Pattern Recognition, pp. 3431–3440 (2015)

13. Sagonas, C., Tzimiropoulos, G., Zafeiriou, S., Pantic, M.: 300 faces in-the-wild challenge: the first facial landmark localization challenge. In: Proceedings of the IEEE International Conference on Computer Vision Workshops, pp. 397–403 (2013)

14. Viola, P., Jones, M.J.: Robust real-time face detection. Int. J. Comput. Vis. **57**, 137–154 (2004)

15. Ren, S., Cao, X., Wei, Y., Sun, J.: Face alignment at 3000 fps via regressing local binary features. In: Proceedings of the IEEE Conference on Computer Vision and Pattern Recognition, pp. 1685–1692 (2014)

16. Belhumeur, P.N., Jacobs, D.W., Kriegman, D.J., Kumar, N.: Localizing parts of faces using a consensus of exemplars. IEEE Trans. Pattern Anal. Mach. Intell. **35**, 2930–2940 (2013)

17. Glorot, X., Bengio, Y.: Understanding the difficulty of training deep feedforward neural networks. In: International Conference on Artificial Intelligence and Statistics, pp. 249–256 (2010)

18. Krizhevsky, A., Sutskever, I., Hinton, G.E.: Imagenet classification with deep convolutional neural networks. In: Advances in Neural Information Processing Systems, pp. 1097–1105 (2012)

19. Girshick, R.: Fast R-CNN. In: Proceedings of the IEEE International Conference on Computer Vision, pp. 1440–1448 (2015)

20. Sun, Y., Chen, Y., Wang, X., Tang, X.: Deep learning face representation by joint identification-verification. In: Advances in Neural Information Processing Systems, pp. 1988–1996 (2014)

21. Szegedy, C., Liu, W., Jia, Y., Sermanet, P., Reed, S., Anguelov, D., Erhan, D., Vanhoucke, V., Rabinovich, A.: Going deeper with convolutions. In: Proceedings of the IEEE Conference on Computer Vision and Pattern Recognition, pp. 1–9 (2015)
22. He, K., Zhang, X., Ren, S., Sun, J.: Deep residual learning for image recognition. arXiv preprint arXiv:1512.03385 (2015)
23. He, K., Zhang, X., Ren, S., Sun, J.: Delving deep into rectifiers: Surpassing human-level performance on imagenet classification. In: Proceedings of the IEEE International Conference on Computer Vision, pp. 1026–1034 (2015)
24. Ioffe, S., Szegedy, C.: Batch normalization: accelerating deep network training by reducing internal covariate shift. In: Proceedings of The 32nd International Conference on Machine Learning, pp. 448–456 (2015)
25. Chen, T., Goodfellow, I., Shlens, J.: Net2net: accelerating learning via knowledge transfer. arXiv preprint arXiv:1511.05641 (2015)
26. Pascanu, R., Mikolov, T., Bengio, Y.: On the difficulty of training recurrent neural networks. In: Proceedings of the 30th International Conference on Machine Learning (ICML-13), pp. 1310–1318 (2013)
27. Le, V., Brandt, J., Lin, Z., Bourdev, L., Huang, T.S.: Interactive facial feature localization. In: Fitzgibbon, A., Lazebnik, S., Perona, P., Sato, Y., Schmid, C. (eds.) ECCV 2012. LNCS, vol. 7574, pp. 679–692. Springer, Heidelberg (2012). doi:10.1007/978-3-642-33712-3_49
28. Zhu, X., Ramanan, D.: Face detection, pose estimation, and landmark localization in the wild. In: 2012 IEEE Conference on Computer Vision and Pattern Recognition (CVPR), pp. 2879–2886. IEEE (2012)
29. Jia, Y., Shelhamer, E., Donahue, J., Karayev, S., Long, J., Girshick, R., Guadarrama, S., Darrell, T.: Caffe: convolutional architecture for fast feature embedding. In: Proceedings of the ACM International Conference on Multimedia, pp. 675–678. ACM (2014)
30. Burgos-Artizzu, X., Perona, P., Dollár, P.: Robust face landmark estimation under occlusion. In: Proceedings of the IEEE International Conference on Computer Vision, pp. 1513–1520 (2013)

Random Forest with Suppressed Leaves for Hough Voting

Hui Liang[1($\boxtimes$)], Junhui Hou[1], Junsong Yuan[1], and Daniel Thalmann[2]

[1] School of Electrical and Electronics Engineering,
Nanyang Technological University, Singapore, Singapore
`hliang1@e.ntu.edu.sg`
[2] Institute for Media Innovation, Nanyang Technological University,
Singapore, Singapore

Abstract. Random forest based Hough-voting techniques have been widely used in a variety of computer vision problems. As an ensemble learning method, the voting weights of leaf nodes in random forest play critical role to generate reliable estimation result. We propose to improve Hough-voting with random forest via simultaneously optimizing the weights of leaf votes and pruning unreliable leaf nodes in the forest. After constructing the random forest, the weight assignment problem at each tree is formulated as a L0-regularized optimization problem, where unreliable leaf nodes with zero voting weights are suppressed and trees are pruned to ignore sub-trees that contain only suppressed leaves. We apply our proposed techniques to several regression and classification problems such as hand gesture recognition, head pose estimation and articulated pose estimation. The experimental results demonstrate that by suppressing unreliable leaf nodes, it not only improves prediction accuracy, but also reduces both prediction time cost and model complexity of the random forest.

1 Introduction

Hough-voting has shown its effectiveness for various classification and regression tasks in computer vision, in which a set of local features are extracted and their votes are cast to provide robust estimation of the target variable (such as object location, pose or category). Some of the commonly used Hough-voting techniques include Hough transform [1] and its generalized version [2], implicit shape model [3] and Hough forest [4]. Recently the combination of random forest and Hough-voting has also proven to be effective in various tasks, like gesture recognition [5], head pose estimation [6], facial landmark localization [7] and articulated pose estimation [8]. These methods can effectively aggregate predictions from spatially-distributed local features and have demonstrated good generalization ability and robustness to partially corrupted observations.

One important factor for Hough-voting to achieve good performance is to determine optimal weight of a vote casted by a local feature in prediction fusion. Due to the complexity of the estimation problem and noisy observations,

© Springer International Publishing AG 2017
S.-H. Lai et al. (Eds.): ACCV 2016, Part III, LNCS 10113, pp. 264–280, 2017.
DOI: 10.1007/978-3-319-54187-7_18

some local features are more informative than others during voting. The unequal contributions of densely sampled local features can be determined with prior spatial-selection, *e.g.*, by constraining them to edge pixels [1,2,9] or certain interest points [3], such that unselected pixels/local features are suppressed during voting. Some empirical strategies are used to represent the unequal contributions of local features, *e.g.*, assigning them with weights determined based on training variances [6,7] or excluding long-range votes [10], but they are not principled solutions. In previous work [11,12], the weights of the voting local features are automatically optimized on the training data in a max-margin framework, while they mainly target at object detection problems.

This paper investigates how to optimize and prune the voting weights of leaf nodes in random forest [13] for Hough voting based regression and classification. The local features are usually densely sampled from images to aggregate local evidence for robust voting. During training, these local features are annotated with ground truth poses or category labels. The forest is learned to make optimal predictions for these annotated local features. During testing, the forest retrieves the votes for the local features sampled from a testing image, and the fused prediction can be obtained via average or mode-seeking with these votes. Considering the densely sampled local features do not contribute equally to the fused prediction, it is non-optimal to treat the leaf nodes the same and assign all local features the same voting weight. Therefore, we derive a formulation to directly optimize the weights of leaf votes in each tree of the forest to improve fused prediction accuracy based on additive fusion of local features for voting. Second, we propose to reduce the model complexity of forest by suppressing leaf nodes with zero weight thus will not participate voting. The problem to fulfill both purposes is formulated as a L0-regularized optimization problem, which can be solved by the Augmented Lagrange Multiplier Method [14]. Particularly, to maintain independence among the trees in the random forest, we choose to

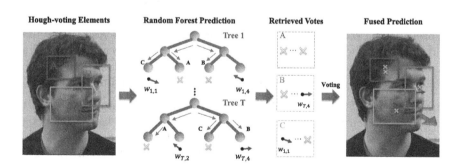

Fig. 1. Random forest with suppressed leaves in an exemplar application of head pose estimation. A, B and C indicate sampled voting local features. Blue arrows at the leaf nodes indicate vote for head orientation and $w_{a,b}$ indicate the corresponding optimized weights, where a and b are tree and leaf indices respectively. $\times$ indicates a suppressed leaf node, and thus patch A does not retrieve any effective votes. The big red arrow denotes fused head orientation prediction. (Color figure online)

optimize the leaf votes at each tree separately. Figure 1 illustrates the proposed optimized random forest in the application of head pose estimation.

The proposed algorithm is motivated by Max-Margin Hough transform [11] and Discriminative Hough forest [12], which associate a weight to the vote casted by each local feature during voting, but with several essential differences. First, unlike [11,12] in which the voting weights are learned to maximize object detection score on correct object locations over incorrect locations, we propose to directly learn the leaf voting weights in random forest to optimize per-image regression or classification accuracy, which improve classification and regression accuracy in Hough voting. Second, different from the max-margin formulation in [11,12], we enforce an additional L0-sparsity constraint on the leaf weights during optimization to suppress non-informative leaves, based on which the forest is further pruned to ignore sub-trees that contain only suppressed leaves. It produces a very compact forest with reduced memory usage and helps to filter out unreliable votes during Hough voting. Third, as model complexity of the forest decreases, the runtime prediction time cost is also largely reduced.

We adapt the proposed method to three different Hough-voting based classification and regression tasks including hand gesture recognition, 2D head pose estimation, and articulated pose estimation on several benchmark datasets [15–17]. The experimental results show that it improves the fused prediction accuracy considerably even with a large portion of suppressed leaf nodes, with the extra benefits of reduced memory usage and running time cost.

2   Literature Review

Hough-voting is widely used in computer vision. During training, local features sampled from training images are coded into a codebook. During testing, the local features sampled from a query image cast their votes for the fused prediction by feature matching in the learned codebook. As each leaf node in random forest can be regarded as a codebook entry to define a mapping between local features and probabilistic votes, the random forest and its variants are also widely used in Hough-voting due to their ability of fast and highly non-linear mapping from feature to Hough votes [5–8,10]. This section reviews previous Hough voting techniques and lays emphasis on relevant random forest based methods.

Hough Voting. The original Hough transform has been proposed in [1] to detect lines and curves. Its generalized version was proposed in [2] to detect arbitrary shapes by encoding the local parts as cookbook entries and use them to vote for the shape parameters. Compared to the original Hough transform, the generalized Hough transform is capable of detecting a variety of shapes in images. However, some local images parts may produce ambiguous predictions, e.g., homogeneous regions. To this end, in [1,2,9] only the edge pixels are allowed to cast votes to detect certain shapes. In [3] the Harris interest points are first detected over the whole image, and the local patches are only extracted around these points for voting-based object localization to avoid using local parts from homogeneous image regions. Nevertheless, the weights of the selected local patches are still determined via their number of activations in the codebook, which can be sub-optimal. To address this issue, in [11] the weights of the local parts are discriminatively optimized in a max-margin formulation to maximize certain object detection scores on the training images. Compared to the methods that rely on uniform weights or Naïve Bayesian weights, this method largely improves the detection performance without incurring computational overhead. A similar idea is proposed in [12], which associates the votes from training local patches with test local patches using a Hough forest and optimizes the weights of training local patches for voting via a max-margin formulation to improve object detection. However, these methods lay more emphasis on object detection, and are not primarily developed to improve regression or classification accuracy by fusing local predictions from Hough voting elements.

Random Forest. Random forests have been extensively studied in literature, and most of its variants aim to improve its performance by learning better split functions or probabilistic leaf votes [18–20]. However, their focus is on general machine learning problems and they do not differentiate between informative and non-informative Hough-voting samples during training. In contrast, in a Hough-voting framework, *e.g.*, body/hand/head pose estimation [6,8,10], face landmark localization [7] and gesture recognition [5], it is favorable that the forest is learned to stress on the informative local parts to produce an optimal fusion strategy. While uniform weights are sometimes assigned to leaf votes in the forest for fusion [5], some other methods have attempted empirical ways to determine the voting weights. For instance, in [10] it is found that leaf nodes containing long range votes are unreliable to predict the body joint positions and should be excluded from fusion. In [7] the leaf vote is assigned voting weight based on trace of the covariance matrix of node sample annotations, and an adaptive threshold is proposed to filter out long-range votes to predict facial point positions. In [6] the leaf nodes that have high training variances are filtered out based on an empirical variance threshold to improve accuracy. However, these strategies cannot guarantee the optimality of the fused prediction.

In [21] a similar quadric form is proposed to optimize the votes stored at leaf nodes in random forest. However, the proposed algorithm differs from their work in the essential spirits. In [21] the purpose is still to improve the prediction accuracy of each individual sample. This is different from the Hough-voting framework, in which we aim to improve the per-frame prediction accuracy by assuming that many voting elements do not have a positive impact on the fused predictions and their corresponding votes in the forest should be filtered out.

3   Hough-Voting with Random Forest

In the proposed algorithm, the random forest [13] is utilized for either classification or regression tasks via Hough-voting given a testing image I. Let Φ denote the value to be predicted, which could be a continuous variable for tasks like object pose estimation or a discrete category label for tasks like gesture recognition. The forest is an ensemble of T randomized trees, each of which is trained independently with a bootstrap training set. A randomized tree consists of a number of intermediate nodes and leaf nodes to produce a nonlinear mapping from feature space to target space. Each intermediate node of the tree has two children nodes and each leaf node stores one or more votes for $\boldsymbol{\Phi}$ and their associated voting weights. Let the set of votes stored at all the leaf nodes of tree t be $\{\hat{\boldsymbol{\Phi}}_{t,k}, w_{t,k}\}_{k=1}^{K_t}$, where $\hat{\boldsymbol{\Phi}}_{t,k}$ is a vote for $\boldsymbol{\Phi}$, $w_{t,k}$ is a value to represent the voting weight and K_t is the number of votes in tree t, which can be equal to or larger than the number of leaf nodes.

Following previous work [5,6,10] on Hough-voting with random forest, the forest is learned to map the features of spatially-distributed voting local parts to the probabilistic votes for $\boldsymbol{\Phi}$ by minimizing the prediction errors for all the training local parts. Let the feature vector of a local part be $\mathcal{F}$. Given a testing image I, a set of voting local parts $\{p_i\}_{i=1}^{N_s}$ are randomly sampled over the region of interest in I and their feature values $\{\mathcal{F}_i\}$ are calculated, which then cast their votes for $\boldsymbol{\Phi}$ independently. To this end, each local part p_i branches down each tree in the forest based on its feature value $\mathcal{F}_i$ until a leaf node is reached, and retrieves the votes stored at the reached leaf nodes. In this framework, we follow the commonly used additive fusion to aggregate these individual votes and the optimal fused prediction is obtained by:

$$\boldsymbol{\Phi}^* = \arg\max_{\boldsymbol{\Phi}} \sum_{i=1}^{N_s} P(\boldsymbol{\Phi}|p_i) = \arg\max_{\boldsymbol{\Phi}} \sum_{t=1}^{T}\sum_{i=1}^{N_s}\sum_{k=1}^{K_t} \delta_i^{t,k} w_{t,k} P(\boldsymbol{\Phi}|\hat{\boldsymbol{\Phi}}_{t,k}), \quad (1)$$

where $\delta_i^{t,k}$ is an indicator function which equals to 1 if a local part p_i retrieves a vote $\hat{\boldsymbol{\Phi}}_{t,k}$ and zero otherwise. As discussed in Sect. 1, the voting weights $w_{t,k}$ are critical for the fused prediction accuracy. The next section demonstrates how it can be optimized and pruned after the forest is training to improve per-image prediction performance in the voting framework in (1). The implementation details of forest training and prediction are presented in Sect. 3.2.

3.1   Leaf Selection and Weight Optimization

A dataset of images $\{I_m\}_{m=1}^{M}$ is needed to train the random forest, each of which is annotated with ground truth $\boldsymbol{\Phi}_m$. Following a general Hough-voting scheme, we randomly sample a fixed number of N training local parts $\{p_{i,m}\}_{i=1}^{N}$ from each image I_m and associate them with image features $\{\mathcal{F}_{i,m}\}$. These local parts are also associated with the ground truth annotations of the images from which they are sampled, i.e., $\{\boldsymbol{\Phi}_{i,m}\}_{i=1}^{N}$, with the detailed description in Sect. 3.2. Traditionally, the forest is trained to learn a nonlinear mapping between $\{\mathcal{F}_{i,m}\}_{i=1,m=1}^{N,M}$ and $\{\boldsymbol{\Phi}_{i,m}\}_{i=1,m=1}^{N,M}$ following the procedures in Sect. 3.2 to minimize the prediction error with the local parts. However, the learned forest will have to contain a large number of leaf nodes to encode the appearance variations of all the training local features, while many of them are not necessarily informative for voting, e.g., homogeneous local parts that tend to produce ambiguous predictions.

Also, minimization of prediction error on the local features cannot guarantee the optimality of the fused prediction via voting.

To maintain the independence of the trees in the forest, we consider only a single tree t next and ignore the suffix t in this section for succinctness. Based on formula (1), the contribution of each local feature to the fused prediction is represented as a linear weighting coefficient w_k associated with the vote it retrieves, which also represents the discriminative power of the local feature. This suggests a way to gap the local-feature based training objective and the Hough-voting based testing objective by adjusting the leaf weights $\{w_k\}_{k=1}^K$. Let $\boldsymbol{w} = [w_1, ..., w_K]^T$ be the concatenated vector of them. Particularly, the purposes of the proposed leaf weight optimization and pruning algorithm are two-fold. On the one hand, the leaf weights are optimized to improve voting-based prediction accuracy, so that the votes corresponding to non-informative local features are assigned small weights and vice versa. On the other hand, a L0-sparsity constraint is enforced on $\boldsymbol{w}$ to reduce model complexity of the forest so that many leaf votes are suppressed. The problem can be formulated as:

$$\boldsymbol{w}^* = \arg \min_{\boldsymbol{w}} \sum_{m=1}^{M} E(\bar{\boldsymbol{\Phi}}_m, \boldsymbol{\Phi}_m)$$

$$s.t. \quad 0 \le \boldsymbol{w} \le u, \|\boldsymbol{w}\|_0 \le S, \tag{2}$$

where $\bar{\boldsymbol{\Phi}}_m$ is the per-image fused prediction, $\boldsymbol{\Phi}_m$ is the ground truth and E is their L2-norm error function. Especially, to handle classification problem in this formulation, E is calculated for the predicted probability histogram and ground truth histogram. $\|\cdot\|_0$ denotes the L0-norm and S is a scalar to control the number of nonzero elements in $\boldsymbol{w}$. Different sparsity ratios of the number of nonzero weights over the number of all weights S/K can adjust the model complexity of the forest. $0 \le \boldsymbol{w} \le u$ enforces all the leaf weights to be non-negative and upper-bounded with u to avoid overfitting.

Given the tree structure and the set of leaf votes $\{\hat{\boldsymbol{\Phi}}_k\}_{k=1}^K$ of tree learned in Sect. 3.2, we first define a per-image fused prediction for each training image based on the sampled local features for voting to approximate that in formula (1) but with only a single tree. As stated before, a fixed number of N local parts are randomly sampled from each training image I_m to construct each tree of the forest. After the tree structure is learned, each of the N local parts reaches one leaf node, and multiple local parts from one image may reach the same leaf node. The fused prediction for each training image can thus be calculated based on the votes retrieved by these N parts from the tree.

By keeping track of the destination leaf nodes of each training local parts, we can define an indicator function $\delta_{i,m}^k$, which equals to 1 if $p_{i,m}$ reaches $\hat{\boldsymbol{\Phi}}_k$ and zero otherwise. We further define a vector $\boldsymbol{q}_m = [q_{m,1}, ..., q_{m,K}]^T$, in which each element $q_{m,k} = \sum_{i=1}^N \delta_{i,m}^k$ represents the number of local parts that are from I_m and reach the k_{th} vote. $\boldsymbol{q}_m$ is normalized to sum to one. Let $\boldsymbol{\alpha} = [\hat{\boldsymbol{\Phi}}_1, ..., \hat{\boldsymbol{\Phi}}_K]^T$ be the concatenation of all leaf votes from a tree. For each training image I_m, its voting based fused prediction $\bar{\boldsymbol{\Phi}}_m$ is defined as the weighted sum of the votes retrieved by its local parts:

$$\bar{\boldsymbol{\Phi}}_m = \sum_{k=1}^{K} q_{m,k} w_k \hat{\boldsymbol{\Phi}}_k = (\boldsymbol{q}_m \circ \boldsymbol{\alpha})^T \boldsymbol{w}, \tag{3}$$

where $\circ$ denotes the Hadamard product. We define two auxiliary matrices $\boldsymbol{Q}, \boldsymbol{A} \in \mathbb{R}^{M \times K}$ and a vector $\boldsymbol{g} \in \mathbb{R}^M$, where $\boldsymbol{Q} = [\boldsymbol{q}_1, ..., \boldsymbol{q}_M]^T$ records the hits on the leaf votes for all training images, each row of which corresponds to one image. $\boldsymbol{A} = [\boldsymbol{\alpha}, ..., \boldsymbol{\alpha}]^T$ stacks vector $\boldsymbol{\alpha}$ for M repetitive rows. $\boldsymbol{g} = [\boldsymbol{\Phi}_1, ..., \boldsymbol{\Phi}_M]^T$ stores ground truths of training images. Let $\boldsymbol{A}_q = \boldsymbol{Q} \circ \boldsymbol{A}$. The prediction error over all training images is represented as $\sum_m E_m = \|\boldsymbol{A}_q \boldsymbol{w} - \boldsymbol{g}\|^2$. Following the augmented Lagrange multiplier method [14], we introduce an auxiliary vector $\boldsymbol{\beta}$ of the same size to $\boldsymbol{w}$, and convert (2) to the following problem:

$$\boldsymbol{w}^*, \boldsymbol{\beta}^* = \arg\min_{\boldsymbol{w}, \boldsymbol{\beta}} \|\boldsymbol{A}_q \boldsymbol{w} - \boldsymbol{g}\|_2^2 \tag{4}$$
$$s.t. \quad 0 \le \boldsymbol{w} \le u, \boldsymbol{w} = \boldsymbol{\beta}, \|\boldsymbol{\beta}\|_0 \le S.$$

Its augmented Lagrangian form is given by:

$$\boldsymbol{w}^*, \boldsymbol{\beta}^* = \arg\min_{\boldsymbol{w}, \boldsymbol{\beta}} \|\boldsymbol{A}_q \boldsymbol{w} - \boldsymbol{g}\|_2^2 + \boldsymbol{y}^T (\boldsymbol{w} - \boldsymbol{\beta}) + \frac{\rho}{2} \|\boldsymbol{w} - \boldsymbol{\beta}\|_2^2 \tag{5}$$
$$s.t. \quad 0 \le \boldsymbol{w} \le u, \|\boldsymbol{\beta}\|_0 \le S,$$

where $\boldsymbol{y}$ is a Lagrange multiplier and ρ is a positive regularization parameter. The solution of (5) can be obtained by solving the following sub-problems for $\boldsymbol{w}^*$ and $\boldsymbol{\beta}^*$ separately in an iterative manner until convergence. To start the iterative optimization, ρ is initialized with a small value and gradually increased to enforce increasingly strong consistency between $\boldsymbol{w}$ and $\boldsymbol{\beta}$. In each iteration, $\boldsymbol{\beta}$ is first fixed to convert (5) to:

$$\boldsymbol{w}^* = \arg\min_{\boldsymbol{w}} \|\boldsymbol{A}_q \boldsymbol{w} - \boldsymbol{g}\|_2^2 + \boldsymbol{y}^T \boldsymbol{w} + \frac{\rho}{2} \boldsymbol{w}^T \boldsymbol{w} - \rho \boldsymbol{\beta}^T \boldsymbol{w}$$
$$= \arg\min_{\boldsymbol{w}} \boldsymbol{w}^T (\boldsymbol{A}_q^T \boldsymbol{A}_q + \frac{\rho}{2} I) \boldsymbol{w} + (\boldsymbol{y}^T - 2\boldsymbol{g}^T \boldsymbol{A}_q - \rho \boldsymbol{\beta}^T) \boldsymbol{w} \tag{6}$$
$$s.t. \quad 0 \le \boldsymbol{w} \le u,$$

which is a standard linearly-constrained quadratic programming problem. The matrix $\boldsymbol{A}_q^T \boldsymbol{A}_q$ can be huge in the applications in this paper, *i.e.* at the magnitude near $10^4 \times 10^4$. However, as the number of training local parts per image is relatively small, and some local parts from the same image may go to the same leaf node, $\boldsymbol{A}_q$ is quite sparse. Thus we utilize the interior-point-convex quadric programming algorithm [22] to solve (6). Next we fix $\boldsymbol{w}$ to solve for $\boldsymbol{\beta}^*$ by converting (5) to:

$$\boldsymbol{\beta}^* = \arg\min_{\boldsymbol{\beta}} -\boldsymbol{y}^T \boldsymbol{\beta} + \frac{\rho}{2} \|\boldsymbol{w} - \boldsymbol{\beta}\|_2^2$$
$$= \arg\min_{\boldsymbol{\beta}} \frac{\rho}{2} \left\| \boldsymbol{w} + \frac{\boldsymbol{y}}{\rho} - \boldsymbol{\beta} \right\|_2^2 \tag{7}$$
$$s.t. \|\boldsymbol{\beta}\|_0 \le S.$$

This problem can be very efficiently solved by element-wise comparison in $\boldsymbol{\beta}$. Finally, we update $\boldsymbol{y}$ and ρ:

$$\boldsymbol{y} \leftarrow \boldsymbol{y} + \rho \times (\boldsymbol{w} - \boldsymbol{\beta})$$
$$\rho \leftarrow \rho\alpha, \tag{8}$$

where $\alpha > 1$ gradually improves the convergence rate. The algorithm stops iteration when $\|\boldsymbol{w} - \boldsymbol{\beta}\|_\infty$ is less than a threshold or a maximum number of iterations is reached. In general, the algorithm takes about 20–30 iterations to converge in the experiments in Sect. 4.

Back pruning of random forest: with weights learned via formula (2), many leaf votes are suppressed to have zero weights, which are non-informative in prediction. Here we define a leaf node to be completely suppressed if the weights of all its votes are zero. For instance, when the forest is trained to predict multiple independent objectives simultaneously, e.g. two Euler angles for head pose estimation, there could be two weights for each leaf vote for these two objectives. In the case of classification, there is only one weight for the class labels in one leaf node. If certain sub-tree in a tree of the forest contains only completely suppressed leaves, it can be removed from the forest to reduce model complexity. This can be efficiently implemented as a post-order tree traversal algorithm. Starting from the root of each tree, each node will check whether all the sub-trees of its child nodes are suppressed in a recursive manner. If yes, the sub-tree with root at this node will be labeled as suppressed, and vice versa.

3.2 Details of Forest Learning and Prediction

A forest is learned with a set of feature and annotation pairs $\{\mathcal{F}_{i,m}, \boldsymbol{\Phi}_{i,m}\}_{i=1,m=1}^{N,M}$. Despite that $\boldsymbol{\Phi}$ can be quite a general term, we are particularly interested its several forms including discrete category labels, Euler angles and 2D/3D positions, which are widely adopted in computer vision problems. Following previous work, the annotation $\boldsymbol{\Phi}_{i,m}$ of a local feature equals to $\boldsymbol{\Phi}_m$ for discrete category labels and Euler angles, and equal to the offset between the center point of $p_{i,m}$ and the ground truth position $\boldsymbol{\Phi}_m$ for 2D/3D positions. To learn the tree structure, each tree is initialized with an empty root node. Starting from the root node, the training samples are split into two subsets recursively to reduce the prediction errors at the child nodes. At the non-leaf nodes, a set of candidate split functions $\{\psi\}$ are randomly generated as the proposals for node splitting, which can take the form $\mathcal{F}_b \leq \tau$, where $\mathcal{F}_b$ is a randomly selected dimension of $\mathcal{F}$ by sampling the feature dimension index, and τ is also a random threshold value to determine whether to branch to the left or right children. Let U be the set of node samples for splitting and $\{\boldsymbol{\Phi}_j | j \in U\}$ be the set of their annotations. The optimal split function is selected based on the following criterion:

$$\psi^* = \arg\max_\psi H(U) - \sum_{s \in \{l,r\}} \frac{|U_s(\psi)|}{|U|} H(U_s(\psi)), \tag{9}$$

where U_l and U_r are the two subsets of U split by ψ. The measure $H(U)$ evaluates the uncertainty of the sample annotations in U. For classification, $H(U)$ is defined as the entropy of the category label distribution of U. For regression of 2D/3D positions, we use the below measure of offset variances:

$$H(U) = \frac{1}{|U|} \sum_{j \in U} \left\| \boldsymbol{\Phi}_j - \bar{\boldsymbol{\Phi}}_j \right\|_2^2. \tag{10}$$

For Euler angles, as they follow circular distribution, we adopt the following measure to evaluate their variances:

$$H(U) = 1 - \sqrt{\left[\frac{\sum_{j \in U} \cos \boldsymbol{\Phi}_j}{|U|} \right]^2 + \left[\frac{\sum_{j \in U} \sin \boldsymbol{\Phi}_j}{|U|} \right]^2}. \tag{11}$$

The samples going to each branch are then used to construct a new tree node by either continuing the splitting procedure or ending up splitting to obtain a leaf node. This is done by checking whether certain stopping criteria are met, *e.g.* the sample annotations are pure enough, or the maximum depth is reached. For each leaf node, its vote can be represented by the average of the ground truth annotations stored at the sample set reaching it. For discrete category labels the vote is a histogram, each bin of which corresponds to the frequency of a category in the node samples. For 2D/3D position, the vote is defined as:

$$\hat{\boldsymbol{\Phi}} = \frac{1}{|U|} \sum_{j \in U} \boldsymbol{\Phi}_j. \tag{12}$$

For Euler angles, the vote is defined as:

$$\hat{\boldsymbol{\Phi}} = \text{atan2} \left[\frac{1}{|U|} \sum_{j \in U} \sin \boldsymbol{\Phi}_j, \frac{1}{|U|} \sum_{j \in U} \cos \boldsymbol{\Phi}_j \right]. \tag{13}$$

More refined votes can also be obtained via clustering node sample annotations [7,10]. As in Sect. 3.1, their weights in Hough voting is obtained via formula (2). If the weight is zero, we simply store an empty vote.

During testing, the optimal fused prediction is obtained by formula (1). For classification, $\boldsymbol{\Phi}$ is a category label and $P(\boldsymbol{\Phi}|\hat{\boldsymbol{\Phi}}_{t,k})$ is a probability distribution histogram stored at the leaf node following [5]. Thus the objective function in (1) is a weighted sum of histograms and the voting-based prediction can be obtained via finding the label with the maximum probability in the fused histogram. For regression, when $\boldsymbol{\Phi}$ is 2D or 3D position, we adopt the weighted Parzen density estimator with a Gaussian kernel to evaluate the objective function in (1) following [7,10], and define $P(\boldsymbol{\Phi}|\hat{\boldsymbol{\Phi}}_{t,k})$ as:

$$P(\boldsymbol{\Phi}|\hat{\boldsymbol{\Phi}}_{t,k}) = \mathcal{N}(\boldsymbol{\Phi}; \hat{\boldsymbol{\Phi}}_{t,k} + \boldsymbol{v}_i, \delta_v^2), \tag{14}$$

where δ_v^2 is an isotropic variance for all dimensions of $\boldsymbol{\Phi}$, and $\boldsymbol{v}_i$ is the center position of the local part as $\hat{\boldsymbol{\Phi}}_{t,k}$ denotes the offset between the target position and local part center. When $\boldsymbol{\Phi}$ is an Euler angle, we utilize 1D wrapped Gaussian kernel [23] to model $P(\boldsymbol{\Phi}|\boldsymbol{\Phi}_{ik})$ within the range $[0, 2\pi]$:

$$P(\boldsymbol{\Phi}|\boldsymbol{\Phi}_{ik}) = \sum_{z \in \mathbb{Z}} \mathcal{N}(\boldsymbol{\Phi} - 2z\pi; \hat{\boldsymbol{\Phi}}_{ik}, \delta_{\theta}^2), \qquad (15)$$

which is infinite wrappings of linear Gaussian within $[0, 2\pi]$, and δ_{θ}^2 is the variance. In practice the summation can be taken over $z \in [-2, 2]$ to approximate the above infinite summation [24]. Therefore, for both regression of 2D/3D position or Euler angle, the objective function in (1) is still sum of Gaussians, and the optimal solution can be efficiently obtained by the Mean-shift algorithm [25].

4   Experiments

In this section we apply random forest with Hough voting to three different problems: (a) hand gesture recognition for classification; (b) head pose estimation for Euler angle regression; (c) articulated hand pose estimation for 3D position regression. In order to better evaluate the effectiveness of the proposed leaf selection and weight optimization algorithm, we test its several variants by (a) assigning uniform weights (UW) to all leaf votes and (b) learning optimized leaf weights via L0-regularized optimization (L0) with different sparsity ratios S/K. All these methods are implemented in C++ and tested on a PC with Intel i7 4710HQ 2.5 GHz CPU without resorting to multi-threading. Their results are also compared to other state-of-the-arts tested on the same datasets.

4.1   Gesture Classification

This experiment is performed on NTU Hand Digits dataset [15], which consists of 1000 depth images of 10 hand digit gestures, *i.e.*, 0–9, from 10 subjects. In implementation the ground truth gesture label is represented as a 10-D vector, with all dimensions equal to 0 except the one corresponding to ground truth label equal to 1. Thus the per-image prediction error in formula (2) is defined as the L2-norm of the fused histogram and this 10-D vector. The measure function $H(U)$ for tree node splitting is taken as the gesture label entropy in the node sample U. Following [26], we perform a 10-fold leave-one-subject-out test in this experiment. The depth difference feature in [27] is adopted for gesture classification with random forest, which is fast to calculate. To train the forest, 100 voting-pixels are sampled from each image with their depth difference feature extracted to train each tree. The tree stops growing in conditions that its depth exceeds 20, the node samples contain only one gesture label or the node sample is less than 50. The number of trees in the forest is set as 4. As the depth difference feature is not rotation-invariant, we utilize the method in [28] to crop the hand region and find the 2D principal orientation of the forearm and rotate the hand region accordingly to perform rough 2D rotation-normalization for all training and testing images. During testing, 500 pixels are randomly sampled within the cropped hand region to vote for hand gesture, and each of them is passed to the random forest to retrieve gesture votes. Their votes are fused via formula (1) to get the gesture prediction of the input image.

Based on the above rotation normalization scheme (RN), we have tested the classification accuracy using random forest with uniform weights and L0-regularized weights learned via different sparsity ratios S/K within $[0.05, 1.0]$. Note that $S/K = 1.0$ means no sparsity constraint. The average number of leaf nodes per tree in the forest is 3678. We first analyze the impact of S/K on the system performance, including memory reduction, per-frame processing time cost and recognition accuracy, as illustrated in Fig. 2. The memory reduction is evaluated in terms of the suppression ratio of tree nodes, $i.e.$, the ratio of the number of suppressed nodes over the number of all tree nodes after forest pruning. The hand segmentation method in [28] incurs an overhead of 11.7 ms to process each frame, which is independent of subsequent gesture classification stage. Thus Fig. 2(b) only includes the time costs for random forest classification to better analyze the effectiveness of leaf suppression. We found that the leaf weights learned via a sparsity ratio S/K of 0.1 already outperforms uniform leaf weights in gesture recognition accuracy, $i.e.$, 97.9% vs 97.0%. At the same time, it reduces the model complexity of the random forest by 83.8% and reduces the classification time cost by 35.5%. Although a small value of S/K has benefit of reduction in both model complexity of random forest and classification time cost, Fig. 2(c) indicates that leaf sparsity cannot be excessively small, which may also suppress some useful voting elements, $e.g.$, $S/K = 0.05$ leads to relative worse prediction accuracy.

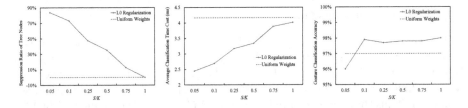

Fig. 2. Influence of sparsity ratio S/K for gesture classification. Left: (a) The suppression ratios of tree nodes. Middle: (b) Average gesture classification time costs. Right: (c) Gesture recognition accuracy.

We also implemented the shape classification forest in [5] for gesture recognition. To this end, we discard the rotation normalization scheme in both the forest training and test stages, and assign uniform weights to all leaf votes. It only achieves 90.0% recognition accuracy on this dataset. Figure 3 illustrates the confusion matrices for gesture recognition using [5], random forest with rotation normalization and uniform leaf weights, and random forest with rotation normalization and leaf weights learned via the proposed L-0 regularized optimization with sparsity ratio 0.1 respectively. In addition, we also compare the

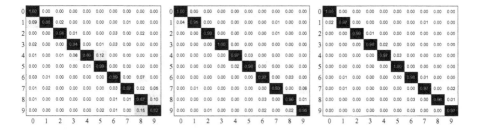

Fig. 3. Confusion matrices of gesture classification using random forest with (a) uniform weights without rotation normalization [5], (b) uniform weights with rotation normalization, and (c) weights obtained via L0-regularized optimization with $S/K = 0.1$ and rotation normalization.

Table 1. Comparison of recognition accuracy on NTU hand digits dataset [15].

Methods	Accuracy	Time cost
Random forest+UW w/o RN [5]	90.0%	14.5 ms
Random forest+UW+RN	97.0%	16.4 ms
Random forest+L0+RN, $S/K = 1.0$	**98.0%**	15.7 ms
Random forest+L0+RN, $S/K = 0.1$	97.9%	14.4 ms
Near-convex decomposition+FEMD [15]	93.9%	4.0 s
Thresholding decomposition+FEMD [15]	93.2%	75.0 ms
H3DF+SVM [26]	95.5%	N.A.
H3DF+SRC [29]	97.4%	N.A.
DCE [30]	97.5%	56.8 ms

proposed method to other state-of-the-art methods, including [15, 26, 29, 30] and the results are summarized in Table 1, which demonstrates that the proposed method is among the state-of-the-art. It is worth noting that [26, 29] adopt much more complex features including the 3D depth normal, while our method only relies on the coarse depth difference feature. Such fine features can be further incorporated into our method to improve gesture recognition. Also, it is worth noting that the most time consuming part of this gesture recognition pipeline is hand segmentation, which can be largely reduced if the body skeleton information is available, *e.g.*, from Kinect SDK.

4.2   Head Pose Estimation

The head pose estimation experiments are performed on Pointing'04 dataset [16], which includes 2790 RGB head images of 15 different subjects, and each image is annotated with bounding boxes and pitch/yaw Euler angles. Both pitch and yaw head rotations are within ±90°. We tested three variants of random forest for head pose estimation: (1) random forest without Hough voting (HV) [13];

(2) random forest with Hough voting using uniform leaf weights; (3) random forest with Hough voting using leaf weights learned via the proposed L0-regularized leaf weight optimization scheme 2 with different sparsity ratios. To implement (1), we crop the head images within annotated bounding boxes, resize them to 64×64 patches and extract HoG descriptors from them at multiple scales. The scales we used include cell size of 8, 16 and 32. The cell blocks are 2×2 and the orientation histograms have 9 bins. We concatenate the HoG features at different scales to form a 2124 dimensional feature vector, which are used for both forest training and testing. To implement (2) and (3), we randomly perturb the low-left and upper-right point of the annotated bounding boxes of human faces by 10 pixels to generate 20 voting image patches for each image. We then extract multiscale HoG features from each of these patches for forest training and testing. The head pose regression forests are trained as in Sect. 3 with prediction goal as two Euler angles, and the measure function $H(U)$ for node splitting is defined as the summation of formula (11) for the two Euler angles.

Performance of these methods are evaluated in terms of mean absolute error in degree between prediction and ground truth rotation angles based on 5-fold cross-validation as in [31,32]. The forest contains 20 trees, each tree stops growing if its depth exceeds 20 or the node samples are less than 10 during training. Similar to Sect. 4.1 we first evaluate the performance of random forest with Hough voting for head pose estimation with uniform weights and L0-regularized weights learned via different sparsity ratios S/K within $[0.05, 1.0]$. The average number of leaf nodes is 7170 when the forest is trained following above Hough voting strategy. Figure 4 illustrates memory reduction, per-frame processing time cost and head pose prediction error for different sparsity ratios. As the multiscale HoG feature needs to be prepared before head pose regression with random forest, this again incurs an overhead of 17.6 ms for each frame independent of different sparsity ratios. Thus Fig. 4 (b) only includes the time costs for head pose regression with random forest to evaluate the impact of different S/K. Again, we observe that the leaf weights learned via $S/K = 0.1$ already outperforms uniform leaf weights in head pose estimation, i.e., $5.2°$ vs $5.3°$ in prediction error, which also reduces the model complexity by 76.7% and reduces the regression time cost by 26.7%. With a larger sparsity ratio of 0.25, the pose estimation error is reduced to only $4.9°$, which still has the benefit of smaller model complexity and running time cost.

In Table 2 we compare the head pose prediction errors of random forest with Hough voting using uniform weights and optimized weights via L0-regularized optimization, random forest without Hough voting [13] and other recent methods tested on the Pointing'04 dataset including [31,32], which indicates the proposed method is among the state-of-the-arts. It can also be observed that Hough voting largely helps head pose estimation, and the leaf weights obtained via L0-regularized optimization further improve the estimation accuracy.

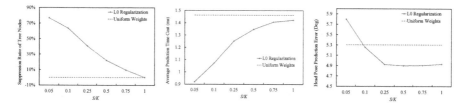

Fig. 4. Influence of sparsity ratio S/K for head pose regression. Left: (a) The suppression ratios of tree nodes. Middle: (b) Average head pose estimation time costs. Right: (c) Head pose estimation error in degrees.

Table 2. Comparison of head pose estimation error on Pointing'04 dataset [16].

Methods	Yaw ($°$)	Pitch ($°$)	Average ($°$)
Random forest w/o HV [13]	7.2	7.6	7.4
Random forest+HV+UW	4.9	5.7	5.3
Random forest+HV+L0, $S/K = 0.1$	4.7	5.7	5.2
Random forest+HV+L0, $S/K = 0.25$	4.4	5.4	**4.9**
Gourier et al. [16]	10.1	15.9	13.0
Class generative model [32]	5.9	6.7	6.3
Partial least squares [31]	11.3	10.5	10.9
Kernel partial least squares [31]	6.6	6.6	6.6

4.3 Articulated Pose Estimation

This experiment is performed for 3D position regression with the proposed algorithm. The MSRA hand pose dataset [17] is adopted for evaluation, which includes 2400 frames of various hand motion captured for six subjects, *i.e.*, 400 frames each. It includes substantial hand posture and scale variation of different subjects. Following [17], the pose estimation accuracy is evaluated in terms of the average error of six hand joints including five fingertips and the wrist, as they are the most flexible hand joints. The depth difference feature is still adopted for regression. We train the forest using images of five subjects and test it on the remaining one, and repeat the test for all subjects. During training, the measure function $H(U)$ for node splitting is defined as the summation of formula (10) for all the hand joints. The forest consists of 3 trees and each tree stops growing if its depth exceeds 15 or the node samples are less than 100. The average leaf number per tree is around $3.5\,K$. To compare to a standard random forest for articulated pose regression [10], we cluster the annotated hand poses of training samples for each hand joint at each leaf node, and refine the leaf vote for one hand joint to be the center of the largest cluster. Note that this experiment is to validate the effectiveness of the proposed method for 3D position regression, we thus predict each hand joint independently and do not consider hand pose constraints adopted in some forests specifically for hand pose estimation [33,34].

Table 3 presents the prediction errors of random forest with different leaf weights for each individual subject and their average. Note the results in [17] are based on model-based hand pose tracking by using temporal smoothing and ground truth pose is used to initialize the first frame during tracking. Therefore, their results are not directly comparable to ours. Again we see that optimized leaf weights via L0-regularized optimization achieves high accuracy compared to uniform weights, *i.e.*, reducing prediction error by 6.4% and pruning 57.3% of the tree nodes with $S/K = 0.2$. Besides, we implemented the regression forest in [10] which utilizes certain threshold to filter out long-range leaf votes. However, we tested it for different thresholds but did not find it improves the prediction accuracy on this dataset.

Table 3. Comparison of random forest with different leaf weights on MSRA hand tracking database [17]. Prediction errors are measured in millimeters.

Subject	1	2	3	4	5	6	Avg.
Random forest+UW	32.0	24.1	32.7	27.7	42.6	33.4	32.1
Random forest+L0, $S/K = 1.0$	29.4	22.0	31.3	25.9	38.9	31.7	29.9
Random forest+L0, $S/K = 0.2$	30.0	22.4	31.4	26.1	39.0	31.5	30.1

5   Conclusion

This paper presents an algorithm to optimize leaf weights and prune tree nodes in random forest to improve classification and regression accuracy during Hough voting. The method not only improves fused prediction accuracy, but also reduces model complexity of random forest and running time cost. In the experiments we show the method achieves good results for several different problems including hand gesture recognition, head orientation estimation and articulated pose estimation. Although the method is developed based on the baseline random forest [13], it is general enough to extend to other more advanced forests, such as [18,19], which is left as our future work.

Acknowledgement. This work is supported in part by Singapore Ministry of Education Academic Research Fund Tier 2 MOE2015-T2-2-114 and Tier 1 RG27/14.

References

1. Duda, R.O., Hart, P.E.: Use of the hough transformation to detect lines and curves in pictures. Commun. ACM **15**, 11–15 (1972)
2. Ballard, D.H.: Generalizing the hough transform to detect arbitrary shapes. Pattern Recogn. **13**, 111–122 (1981)
3. Leibe, B., Leonardis, A., Schiele, B.: Combined object categorization and segmentation with an implicit shape model. In: ECCV Workshop on Statistical Learning in Computer Vision, pp. 17–32 (2004)

4. Gall, J., Lempitsky, V.: Class-specific hough forests for object detection. In: IEEE Conference on Computer Vision and Pattern Recognition, pp. 1022–1029. IEEE (2009)
5. Keskin, C., Kıraç, F., Kara, Y.E., Akarun, L.: Hand pose estimation and hand shape classification using multi-layered randomized decision forests. In: Fitzgibbon, A., Lazebnik, S., Perona, P., Sato, Y., Schmid, C. (eds.) ECCV 2012. LNCS, vol. 7577, pp. 852–863. Springer, Heidelberg (2012). doi:10.1007/978-3-642-33783-3_61
6. Fanelli, G., Dantone, M., Gall, J., Fossati, A., Van Gool, L.: Random forests for real time 3D face analysis. Int. J. Comput. Vis. **101**, 437–458 (2013)
7. Yang, H., Patras, I.: Sieving regression forest votes for facial feature detection in the wild. In: IEEE International Conference on Computer Vision, pp. 1936–1943. IEEE (2013)
8. Xu, C., Cheng, L.: Efficient hand pose estimation from a single depth image. In: IEEE International Conference on Computer Vision, pp. 3456–3462. IEEE (2013)
9. Sirmaçek, B., Ünsalan, C.: Road network extraction using edge detection and spatial voting. In: International Conference on Pattern Recognition, pp. 3113–3116. IEEE (2010)
10. Girshick, R., Shotton, J., Kohli, P., Criminisi, A., Fitzgibbon, A.: Efficient regression of general-activity human poses from depth images. In: IEEE International Conference on Computer Vision, pp. 415–422. IEEE (2011)
11. Maji, S., Malik, J.: Object detection using a max-margin hough transform. In: IEEE Conference on Computer Vision and Pattern Recognition, pp. 1038–1045. IEEE (2009)
12. Wohlhart, P., Schulter, S., Köstinger, M., Roth, P.M., Bischof, H.: Discriminative hough forests for object detection. In: British Machine Vision Conference, pp. 1–11(2012)
13. Breiman, L.: Random forests. Mach. Learn. **45**, 5–32 (2001)
14. Lin, Z., Chen, M., Ma, Y.: The augmented lagrange multiplier method for exact recovery of corrupted low-rank matrices. arXiv preprint arXiv:1009.5055 (2010)
15. Ren, Z., Yuan, J., Meng, J., Zhang, Z.: Robust part-based hand gesture recognition using kinect sensor. IEEE Trans. Multimedia **15**, 1110–1120 (2013)
16. Gourier, N., Hall, D., Crowley, J.L.: Estimating face orientation from robust detection of salient facial features. In: ICPR International Workshop on Visual Observation of Deictic Gestures, Citeseer (2004)
17. Qian, C., Sun, X., Wei, Y., Tang, X., Sun, J.: Realtime and robust hand tracking from depth. In: IEEE Conference on Computer Vision and Pattern Recognition, pp. 1106–1113. IEEE (2014)
18. Schulter, S., Leistner, C., Wohlhart, P., Roth, P.M., Bischof, H.: Alternating regression forests for object detection and pose estimation. In: 2013 IEEE International Conference on Computer Vision (ICCV), pp. 417–424. IEEE (2013)
19. Rota Bulo, S., Kontschieder, P.: Neural decision forests for semantic image labelling. In: IEEE Conference on Computer Vision and Pattern Recognition, pp. 81–88. IEEE (2014)
20. Hara, K., Chellappa, R.: Growing regression forests by classification: applications to object pose estimation. In: Fleet, D., Pajdla, T., Schiele, B., Tuytelaars, T. (eds.) ECCV 2014. LNCS, vol. 8690, pp. 552–567. Springer, Cham (2014). doi:10. 1007/978-3-319-10605-2_36
21. Ren, S., Cao, X., Wei, Y., Sun, J.: Global refinement of random forest. In: The IEEE Conference on Computer Vision and Pattern Recognition (2015)
22. Gould, N., Toint, P.L.: Preprocessing for quadratic programming. Math. Program. Series B **100**, 95–132 (2004)

23. Fisher, N.I.: Statistical Analysis of Circular Data. Cambridge University Press, Cambridge (2000)
24. Herdtweck, C., Curio, C.: Monocular car viewpoint estimation with circular regression forests. In: IEEE Intelligent Vehicles Symposium (2013)
25. Comaniciu, D., Meer, P.: Mean shift: a robust approach toward feature space analysis. IEEE Trans. PAMI **24**, 603–619 (2002)
26. Zhang, C., Yang, X., Tian, Y.: Histogram of 3D facets: a characteristic descriptor for hand gesture recognition. In: IEEE International Conference and Workshops on Automatic Face and Gesture Recognition, pp. 1–8. IEEE (2013)
27. Shotton, J., Girshick, R., Fitzgibbon, A., Sharp, T., Cook, M., Finocchio, M., Moore, R., Kohli, P., Criminisi, A., Kipman, A., et al.: Efficient human pose estimation from single depth images. IEEE Trans. Pattern Anal. Mach. Intell. **35**, 2821–2840 (2013)
28. Liang, H., Yuan, J., Thalmann, D.: 3D fingertip and palm tracking in depth image sequences. In: Proceedings of the 20th ACM International Conference on Multimedia, pp. 785–788. ACM (2012)
29. Zhang, C., Tian, Y.: Histogram of 3D facets: a depth descriptor for human action and hand gesture recognition. Comput. Vis. Image Underst. **139**, 29–39 (2015)
30. Lai, Z., Yao, Z., Wang, C., Liang, H., Chen, H., Xia, W.: Fingertips detection and hand gesture recognition based on discrete curve evolution with a kinect sensor. In: Visual Communications and Image Processing (2016)
31. Haj, M.A., Gonzalez, J., Davis, L.S.: On partial least squares in head pose estimation: How to simultaneously deal with misalignment. In: 2012 IEEE Conference on Computer Vision and Pattern Recognition, pp. 2602–2609. IEEE (2012)
32. Fenzi, M., Leal-Taixé, L., Rosenhahn, B., Ostermann, J.: Class generative models based on feature regression for pose estimation of object categories. In: Proceedings of the IEEE Conference on Computer Vision and Pattern Recognition, pp. 755–762 (2013)
33. Sun, X., Wei, Y., Liang, S., Tang, X., Sun, J.: Cascaded hand pose regression. In: Proceedings of the IEEE Conference on Computer Vision and Pattern Recognition, pp. 824–832 (2015)
34. Tang, D., Chang, H.J., Tejani, A., Kim, T.K.: Latent regression forest: structured estimation of 3D articulated hand posture. In: Proceedings of the IEEE Conference on Computer Vision and Pattern Recognition, pp. 3786–3793. IEEE (2014)

Sign-Correlation Partition Based on Global Supervised Descent Method for Face Alignment

Yongqiang Zhang[1]([✉]), Shuang Liu[2], Xiaosong Yang[2], Daming Shi[1], and Jian Jun Zhang[2]

[1] Harbin Institute of Technology, Harbin, China
seekever@foxmail.com
[2] Bournemouth University, Poole, UK

Abstract. Face alignment is an essential task for facial performance capture and expression analysis. As a complex nonlinear problem in computer vision, face alignment across poses is still not studied well. Although the state-of-the-art Supervised Descent Method (SDM) has shown good performance, it learns conflict descent direction in the whole complex space due to various poses and expressions. Global SDM has been presented to deal with this case by domain partition in feature and shape PCA spaces for face tracking and pose estimation. However, it is not suitable for the face alignment problem due to unknown ground truth shapes. In this paper we propose a sign-correlation subspace method for the domain partition of global SDM. In our method only one reduced low dimensional subspace is enough for domain partition, thus adjusting the global SDM efficiently for face alignment. Unlike previous methods, we analyze the sign correlation between features and shapes, and project both of them into a mutual sign-correlation subspace. Each pair of projected shape and feature keep sign consistent in each dimension of the subspace, so that each hyperoctant holds the condition that one general descent exists. Then a set of general descent directions are learned from the samples in different hyperoctants. Our sign-correlation partition method is validated in the public face datasets, which includes a range of poses. It indicates that our methods can reveal their latent relationships to poses. The comparison with state-of-the-art methods for face alignment demonstrates that our method outperforms them especially in uncontrolled conditions with various poses, while keeping comparable speed.

1   Introduction

Face alignment is an important computer vision task, and plays a key role in many facial analysis applications, such as face recognition, performance-based facial animation, and expression analysis. It aims at locating predefined facial

---

Electronic supplementary material The online version of this chapter (doi:10. 1007/978-3-319-54187-7_19) contains supplementary material, which is available to authorized users.

S.-H. Lai et al. (Eds.): ACCV 2016, Part III, LNCS 10113, pp. 281–295, 2017.
DOI: 10.1007/978-3-319-54187-7_19

landmarks (such as eye corners, nose tip, mouth corners) in face images automatically. Face alignment usually takes a face bounding box from a face detector as input, and fits initial landmarks positions into optimal locations.

Since the ground truth of shape is unknown during test, how to predict the shape increment from initial shape to real shape is a hard problem. Generally, The global or local face appearance is considered as extra constraints for optimization. Sufficient labeled face images are also very important for learning a reliable face alignment model. Recently there have been many methods proposed to face alignment. Most of them can be categorized into two groups according to the underlying model: generative models and discriminative models.

Typical generative models include Active Shape Model (ASM) [1], Active Appearance Model (AAM) [2], and their extensions [3–6]. In this type of methods, the optimization target is model parameters. It means searching the best parameters to generate the most fitting shape (facial landmarks). These methods mitigate the influence of various poses and illumination, but due to sub-optimization problem, they are sensitive to initialization and often tend to fail in the wild condition.

Recently discriminative models have shown better performance for face alignment. One discriminative method directly learns a mapping between facial appearance features and shapes. A kind of discriminative methods are based on local classifiers or response maps for landmarks [7,8]. These methods deal with each landmark independently, and ignore the their relationship. As a popular discriminative model, current cascaded regression-based methods take all the landmarks as a whole, and solve a nonlinear optimization problem by cascaded regression theory. The main difference between cascaded regression and related boosting regression is that cascaded regression uses shape-indexed features extracted from the image according to the current estimated shape. Cascaded Pose Regression (CPR) [9] for pose estimation has been widely extended into face alignment in current works, represented by Explicit Shape Regression (ESR) [10], and Supervised Descent Method (SDM) [11]. It is noticed that SDM provides a theoretical explanation of the cascaded regression from the point of view of optimizing a nonlinear problem, as a significant achievement in cascaded regression-based methods.

Following the cascaded regression framework, many researchers focus on improving its efficiency and accuracy in uncontrolled conditions, including various poses, expressions, lighting and partial occlusions. Some of them can handle partial occlusions [12–14]. Some works mainly aim at speeding up the prediction process while keeping high accuracy [13,15]. The choice and learning of shape-indexed features are also studied [15–17]. A series of regression methods have been employed into cascaded regression framework to deal with over-fitting and local minima problems in the wild condition, including ridge regression [18], Support Vector [19], Gaussian process [20,21], Random Forest voting [14,22,23], Deep Neural Nets [16,24,25], and project-out cascaded regression [26].

Although these works have produced remarkable results on nearly frontal face alignment, it is still hard to locate landmarks across large poses and expressions under uncontrolled conditions. The variation of poses leads to non-convex and multiple local minima problems. Especially Xiong et al. [27] theoretically addresses the limitation of SDM and proposes descent domain partition in feature and shape PCA space separately. Though their scheme works well for face tracking and pose estimation, it is not suitable for face alignment across various poses, because the ground truth of shapes or features are unknown and it is not able to find approximation due to lack of previous frame. A few recent works [12,13,28–31] begin to consider the influence of multiple poses. Most of them deal with the problem indirectly by random schemes or data augment, and they can only handle small changes in poses. How to solve non-convex and multiple local minima problems caused by large poses is still not well studied.

Inspired by Xiongs work [27], we proposed a novel sign-correlation subspace method for partitioning descent domains to achieve robust face alignment across poses. The main contributions of our work are: (1) The inherent relationship between poses space and appearance features or shapes space is explicitly obtained by sign-correlation reduced dimension strategy. The whole features and shapes spaces are projected into a mutual sign-correlation subspace, which mainly represents the variation of poses. (2) The decent domains partition is produced according to the signs of each dimension in this sign-correlation subspace. Since the face appearance features during cascaded regression, what we need to do for decent domains partition is to project features space into joint sign-correlation subspace and split whole sample space into different hyperoctants as decent domains. (3) Our method is validated on challenging face datasets, which includes face images from different poses. The results show that it can split complex sample space into homogeneous domains related to poses, thus a mutual manifold of feature and shape spaces are obtained. The experiments for face alignment indicate that our method get state-of-the-art performance for nearly frontal face images, and it is more robust on datasets with multiple poses, compared with current methods.

2   Related Work

2.1   Cascaded Regression to Face Alignment

Both of generative models and discriminative models have been studied for face alignment. As a typical generative model, ASM [1] is proposed to take advantage of prior knowledge from training datasets, which is one of the earliest data-driven model for shape fitting. PCA is used to build a linear combination model of major shape basis, and local textures around control points are also used for fitting the shape well. AAM [2] considers the global appearance rather than only local textures in ASM, and a PCA model is trained for global appearance while the shape PCA is trained at the same time. AAM can warp the initial shape and appearance into the current face very well due to both of its shape constraint and appearance constraint. There are also many methods based on them, like multi-view ASM [3], CLM [4], bilinear AAM [5] and tensor-based

AAM [6]. Since these methods are parametric models, it is hard to avoid a sub-optimization problem. Unexpected results often occur in the wild condition due to a inappropriate initialization.

Among discriminative models, cascaded regression based methods have shown more promising performance than local classifiers or response map based methods [7,8] and generative models. ESR [10] uses shape indexed intensity difference features for face alignment based on CPR [9]. Moreover, SDM extracts shape-indexed SIFT features and learns a sequence of general descent maps from supervised training data, providing a solution when Newton Descent method is hard to be utilized for a not analytically differentiable nonlinear function or Hessian matrix is too large and not positive definite. Since SDM tends to average conflict descent directions over whole non-convex space, it is still limited in the wild scenes, like large poses, extreme expressions and partial occlusions.

Later research mainly focuses on improving performance based on ESR and SDM. Burgos-Artizzu et al. [12] integrate part visibility term into landmarks and presents interpolated shape-indexed features to tackle with occlusions and high shape variances. Kazemi et al. [32] estimate facial landmarks by learning an ensemble of regression trees (ERT) directly from a sparse subset of pixel intensities. Their ERT achieves millisecond performance and can handle partial or uncertain labels, but the correlation of shape parameters is little taken into account. Instead of least squares regression, Xing et al. [13] learn sparse Stagewise Relational Dictionary (SRD) between facial appearances and shapes, which improves the robustness under different views and severe occlusions. Some recent research aims at choosing or learning shape-indexed features. Yan et al. [15] compare the performance of different local feature descriptors for face alignment, including SIFT, HOG, LBP and Gabor, and HOG shows the best results in their experiments. Ren et al. [33] build local binary features by learning regression random forest for each landmark independently, and then learn a global cascaded linear regressor with pre-built binary features. Deep Neural Networks [16,24,25] have also been studied for face landmark detection. DNNs-based methods fuse the feature description and networking training in a unified framework, but it is still a very hard task to tune many free parameters.

2.2 Multi-pose Face Alignment

More recently Xiong et al. [27] analyze the drawbacks of SDM and splits whole sample space into descent domains by PCA in both of feature and shape spaces, but it can not be used for face alignment across various poses due to unknown real shapes or features, which can be approximated by previous frame for face tracking and pose estimation tasks. There have been some works done to improve the SDM for face alignment. Feng et al. [28] propose random cascaded-regression copse, learning a sets of cascaded strong regressors corresponding to different subsets of samples and averaging all predictions of them as final output. Similarly, Yang et al. [29] propose random subspace SDM, randomly selecting a small number of dimensions from the whole feature space and training an ensemble of regressors in several feature subspaces. Liu et al. [18] modify traditional

SDM with multi-scale HOG features, global to local regression of features and rigid regularization to improve the accuracy and robustness. L2,1 norm based kernel SVR is presented by Martinez et al. [19] to substitute the commonly used least squares regressor, which improves the performance of face alignment across views. Gaussian process [20,21] and Random Forest voting [14,22,23] are also introduced into cascaded regression framework. Zhang et al. [34] and Zhu et al. [30] further study hierarchical or coarse-to-fine searching for face alignment. Feng et al. [31] combine synthetic images with real images to train cascaded collaborative regression with dynamic weighting, handling the pose variations better. Fan et al. [17] combine projective invariant characteristic number with appearance based constraints and solve a quadratic optimization by the standard gradient descent. Though their method show well pose invariant, it can only handle a small number of landmarks. Tzimiropoulos presents project-out cascaded regression (PO-CR) [26] and extend the learn-based Newtons method further: Instead of learning directly a mapping from appearance features to non-parametric shapes, PO-CR learns a sequence of Jacobian and Hessian matrices based on parametric shape model. It shows noticeable results on the challenging datasets.

Some methods among them have begun to deal with the impact of poses, like RPCR [12], SRD [13], CCR [31], hierarchical localization [34] and coarse-to-fine searching [30]. However, few of them can give a clear interpretation for the correlation between poses and feature or shapes. Most of these methods loose the problem by different strategies, but how to achieve robust and accurate face alignment across poses is still a challenging task. Following previous methods, our work focuses on analyzing the underlying relationship between feature space and shape space, and finding a joint pose-related subspace for global supervised descent domains partition, to improve the performance of face alignment under challenging multi-pose conditions.

3 Sign-Correlation Subspace for Descent Domains Partition

3.1 Descent Domain Partition Problem in SDM

In order to keep the completeness of our work, we first review the problem of SDM and the sign-correlation condition for existence of a supervised descent domain. According to SDM, setting one image $\mathbf{d}$, p landmarks $\boldsymbol{x} = [x_1, y_1, \ldots, x_p, y_p]$, a feature mapping function $\boldsymbol{h}(\mathbf{d}(\boldsymbol{x}))$ corresponding to image $\mathbf{d}$, where $\mathbf{d}(\boldsymbol{x})$ indexes landmarks in the image $\mathbf{d}$, the face alignment problem can be regarded as a optimization problem,

$$f(\boldsymbol{x}_0 + \Delta\boldsymbol{x}) = \|\boldsymbol{h}(\mathbf{d}(\boldsymbol{x}_0 + \Delta\boldsymbol{x})) - \phi_*\|_2^2 \tag{1}$$

where $\phi_* = \boldsymbol{h}(\mathbf{d}(\boldsymbol{x}_*))$ represents the feature extracted according to correct landmarks $\boldsymbol{x}_*$, which is known in the training images, but unknown in the testing images. For initial locations of landmarks $\boldsymbol{x}_0$, we solve $\Delta\boldsymbol{x}$, which minimizes the feature alignment error $f(x_0 + \Delta x)$. Since the feature function is usually not analytically differentiable, it is hard to solve the problem with traditional Newtons methods. Alternatively, a general descent mapping can be learned from training datasets. The supervised descent method form is,

$$x_k = x_{k-1} - \mathbf{R}_{k-1}(\phi_{k-1} - \phi_*) \tag{2}$$

Since ϕ_* of a testing image is unknown but constant, SDM modifies the objective to align with respect to the average one $\overline{\phi}_*$ over training set, the update rule then is modified,

$$\Delta x = \mathbf{R}_k(\overline{\phi_*} - \phi_k) \tag{3}$$

Instead of learning only one $\mathbf{R}_k$ over all samples during one updating step, The Global SDM learns a series of $\mathbf{R}^t$, one for a subset of samples S^t, where the whole samples are divided into T subsets $S = \{S^t\}_1^T$.

A generic DM exists under these two conditions: (1) $\mathbf{R}h(x)$ is a strictly locally monotone operator anchored at the optimal solution (2) $h(x)$ is locally Lipschitz continuous anchored at x_*. For a function with only one minimum, these normally hold. But a complex function might have several local minima in a relatively small neighborhood, thus the original SDM tends to average conflicting gradient directions. Therefore, the Global SDM proves that if the samples are properly partitioned into a series of subsets, there is a DM in each of the subsets. The $\mathbf{R}_t$ for subset S_t can be solved with a constrained optimization form,

$$\min_{S,\mathbf{R}} \sum_{t=1}^{T} \sum_{i \in S^t} \|\Delta x_* - \mathbf{R}_t \Delta \phi^{i,t}\|^2 \tag{4}$$

$$s.t. \ \Delta x_*^i \mathbf{R}_t \Delta \phi^{i,t} > 0, \forall \ t, i \in S^t \tag{5}$$

where $\Delta x_*^i = x_*^i - x_k^i$, $\Delta \phi^{i,t} = \overline{\phi}_*^t - \phi^i$, and where $\overline{\phi}_*^t$ – average all ϕ_* over the subset S^t. Equation 5 guarantees that the solution satisfies DM condition 1. It is NP-hard to solve Eq. 4, so a deterministic scheme is proposed to approximate the solution. A set of sufficient conditions for Eq. 5 is given:

$$\Delta x_*^{i\,T} \Delta \mathbf{X}_*^t > \mathbf{0}, \forall \ t, i \in S^t \tag{6}$$

$$\Delta \Phi^{t\,T} \Delta \phi^{i,t} > \mathbf{0}, \forall \ t, i \in S^t \tag{7}$$

where $\Delta \mathbf{X}_*^t = [\Delta x_*^{1,t}, \dots, \Delta x_*^{i,t}, \dots]$, each column is $\Delta x_*^{i,t}$ from the subset S^t; $\Delta \Phi^t = [\Delta \phi^{1,t}, \dots, \Delta \phi^{i,t}, \dots]$, each column is $\Delta \phi^{i,t}$ from the subset S^t.

Since the dot product of any two vectors within the same hyperoctant (the generalization of quadrant) is positive, an ideal sufficient partition can be regarded as that each subset S^t occupies a hyperoctant both in the parameter space Δx and feature space $\Delta \phi$. However, this leads to exponential number of DMs. Assuming Δx is n-dimension, and $\Delta \phi$ is m-dimension, the number of subsets will be 2^{n+m}. Moreover, if the number of all samples is small, there will be many empty subsets, and also the volume of some subsets will be too small to train.

It's known that as Δx and $\Delta \phi$ are embedded in a lower dimensional manifold for human faces. So dimension reduction methods (e.g. PCA) on the whole training set Δx and $\Delta \phi$ can be used for approximation. The Global SDM authors project Δx onto the subspace expended by the first two components of Δx space, and project $\Delta \phi$ onto the subspace by the first component of $\Delta \phi$ space. So there are 2^{2+1} subsets in their work. It is a very naive scheme and not suitable for face alignment. Correlation-based dimension reduction theory can be introduced to develop a more practical and efficient strategy for low-dimension approximation of the high dimensional partition problem.

3.2   Sign-Correlation Subspace Partition

Interestingly, Xiong et al. [27] have proven that if one subset S^t satisfies: For any two samples $\{\Delta x^{i,t}, \Delta \phi^{i,t}\}$, $\{\Delta x^{k,t}, \Delta \phi^{k,t}\}$ within S^t, the signs of each corresponding $j-th$ dimension $\{\Delta x_j^{i,t}, \Delta \phi_j^{i,t}\}$ between the samples keep the same,

$$sign(\Delta x_j^{i,t}, \Delta \phi_j^{i,t}) = sign(\Delta x_j^{k,t}, \Delta \phi_j^{k,t}), \forall i, k \in S^t, j = 1 : min(n, m) \qquad (8)$$

Then there must exist a DM $\mathbf{R}^t$ in one updating step. Equation 8 provides a possible partition strategy: all the samples that follow Eq. 8 can be put into a subset, and there would be $2^{min(n,m)}$ subsets in total. It is noticed that there are two limits of this partition strategy: (1) It can not guarantee the samples lie in the same small neighborhood. In other words, even if $\{\Delta x^{i,t}, \Delta \phi^{i,t}\}$, $\{\Delta x^{k,t}, \Delta \phi^{k,t}\}$ keep Eq. 8, the $\Delta x^{i,t}$, $\Delta x^{k,t}$ may be very far from each other. (2) It only considers the dim-to-dim correlation of the first $min(n, m)$ dimensions in the Δx space and $\Delta \phi$ space, and other dimensions are ignored. The correlation of any $j-th$ dimension of Δx with a non-corresponding $j'-th$ dimension $j' \neq j$ of $\Delta \phi$ is also ignored.

Considering the low dimensional manifold, the Δx space and $\Delta \phi$ space can be projected onto a medium low dimensional space with projection matrix $\mathbf{Q}$ and $\mathbf{P}$, respectively, which keeps the projected vectors $v = \mathbf{Q}\Delta x$, $u = \mathbf{P}\Delta \phi$ correlated enough: (1) v, u lie in the same low dimensional space. (2) For each $j-th$ dimension, $sign(v_j, u_j) = 1$. If the projection holds these two conditions, the projected samples $\{u^i, v^i\}$ can be partitioned into different hyperoctants in the medium space only according to the signs of u^i, due to condition 2. Since samples in a hyperoctant are close enough to each other, this partition can hold the small neighborhood better. It is also a compact low dimensional approximation of the high dimensional hyperoctant-based partition strategy in both Δx space and $\Delta \phi$ space, which is an sufficient condition for the existence of a generic DM, as mentioned above.

For convenience, we re-denote Δx as $y \in \Re^n$, re-denote $\Delta \phi$ as $x \in \Re^m$, $\mathbf{Y}_{s \times n} = [y^1, \ldots, y^i, \ldots, y^s]$ is all the y^i of training set. $\mathbf{X}_{s \times m} = [x^1, \ldots, x^i, \ldots, x^s]$ is all the x^i of training set. The projection matrices are

$$\mathbf{Q}_{r \times n} = [q_1, \ldots, q_j, \ldots, q_r]^T, \; q_j \in \Re^n,$$
$$\mathbf{P}_{r \times m} = [p_1, \ldots, p_j, \ldots, p_r]^T, \; p_j \in \Re^m,$$

Projection vectors are $v = Qy$, $u = Px$. Here we denote projection vectors w_j, z_j along the sample space: $w_j = Yq_j = [v_j^1, \ldots, v_j^i, \ldots, v_j^s]^T$, $z_j = Xp_j = [u_j^1, \ldots, u_j^i, \ldots, u_j^s]^T$. This problem can be formulated as a constrained optimization form,

$$\min_{P,Q} \sum_{j=1}^{r} \|Yq_j - Xp_j\|^2 = \min_{P,Q} \sum_{j=1}^{r} \sum_{i=1}^{s} (v_j^i - u_j^i)^2 \tag{9}$$

$$s.t. \sum_{j=1}^{r} \sum_{i=1}^{s} sign(v_j^i u_j^i) = sr \tag{10}$$

It can be seen that w_j and z_j are the projected values of all the samples Y or X along a special direction q_j or p_j. For a fixed projected $j-th$ dimension, assuming the w_j and z_j is normalized, which means that the mean of $\{v_j^i\}_{i=1:s}$ is zero, and the standard deviation of it is $1/s$, so is $\{u_j^i\}_{i=1:s}$. Thus $w_j^T w_j = 1$, $z_j^T z_j = 1$, $w_j^T e = 0$, $z_j^T e = 0$, where $e = [1,1,\ldots,1]^T$, then Eq. 9 can be simplified as,

$$\min_{P,Q} \sum_{j=1}^{r} \|w_j - z_j\|^2 = \max_{P,Q} \sum_{j=1}^{r} w_j^T z_j \tag{11}$$

For a fixed projected $j-th$ dimension, the constraint $\sum_{i=1}^{s} sign(v_j^i u_j^i) = s$ means that all the pairs $\{v_j^i, u_j^i\}$ of samples in $j-th$ dimension keeps the consistence of sign. There is a fact: if the angle θ_j between w_j and z_j is 0, the term $w_j^T z_j$ will reach maximum, so the sign condition must hold; and if the angle θ_j is $\pi/2$, the term $w_j^T z_j$ will reach 0, so the sign condition will fail completely. Moreover, fixing the $|v_j^i u_j^i|$, the $\cos \theta_j$ will get larger while the $\sum_{i=1}^{s} sign(v_j^i u_j^i)$ rises, and $\sum_{i=1}^{s} sign(v_j^i u_j^i)$ tends to be larger with the $\cos \theta_j$ growing. Given some constraints, it can be proved that the $\cos \theta_j$ can be taken as an approximation of the sign summation function for optimization,

$$\frac{1}{s} \sum_{i=1}^{s} sign(v_j^i u_j^i) \approx \cos \theta_j = w_j^T z_j \tag{12}$$

When the samples $\{y^i\}_{i=1:s}$ and $\{x^i\}_{i=1:s}$ are normalized (removing means and dividing standard deviation during pre-processing), the sign-correlation constrained optimization problem will be solved with the standard Canonical-Correlation Analysis (CCA). The CCA problem for normalized $\{y^i\}_{i=1:s}$ and $\{x^i\}_{i=1:s}$ is,

$$\max_{(p)_j, q_j} q_j^T cov(Y, X) p_j \tag{13}$$

$$s.t. \ q_j^T var(Y, Y) q_j = 1, \ p_j^T var(X, X) p_j = 1 \tag{14}$$

Following CCA algorithm, the max sign-correlation dimension p_1 and q_1 is solved at first. Then one seeks p_2 and q_2 by maximizing the same correlation subject to the constraint that they are to be uncorrelated with the first pair w_1, z_1 of canonical variables; This procedure may be continued up to r times until p_r and q_r is solved.

After all p_j and q_j is solved, we only need the projection matrix $\mathbf{P}$ in Δx space. Then we project each Δx^i into the sign-correlation subspace and get reduced feature $u^i = \mathbf{P}\Delta x^i$. Then we partition the whole sample space into independent descent domains by judging the sign of each dimension of u^i and group it into corresponding hyperoctant. Finally, in order to solve Eq. 4 at each iterative step, we learn a descent mapping for every subset at each iterative step with the ridge regression algorithm. When testing a face image, we also use the projection matrix $\mathbf{P}$ to find its corresponding decent domain and predict its shape increment at each iterative step.

4 Experiments and Evaluation

Our work mainly focuses on face alignment across poses, so we conduct experiments especially for this task to analyze and evaluate our sign-correlation partition method. Firstly, we validate our method on multi-pose datasets, and the PCA partition scheme is also compared to our sign-correlation partition method. Then we also test our method on common datasets for general face alignment and compare it with state-of-the-art methods. According to Yans research [15], the multi-scale HOG outperforms multi-scale SIFT and other typical local descriptors HOG, SIFT, LBP and Gabor. We adopt multi-scale HOG as feature mapping function in our sign-correlation partition SDM algorithm. The two domains are enough for partition, so we only use the first sign-correlation projection component in appearance feature space.

4.1 Sign-Correlation Partition Validation

In this section, we validate the underlying relationship between our sign-correlation partition and the variation of poses. Two widely used benchmark datasets are used in our validation: MTFL [35] and 300W [36]. MTFL dataset contains labeled face images from AFLW [37], LFW [38] and Internet. This dataset annotates 5 landmarks and labels 5 different left-right poses with the flags of gender, smile and glasses. Here we only focus on non-frontal poses to verify our partition method. There are 2550 non-frontal face images in original MTFL dataset, and the number of left ones is not equal to the number of right ones. For fairness, we augment these non-frontal images by a horizontal flip, so that we get the same numbers of left and right images. 300W dataset is mainly made up of images from LFPW [39], HELEN [40], AFW [41] with 68 re-annotated landmarks. The 3148 images from training dataset are selected in our validation. The flip augment is also used for obtaining the same number of left and right images. The left or right poses are estimated by a typical pose estimation taking known landmarks as input.

We partition the multi-poses images into two domains by the first sign-correlation projected dimension. The PCA partition by the first principle component is also tested as comparison. The results in Figs. 1 and 2 show that

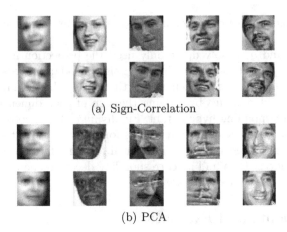

Fig. 1. Pose validation on MTFL. In each subfigure: first column shows average faces of two subsets, and first or second row shows samples in subset 1 or 2.

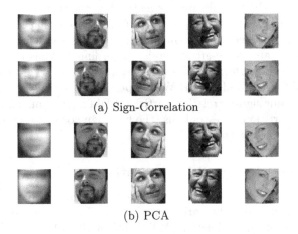

Fig. 2. Pose validation on 300W. In each subfigure: first column shows average faces of two subsets, and first or second row shows samples in subset 1 or 2.

each sign-correlation domain mainly contain left or right pose images, and the accuracy of pose partition is high, as shown in Table 1. It indicates that our sign-correlation partition method can construct descent domains highly related to pose variations only with face appearance features. On the contrary, PCA partition only with face appearance features can not capture the pose variation well, and the partition result is nearly random.

4.2   Comparison of Face Alignment

We evaluate the proposed sign-correlation partition SDM method on the challenging 300W dataset, and compare it with state-of-the-art methods ESR [10],

Table 1. Pose accuracy validation on MTFL and 300W

Datasets	Left/right number	PCA	Sign-Corr
MTFW	2550	0.7780	**0.9275**
300W	3148	0.5179	**0.9319**

Table 2. Comparison with current methods on 300W dataset

Datasets	Full	Common	Challenging
ESR	7.58	5.28	17.00
SDM	7.52	5.60	15.40
ERT	6.41	5.22	13.03
LBF	6.32	4.95	11.98
Ours	**5.88**	5.07	**10.79**

SDM [11], ERT [32], and LBF [33]. As mentioned above, there are 68 labeled landmarks in this dataset. Its training part contains 3148 images form AFW and training parts of LFPW, and HELEN dataset, and its testing part contains 689 images from testing parts of LFPW, and HELEN and IBUG. Among them, the LFPW dataset, although more challenging than other near-frontal datasets, is mainly made up of small pose variations, and the result on it nearly reaches limitation. The HELEN dataset contains faces of different genders, poses, and expressions. The IBUG testing dataset is the most challenging one due to extreme poses, expressions and lighting.

We conduct three experiments by testing different parts of the whole 300W datset: common subset: LFPW and HELEN, challenging dataset: IBUG, and full

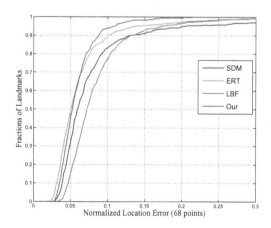

Fig. 3. CED curves over 300W

dataset. Following the standard [38], the normalized inner-pupil distance land-mark error is used in our evaluation. The inner-pupil errors of different methods are given in Table 2. The cumulative error distribution (CED) curves are also plotted, as shown in Fig. 3. The results illustrate that our method outperforms most of current methods over the full datasets. We also get comparable results on common LFPW and HELEN datasets. Our method works better especially on the challenging IBUG dataset with large variations of poses.

5   Conclusion

In this paper we propose a novel sign-correlation partition method based on global SDM, which achieves promising performance for face alignment on the challenging datasets. We analyze the underlying relationship between shape/feature space and pose space by sign-correlation reduced dimensional projection. Taking the advantage of the inherent connection of shapes with features within a mutual pose-related subspace, the global descents partition can be operated according to different hyperoctants in the projected sign-correlation subspace. Due to the high consistence of sign between shapes and features in this subspace, it is able to partition the descent domains only depending on features and learned sign-correlation projection components. Our method adjusts the global SDM efficiently for face alignment, the original partition scheme of which is not suitable for face alignment. Moreover, we provide a clearer explanation about the influence of poses on the problems of multiple local minima and global descent direction conflict, and tackle with the robust face alignment across poses in a direct way. The experiment on the widely used multi-pose dataset indicates that our sign-correlation partition method can divide global complex space into several pose-related descent domains only with appearance features, while PCA-based partition only in feature space does not work. Our method also achieves noticeable Search Results performance for face alignment on challenging datasets, compared with popular methods. Although our sign-correlation subspace method improves the robustness in extreme conditions, there are still some parts we need to study in the future: The number of sign-correlation dimensions is needed to be chosen more carefully. On the other side, since the partition accuracy is limited by linear reduced dimensional projection, the kernel method can also be introduced into sign-correlation analysis.

Acknowledgement. This work was supported by Harbin Institute of Technology Scholarship Fund 2016 and National Centre for Computer Animation, Bournemouth University.

References

1. Cootes, T.F., Taylor, C.J., Cooper, D.H., Graham, J.: Active shape models—their training and application. Comput. Vis. Image Underst. **61**, 38–59 (1995)
2. Cootes, T.F., Edwards, G.J., Taylor, C.J.: Active appearance models. IEEE Trans. Pattern Anal. Mach. Intell. **23**, 681–685 (2001)
3. Romdhani, S.: A multi-view nonlinear active shape model using kernel PCA. In: British Machine Vision Conference, pp. 483–492 (1999)
4. Cristinacce, D., Cootes, T.F.: Feature detection and tracking with constrained local models. BMVC **41**, 929–938 (2006)
5. Gonzalezmora, J., Torre, F.D.L., Murthi, R., Guil, N., Zapata, E.L.: Bilinear active appearance models. In: IEEE International Conference on Computer Vision, pp. 1–8 (2007)
6. Lee, H.S., Kim, D.: Tensor-based AAM with continuous variation estimation: application to variation-robust face recognition. IEEE Trans. Pattern Anal. Mach. Intell. **31**, 1102–1116 (2009)
7. Saragih, J.M., Lucey, S., Cohn, J.F.: Deformable model fitting by regularized landmark mean-shift. Int. J. Comput. Vision **91**, 200–215 (2011)
8. Valstar, M., Martinez, B., Binefa, X., Pantic, M.: Facial point detection using boosted regression and graph models, pp. 2729–2736 (2010)
9. Dollar, P., Welinder, P., Perona, P.: Cascaded pose regression, vol. 238, pp. 1078–1085. IEEE (2010)
10. Cao, X., Wei, Y., Wen, F., Sun, J.: Face alignment by explicit shape regression US Patent Application number 13/728,584 (2012)
11. Xiong, X., De, la Torre, F.: Supervised descent method and its applications to face alignment. In: IEEE Conference on Computer Vision and Pattern Recognition, pp. 532–539 (2013)
12. Burgosartizzu, X.P., Perona, P., Dollar, P.: Robust face landmark estimation under occlusion. In: IEEE International Conference on Computer Vision, pp. 1513–1520 (2013)
13. Xing, J., Niu, Z., Huang, J., Hu, W., Yan, S.: Towards multi-view and partially-occluded face alignment. In: Computer Vision and Pattern Recognition, pp. 1829–1836 (2014)
14. Yang, H., He, X., Jia, X., Patras, I.: Robust face alignment under occlusion via regional predictive power estimation. IEEE Trans. Image Process. **24**, 2393–2403 (2015)
15. Yan, J., Lei, Z., Yi, D., Li, S.Z.: Learn to combine multiple hypotheses for accurate face alignment. In: IEEE International Conference on Computer Vision Workshops, pp. 392–396 (2013)
16. Zhang, J., Shan, S., Kan, M., Chen, X.: Coarse-to-fine auto-encoder networks (CFAN) for real-time face alignment. In: Fleet, D., Pajdla, T., Schiele, B., Tuytelaars, T. (eds.) ECCV 2014. LNCS, vol. 8690, pp. 1–16. Springer, Heidelberg (2014). doi:10.1007/978-3-319-10605-2_1
17. Fan, X., Wang, H., Luo, Z., Li, Y., Hu, W., Luo, D.: Fiducial facial point extraction using a novel projective invariant. IEEE Trans. Image Process. **24**, 1164–1177 (2015)
18. Liu, L., Hu, J., Zhang, S., Deng, W.: Extended supervised descent method for robust face alignment. In: Jawahar, C.V., Shan, S. (eds.) ACCV 2014. LNCS, vol. 9010, pp. 71–84. Springer, Cham (2015). doi:10.1007/978-3-319-16634-6_6

19. Martinez, B., Valstar, M.F.: L 2,1-based regression and prediction accumulation across views for robust facial landmark detection. Image Vis. Comput. **45**, 371–382 (2015)

20. Lee, D., Park, H., Yoo, C.D.: Face alignment using cascade Gaussian process regression trees. In: IEEE Conference on Computer Vision and Pattern Recognition, pp. 4204–4212 (2015)

21. Martinez, B., Pantic, M.: Facial landmarking for in-the-wild images with local inference based on global appearance. Image Vis. Comput. **36**, 40–50 (2015)

22. Lindner, C., Bromiley, P.A., Ionita, M.C., Cootes, T.F.: Robust and accurate shape model matching using random forest regression-voting. IEEE Trans. Pattern Anal. Mach. Intell. **37**, 1862–1874 (2015)

23. Yang, H., Patras, I.: Fine-tuning regression forests votes for object alignment in the wild. IEEE Trans. Image Process. **24**, 619–631 (2014). A Publication of the IEEE Signal Processing Society

24. Sun, Y., Wang, X., Tang, X.: Deep convolutional network cascade for facial point detection. In: Conference on Computer Vision and Pattern Recognition, pp. 3476–3483 (2013)

25. Zhou, E., Fan, H., Cao, Z., Jiang, Y., Yin, Q.: Extensive facial landmark localization with coarse-to-fine convolutional network cascade. In: IEEE International Conference on Computer Vision Workshops, pp. 386–391 (2013)

26. Tzimiropoulos, G.: Project-out cascaded regression with an application to face alignment. In: IEEE Conference on Computer Vision and Pattern Recognition (2015)

27. Xiong, X., De la Torre, F.: Global supervised descent method. In: IEEE Conference on Computer Vision and Pattern Recognition, pp. 2664–2673 (2015)

28. Feng, Z.H., Huber, P., Kittler, J., Christmas, W., Wu, X.J.: Random cascaded-regression copse for robust facial landmark detection. IEEE Sig. Process. Lett. **22**, 76–80 (2015)

29. Yang, H., Jia, X., Patras, I., Chan, K.P.: Random subspace supervised descent method for regression problems in computer vision. IEEE Sig. Process. Lett. **22**, 1816–1820 (2015)

30. Zhu, S., Li, C., Loy, C.C., Tang, X.: Face alignment by coarse-to-fine shape searching. In: CVPR, pp. 4998–5006 (2015)

31. Feng, Z.H., Hu, G., Kittler, J., Christmas, W., Wu, X.J.: Cascaded collaborative regression for robust facial landmark detection trained using a mixture of synthetic and real images with dynamic weighting. IEEE Trans. Image Process. **24**, 3425–3440 (2015)

32. Kazemi, V., Sullivan, J.: One millisecond face alignment with an ensemble of regression trees. In: Computer Vision and Pattern Recognition, pp. 1867–1874 (2014)

33. Ren, S., Cao, X., Wei, Y., Sun, J.: Face alignment at 3000 fps via regressing local binary features. IEEE Trans. Image Process. **25**, 1685–1692 (2014)

34. Zhang, Z., Zhang, W., Ding, H., Liu, J., Tang, X.: Hierarchical facial landmark localization via cascaded random binary patterns. Pattern Recogn. **48**, 1277–1288 (2014)

35. Zhang, Z., Luo, P., Loy, C.C., Tang, X.: Facial landmark detection by deep multi-task learning. In: European Conference on Computer Vision, pp. 94–108 (2014)

36. Sagonas, C., Tzimiropoulos, G., Zafeiriou, S., Pantic, M.: A semi-automatic methodology for facial landmark annotation. In: IEEE Conference on Computer Vision and Pattern Recognition Workshops, pp. 896–903 (2013)

37. Kostinger, M., Wohlhart, P., Roth, P.M., Bischof, H.: Annotated facial landmarks in the wild: a large-scale, real-world database for facial landmark localization. In: IEEE International Conference on Computer Vision Workshops, ICCV 2011 Workshops, Barcelona, Spain, pp. 2144–2151, 6–13 November 2011
38. Huang, G.B., Mattar, M., Berg, T., Learned-Miller, E.: Labeled faces in the wild: a database for studying face recognition in unconstrained environments (2008)
39. Belhumeur, P.N., Jacobs, D.W., Kriegman, D.J., Kumar, N.: Localizing parts of faces using a consensus of exemplars. IEEE Trans. Pattern Anal. Mach. Intell. **35**, 2930–2940 (2013)
40. Le, V., Brandt, J., Lin, Z., Bourdev, L., Huang, T.S.: Interactive facial feature localization. In: European Conference on Computer Vision, pp. 679–692 (2012)
41. Ramanan, D.: Face detection, pose estimation, and landmark localization in the wild. In: IEEE Conference on Computer Vision and Pattern Recognition, pp. 31–37 (2015)

Deep Video Code for Efficient Face Video Retrieval

Shishi Qiao[1,2], Ruiping Wang[1,2,3(✉)], Shiguang Shan[1,2,3], and Xilin Chen[1,2,3]

[1] Key Laboratory of Intelligent Information Processing of Chinese Academy
of Sciences (CAS), Institute of Computing Technology, CAS, Beijing 100190, China
shishi.qiao@vipl.ict.ac.cn, {wangruiping,sgshan,xlchen}@ict.ac.cn
[2] University of Chinese Academy of Sciences, Beijing 100049, China
[3] Cooperative Medianet Innovation Center, Beijing, China

Abstract. In this paper, we address the problem of face video retrieval. Given one face video of a person as query, we search the database and return the most relevant face videos, *i.e.*, ones have same class label with the query. Such problem is of great challenge. For one thing, faces in videos have large intra-class variations. For another, it is a retrieval task which has high request on efficiency of space and time. To handle such challenges, this paper proposes a novel Deep Video Code (*DVC*) method which encodes face videos into compact binary codes. Specifically, we devise a multi-branch CNN architecture that takes face videos as training inputs, models each of them as a unified representation by temporal feature pooling operation, and finally projects the high-dimensional representations into Hamming space to generate a single binary code for each video as output, where distance of dissimilar pairs is larger than that of similar pairs by a margin. To this end, a smooth upper bound on triplet loss function which can avoid bad local optimal solution is elaborately designed to preserve relative similarity among face videos in the output space. Extensive experiments with comparison to the state-of-the-arts verify the effectiveness of our method.

1 Introduction

Face video retrieval in general is to retrieve shots containing a particular person given one video clip of him/her [1]. As depicted in Fig. 1, given one face clip as query, we search the database and return the most relevant face clips according to their distance to the query. It is a promising research area with wide range of applications, such as: 'intelligent fast-forwards' - where the video jumps to the next shot containing the specific actor; retrieval of all the shots containing a particular family member from thousands of short videos; and locating and tracking criminal suspects from masses of surveillance videos [2].

While face video retrieval is in great demand, it is still of great challenge. The video data usually tends to have unconstrained recording environment, which

Electronic supplementary material The online version of this chapter (doi:10.1007/978-3-319-54187-7_20) contains supplementary material, which is available to authorized users.

© Springer International Publishing AG 2017
S.-H. Lai et al. (Eds.): ACCV 2016, Part III, LNCS 10113, pp. 296–312, 2017.
DOI: 10.1007/978-3-319-54187-7_20

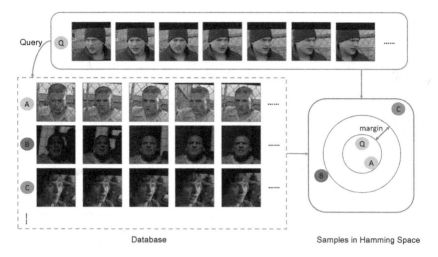

Database   Samples in Hamming Space

Fig. 1. Illustration of face video retrieval and the motivation of our method. Q, A, B and C denote face clips and the colors around them represent their class labels. Given the query Q, we search the database to find the most relevant face clips according to their distance to Q. It is observed that Q, A, B and C have large intra and inter class variations caused by illumination (B), pose (Q), expressions (C), occlusion (A), etc. We aim to project them into a common Hamming space where distance of dissimilar pairs is larger than that of similar pairs by a margin.

leads to large intra-class variations caused by illumination, pose, expressions, resolution and occlusion, as is shown in Fig. 1. Fortunately, videos provide multiple consecutive faces of one person and each frame forms a part of the person's story. One can mine complementary information from each frame to obtain a comprehensive representation for the video. However, modeling face video as a whole is nontrivial. To address this issue, a typical class of video-based face recognition methods [3–10] put the sequential dynamic information of video aside, and simply treat the video as a set of images (*i.e.* frames) and then formulate the problem as image set classification. While encouraging performance has been gained in such works, most of them utilize hand-crafted features which cannot well capture the semantic information and are hard to deal with the challenging intra-class variations. To overcome such limitations, we devise a multi-branch CNN architecture to model face video as a whole by exploiting the convolutional temporal feature pooling scheme which has been proved effective and efficient for video classification task [11]. Another challenge of face video retrieval task is the demand for low memory cost and efficient distance calculation, especially in large scale data scenarios. Obviously, high-dimensional representation of the video is not the best choice. Here we resort to hashing method which is a popular solution for approximate nearest neighbor search. In general, hashing is an approach of transforming the data item to a low-dimensional representation, or equivalently a short code consisting of a sequence of bits [12]. Therefore, we hope the designed

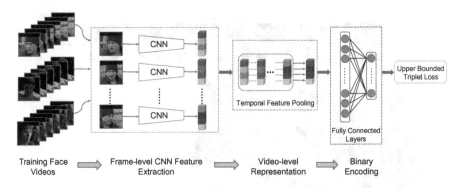

Fig. 2. Framework of the proposed DVC method. We integrate frame-level feature extraction, video-level representation and binary encoding in a unified framework by designing an end-to-end multi-branch CNN architecture. Taking face videos with their class labels as training inputs, DVC first extracts convolutional features for each frame and then utilizes temporal feature pooling on all frames belonging to the same video to produce video-level representation. Finally the fully connected layers project the feature representations of all face videos from the high-dimensional Euclidean space into a much lower-dimensional Hamming space, using the elaborately designed upper bounded triplet loss function.

CNN architecture can further project the high-dimensional representation into compact binary code, which we name as Deep Video Code (*DVC*), to satisfy the aforementioned demand with ease in an end-to-end learning manner.

Figure 2 shows the framework of our method. The multi-branch network takes face videos instead of face images as training inputs. Then each branch (weights of branches are shared) of the network extracts convolutional features for each frame simultaneously. In order to further obtain a unified representation for each video, we regard the convolution filter as a local concept classifier or detector [13] and pool the classification or detection results of all frames temporally to mine complementary information of frames within one video. With such representation, we further elaborately design a triplet loss function which aims to separate the positive sample pair (a pair of samples coming from the same class) from the negative pair by a distance margin to generate a single binary code for each video as output. To avoid converging to a bad local optimum, the gradients of the loss function should descend stably. To this end, we turn to optimize a smooth upper bound on the loss function inspired by a recent metric learning method [14]. Extensive experiments with comparison to the state-of-the-arts on two challenging TV-Series datasets released in [15] verify the effectiveness of our method on face video retrieval task.

The rest of this paper is organized as follows: We start with related works to our method in Sect. 2. Section 3 describes DVC in detail. Section 4 evaluates the proposed method with comparison to the state-of-the-arts extensively. Section 5 ends the paper with conclusions.

2   Related Work

As introduced above, our method treats face video as a whole and aims to encode high-dimensional video representations into compact binary codes. Hence, in this section, we give a brief review of related works including face video retrieval, image set and video analysis, and hashing methods.

Face Video Retrieval. More and more works on face video retrieval have been published in recent years [1,2,15–20]. Arandjelović and Zisserman [16,17] built an end-to-end system to retrieve shots in feature-length films. They proposed a cascade of processing steps to normalize the effects of the changing image environment and use the signature image to represent face shot. However, they did not make full use of information provided by multiple frames. To take advantage of rich information of videos, [2] developed a video shot retrieval system which represents each face video as distributions of histograms and measures them by chi-square distance. These early works utilize high-dimensional features which are not appropriate for efficient retrieval task. [19] is probably the first work which proposed to compress face video into compact binary code by means of learning to hash. [15] further improved the video modeling procedure by representing face video as the set covariance matrix with Fisher Vector as frame feature. [20] proposed a hashing method across Euclidean space and Riemannian manifold to solve the problem of face video retrieval with image query. While certain successes have been achieved in these works, they make little effort to frame-level feature extraction by merely relying on hand-crafted features, which are unfavorable of handling realistic challenging image variations. [21] thus made an early attempt to employ a typical standard deep CNN network to extract frame-level features and hashing codes separately for each single frame. Since their learning totally ignores the correlation information among consecutive frames and the resulting frame-level hashing codes need to be further aggregated to form the final video-level hashing code, the ordinary CNN network framework seems not an optimal solution to efficient video hashing.

Image Set and Video Analysis. Recent years have witnessed an increasing works on video-based face recognition, and among them, a typical class of methods [3–10] simply treat the problem as image set (formed by frames) classification and focus on modeling image set with different representations and measuring their similarities. However, the frame-level features they use lack strong representation power. More recently, deep CNN based methods for video classification [11,13,22,23], event detection [13] and pose estimation [24] have achieved outstanding performance compared with conventional hand-crafted features based methods. [22,23] proposed to utilize 3D CNN to capture spatial appearance and temporal motion information of video sequence. [13] discovered that simply aggregating the frame-level latent concept descriptors can achieve promising performance. However, the extraction and aggregation of frame-level descriptors are separated which may degrade the performance. To tackle this problem, [11] experimented with different frame-level feature aggregation methods in end-to-end CNNs and found that the performance of simple temporal feature pooling has been quite comparable with other more complex alternatives, verifying the usefulness of joint learning of image representation and feature aggregation.

Hashing Methods. Hashing is widely applied in retrieval area especially for large-scale approximate nearest neighbor (ANN) search problem. Compared with retrieval methods using real-valued features such as [25–27], hashing is more space and time efficient. In early years, studies mainly focus on data-independent hashing methods, such as a family of methods known as Locality Sensitive Hashing (LSH) [28] and random matrix factorization based method [10]. However, LSH methods usually require long codes to achieve satisfactory performance. To overcome this limitation, data-dependent hashing methods attempt to learn similarity-preserving compact binary codes using training data. Such methods can be further divided into unsupervised [29–31] and (semi-)supervised methods [31–44]. Since unsupervised methods cannot take advantage of label information, their performances are usually inferior to supervised methods. In supervised methods, usually an objective function in the form of point-wise [41], pairwise [32–36,39,42] or triplet [37,38,40,43–45] loss is designed. Compared with point-wise and pairwise loss, the objective of triplet loss is to preserve rank order among samples which is very well suited to the preservation of semantic similarity on challenging datasets [45]. In light of the recent progress of deep CNN network in learning robust image representation, there are also growing interests in developing hashing methods using CNN architecture for traditional image retrieval task, such as DLBHC [41] and DNNH [43]. However, as noted in [14], when using the triplet loss function in deep CNN network, it is better to make use of "difficult" triplets and straightforwardly optimizing such triplet loss with mini-batch gradient descent algorithm would probably lead to a bad local optimum [46,47]. This observation encourages us to elaborately optimize a smooth upper bound on the triplet loss function in our devised multi-branch CNN architecture, which aims to preserve relative similarity among face videos in the output Hamming space and simultaneously avoids bad local optimal solution.

3   Approach

Our goal is to learn compact binary codes for face videos such that: (a) each face video should be treated as a whole, *i.e.*, we should learn a single binary code for each video; (b) the binary codes should be similarity-preserving, *i.e.*, the Hamming distance between similar face videos should be smaller than that between dissimilar face videos by a margin. To fulfill the task, as demonstrated in Fig. 2, our method mainly involves two steps: (1) video modeling, which extracts frame-level features and aggregates them for single video-level representation, and (2) binary encoding, which learns the optimal binary codes for face videos under the designed upper bounded triplet loss function.

3.1   Video Modeling

In this step, what we need is to learn a powerful representation for each face video with the help of CNN. A straightforward method is to train a CNN model first, extract CNN features for each frame with the model, and then aggregate them into a unified representation (e.g. by averaging features of all frames). However, such method separates frame-level feature extraction and video-level representation, which may lead to poor coupling between the two steps and thus poor performance.

Another solution widely used in video modeling with CNNs is 3D Convolution Neural Network (3D CNN). To cope with video data, the 3D CNN takes the motion variation in temporal dimension into consideration. In [22], the 3D convolution is implemented by stacking image frames to construct a cube and then convolving the 3D kernels on the cube. Since 3D CNN connects each feature map to multiple contiguous feature maps (frames for the first convolution layer) in the previous layer, it has the power of making use of more frames' information in videos. However, the goal of 3D CNN is to capture motion information from frame sequences and appearance information from each frame simultaneously, which has high request on the network. Consequently learning of both motion and appearance information is degraded. As far as the face video retrieval task concerns, the temporal motion information among frames contributes little to distinguishing one face video from another for that we care more about the appearance of the faces in video than their heads' motion. Therefore 3D CNN is not the best choice for face video modeling.

For the sake of extracting appearance information and modeling the video as a whole simultaneously, we devise the multi-branch CNN architecture, as illustrated in Fig. 2. Let $F = [f_1, f_2, ..., f_n]$ be a face video with n frames, where f_i denotes the i-th frame. We first propagate each frame through the stacked convolution layers, pooling layers and ReLU [48] non-linear activation layers in one of the multiple branches. By doing this, the CNN features $\mathbf{d}_i \in \mathbb{R}^m$ for each frame are produced. Next, to mine complementary information from these frame-level CNN activations, we adopt the temporal feature pooling methods. Temporal feature pooling has been extensively used for video classification, the resulting vector of which can be used to make video-level predictions [11]. Generally speaking, faces in a video have large variations. Each of them may only carry partial but complementary information. Specifically, the convolution kernel can be regarded as a local concept classifier or detector. While some concepts only exist on one or a few faces in the video, the detectors will only have large responses on some of these faces. By using back-propagation algorithm during training the CNN, gradients coming from the top layers help learn useful local concept detectors, while allowing the network to decide which input frame is relevant to the video representation. In this paper, we test two kinds of temporal feature pooling, *i.e.*, the max-pooling and average pooling.

Until now, we have obtained the m-dimensional vector which serves as the real-valued representation of the face video. In the following we will continue to prorogate the network activations to encode the representation into compact and similarity-preserving binary codes.

3.2 Binary Encoding

We can obtain a representation for each face video using the temporal feature pooling as introduced above. However, such representation suffers from high dimensionality which leads to high space and time complexity when applied to retrieval task straightforwardly. Besides, we still need to devise an objective function to supervise the learning of video representation. To address these problems, we propose a hashing method with triplet loss function. By doing this, the high-dimensional representation is further projected into a much lower-dimensional Hamming space.

To guarantee the similarity-preserving power of learned hashing functions, several kinds of objectives are proposed. Among them, the triplet ranking loss based hash learning methods are very promising because the objective of triplet constraints is to preserve relative rank order among samples, which is agree with the objective of retrieval task. The triplet constraints can be described as the form: "image i is more similar to image j than to image k" [43]. When the dataset is challenging, such constraints are easier to be satisfied than point-wise or pairwise constraints. Apart from that, such form of triplet-based relative similarities are easier to be constructed than others (e.g., for two images with multiple attributes/tags, simply count the number of common attributes/tags as the similarity metric between them). For better understanding of the triplet ranking loss in hash learning, let i, j, k be three samples and i is more similar to j than to k, our goal is to map these three samples into Hamming space where their relative similarity or distance can be well preserved, as illustrated in Fig. 1. Otherwise, punishment should be put on them, defined by:

$$J_{i,j,k} = \max(0, \alpha + D_h(\mathbf{b}_i, \mathbf{b}_j) - D_h(\mathbf{b}_i, \mathbf{b}_k)).$$
$$s.t. \ \mathbf{b}_i, \mathbf{b}_j, \mathbf{b}_k \in \{0, 1\}^c \tag{1}$$

where $D_h(\cdot)$ denotes the Hamming distance between two binary vectors and $\alpha > 0$ is a margin threshold parameter. $\mathbf{b}_i$, $\mathbf{b}_j$ and $\mathbf{b}_k$ are the c-bit binary codes of sample i, j and k, respectively.

Generally we use the gradient descent algorithm with mini-batch to train the CNN with the aforementioned triplet ranking loss. In this case, triplets are constructed at random. Therefore, a substantial part of them contribute little to the convergence of the network during each iteration as they already meet the triplet constraint as described in Eq. (1) or their loss is quite small. To cope with this problem, our approach tends to make use of "difficult" triplets, $i.e.$, given a pair of similar samples, we actively find the dissimilar neighbor closest to them in current learned Hamming space. Based on this idea, we rewrite Eq. (1) and give the overall loss function per batch as:

$$J = \frac{1}{2|\widehat{\mathcal{P}}|} \sum_{(i,j) \in \widehat{\mathcal{P}}} \max(0, J_{i,j}),$$

$$J_{i,j} = \max \left(\max_{(i,k) \in \widehat{\mathcal{N}}} \{\alpha - D_h(\mathbf{b}_i, \mathbf{b}_k)\}, \right.$$

$$\left. \max_{(j,l) \in \widehat{\mathcal{N}}} \{\alpha - D_h(\mathbf{b}_j, \mathbf{b}_l)\} \right) + D_h(\mathbf{b}_i, \mathbf{b}_j). \tag{2}$$

$$s.t. \ \mathbf{b}_i, \mathbf{b}_j, \mathbf{b}_k, \mathbf{b}_l \in \{0, 1\}^c$$

where $\widehat{\mathcal{P}}$ and $\widehat{\mathcal{N}}$ are the set of positive and negative pairs (*i.e.*, similar and dissimilar pairs) in the training mini-batch, respectively. Note that here we allow both sample i and j play the role of anchor point in the triplet structure in order to make full use of samples in the batch.

However, such loss function is non smooth, which causes the network converge unstably and is very likely to get into a bad local optimum. Inspired by [14], we turn to optimize a smooth upper bound on Eq. (2), defined as:

$$\tilde{J} = \frac{1}{2|\widehat{\mathcal{P}}|} \sum_{(i,j)\in\widehat{\mathcal{P}}} \max(0, \tilde{J}_{i,j}),$$

$$\tilde{J}_{i,j} = \log\left(\sum_{(i,k)\in\widehat{\mathcal{N}}} \exp\{\alpha - D_h(\mathbf{b}_i, \mathbf{b}_k)\} \right.$$

$$\left. + \sum_{(j,l)\in\widehat{\mathcal{N}}} \exp\{\alpha - D_h(\mathbf{b}_j, \mathbf{b}_l)\} \right) + D_h(\mathbf{b}_i, \mathbf{b}_j). \tag{3}$$

$$s.t. \ \mathbf{b}_i, \mathbf{b}_j, \mathbf{b}_k, \mathbf{b}_l \in \{0, 1\}^c$$

From Eq. (3), we can see that the triplet loss $\tilde{J}_{i,j}$ takes all dissimilar pairs into consideration which makes the overall loss $\tilde{J}$ for each batch more smooth. At the same time, to make use of "difficult" triplets, the exp operator strengthens the contributions of such "difficult" triplets while weakens other "easy" ones in summation terms.

Unfortunately, it is infeasible to optimize Eq. (3) directly because the binary constraints require discretizing the real-valued output of the network (e.g. with signum function) and will make it intractable to train the network with back propagation algorithm. For ease of optimization, we adopt the strategy in [43], *i.e.*, replace the Hamming distance $D_h(\cdot)$ with square of Euclidean distance and relax the binary constraints on $\mathbf{b}$ to range constraints. We formulate the relaxed overall loss function as follows:

$$\tilde{J} = \frac{1}{2|\widehat{\mathcal{P}}|} \sum_{(i,j)\in\widehat{\mathcal{P}}} \max(0, \tilde{J}_{i,j}),$$

$$\tilde{J}_{i,j} = \log\left(\sum_{(i,k)\in\widehat{\mathcal{N}}} \exp\{\alpha - D_e^2(\mathbf{b}_i, \mathbf{b}_k)\} \right.$$

$$\left. + \sum_{(j,l)\in\widehat{\mathcal{N}}} \exp\{\alpha - D_e^2(\mathbf{b}_j, \mathbf{b}_l)\} \right) + D_e^2(\mathbf{b}_i, \mathbf{b}_j). \tag{4}$$

$$s.t. \ \mathbf{b}_i, \mathbf{b}_j, \mathbf{b}_k, \mathbf{b}_l \in (0, 1)^c$$

where D_e denotes the Euclidean distance and the binary constraints on $\mathbf{b}_i, \mathbf{b}_j,$ $\mathbf{b}_k$ and $\mathbf{b}_l$ are relaxed to range constraints of 0 to 1.

With Eq. (4), back-propagation algorithm with mini-batch gradient descent method is applied to train the network. Specifically, we give the gradients of Eq. (4) with respect to the relaxed binary vectors as follows:

$$
\begin{aligned}
\frac{\partial \tilde{J}}{\partial D_e^2(\mathbf{b}_i, \mathbf{b}_j)} &= \frac{1}{2|\widehat{\mathcal{P}}|} \mathbb{1}[\tilde{J}_{i,j} > 0] \\
\frac{\partial \tilde{J}}{\partial D_e^2(\mathbf{b}_i, \mathbf{b}_k)} &= \frac{1}{2|\widehat{\mathcal{P}}|} \mathbb{1}[\tilde{J}_{i,j} > 0] \frac{-\exp\{\alpha - D_e^2(\mathbf{b}_i, \mathbf{b}_k)\}}{\exp\{\tilde{J}_{i,j} - D_e^2(\mathbf{b}_i, \mathbf{b}_j)\}} \\
\frac{\partial \tilde{J}}{\partial D_e^2(\mathbf{b}_j, \mathbf{b}_l)} &= \frac{1}{2|\widehat{\mathcal{P}}|} \mathbb{1}[\tilde{J}_{i,j} > 0] \frac{-\exp\{\alpha - D_e^2(\mathbf{b}_j, \mathbf{b}_l)\}}{\exp\{\tilde{J}_{i,j} - D_e^2(\mathbf{b}_i, \mathbf{b}_j)\}} \\
\frac{\partial D_e^2(\mathbf{b}_i, \mathbf{b}_j)}{\partial \mathbf{b}_i} &= 2(\mathbf{b}_i - \mathbf{b}_j), \quad \frac{\partial D_e^2(\mathbf{b}_i, \mathbf{b}_k)}{\partial \mathbf{b}_i} = 2(\mathbf{b}_i - \mathbf{b}_k) \\
\frac{\partial D_e^2(\mathbf{b}_i, \mathbf{b}_j)}{\partial \mathbf{b}_j} &= 2(\mathbf{b}_j - \mathbf{b}_i), \quad \frac{\partial D_e^2(\mathbf{b}_j, \mathbf{b}_l)}{\partial \mathbf{b}_j} = 2(\mathbf{b}_j - \mathbf{b}_l) \\
\frac{\partial D_e^2(\mathbf{b}_i, \mathbf{b}_k)}{\partial \mathbf{b}_k} &= 2(\mathbf{b}_k - \mathbf{b}_i), \quad \frac{\partial D_e^2(\mathbf{b}_j, \mathbf{b}_l)}{\partial \mathbf{b}_l} = 2(\mathbf{b}_l - \mathbf{b}_j)
\end{aligned}
\tag{5}
$$

where, $\mathbb{1}[\cdot]$ is the indicator function which equals 1 if the expression in the bracket is true and 0 otherwise. As is shown in Eq. (5), the gradients of each iteration contain all negative pairs' information which makes the optimization more stable.

With these computed gradients over mini-batches, the rest of back-propagation can be run in standard manner.

3.3   Implementation Details

Network parameters: We implement our DVC method with Caffe platform[1] [49]. Due to memory limitation, we resize each face image to 100×100. In frame-level feature extraction procedure, each branch consists of three convolution-pooling layers. The convolution layers include 32, 32 and 64 5×5 filters with stride 1 respectively, and the size of pooling window is 3×3 with stride 2. Such branch is duplicated multiple times (both the configuration and the weights are shared) and they work side by side. Following the frame-level feature extraction, we implement the temporal feature pooling layer using the Eltwise layer provided by Caffe. Next is two fully connected layers for transforming the high-dimensional representation to low-dimensional real-valued vector. The first fully connected layer contains 500 nodes and the second contains c nodes, where c is the length of the final binary codes. To satisfy the range constraints, we append a sigmoid layer after the last fully connected layer. Moreover, all the convolution layers and the first fully connected layer are equipped with the ReLU activation function.

During training, the weights of each layer were initialized with "Xavier" method [50]. We set batch size to 200, momentum to 0.9 and weight decay to 0.004. We adopt the fixed learning rate policy with 10^{-4} and train the network with $150,000$ iterations. The margin α in Eq. (4) is empirically set to 1.

---

[1] The source code of DVC is available at http://vipl.ict.ac.cn/resources/codes.

Training methodology: To speed up the converging of the network, we generate triplet samples online as [14] does, *i.e.*, the input of network is not in triplet form but in point-wise form. Triplets are constructed from each batch at loss layer according to their class labels. By doing this, much more triplets can be utilized during each iteration. As a result, the network will converge faster and the structure of the whole network can be simplified. Therefore the computational resources and storage space can be used more efficiently. However, when distribution of different classes in dataset is uneven, if we draw samples randomly, we may fail to construct a triplet in a mini-batch. Worse still, training in such manner would inevitably cause bad result where classes with large number of samples are trained well and the others may be disappointing. To make sure that face videos of each class distribute uniformly in each batch, we randomly select a few (e.g. 10) class labels first, and then load same number of face videos for each selected class. With such processing, the number of triplets in each batch can be guaranteed and the uneven distribution of different classes as will be introduced in Sect. 4.1 can be alleviated.

Another trick for training DVC is the use of finetuning. Since the frame-level feature extraction unit of the designed CNN consists of multiple duplicated branches and the convolution operations are time-consuming, training the multi-branch network straightforwardly is not recommended. Hence we turn to first train the network thoroughly with single frame as input (only preserve one branch in frame-level feature extraction unit). Then we expand the single-branch model to the multi-branch model. In this way, we achieve significant speedup compared to training it from scratch. In a similar way, as networks with different code lengths share the same configurations except the last fully connected layer, we train the long code network by finetuning it with short code network. Due to the limited space, evaluation of the effectiveness of finetuning policy on training DVC is introduced in our supplementary materials.

4 Experiments

In this section, we first evaluate the power of different CNN architectures discussed in Sect. 3.1 for video modeling. Then comparison with state-of-the-art hashing methods is conducted to illustrate the effectiveness of the proposed method.

4.1 Datasets and Evaluation Protocols

We carry face video retrieval experiments on the ICT-TV dataset [15]. ICT-TV dataset contains two face video collections clipped from two popular American TV-Series, *i.e.*, the Big Bang Theory (BBT) and Prison Break (PB). These two TV-Series are quite different in their filming style and therefore pose different challenges. BBT is a sitcom and most stories take place indoors. Each episode contains 5–8 characters. By contrast, many scenes of PB are taken outdoors with a main cast list around 19 characters. Consequently face videos from PB have larger variation of illumination. All the face videos are extracted from the whole first season of each TV-Series, *i.e.*, 17 episodes of BBT and 22 episodes of PB. Each frame of face video has been cropped to the size of 150×150. The numbers of face videos of these two datasets are 4667 and 9435 respectively. Actually the numbers of face videos of different characters are quite different from each other, which range from 11 to 1528 and 49 to 1965 per character in BBT and PB respectively. To tackle this problem, we adopt the training methodology as introduced in Sect. 3.3 (due to the limited space, the distribution of face videos per character and validation of the effectiveness of the trick we adopt can be found in the supplementary materials). We use the extracted block discrete cosine transformation features of face images as used in [19] for all traditional methods, *i.e.*, methods using hand-crafted features.

We abandon the "Unknown" class in both collections, then randomly select $\frac{2}{3}$ of each character's face videos for training and leave the rest for evaluating the retrieval performance. In this way, we obtain 2971 and 5001 face videos for training, and leave 1487 and 2499 face videos for evaluation on BBT and PB respectively. We will introduce how to use the split for different methods in Sects. 4.2 and 4.3 in detail.

Following previous works [19,21,43], we adopt the mean Average Precision (mAP) and precision recall curves calculated among the whole test set of each dataset for quantitative evaluation.

4.2 Evaluation of Video Modeling

In this part, we validate the effectiveness of the proposed temporal feature pooling. We mainly evaluate different video modeling solutions discussed in Sect. 3.1, *i.e.*, single-frame model, 3D CNN model, temporal max-pooling model and temporal average pooling model. We denote them as *Single*, *3DCNN*, *T-max* and *T-avg* respectively for convenience. The CNN architectures of the *T-max*, *T-avg* and *Single* have been discussed in Sect. 3.3. *3DCNN*'s network configuration is same with *Single* except that the convolution kernels of the first convolution layer are $3 \times 3 \times 3T$ where T is the number of frames of a face video, each frame has 3 channels and totally $3T$ channels in temporal dimension. We can observe that the scale of training set on both dataset is too small to train deep CNNs from scratch. To augment data, we fix the number of frames for each face videos to 10 and segment each face video into multiple sub-videos with such size. Specifically, we slide the segment window 5 frames after each segmentation until the window gets to the last frame. By doing this, we expand the training set from 2971 and 5001 to 25590 and 40941, and the test set from 1487 and 2499 to 12735 and 20357 for BBT and PB respectively. The expanded data is supplied to train *3DCNN*, *T-max* and *T-avg*. For *Single* network, all frames of all face videos in training set are used for training. To represent each face video as a whole, we simply average the representations of the segmented sub-videos or all frames belonging to that face video.

The retrieval mAP of different models are listed in Table 1. From this table, we can reach three conclusions: (1) *3DCNN* performs worst both on BBT and PB.

Table 1. Comparison of retrieval mAP of different video modeling methods on BBT and PB with 12-bit binary codes.

	BBT				PB			
	3DCNN	*Single*	*T-avg*	*T-max*	*3DCNN*	*Single*	*T-avg*	*T-max*
mAP	0.9759	**0.9853**	0.9791	0.9808	0.8573	0.9547	0.9552	**0.9590**

The reason is that *3DCNN* aims to learn both spatial and temporal information, which has high request on the network. Besides, it mixes appearance information of multiple frames together in the first convolution layer which causes the network fail to capture the appearance information of each frame in the latter layers. Therefore the appearance information of face is not learned well. However, the performance of *Single* method demonstrates the importance of appearance information for face video retrieval. In spite of local motion information, just extracting the appearance information from each single face by *Single* method can achieve promising even best (on BBT) performance. (2) *T-max* and *T-avg* achieve comparable performances with the *Single* method. The reason why *T-max* and *T-avg* perform slight worse than *Single* on BBT is that BBT only has 25590 sub-videos while the number of frames is 136788 for training, and the numbers of network weights in *Single*, *T-avg* and *T-max* are equal. Therefore the sub-videos in training set of BBT for *T-max* and *T-avg* may be not enough. To verify the assumption, we enlarge the training set to 30529 (reduce test set correspondingly) and find that mAP of *T-max* increases to 0.9917 which outperforms *Single* with 0.9892. Besides, the *Single* method needs more computation of projection through fully connected layers for all frames. *T-max* and *T-avg* methods only need to project sub-videos into binary codes, the computation cost of which is much smaller than the former. For PB, sub-videos for training is about 1.6 times as many as that of BBT which is enough to train *T-max* and *T-avg*. Moreover, samples in PB have larger intra and inter class variations than BBT. Hence information carried by different frames is more complementary and the temporal feature pooling operation can mine such information to represent face videos more stably. Therefore, *T-max* and *T-avg* tend to outperform *single* on PB. (3) *T-max* performs better than *T-avg* both on BBT and PB. Proceed from the procedure of convolution operation, each convolutional filter can be regarded as a local concept classifier or detector. When appearance of frames in a video changes fiercely, some local concepts may only exist on one of those faces. Thus, it is better to use the largest activation of convolutional results than the average one.

4.3   Comparison with the State-of-the-Art

Comparative methods: We compare our DVC with LSH [28], SH [29], BRE [32], ITQ [31], CCA-ITQ [31], MLH [34], KSH [35], CVC [19], DLBHC [41] and DNNH [43]. Strictly speaking, the ten compared hashing methods except CVC are not specifically designed for face video retrieval task. To conduct face video retrieval experiments on these methods, we trained them by treating a face image as a sample and finally all the frame-level binary representations are fused by hard-voting method as the representation of the face video. For fair comparison, DLBHC and DNNH used the same network structure, as the *Single* network in Sect. 4.2. Note that in this case, the DNNH actually is the fully connected version described in [43]. For evaluating the loss function, we also implement the *Single* version of DVC, which is named as **DVC-s**. The *T-max* version of DVC is named as **DVC-m**.

Training set: If possible, we would like to use the whole training data to train all methods. However, MLH and KSH cost too much memory. Hence we had to randomly select 5K and 10K frames from all training face videos for MLH and KSH respectively, which costs more than 8 GB of memory. Parameters of the compared methods were all set based on the authors' suggestions in their original publications.

Results: Table 2 shows the retrieval performance comparison and Fig. 3 gives the precision recall curves on two datasets with 48-bit binary codes (more results with other code lengths can be found in supplementary materials). In general, supervised hashing methods perform better than unsupervised methods, validating the importance of label information for learning similarity-preserving Hamming space. In addition, those CNN-based methods outperform conventional hashing methods by a large margin, demonstrating the advantage of joint feature learning and binary coding. Apart from that, we also attempted to train some conventional hashing methods with CNN features, their performances improved

Table 2. Comparison of retrieval mAP of our DVC method and the other hashing methods on BBT and PB.

Method	BBT				PB			
	12-bit	24-bit	36-bit	48-bit	12-bit	24-bit	36-bit	48-bit
LSH [28]	0.2778	0.3062	0.3171	0.3679	0.1259	0.1360	0.1375	0.1412
SH [29]	0.3745	0.3652	0.3473	0.3329	0.1403	0.1496	0.1504	0.1504
ITQ [31]	0.4771	0.4928	0.4924	0.4968	0.1414	0.1525	0.1563	0.1608
CCA-ITQ [31]	0.7159	0.8141	0.8406	0.8547	0.1819	0.2312	0.2595	0.2814
BRE [32]	0.4275	0.4810	0.4869	0.4860	0.1423	0.1468	0.1501	0.1510
MLH [34]	0.7670	0.8058	0.8141	0.8402	0.2294	0.2325	0.2550	0.2783
KSH [35]	0.8819	0.8830	0.8814	0.8856	0.3405	0.3840	0.3993	0.4086
CVC [19]	0.7784	0.8121	0.8158	0.8166	0.2767	0.3314	0.3554	0.3648
DLBHC [41]	0.9870	0.9914	0.9922	0.9922	0.9476	0.9498	0.9521	0.9602
DNNH [43]	**0.9878**	0.9884	**0.9927**	0.9909	0.9262	0.9306	0.9335	0.9262
DVC-s	0.9853	**0.9933**	**0.9927**	**0.9941**	0.9547	0.9663	**0.9785**	**0.9788**
DVC-m	0.9808	0.9926	0.9917	0.9915	**0.9590**	**0.9707**	0.9741	0.9727

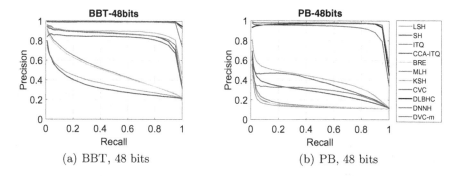

Fig. 3. Comparison with the state-of-the art hash learning methods with precision recall curves on two databases.

significantly, but still inferior to our DVC. Detailed results are introduced in supplementary materials.

It is observed that performances of CNN-based methods on BBT are almost the same because face videos in BBT have small variations. However, when it comes to PB which is more challenging than BBT as described in Sect. 4.1, the performance gap between our method and others especially DNNH becomes larger. We attribute it to the designed loss functions and video representation policies. On one hand, DVC is based on triplet constraints, which aims to optimize the relative rank order among samples. Hence, DVC is very well suited to the preservation of semantic similarity on challenging datasets. Besides, we optimize the smooth upper bound on triplet loss function which biases towards triplets with large loss and simultaneously leads the network to converge stably. However, DLBHC aims to embed label information into binary codes with pointwise loss which neglects the relative similarity among samples, thus encoding dissimilar images to similar codes would not be punished as long as the classification accuracy is unaffected. Though DNNH is based on triplet constraints, most triplets in a batch contribute little to the convergence of the network for that they already meet the constraints and they overwhelm triplets that violate the constraints. Therefore, our DVC-s outperforms them. On the other hand, DVC-m utilizes the temporal max-pooling network to represent face video which has the ability of making use of more complementary information, while DLBHC, DNNH and DVC-s simply average all frame-level binary codes, the final video code is unreliable when faces have large variations in the video. Moreover, the performance of DVC-m tends to be inferior to DVC-s especially when codes become longer on PB, the reason for it is same with that explained in Sect. 4.2, *i.e.*, the number of training sub-videos become insufficient when complexity of network becomes high.

5   Conclusion

In this paper, we propose a multi-branch CNN architecture, which takes face videos as inputs and outputs compact binary codes. The learned DVC achieves promising performance compared with state-of-the-art hashing methods on two challenging face video datasets for face video retrieval. We owe it to two aspects: **First**, the integration of frame-level non-linear convolutional feature learning, video-level modeling by temporal feature pooling and hash coding for extracting compact video code. **Second**, the optimization of a smooth upper bound on triplet loss function for hash learning. In the future, we would construct a larger and more challenging face video dataset to train more complicated CNNs.

Acknowledgements. This work is partially supported by 973 Program under contract No. 2015CB351802, Natural Science Foundation of China under contracts Nos. 61390511, 61379083, 61272321, and Youth Innovation Promotion Association CAS No. 2015085.

References

1. Shan, C.: Face recognition and retrieval in video. In: Schonfeld, D., Shan, C., Tao, D., Wang, L. (eds.) Video Search and Mining. Springer, Heidelberg (2010)
2. Sivic, J., Everingham, M., Zisserman, A.: Person spotting: video shot retrieval for face sets. In: Leow, W.-K., Lew, M.S., Chua, T.-S., Ma, W.-Y., Chaisorn, L., Bakker, E.M. (eds.) CIVR 2005. LNCS, vol. 3568, pp. 226–236. Springer, Heidelberg (2005). doi:10.1007/11526346_26
3. Yamaguchi, O., Fukui, K., Maeda, K.: Face recognition using temporal image sequence. In: FG (1998)
4. Cevikalp, H., Triggs, B.: Face recognition based on image sets. In: CVPR (2010)
5. Hu, Y., Mian, A.S., Owens, R.: Sparse approximated nearest points for image set classification. In: CVPR (2011)
6. Kim, T.K., Kittler, J., Cipolla, R.: Discriminative learning and recognition of image set classes using canonical correlations. IEEE TPAMI **29**, 1005–1018 (2007)
7. Wang, R., Chen, X.: Manifold discriminant analysis. In: CVPR (2009)
8. Wang, R., Shan, S., Chen, X., Gao, W.: Manifold-manifold distance with application to face recognition based on image set. In: CVPR (2008)
9. Wang, R., Guo, H., Davis, L.S., Dai, Q.: Covariance discriminative learning: a natural and efficient approach to image set classification. In: CVPR (2012)
10. Parkhi, O., Simonyan, K., Vedaldi, A., Zisserman, A.: A compact and discriminative face track descriptor. In: CVPR (2014)
11. Ng, J.Y.H., Hausknecht, M., Vijayanarasimhan, S., Vinyals, O., Monga, R., Toderici, G.: Beyond short snippets: Deep networks for video classification. In: CVPR (2015)
12. Wang, J., Shen, H.T., Song, J., Ji, J.: Hashing for similarity search: a survey. arXiv preprint arXiv:1408.2927 (2014)
13. Xu, Z., Yang, Y., Hauptmann, A.G.: A discriminative CNN video representation for event detection. In: CVPR (2015)
14. Song, H.O., Xiang, Y., Jegelka, S., Savarese, S.: Deep metric learning via lifted structured feature embedding. arXiv preprint arXiv:1511.06452 (2015)

15. Li, Y., Wang, R., Shan, S., Chen, X.: Hierarchical hybrid statistic based video binary code and its application to face retrieval in TV-series. In: FG (2015)
16. Arandjelović, O., Zisserman, A.: Automatic face recognition for film character retrieval in feature-length films. In: CVPR (2005)
17. Arandjelović, O., Zisserman, A.: On film character retrieval in feature-length films. In: Hammoud, R.I. (ed.) Interactive Video. Springer, Heidelberg (2006)
18. Everingham, M., Sivic, J., Zisserman, A.: Hello! My name is... Buffy-automatic naming of characters in TV video. In: BMVC (2006)
19. Li, Y., Wang, R., Cui, Z., Shan, S., Chen, X.: Compact video code and its application to robust face retrieval in TV-series. In: BMVC (2014)
20. Li, Y., Wang, R., Huang, Z., Shan, S., Chen, X.: Face video retrieval with image query via hashing across Euclidean space and Riemannian manifold. In: CVPR (2015)
21. Dong, Z., Jia, S., Wu, T., Pei, M.: Face video retrieval via deep learning of binary hash representations. In: AAAI (2016)
22. Ji, S., Xu, W., Yang, M., Yu, K.: 3D convolutional neural networks for human action recognition. IEEE TPAMI **35**, 221–231 (2013)
23. Karpathy, A., Toderici, G., Shetty, S., Leung, T., Sukthankar, R., Li, F.F.: Large-scale video classification with convolutional neural networks. In: CVPR (2014)
24. Pfister, T., Charles, J., Zisserman, A.: Flowing convnets for human pose estimation in videos. In: ICCV (2015)
25. Chatfield, K., Arandjelović, R., Parkhi, O., Zisserman, A.: On-the-fly learning for visual search of large-scale image and video datasets. Int. J. Multimedia Inf. Retrieval **4**, 75–93 (2015)
26. Crowley, E.J., Parkhi, O.M., Zisserman, A.: Face painting: querying art with photos. In: BMVC (2015)
27. Ghaleb, E., Tapaswi, M., Al-Halah, Z., Ekenel, H.K., Stiefelhagen, R.: ACCIO: a data set for face track retrieval in movies across age. In: ACM International Conference on Multimedia Retrieval (2015)
28. Gionis, A., Indyk, P., Motwani, R.: Similarity search in high dimensions via hashing. In: VLDB (1999)
29. Weiss, Y., Torralba, A., Fergus, R.: Spectral hashing. In: NIPS (2009)
30. Liu, W., Wang, J., Kumar, S., Chang, S.F.: Hashing with graphs. In: ICML (2011)
31. Gong, Y., Lazebnik, S.: Iterative quantization: a procrustean approach to learning binary codes. In: CVPR (2011)
32. Kulis, B., Darrell, T.: Learning to hash with binary reconstructive embeddings. In: NIPS (2009)
33. Wang, J., Kumar, S., Chang, S.F.: Semi-supervised hashing for scalable image retrieval. In: CVPR (2010)
34. Norouzi, M., Fleet, D.J.: Minimal loss hashing for compact binary codes. In: ICML (2011)
35. Liu, W., Wang, J., Ji, R., Jiang, Y.G., Chang, S.F.: Supervised hashing with kernels. In: CVPR (2012)
36. Rastegari, M., Farhadi, A., Forsyth, D.: Attribute discovery via predictable discriminative binary codes. In: Fitzgibbon, A., Lazebnik, S., Perona, P., Sato, Y., Schmid, C. (eds.) ECCV 2012. LNCS, vol. 7577, pp. 876–889. Springer, Heidelberg (2012). doi:10.1007/978-3-642-33783-3_63
37. Wang, J., Liu, W., Sun, A., Jiang, Y.G.: Learning hash codes with listwise supervision. In: ICCV (2013)
38. Wang, J., Wang, J., Yu, N., Li, S.: Order preserving hashing for approximate nearest neighbor search. In: ACM International Conference on Multimedia (2013)

39. Xia, R., Pan, Y., Lai, H., Liu, C., Yan, S.: Supervised hashing for image retrieval via image representation learning. In: AAAI (2014)
40. Zhang, R., Lin, L., Zhang, R., Zuo, W., Zhang, L.: Bit-scalable deep hashing with regularized similarity learning for image retrieval and person re-identification. IEEE TIP **24**, 4766–4779 (2015)
41. Lin, K., Yang, H.F., Hsiao, J.H., Chen, C.S.: Deep learning of binary hash codes for fast image retrieval. In: CVPRW (2015)
42. Liong, V.E., Lu, J., Wang, G., Moulin, P., Zhou, J.: Deep hashing for compact binary codes learning. In: CVPR (2015)
43. Lai, H., Pan, Y., Liu, Y., Yan, S.: Simultaneous feature learning and hash coding with deep neural networks. In: CVPR (2015)
44. Zhao, F., Huang, Y., Wang, L., Tan, T.: Deep semantic ranking based hashing for multi-label image retrieval. In: CVPR (2015)
45. Norouzi, M., Fleet, D.J., Salakhutdinov, R.R.: Hamming distance metric learning. In: NIPS (2012)
46. Schroff, F., Kalenichenko, D., Philbin, J.: FaceNet: a unified embedding for face recognition and clustering. In: CVPR (2015)
47. Parkhi, O.M., Vedaldi, A., Zisserman, A.: Deep face recognition. In: BMVC (2015)
48. Nair, V., Hinton, G.E.: Rectified linear units improve restricted Boltzmann machines. In: ICML (2010)
49. Jia, Y., Shelhamer, E., Donahue, J., Karayev, S., Long, J., Girshick, R., Guadarrama, S., Darrell, T.: Caffe: convolutional architecture for fast feature embedding. In: ACM International Conference on Multimedia (2014)
50. Glorot, X., Bengio, Y.: Understanding the difficulty of training deep feedforward neural networks. In: AISTATS (2010)

From Face Images and Attributes to Attributes

Robert Torfason, Eirikur Agustsson, Rasmus Rothe, and Radu Timofte[(✉)]

Computer Vision Laboratory, ETH Zurich, Zurich, Switzerland
radu.timofte@vision.ee.ethz.ch

Abstract. The face is an important part of the identity of a person. Numerous applications benefit from the recent advances in prediction of face attributes, including biometrics (like age, gender, ethnicity) and accessories (eyeglasses, hat). We study the attributes' relations to other attributes and to face images and propose prediction models for them. We show that handcrafted features can be as good as deep features, that the attributes themselves are powerful enough to predict other attributes and that clustering the samples according to their attributes can mitigate the training complexity for deep learning. We set new state-of-the-art results on two of the largest datasets to date, CelebA and Facebook BIG5, by predicting attributes either from face images, from other attributes, or from both face and other attributes. Particularly, on Facebook dataset, we show that we can accurately predict personality traits (BIG5) from tens of 'likes' or from only a profile picture and a couple of 'likes' comparing positively to human reference.

1 Introduction

Attributes are semantic features allowing for mid-level representations, between the low-level features and the high-level labels, with application to recognition of people, objects, and activities [1,2]. Face recognition and verification [3,4] are two domains where attributes have been applied successfully. The facial attributes (see Fig. 1) can include biometrics such as age, ethnicity, gender, but also particular categories distinguishing among, for example, hair colors, types of beards, or hair styles. Other attributes can capture the presence of accessories (such as earrings, eyeglasses, hat), makeup changes (lipstick), and mood/facial expressions. Behind each face is a distinct individual with attributes derived from preferences ('likes' for certain items from Facebook), CV (education, experience), habits, geolocations, and even personality traits. The accurate prediction of such attributes is the key for the success of HCI applications where the computer application needs to 'read' the human by estimation of age, gender, and facial expression to interact accordingly. Visual data can be stored and searched using attributes [5], moreover, in cases when the visual data is insufficient the attributes might bring the extra information for more accurate recognition.

Attribute prediction is a challenging problem and while for biometrics the research spans over decades, the other attributes have been investigated less and, moreover, most of the recent research treats each attribute individually as

© Springer International Publishing AG 2017
S.-H. Lai et al. (Eds.): ACCV 2016, Part III, LNCS 10113, pp. 313–329, 2017.
DOI: 10.1007/978-3-319-54187-7_21

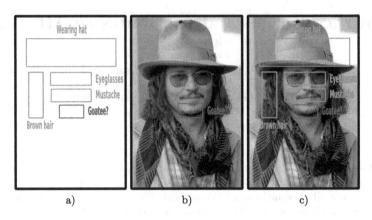

a)   b)   c)

Fig. 1. Attribute completion task: 'Goatee' attribute prediction from (a) other attributes, (b) face image, and (c) combined face image and other attributes.

a prediction from face image domain to the attribute label. The recent release of large-scale attribute datasets such as CelebA [6] (with face attributes) and Facebook BIG5 (with personality traits and 'likes') [7] fueled the recent advances and the interest resurgence for attributes as research topic. Convolutional Neural Networks (CNNs) proved their effectiveness on many computer vision tasks such as classification and end-to-end mappings [8–10]. The same is valid for attribute prediction where the recent top solutions rely on deep learning for representations and/or for direct attribute prediction [1,6,11–14]. We observe that generally the methods treat the prediction of each attribute individually starting from the face image even if there are clear dependencies between some attributes ('males' are unlikely to wear 'earrings' or 'lipsticks', 'females' to be 'bold' and have a 'goatee'). This leads to expensive solutions as the number of models and the training time are linear in the number of attributes and underachievement because of discarding important information – the strong dependencies.

We propose to make use of the strong dependencies among attributes. By doing so, we can robustly train deep features meaningful to all attributes. Furthermore, we can use attributes to predict other attributes and thus to solve the attribute completion task (see Fig. 1). Note that with most of our settings we are capable to achieve state-of-the-art performance for most attributes of two recent large-scale datasets: CelebA [6] and Facebook BIG5 [7].

Our main contributions are as follows:

1. we propose clustering of attributes for efficient CNN representation learning;
2. we show that classic hand-crafted features can be as good as deep features and complementary;
3. we study the correlations between attributes for attribute completion with and without visual features;
4. we achieve state-of-the-art performance for most attributes on CelebA;

5. we achieve better than human references performance for personality traits prediction on Facebook BIG5 dataset by using a history of at least 20 'likes' and surprisingly good accuracy when using solely the profile picture;

The paper is structured as follows. In Sect. 2 we review related works. In Sect. 3 we describe the datasets and experimental setup. In Sect. 4 we study the prediction of attributes when other attributes are known. In Sect. 5 we study the direct prediction of (facial) attributes from face images and propose clustering of attributes for representation learning. In Sect. 6 we analyze the performance when both visual and attributes are used for predicting other attributes. We discuss the results and conclude the paper in Sect. 7.

2   Related Work

Visual attributes in their simplest and most common form are binary properties describing the image contents (see Fig. 1). In general, any (image) property that can be quantified can be an attribute. For example, 'age' as biometric and 'personality trait score' of an user with an image.

Very recently, Liu *et al.* [6] show remarkable attribute prediction results using deep learning on large datasets (**CelebA** and LFWA) with thousands of face images annotated with 40 binary attributes. Youyou *et al.* [7] show that Facebook 'likes' can be used to predict the scores of **BIG5** personality traits. Note that in this case the number of items/attributes that can be 'liked' is very large, while the 'likes' are very sparse and binary. One item is informative for a Facebook user only when is liked, otherwise is difficult to interpret as it needs additional information (did the user not like it on purpose? or was the user not aware of it?). The personality traits scores are real values. Both likes and personality scores are attributes for the Facebook user represented by a profile picture. We denote Youyou *et al.*'s dataset as **Facebook BIG5** dataset. Prior Youyou *et al.*, Kosinski *et al.* [15] successfully predicted different user attributes (biometrics and traits) using Facebook data.

Visual attribute prediction is a **multi-label classification (MLC)** problem, where each image has multiple class labels/attributes. Most of the literature treats binary attributes individually (*e.g.* biometrics such as gender and ethnic group) in both modeling and prediction phases. This is known as Binary Relevance (BR) [16] which does not model correlations between binary labels and uses a number of models/classifiers in the number L of labels/attributes. Pairwise classification (PW) [17] with $\frac{L(L-1)}{2}$ models/classifiers and label combination (LC) [18] with $2^L - 1$ labels are other typical approaches which account for correlations but suffer from time complexity for large L values. Recently, classifier chains (CC) [19] were shown to improve over PW while keeping in check the complexity.

Fine-Grained Image Classification (FGIC) is another direction closely related to our work. FGIC deals with class labels that are visually very similar, equivalent to, in attributes terms, images which differs in as few as 1 attribute when a (very) large number of attributes are available [20].

Multi-task learning (MTL) aims at solving problems with shared and/or related information [21] and is a very active direction in attribute prediction. Wang and Forsyth [22] learn jointly visual attributes, object classes and visual saliency in images with an iterative procedure and multiple instance learning. Wang and Mori [23] learn discriminative latent models for objects and attributes. Parikh and Grauman [24] propose to model relative attributes instead of binary attributes and to relate unseen object categories to seen objects through attributes, which is a form of zero-shot learning.

The visual attributes once predicted are generally used as medium-level semantic features with applications such as indexing and retrieval [3,5], zero-shot classification of unseen objects based on attribute description [25,26], and recognition [4].

3   Experimental Setup

In our work we employ two complementary datasets and corresponding benchmarks as described below.

3.1   CelebA Face Images and Attributes Dataset

CelebA dataset was recently introduced by Liu *et al.* [6] for studying facial attributes in the wild. CelebA contains images of ten thousand celebrities, each with ~20 different images for a total of 202,599 images. Each image in CelebA is annotated with 40 binary facial attributes (such as 'Attractive', 'Male', 'Gray hair', 'bald', 'has beard', 'wearing lipstick', see Fig. 2). Additionally, each image has 5 facial landmarks annotated, 2 for eyes, 1 for the nose and 2 for the corners of the person's mouth. In all our experiments we adhere to the benchmark setup from [6] such that we can directly compare our results on the test partition of the CelebA dataset with other found in the literature.

3.2   Facebook Profile Images, Likes and Personality Traits Dataset

Facebook BIG5 personality traits dataset was recently introduced by Youyou *et al.* [7] for studying the prediction of personality traits from the 'likes' history of the users of the Facebook social network. A 'like' shows a positive association of a user with online and offline items/subjects such as sport, musicians, products, celebrities, etc. However, the absence of a 'like' of a user for an item is not informative because the user equally could have a negative position, neutral position, positive but not expressed, or was not aware of that item within Facebook framework. The dataset has been collected by the myPersonality project[1] and targets Facebook users and their attributes. Thousands of Facebook users willingly agreed to provide their 'likes' history and profile pictures, and to fill in a standard 100-item International Personality Item Pool

---

[1] http://mypersonality.org.

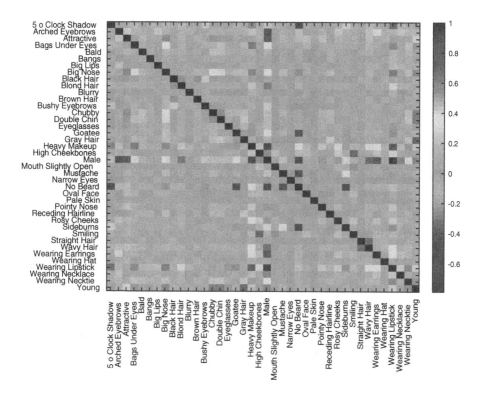

Fig. 2. Correlation between individual attributes in the CelebA dataset.

(IPIP) Five-Factor Model of personality [27] questionnaire [28]. The personality traits (aka BIG5) measured by the 100 IPIP are: openness, conscientiousness, extraversion, agreeableness and neuroticism. Each trait gets a score on a fixed scale. The number of users/samples that have results from the personality test, a history of likes and have profile images available make up the dataset 51,617 samples as used in our experiments. Note that in the original work of Youyou *et al.* [7] were used 70,520 samples (many without profile images) for experimental evaluation. For our experiments we keep the reporting setup from Youyou *et al.* [7] scaled to our lower number of samples. That is, we report Pearson correlation results computed with 10 fold cross-validation. The Pearson correlation is computed between the predicted personality trait score and the ground truth score as obtained based on the filled in 100 IPIP questionnaire by the Facebook user. The self-other agreement, human judgment references for work colleague, friend, family, spouse, humans' average are all reported as in [7].

4   From Attributes to Attribute Prediction

When analyzing a face image, we may already know some of the attributes, *e.g.* based on the identity of the subject, descriptive text or other metadata. In such

Table 1. CelebA prediction results (average accuracy [%] and standard deviation) from subsets of attributes.

# train attr	39	30	20	10	1	Flat prediction
SVM	**87.44 ± 7.77**	86.76 ± 8.03	85.98 ± 8.65	84.23 ± 10.50	80.76 ± 14.27	80.04 ± 15.35
LASSO	87.18 ± 7.72	87.11 ± 8.18	85.92 ± 8.95	84.20 ± 10.32	80.76 ± 14.27	80.04 ± 15.35

cases, we could leverage this information to predict the other attributes, *i.e.* solve the attribute completion task. This is one of the many cases and potential applications when it is desirable to be able to predict attributes or a set of attributes by using a limited set of other attributes.

4.1   CelebA Face Attributes

Each CelebA image is labeled with 40 facial attributes. These attributes have predictive quality for each other due to their dependencies, *e.g.* 'wearing lipstick' correlates with 'heavy makeup' and 'young' with 'attractive'. The amount of annotated attributes and the total number of samples makes CelebA an ideal dataset to analyze the attribute completion task when starting from other attributes. In Fig. 2 we show the matrix of correlations for the facial attributes of CelebA as found in the training data. We represent the samples by binary vectors corresponding to their known attributes at test. We use LASSO regression [29] followed by thresholding, or a linear SVM [30,31] for binary attribute prediction. We consider 5 cases/setups corresponding to the number of attributes that are known and used to build prediction models for the remaining attributes. We train models using 1, 10, 20, 30, and 39 attributes, resp., for predicting the remaining attributes, individually. For the case with 39 known attributes we train 40 SVM models and achieve a 87.44% average accuracy and 7.77 standard deviation at test. Table 1 summarizes how the predictive quality diminishes when smaller subsets of attributes are used to predict other attributes. The reported results were generated by taking 1500 random subsets of 10 attributes, 1000 random subsets of 20 attributes, and 500 random subsets of 30 attributes and the predictive results for the remaining 30, 20, and 10 corresponding attributes were averaged. When predicting 1 attribute from 39 and 39 attributes from 1, the number of permutations is computationally manageable and all different permutations are averaged.

In the setting where 1 attribute is predicted from 1 other attribute, the classifier in general learns the flat prediction. This is the case except for highly correlated features where the prediction improves slightly from the flat baseline prediction.

The average prediction accuracy of 87.44% achieved using solely attributes (39) to predict the missing attributes is remarkable as it compares favorably with many results from Table 2.

4.2   Facebook BIG5 Personality Traits

Each user's history of likes can be viewed as a set of 'attributes' and has predictive quality for each user's personality traits (as measured by the BIG5 personality test). The 'likes' show affinity to certain topics and can therefore have predictive quality. As an example, a user that has 'liked' a Facebook page for 'meditation' or 'yoga' is probably likely to score high on 'openness' in the BIG5 personality test. The data for the 'likes' is very sparse, millions of pages have 'likes' from users but on average for our dataset a user has 140 'likes'. Therefore only a limited subset of the 'likes' categories have any predictive value. Because of this inherent sparsity of the data, LASSO-regression [29] was used for prediction.

First a feature elimination is performed, where pages with less than 10 likes in total are removed, reducing the number of features from 10^6 to 10^4. This subset of features is then regressed on using lasso-regression. In order for the results to be comparable with those gathered in [7] the same methodology and metrics are used. 10-fold cross-validation is performed on the dataset, and the Pearson correlation coefficient is calculated for each split and then averaged over all the splits. For plotting, the data is partitioned based on how many 'likes' a user has, the correlation is plotted as a function of the number of 'likes' a user has. The average results over the 5 personality traits are depicted in Fig. 6. The number of users that have 'likes' below 10 is small, so to estimate the predictive quality of users having 1 to 9 likes, we randomly query users with 10–20 'likes' and

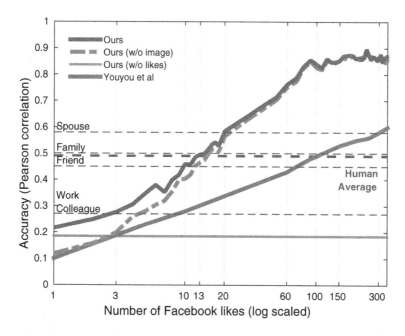

Fig. 3. Average accuracy of personality traits prediction. Human references included.

Table 2. CelebA prediction results (accuracy [%]).

Attribute \ Method	All 0 prediction	Flat prediction	FaceTracer [32]	PANDA-w [1]	PANDA-1 [1]	[33]+ANet [6]	LNets+ANet(w/o) [6]	LNets+ANet [6]	Attributes (ours)	fc6 ft (ours)	fc6 ft + attr. (ours)	fc6 ft + hc (ours)	fc6 ft+hc+attr.(ours)
5 o'Clock Shadow	90.01	90.01	85	82	88	86	88	91	93.33	93.80	94.66	94.16	**94.87**
Arched Eyebrows	71.56	71.56	76	73	78	75	74	79	80.59	79.91	83.39	80.99	**84.08**
Attractive	50.42	50.42	78	77	81	79	77	81	78.20	81.04	81.87	82.10	**82.62**
Bags Under Eyes	79.74	79.74	76	71	79	77	73	79	83.19	82.82	85.40	83.61	**85.79**
Bald	97.88	97.88	89	92	96	92	95	98	97.89	98.80	**98.85**	98.72	98.78
Bangs	84.43	84.43	88	89	92	94	92	95	84.45	94.42	94.55	95.02	**95.14**
Big Lips	67.30	67.30	64	61	67	63	66	68	71.32	68.75	73.08	70.85	**73.39**
Big Nose	78.80	78.80	74	70	75	74	75	78	83.55	81.95	84.57	82.12	**85.35**
Black Hair	72.84	72.84	70	74	85	77	84	88	76.90	87.37	88.27	87.45	**88.60**
Blonde Hair	86.67	86.67	80	81	93	86	91	95	86.65	95.30	95.63	95.33	**95.77**
Blurry	94.94	94.94	81	77	86	83	80	84	94.94	95.91	96.04	96.13	**96.24**
Brown Hair	82.03	82.03	60	69	77	74	78	80	82.44	86.61	87.44	87.22	**88.04**
Bushy Eyebrows	87.05	87.05	80	76	86	80	85	90	87.28	89.32	89.55	90.65	**90.75**
Chubby	94.70	94.70	86	82	86	86	86	91	95.78	95.03	95.95	95.26	**96.16**
Double Chin	95.43	95.43	88	85	88	90	88	92	96.47	95.79	96.68	96.15	**96.80**
Eyeglasses	93.54	93.54	98	94	98	96	96	**99**	93.56	98.69	98.72	98.92	98.89
Goatee	95.42	95.42	93	86	93	92	92	95	95.83	96.77	97.08	96.93	**97.15**
Gray Hair	96.81	96.81	90	88	94	93	93	97	96.90	97.95	98.03	98.16	**98.29**
Heavy Makeup	59.50	59.50	85	84	90	87	85	90	88.75	89.92	91.19	90.61	**91.78**
High Cheekbones	51.82	51.82	84	80	86	85	84	**87**	84.20	85.65	87.00	86.21	87.41
Male	61.35	61.35	91	93	97	95	94	**98**	93.56	97.13	97.61	97.50	97.74
Mouth Slightly Open	50.49	50.49	87	82	**93**	85	86	92	77.00	87.70	87.79	89.35	89.27
Mustache	96.13	96.13	91	83	93	87	91	95	96.74	96.62	**97.31**	96.67	97.25
Narrow Eyes	85.13	85.13	82	79	84	83	77	81	85.13	85.21	86.36	85.52	**85.99**
No Beard	14.63	85.37	90	87	93	91	92	95	94.87	95.25	96.31	95.34	**96.38**
Oval Face	70.44	70.44	64	62	65	65	63	66	75.83	74.06	77.71	75.00	**78.33**
Pale Skin	95.79	95.79	83	84	91	89	87	91	95.79	95.81	95.83	96.71	**96.81**
Pointy Nose	71.43	71.43	68	65	71	67	70	72	72.87	74.45	74.74	75.13	**75.60**
Receding Hairline	91.51	91.51	76	82	85	84	85	89	91.65	92.13	92.52	92.54	**92.67**
Rosy Cheeks	92.83	92.83	84	81	87	85	87	90	92.84	93.95	94.26	94.43	**94.82**
Sideburns	95.36	95.36	94	90	93	94	91	96	96.04	97.30	97.53	97.54	**97.58**
Smiling	49.97	50.03	89	89	92	92	88	92	84.76	91.59	92.22	92.05	**92.65**
Straight Hair	79.01	79.01	63	67	69	70	69	73	79.29	82.59	83.00	82.58	**83.21**
Wavy Hair	63.60	63.60	73	76	77	79	75	80	76.09	82.58	83.90	82.40	**84.17**
Wearing Earrings	79.34	79.34	73	72	78	77	78	82	81.97	86.47	86.60	86.90	**87.25**
Wearing Hat	95.80	95.80	89	91	96	93	96	**99**	95.79	98.45	98.59	98.76	98.89
Wearing Lipstick	47.81	52.19	89	88	93	91	90	93	92.92	93.15	94.11	93.46	**94.13**
Wearing Necklace	86.21	86.21	68	67	67	70	68	71	86.31	86.47	**86.80**	86.38	86.80
Wearing Necktie	92.99	92.99	86	88	91	90	86	93	93.23	95.69	**95.91**	95.76	95.82
Young	24.29	75.71	80	77	84	81	83	87	82.69	88.36	88.72	88.29	**88.92**
Average	76.88	80.04	81	79	85	83	83	87	87.44	89.77	90.62	90.22	**91.00**

randomly subsample their likes to create user queries with the desired number of likes. The number of 'likes' a user has is a predictive indicator of personality, so to minimize this bias we restrict ourselves to query the users that are most similar in number of likes, *i.e.* the users that have 10–20 likes.

As shown in Fig. 3 our average prediction of personality traits based solely on likes improves with the number of 'likes' available per user. This is expected as more information is known about the user. Note that our solution improves greatly over the results reported by Youyou *et al.* [7] for 4 or more likes, and that above 20 likes exceeds the performance of the spouse (0.58) to reach a plateau at more than 100 likes (0.86). Also, Youyou *et al.* LOWESS smooth and disattenuate their results, while we report the raw results.[2]

In the remainder of the paper, if not mentioned otherwise, we will use SVM prediction for CelebA experiments as the features are medium to low dimensional and the LASSO prediction is used for Facebook BIG5 due to the high dimensionality and sparsity of the data.

5   From Face Image to Attribute Prediction

In this section we study attribute prediction task when starting from a face image, which is the traditional way of visual attribute prediction. In particular we evaluate different handcrafted and deep learned features on CelebA to then validate and report prediction results on both CelebA and Facebook BIG5 datasets. We used the off-the-shelf face detector of Mathias *et al.* [34] for detection and alignment of the faces as done for the DEX Network in [11,12].

a) For the deep features descriptors the face is located and then a 40% padding is added around the face to get more context from the surrounding region

b) After the face is located it is split into blocks and handcrafted features are extracted for each block

c) Using relative location of facial landmarks, regions of interest are extracted from the face and features extracted individually for each region

Fig. 4. Extraction sections for visual features.

5.1   Handcrafted Features

The recent years showed deep learned features to be generally more robust and to provide the best results for many computer vision tasks. However, they suffer from complexity as they usually require large datasets for training and have higher time and memory complexities for both training and testing than many handcrafted features (hc). Therefore, in the following we study three typical handcrafted features to embed the visual information from face images: LBP, SIFT and Color histograms. They are extracted in all the sections marked in Fig. 4 and concatenated per each feature type.

---

[2] Disattenuation always leads to equal or better results.

Locally binary patterns (LBP) [35] were proposed by Ojala *et al.* For each pixel in the image, its surrounding pixels are thresholded based on the pixel-value, which results in a binary number. A histogram of these numbers over an image or a section of an image is a powerful texture descriptor. In [35] a generalized gray-scale and rotation invariant version of the locally-binary-patterns is proposed, so called 'uniform' locally-binary-patterns. Here a 8 point LBP with radius 1 is used which results in a 10 bin histogram. This histogram is l2-normalized resulting in a 10-dimensional feature vector extracted for each section marked in Fig. 4.

Scale-invariant feature transform (SIFT). For each section in Fig. 4 a SIFT descriptor [36] is extracted. A neighborhood around the center of the image (or center of the image subsection) is taken. This neighborhood is divided into 16 sub-blocks and for each sub-block a 8 bin orientation histogram is created. This descriptor is then l2-normalized and used as a 128-dimensional feature vector.

Color histograms (Color). The color histograms have 64 uniform bins $(4 \times 4 \times 4)$ in the RGB color-space. These histograms are computed per each face image section and l2-normalized which results in a 64-dimensional feature vector.

5.2   Deep Features

For the deep features, we extract the **fc6** and **fc7** 4096-dimensional layers of the DEX Network [11], which is a variant of the VGG-16 [37] trained for age estimation on hundreds of thousands of face images from IMDB-WIKI dataset [11,12]. Since most facial attributes are relevant signals for age estimation, these features work quite well off-the-shelf.

To further improve performance, we can *finetune (ft)* the network. However this poses a challenge, since attribute prediction is not a standard classification problem, but a multi-label one. While the network could be adapted for direct attribute prediction, it would result in a mismatch between the pre-trained network, trained for age classification, and the target task, attribute prediction. Instead, we cast the attribute prediction into a classification problem by predicting *attribute configurations*. To do this directly would be infeasible, since L binary attributes have 2^L distinct possible configurations. However, due to the inter-dependencies between attributes, many of these configurations are unlikely to occur. We exploit this, and quantize the configuration space, by partitioning the training labels into K clusters (using standard k-means clustering).

This transformation of the labels encodes the inter-dependencies of the attributes into a compact representation, for which we can learn fine-tuned features by treating as a a K-class classification task over the images. We thus only need to fine-tune the network once, obtaining a feature representation suitable for predicting jointly *all the attributes*, by discarding the last layer(s) of the network. For the final model, the quantized labels are discarded and an independent linear classifier is trained for each original attributes.

Compared to fine-tuning a separate binary classification network for each attribute, this reduces the fine-tuning time by a factor L. In all our fine-tuning experiments we set $K = 50$, and iterate over 15000 minibatches with 10 images per batch, based on monitoring the validation error on a 90%/10% split of the training data.

In the same way as for the handcrafted features, the deep features are l2-normalized and when concatenated with other features, the total concatenation is also l2-normalized.

5.3   Prediction Results

CelebA. In Table 3 (w/o attributes) we report the performance on CelebA using various combinations of handcrafted and deep features extracted from the face image. We report the numbers for each of these settings using standard accuracy, positive rate $P = tp/(tp + fp)$, negative rate $N = tn/(tn + fn)$ and the average $(P+N)/2$, where tp, fp are the number of true and false positives, and tn, fn the number of true and false negatives. The average $(P+N)/2$ provides an unbiased (ub.) metric for highly unbalanced class predictions as is the case for the CelebA dataset. The difference between the baseline and any of the classification methods becomes more pronounced and the numbers more illustrative of how the true predictions are split between the positive and the negative labels.

The focus in the discussion is on the accuracy numbers as they are the metric to compare to previous work but in brackets behind those numbers the unbiased metric will be reported.

Table 3. CelebA prediction results. Acc: Standard accuracy. P: Positive rate $tp/(tp + fp)$. N: Negative rate $tn/(tn + fn)$. Unbiased metric $(P + N)/2$.

| | Prediction method | | Average performance [%] | | | | | | |
| | | | w/o attributes | | | | w/ attributes | | |
	Metric	Acc	P	N	$\frac{P+N}{2}$	Acc	P	N	$\frac{P+N}{2}$
	Flat prediction	80.04	6.58	73.46	40.02	–	–	–	–
	Attributes (39)	–	–	–	–	87.44	62.05	88.29	75.17
handcraft	Color	84.52	55.48	84.33	69.90	88.58	71.65	89.56	80.60
	LBP	84.50	46.89	84.06	65.47	88.34	70.30	89.32	79.81
	SIFT	87.02	65.24	87.13	76.18	89.21	72.30	90.15	81.23
	Color + LBP + SIFT	88.27	71.35	88.73	80.04	89.90	75.49	90.90	83.19
deep features	fc6	88.76	71.93	89.47	80.70	90.02	75.40	91.01	83.21
	fc7	88.12	70.66	88.75	79.70	89.65	73.89	90.50	82.20
	PCA (fc6 + fc7)	88.80	72.68	89.48	81.08	90.06	75.73	91.03	83.38
	fc6 ft	89.77	74.23	90.56	82.40	90.62	76.60	91.53	84.07
	fc7 ft	89.41	72.91	90.12	81.52	90.27	75.94	91.14	83.54
combined	fc6 +Color+LBP+SIFT	89.81	74.91	90.68	82.79	90.72	77.75	91.65	84.70
	fc7 +Color+LBP+SIFT	89.55	74.96	90.18	82.57	90.49	77.52	91.25	84.38
	PCA +Color+LBP+SIFT	89.82	75.31	90.69	83.00	90.69	77.22	91.71	84.47
	fc6 ft +Color+LBP+SIFT	**90.22**	**76.07**	**91.01**	**83.54**	**91.00**	**78.35**	**91.89**	**85.12**
	fc7 ft +Color+LBP+SIFT	90.01	75.51	90.90	83.21	90.78	78.08	91.64	84.86
	LNets+ANet [6]	87	–	–	–	–	–	–	–

Since most attributes in CelebA are unlikely to occur, the Flat prediction (that is predicting always the most often occurring 0 or 1 for each attribute) gives a powerful baseline of 80.04% (40.02% for ub. metric $(P+N)/2$). The handcrafted features (Color, LBP, SIFT) boost the performance significantly over Flat, with their combination reaching 88.27% (80.04% ub.). As often reported, using off-the-shelf deep features (Deep fc6 from DEX Network) beats the handcrafted ones by a small margin, +0.49% (+0.66% ub.). Surprisingly though, the handcrafted and deep features are quite complementary, giving 89.81% (82.79% ub.) in combination.

The finetuned features (Deep fc6 - finetuned) significantly improve the performance, boosting the non-finetuned by +1.01% (1.70% ub.) and giving 90.22% (83.54% ub.) in combination with the handcrafted ones. In our setups the deep fc6 features always led to better results than the deep fc7 features, while the combination of fc6 and fc7 features through a PCA projection preserving 99.9% of the energy just marginally improves over the results using fc6 alone.

Facebook BIG5. In Fig. 3 we show the accuracy of personality trait prediction on Facebook BIG5 dataset when using solely the profile image information, the 'Ours (w/o likes)' plot. It is remarkable that the performance of 0.18, achieved using deep fc6 features from DEX Network, is better than that using solely up to 3 likes and not far from the work colleague reference of 0.27.

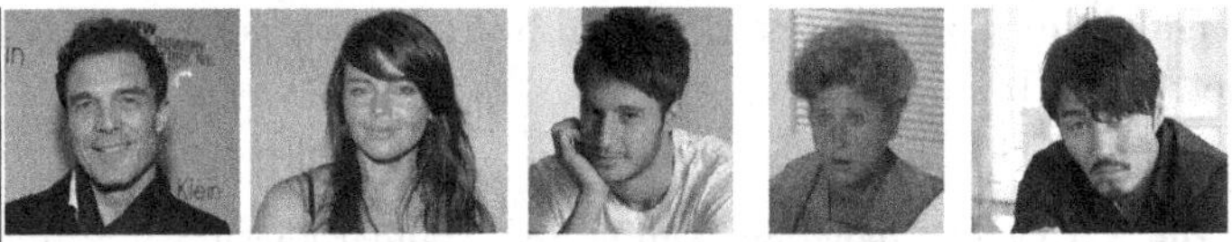

	Image Predictor	A I C	A I C	A I C	A I C	A I C
attributes	Arched Eyebrows	0 0 0	0 0 0	0 1 1	0 0 0	0 0 0
	Bangs	0 0 0	0 1 1	0 0 0	0 0 0	0 0 0
	Black Hair	0 1 1	0 1 1	0 0 0	0 0 0	1 1 1
	Male	1 1 1	0 0 0	0 0 0	1 1 1	1 1 1
	No Beard	1 0 1	1 1 1	1 1 1	1 1 1	0 0 0
	Young	1 1 1	1 1 1	1 1 1	1 1 1	1 1 1

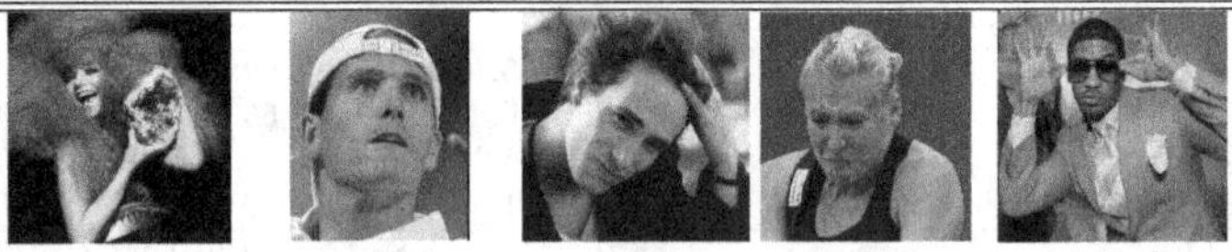

	Image Predictor	A I C	A I C	A I C	A I C	A I C
attributes	Balds	0 0 0	0 0 0	0 0 0	0 0 0	0 0 0
	Eyeglasses	0 0 0	0 0 0	0 0 0	0 0 0	0 1 1
	High Cheekbones	1 0 0	0 0 0	1 0 0	0 0 0	0 0 0
	Mustache	0 0 0	0 0 0	0 0 0	0 0 0	1 0 1
	Wavy Hair	0 1 1	0 0 0	0 0 0	0 0 0	0 0 0
	Wearing Hat	0 0 0	0 1 1	0 0 0	0 0 0	0 0 0

Fig. 5. Examples of correct (green) and wrong (red) attribute predictions from (**A**)ttributes, (**I**)mages (fc6 ft + hc), and (**C**)ombined attributes and images (fc6 ft + hc + attr). (Color figure online)

6   From Face Images and Attributes to Attributes

In previous sections we achieved top attribute prediction accuracies when starting from either i) other attributes, or ii) handcrafted features and/or deep features extracted from face images. In this section we combine and discuss the combination of face features and attributes for improved attribute prediction.

CelebA. The image features and the attributes are complimentary. Image features provide visual information while the attributes can provide information that is otherwise not easy to gather from images. Thus, we predict attributes by using image features and the rest of the attributes (concatenate image features with attributes). As can be seen in Table 2 (columns with +attr), concatenating the attributes with image features gives a boost to the average accuracy, up to 91.00% (85.12% ub.) when using the Deep fc6 and handcrafted features. In Fig. 5 we show several examples of correct and wrong attribute prediction when starting from attributes (39), from image features (fc6 ft + hc), and from the combined attributes and images (fc6 ft + hc + attr). There are a couple of examples (Male, Bangs and Young attributes) where the failure is shared by all 3 predictors, while in most other cases at least 1 of the 3 predictors is correct (Fig. 5).

Facebook BIG5. The likes provide a reliable way to predict the results of the BIG5 personality tests (in terms of Pearson correlation coefficient) but when the number of likes is reduced this prediction obviously breaks down. Therefore

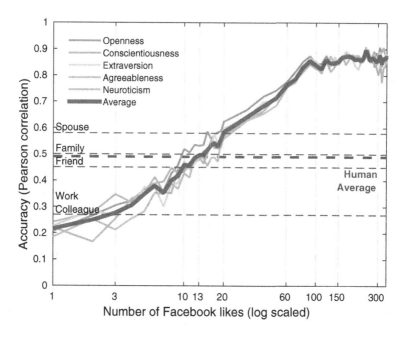

Fig. 6. Accuracy of personality traits prediction using 'likes' and image information.

we employ the profile image as additional information to the likes (see Fig. 3). In the 'region of few likes' the profile images from the Facebook BIG5 dataset provide a way to augment the prediction so that instead of starting from 0, one can obtain a good prediction using only the image (a sound 0.18 Pearson correlation). When the number of likes increases the prediction becomes more reliable. With 3 likes and a profile face image the prediction performance surpasses the work colleague reference. With 10 likes and a picture we get comparable to the friend performance, while above 20 likes we exceed the spouse in predicting the personality traits of the Facebook user. Above 20 likes there is just a marginal benefit from using the profile image. Thus, the importance of the profile image is critical when very few or no likes are available and diminishes with the increase in the number of likes available for a user.

7   Conclusion

In this paper we studied the prediction of attributes from other attributes, from face images, and from both attributes and face images. At the same time we analyzed handcrafted and deep visual features. Our validation was done on very recent large and complementary datasets: CelebA and Facebook BIG5. Our main findings are as follows:

1. The attributes to attribute prediction is a powerful technique not only on facial attribute prediction (CelebA) but also for predicting personality attributes based on sparse attributes such as likes in Facebook.
2. The handcrafted features, while a bit overlooked by the most recent works, are capable to capture the visual information and to lead to comparable performance to methods based on deep features. Furthermore, they are complementary when combined with the deep features.
3. The combination of visual and attribute information leads to better results than when using each individually.
4. For efficient deep learning we propose clustering of samples according to the attributes and, thus, achieve significant time savings when compared to per attribute training (a factor of 40 for CelebA).
5. We achieve state-of-the-art prediction for most facial attributes from CelebA.
6. With as few as 20 likes we are able to predict as accurate as a spouse the personality traits for a Facebook user, with only 3 likes and a profile picture we reach the level of accuracy of a work colleague, while at 100 likes we achieve a 0.86 Pearson correlation of the prediction with the ground truth.
7. We require 3 up to 10 times fewer likes for better BIG5 prediction accuracy when compared with previous results reported on Facebook BIG5 dataset.

Some observations are harsh but necessary:

(i) The flat prediction, without even touching the test samples, is comparable to a number of (involved) methods from the literature.

(ii) Common standard handcrafted features combined with standard linear classifiers can achieve top performance comparable or better than existing deep learned top methods, and are complementary to the deep features.

(iii) People are amazingly easy to 'read' by knowing just a profile picture and/or a couple of likes in a social network as shown by our experiments.

As future work we plan to further explore the relationships between images, attributes and any other information connected to an individual, object, or simple conceptualized instance and to be able to bridge them such that to not only predict attributes but to embed knowledge and transfer it.

Acknowledgement. This work was supported by the ETH General Fund (OK) and by a K40 GPU grant from NVidia. We thank Michal Kosinski and David Stillwell for providing the Facebook BIG5 dataset.

References

1. Zhang, N., Paluri, M., Ranzato, M., Darrell, T., Bourdev, L.: PANDA: pose aligned networks for deep attribute modeling. In: The IEEE Conference on Computer Vision and Pattern Recognition (CVPR) (2014)
2. Liu, J., Kuipers, B., Savarese, S.: Recognizing human actions by attributes. In: 2011 IEEE Conference on Computer Vision and Pattern Recognition (CVPR), pp. 3337–3344 (2011)
3. Kumar, N., Berg, A., Belhumeur, P.N., Nayar, S.: Describable visual attributes for face verification and image search. IEEE Trans. Pattern Anal. Mach. Intell. **33**, 1962–1977 (2011)
4. Layne, R., Hospedales, T.M., Gong, S.: Person re-identification by attributes. In: BMVC (2012)
5. Siddiquie, B., Feris, R.S., Davis, L.S.: Image ranking and retrieval based on multi-attribute queries. In: 2011 IEEE Conference on Computer Vision and Pattern Recognition (CVPR), pp. 801–808 (2011)
6. Liu, Z., Luo, P., Wang, X., Tang, X.: Deep learning face attributes in the wild. In: The IEEE International Conference on Computer Vision (ICCV) (2015)
7. Youyou, W., Kosinski, M., Stillwell, D.: Computer-based personality judgments are more accurate than those made by humans. Proc. Natl. Acad. Sci. **112**, 1036–1040 (2015)
8. Ciresan, D., Meier, U., Schmidhuber, J.: Multi-column deep neural networks for image classification. In: 2012 IEEE Conference on Computer Vision and Pattern Recognition (CVPR), pp. 3642–3649 (2012)
9. Krizhevsky, A., Sutskever, I., Hinton, G.E.: Imagenet classification with deep convolutional neural networks. In: Pereira, F., Burges, C.J.C., Bottou, L., Weinberger, K.Q. (eds.) Advances in Neural Information Processing Systems, vol. 25, pp. 1097–1105. Curran Associates, Inc., New York (2012)
10. He, K., Zhang, X., Ren, S., Sun, J.: Deep residual learning for image recognition. In: 2016 IEEE Conference on Computer Vision and Pattern Recognition (CVPR) (2016)
11. Rothe, R., Timofte, R., Van Gool, L.: DEX: deep expectation of apparent age from a single image. In: The IEEE International Conference on Computer Vision (ICCV) Workshops. (2015)

12. Rothe, R., Timofte, R., Van Gool, L.: Deep expectation of real and apparent age from a single image without facial landmarks. Int. J. Comput. Vis., 1–14 (2016). doi:10.1007/s11263-016-0940-3
13. Rothe, R., Timofte, R., Van Gool, L.: Some like it hot - visual guidance for preference prediction. In: 2016 IEEE Conference on Computer Vision and Pattern Recognition (CVPR) (2016)
14. Uricar, M., Timofte, R., Rothe, R., Matas, J., Van Gool, L.: Structured output SVM prediction of apparent age, gender and smile from deep features. In: Computer Vision and Pattern Recognition (CVPR) Workshops (2016)
15. Kosinski, M., Stillwell, D., Graepel, T.: Private traits and attributes are predictable from digital records of human behavior. Proc. Natl. Acad. Sci. **110**, 5802–5805 (2013)
16. Tsoumakas, G., Katakis, I.: Multi-label classification: an overview. Department of Informatics, Aristotle University of Thessaloniki, Greece (2006)
17. Fürnkranz, J., Hüllermeier, E., LozaMencía, E., Brinker, K.: Multilabel classification via calibrated label ranking. Mach. Learn. **73**, 133–153 (2008)
18. Boutell, M.R., Luo, J., Shen, X., Brown, C.M.: Learning multi-label scene classification. Pattern Recogn. **37**, 1757–1771 (2004)
19. Read, J., Pfahringer, B., Holmes, G., Frank, E.: Classifier chains for multi-label classification. Mach. Learn. **85**, 333–359 (2011)
20. Akata, Z., Reed, S., Walter, D., Lee, H., Schiele, B.: Evaluation of output embeddings for fine-grained image classification. In: The IEEE Conference on Computer Vision and Pattern Recognition (CVPR) (2015)
21. Caruana, R.: Multitask learning. Mach. Learn. **28**, 41–75 (1997)
22. Wang, G., Forsyth, D.: Joint learning of visual attributes, object classes and visual saliency. In: 2009 IEEE 12th International Conference on Computer Vision, pp. 537–544 (2009)
23. Wang, Y., Mori, G.: A discriminative latent model of object classes and attributes. In: Daniilidis, K., Maragos, P., Paragios, N. (eds.) ECCV 2010. LNCS, vol. 6315, pp. 155–168. Springer, Heidelberg (2010). doi:10.1007/978-3-642-15555-0_12
24. Parikh, D., Grauman, K.: Relative attributes. In: 2011 International Conference on Computer Vision, pp. 503–510 (2011)
25. Lampert, C.H., Nickisch, H., Harmeling, S.: Learning to detect unseen object classes by between-class attribute transfer. In: 2009 IEEE Conference on Computer Vision and Pattern Recognition, CVPR 2009, pp. 951–958 (2009)
26. Palatucci, M., Pomerleau, D., Hinton, G.E., Mitchell, T.M.: Zero-shot learning with semantic output codes. In: Bengio, Y., Schuurmans, D., Lafferty, J.D., Williams, C.K.I., Culotta, A. (eds.) Advances in Neural Information Processing Systems, vol. 22, pp. 1410–1418. Curran Associates, Inc., New York (2009)
27. Costa, P.T., McCrae, R.R.: Revised NEO personality inventory (NEO PI-R) and NEP five-factor inventory (NEO-FFI): professional manual. Psychological Assessment Resources Lutz, FL (1992)
28. Goldberg, L.R., Johnson, J.A., Eber, H.W., Hogan, R., Ashton, M.C., Cloninger, C.R., Gough, H.G.: The international personality item pool and the future of public-domain personality measures. J. Res. Pers. **40**, 84–96 (2006). Proceedings of the 2005 Meeting of the Association of Research in PersonalityAssociation of Research in Personality
29. Tibshirani, R.: Regression shrinkage and selection via the lasso. J. Roy. Stat. Soc. Ser. B (Methodol.) **58**, 267–288 (1996)
30. Cortes, C., Vapnik, V.: Support-vector networks. Mach. Learn. **20**, 273–297 (1995)

31. Chang, C.C., Lin, C.J.: LIBSVM: a library for support vector machines. ACM Trans. Intell. Syst. Technol. **2**, 27:1–27:27 (2011)
32. Kumar, N., Belhumeur, P., Nayar, S.: FaceTracer: a search engine for large collections of images with faces. In: Forsyth, D., Torr, P., Zisserman, A. (eds.) ECCV 2008. LNCS, vol. 5305, pp. 340–353. Springer, Heidelberg (2008). doi:10.1007/978-3-540-88693-8_25
33. Li, J., Zhang, Y.: Learning surf cascade for fast and accurate object detection. In: 2013 IEEE Conference on Computer Vision and Pattern Recognition (CVPR), pp. 3468–3475 (2013)
34. Mathias, M., Benenson, R., Pedersoli, M., Gool, L.: Face detection without bells and whistles. In: Fleet, D., Pajdla, T., Schiele, B., Tuytelaars, T. (eds.) ECCV 2014. LNCS, vol. 8692, pp. 720–735. Springer, Heidelberg (2014). doi:10.1007/978-3-319-10593-2_47
35. Ojala, T., Pietikainen, M., Maenpaa, T.: Multiresolution gray-scale and rotation invariant texture classification with local binary patterns. IEEE Trans. Pattern Anal. Mach. Intell. **24**, 971–987 (2002)
36. Lowe, D.G.: Object recognition from local scale-invariant features. In: Proceedings of the International Conference on Computer Vision, ICCV 1999, vol. 2, p. 1150. IEEE Computer Society, Washington, DC (1999)
37. Simonyan, K., Zisserman, A.: Very deep convolutional networks for large-scale image recognition. CoRR abs/1409.1556 (2014)

Learning with Ambiguous Label Distribution
for Apparent Age Estimation

Ke Chen$^{(\boxtimes)}$ and Joni-Kristian Kämäräinen

Department of Signal Processing, Tampere University of Technology,
33720 Tampere, Finland
{ke.chen,joni.kamarainen}@tut.fi

Abstract. Annotating age classes for humans' facial images according to their appearance is very challenging because of dynamic person-specific ageing pattern, and thus leads to a set of unreliable apparent age labels for each image. For utilising ambiguous label annotations, an intuitive strategy is to generate a *pseudo* age for each image, typically the average value of manually-annotated age annotations, which is thus fed into standard supervised learning frameworks designed for chronological age estimation. Alternatively, inspired by the recent success of label distribution learning, this paper introduces a novel concept of ambiguous label distribution for apparent age estimation, which is developed under the following observations that (1) soft labelling is beneficial for alleviating the suffering of inaccurate annotations and (2) more reliable annotations should contribute more. To achieve the goal, label distributions of sparse age annotations for each image are weighted according to their reliableness and then combined to construct an ambiguous label distribution. In the light, the proposed learning framework not only inherits the advantages from conventional learning with label distribution to capture latent label correlation but also exploits annotation reliableness to improve the robustness against inconsistent age annotations. Experimental evaluation on the FG-NET age estimation benchmark verifies its effectiveness and superior performance over the state-of-the-art frameworks for apparent age estimation.

1 Introduction

Chronological age estimation [1–7] is to predict persons' true age given their facial images, which is a hot yet challenging topic in computer vision. In supervised learning based frameworks for age estimation, the unique chronological age labels are provided to supervise model training. However, due to inherent ambiguities in age annotation, a large number of facial images can be readily found on the Internet, but reliable annotations of the exact age of images are usually lacking, which leads to sparsely distributed data [1] in the public benchmarks such as the FG-NET and MORPH datasets. More challengingly, apparent age estimation investigated in this paper is to estimate apparent ages of human faces (intuitively, how old the persons look like) from apparent age annotations instead of their chronological age. Apparent age estimation can be categorised

© Springer International Publishing AG 2017
S.-H. Lai et al. (Eds.): ACCV 2016, Part III, LNCS 10113, pp. 330–343, 2017.
DOI: 10.1007/978-3-319-54187-7_22

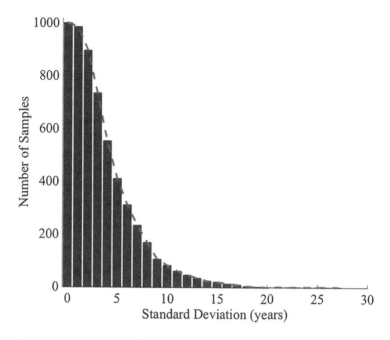

Fig. 1. Number of samples with larger standard deviation than coordinates in the horizontal axis to reflect label ambiguity on the FG-NET benchmark with manual annotations provided by Han *et al.* [8]. The maximum of standard deviation for age annotations is 27.14; standard deviation of 41.12% samples is larger than 5 years, while that of 8.38% samples is larger than 10 years.

into a weakly-supervised learning paradigm, as the supervised information conveyed in such a problem are implicit. Learning a mapping function between imagery feature representation and a set of age label annotations for each facial image is made even more difficult due to large variation of persons' appearance caused by both intrinsic and extrinsic factors and label ambiguity.

On one hand, low-level feature extracted from facial images is largely varied caused by intrinsic and extrinsic factors. Person-specific ageing procedure generally lies in the changes of shape (during childhood) and texture (during adulthood). In this sense, visual appearance of faces varies a lot across individuals because of different gender, hairstyle, ethnicity *etc.* In addition, changing illumination conditions and head poses of human faces also affect the extracted features, which further increases feature inconsistency and thus the difficulty in age estimation. On the other hand, label inconsistency intrinsically caused by manual annotations is the main challenge in apparent age estimation, which has been investigated in very few existing work. As shown in Fig. 1, standard deviation of apparent age annotations for each instance is first calculated and the cumulative size of samples larger than standard deviation coordinates is reported, for the purpose of visualising the uncertainty of manual annotations. Each image in apparent age estimation is associated with a number of uncertain apparent age annotations instead of a unique chronological age. The straight-forward solution is to average annotated age classes to generate a pseudo age [9–13], which is readily applied to the existing supervised learning frameworks for chronological age estimation. However, the induced uncertainty has not been exploited in such a setting. We observe that the mean of apparent age classes could miss reflecting label variation across annotations. Such an observation motivates us to take incorporating annotation reliableness explicitly in the label representation into account to achieve more robust performance.

We consider that *latent label correlation mining* and *reliable annotation exploiting* are two key factors for accurate and robust apparent age estimation. To this end, we propose a novel framework based on the recent label distribution learning paradigm to combine label distribution from ambiguous annotations to alleviate feature and label inconsistency. Compared to the mean and/or standard deviation over all annotated age labels in [9–13], the proposed ambiguous label distribution learning aims to construct a weighted label density space, which is then mapped from low-level feature space. The weighting strategies according to the annotations' reliableness play a vital role in generating such an ambiguous label distribution. On one hand, more reliable age annotations (determined by reliableness) with higher weights will contribute more to achieving robust performance from label uncertainty. On the other hand, combining a number of distribution from different annotations will have richer description degree in label distribution in comparison with single label distribution generated by mean and standard deviation of all apparent age. Specifically, the proposed ambiguous label distribution usually have more than one peak and asymmetric distribution rather than single peak and symmetric structure in original label distribution learning [4].

2   Related Work

Chronological Age Estimation – The recent frameworks for estimating persons' chronological ages given facial images can be categorised into three groups: classification based [3, 4, 14–16], regression based [1, 7, 17, 18], and ranking based [19, 20]. Considering cumulative dependent nature across age classes (*i.e.* the closer age labels of facial images are, the more visual similarity they share), the frameworks [1, 3, 4, 7, 17, 18] explicitly or implicitly mining latent label correlation are more favourable for facial age estimation problem. Chen *et al.* [1] exploited the cumulative dependency across age classes to achieve robust performance in a two-layer attribute learning framework. Geng *et al.* [4] designed a framework by learning from label distribution in the manner of multi-label learning instead of a single independent class label to capture latent age class correlation. Specifically, for each instance, a label distribution vector (whose size is equal to the age range, typically [0, 100]) gives description degree to each element, which reflects its describing capability. The maximum description degree is allocated to the element having the relative position of the chronological age in the label distribution vector. The label distribution vectors are then mapped from imagery feature vectors. Evidently, for any two age classes, the design of label distribution learning captures their correlation via the values of their corresponding positions in the label distribution vectors. An advanced attempt of label distribution learning with adaptively updating label distribution for each age group was introduced to mitigate the suffering of dynamic ageing procedure during different period (*i.e.* childhood and adulthood) [16].

Apparent Age Estimation – In apparent age estimation, each training facial image is associated with a number of inconsistently annotated age labels. Such a problem was cast into a partial label learning paradigm [21–23], which assumes that only one annotation is valid among a set of candidate annotations. However, the existing partial label learning frameworks were designed for classification problems without considering ordinal dependency across age classes, which are less suitable for apparent age estimation. An intuitive strategy is to construct a pseudo age label for each sample from a number of manual annotations, which can be directly applied and incorporated to the existing supervised learning frameworks originally developed for chronological age estimation. The typical strategy for pseudo age generation is to use the mean value of apparent age. Since the competition organised by ChaLearn [24] is popular, apparent age estimation has attracted wide attention in the field and a number of recent frameworks [9–13] based on Convolutional Neural Networks (CNN) [25] were proposed, which concerned mainly on training and/or fine-tuning deep CNN models for better imagery representation. Inspired by recent success of the existing frameworks such as cumulative attributes [1] and label distribution learning [4] designed for chronological age estimation, their concept have inspired to design DeepCodeAge [11] and deep label distribution learning [9] respectively. The framework proposed by Yang *et al.* [9] was one of the first attempt to handle the uncertainty of apparent age, which shares similar script as our learning with ambiguous label distribution (LALD). Nevertheless, the differences of our method lie in the utilisation of combined label distribution from independent apparent age annotations for each image instead of a single label distribution based on global statistics (*e.g.* , mean and standard deviation [9]). Moreover, the proposed method in this paper also takes the reliableness of every annotation into account, which further boost the estimation performance. Consequently, owing to the introduction of label distribution combination and reliableness enhancing, our ambiguous label distribution is more informative and robust than the direct utilisation of label distribution with global statistics.

Contributions – The contributions and novelties of this paper are three-fold as:

- To the best knowledge of authors, this paper is the first attempt for apparent age estimation to exploit annotation reliableness to handle with label uncertainty and reduce the negative effect caused by annotation outliers.

- Compared to the mean and standard deviation over all apparent age annotations, the proposed ambiguous label distribution owing to label distribution combination and reliableness enhancing is more informative and robust in view of using annotation density of each image.
- Extensive experiments on the public FG-NET benchmark gain notable advantage on accuracy of ambiguous label distribution learning for apparent age estimation to tackle with both feature inconsistency and label ambiguity.

3   Methodology

Given imagery feature representation x and its corresponding age annotations $y \in \mathbb{R}^D$, training samples consist of $\{x_i, y_i\}^{i=1,2,\cdots,N}$, where N denotes the number of training samples. The pipeline of the proposed algorithm illustrated in Fig. 2 is given in details as the following:

- For ith training sample, we first generate vector-formed ordinary label distribution $l_1, l_2, \cdots l_D$ of apparent age according to their relative positions y_i in a chronological age range. Label distributions are then combined together according to their reliableness, i.e. the distance to the pseudo age, to construct ambiguous label distribution a_i (see Sect. 3.1).
- Learning the mapping between imagery feature representation x and an ambiguous label distribution a is achieved by adopting label distribution learning [3, 4] (see Sect. 3.2).

During testing, imagery feature of an unseen image are fed into the trained model to predict the person's age, i.e. the age class having the maximum predicted description degree in the ambiguous label distribution.

3.1   Ambiguous Label Distribution Construction

The concept of label distribution was firstly introduced by Geng et al. [4] for chronological age estimation, which is investigated briefly here and named as ordinary label distribution to distinguish from the proposed ambiguous label distribution. Given a scalar-valued age label $y \in \mathbb{R}$ for each image instance, a label distribution vector $l \in \mathbb{R}^K$ is generated, whose dimension K is equal to the size of age range. Each dimension in such a label distribution corresponds to an age class according to its relative positioning in value, which has a description degree $d_x \in [0, 1]$ to indicate the capability to describe the proportion of the samples. In mathematics, description degree d_x^k possibly represented as conditional probability $P(k|x)$ indicates that age label $k \in [0, K-1]$ describe the proportion d_x^k of the sample. The sum of real-valued description degree d_x^k of all elements in the label distribution vector is equal to one. The assumption of label distribution is two-fold: a) true labels have the highest description degree in l; and b) the farther labels are away from chronological ages, the lower description degree they have. As label distribution changing along ordinal age classes continuously and cumulatively, description degree also reflects the support of

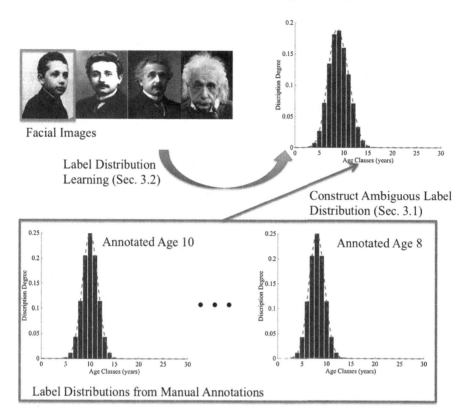

Fig. 2. The pipeline of the proposed learning with ambiguous label distribution.

neighbouring labels contributing to the exact label y associated to instance x. Consequently, all age classes k having positive values are assumed to contribute to discriminating training samples to the age class y. Typical label distributions are Gaussian and triangle distributions [4] anchored in chronological age class. In the light of Gaussian distribution consistently superior to triangle distribution, which is thus adopted in the experiments of this paper.

The setting of original label distribution designed for chronological age estimation can be readily employed for apparent age estimation by obtaining a pseudo age label for each instance from a set of unreliable annotations. Ordinary label distribution only has a single peak and the symmetric distribution. We aim to enrich the capability of label representation by proposing a novel ambiguous label distribution (ALD) $a = [d_x^0, d_x^1, \cdots, d_x^{K-1}] \in \mathbb{R}^K$. For each image, it combines label distribution l from each apparent age annotation, which is then normalised to satisfy that each element of description degree $d_x^k \in [0,1]$ in a and $\sum^K d_x^k = 1$.

The strategies to construct the ambiguous label distribution play an important role and are sensitive to the estimation performance, which are investigated here and will be experimentally evaluated and compared. To this end, two factors need to be concerned for an informative and robust ambiguous label distribution: (1) pseudo age acquisition and (2) combination of label distribution l. Evidently, averaging all annotated age labels and maximum majority voting are two types of intuitive strategies for determining pseudo ages for apparent age estimation problem. Moreover, we also introduce the third type by averaging annotated age classes, *i.e.* without counting the repeated annotations for the identical age classes. The third one consistently achieves superior performance to the rest two in our experiments with more detailed analysis given in the experimental part (Table 2). The solution to the second question is to incorporate their reliableness (*e.g.* the first- or second- order statistics between apparent age and the pseudo age) as the weights for the corresponding label distribution. Adopting the weighting strategies is aimed to improve the robustness, as the less reliable annotation far away from the pseudo age should be given lower weights to reduce their negative effect. For each image, the distance measure between annotations and pseudo age is thus employed as weights of reliableness, which are respectively multiplied with their corresponding label distributions and then summed up to construct the proposed ambiguous label distribution. Our framework achieves better performance than non-weighted combination in our evaluative verification (Table 3).

3.2 Mapping from Feature Input to Distribution Output

With the generated ambiguous label distribution for each image, the training set becomes $\{x, a\}_i, i = 1, 2, \cdots, N$. Entry $a_j, j = 1, 2, \cdots, K$ of $a \in \mathbb{R}^K$ denotes description degree for the jth age class, where age label for the jth age class is $k = j - 1$. The aim is to learn a conditional density function $p(y|x; \theta)$ to minimise the distance between the predicted $\hat{a}$ generated by θ and the ground truth a, where θ is the parameter vector to be optimised. Evidently, the problem is cast as a label distribution learning problem, which has been well presented in [4]. The object function for ambiguous label distribution learning can be written as:

$$\min_{\theta} \quad \sum_i P(a_i || p(y_i | x_i; \theta)),$$ (1)

where $P(a^h || a^w)$ is to measure the similarity between two distributions a^h and a^w. In this paper, Kullback-Leibler divergence [26] is employed, which can be mathematically depicted as the following:

$$P(a^h || a^w) = \sum_j (a_j^h \ln \frac{a_j^h}{a_j^w}),$$ (2)

where a_j^h and a_j^w denote the jth element in a^h and a^w respectively. Substituting Eq. (2) into (1), object function can thus be formulated as:

$$\min_{\theta} \quad \sum_i \sum_j (d_{x_i}^{j-1} \ln \frac{d_{x_i}^{j-1}}{p((j-1)|x_i; \theta)}).$$ (3)

Fig. 3. Illustrative example images from the FG-NET dataset.

As a result, the optimised parameter θ^* can be determined by

$$\theta^* = \underset{\theta}{\arg\min} \sum_i \sum_j (d_{\boldsymbol{x}_i}^{j-1} \ln \frac{d_{\boldsymbol{x}_i}^{j-1}}{p((j-1)|\boldsymbol{x}_i; \boldsymbol{\theta})})$$

$$= \underset{\theta}{\arg\max} \sum_i \sum_j d_{\boldsymbol{x}_i}^{j-1} \ln p((j-1)|\boldsymbol{x}_i; \boldsymbol{\theta}). \qquad (4)$$

Let us assume a maximum entropy model [27] as

$$p((j-1)|\boldsymbol{x}_i; \boldsymbol{\theta}) = \frac{\exp(\sum_r \boldsymbol{\theta}_{j-1,r} \boldsymbol{x}_i^r)}{\sum_j \exp(\sum_r \boldsymbol{\theta}_{j-1,r} \boldsymbol{x}_i^r)}, \qquad (5)$$

where $\boldsymbol{x}_i^r$ denotes the rth entry of feature $\boldsymbol{x}_i$ and $\boldsymbol{\theta}_{j-1,r}$ is the element of $\boldsymbol{\theta}$ associated to the jth label (*i.e.* age $j-1$ class in the light of starting from age 0 in the age range) and rth feature element. Substituting Eq. (5) into (4) yields the object function as

$$F(\theta) = \sum_{i,j} d_{\boldsymbol{x}_i}^{j-1} \ln p((j-1)|\boldsymbol{x}_i; \boldsymbol{\theta})$$

$$= \sum_{i,j} d_{\boldsymbol{x}_i}^{j-1} \sum_r \boldsymbol{\theta}_{j-1,r} \boldsymbol{x}_i^r - \sum_i \ln \sum_j \exp(\sum_r \boldsymbol{\theta}_{j-1,r} \boldsymbol{x}_i^r). \qquad (6)$$

A number of optimisation algorithms such as improved iterative scaling (IIS) [28], Conditional Probability Neural Network (CPNN) [4], quasi-Newton method BFGS [29] have been investigated and evaluated to address object function (6) in [3,4]. In the light of stable performance of BFGS [30] as well as its high computational efficiency [3], we adopt BFGS algorithm to optimise object function (6).

4 Experiments

4.1 Datasets and Settings

Datasets – We evaluate the proposed framework on the public benchmark FG-NET [1,4,7,14,17,19], which is the only dataset for age estimation having human annotations provided by Han *et al.* [8]. Specifically, the FG-NET dataset contains 82 persons varying from age 0 to age 69 with 1002 images in total and Fig. 3 shows all the images of the first identity. Evidently, the appearance of example faces are largely varied because of hairstyle, expression, beard style, whether wearing glasses and head poses, which makes the FG-NET dataset common and difficult for evaluating age estimation algorithm. Manual annotations for the FG-NET dataset have large uncertainty illustrated in Fig. 1, which makes apparent age estimation more challenging.

Features – Active Appearance Model (AAM) feature [31] is adopted as low-level imagery features because of its popularity in the recent works [1,7,14,17, 19,32,33]. In details, the parameters of AAM model including visual appearance, shape, and texture cues to form a 200-dimensional feature vector.

Settings – Two experiments are conducted according to the settings of data split. In the first experiment, we followed the same leave-one-person-out setting as in [1,7,17,19,32,33], whose testing images for each fold belong to an unseen person identity. In the second experiment, the total images of the FG-NET dataset was randomly split into 80% data for training and the remaining 20% for testing (*i.e.* 800 images for training and 202 images for testing) and we repeated the experiment 30 times.

Comparative Methods – We compare four algorithms with the proposed LALD framework, namely Instance-based PArtial Label learning (IPAL) [21], support vector regression with linear kernel (SVR) [34], two label distribute learning methods: CPNN [4] and BFGS-LLD [3]. IPAL can directly be applied to apparent age estimation with the capability of coping with multiple annotations for one instance, SVR, CPNN and BFGS-LLD employs pseudo age by averaging apparent age annotations to replace the true chronological ages. In IPAL, the number of nearest neighbours and the balancing coefficient are set to 5 and 0.45 respectively. Free parameter C in SVR to trade off the loss function and regularised term is tuned by four-fold cross-validation with $[10^{-5} : 10 : 10^5]$[1]. We set the size of hidden layers in CPNN to be 400.

Evaluation Metrics – Two evaluation metrics for chronological age estimation, namely *mean absolute error* (mae) and *cumulative score* (cs) [14] are not suitable for apparent age estimation because of unavailable unique chronological age labels for training and testing samples. In view of this, we adopt the evaluation metric $\in [0, 1]$ for each testing instance, introduced in [24]:

$$\epsilon = 1 - \exp\left(-\frac{(\hat{y} - \mu)^2}{2\sigma^2}\right)$$

---

[1] Following the usage in Matlab, the notation $[x : y : z]$ represents an array starting from x to z with the step of y.

where $\hat{y}$ denotes the predicted age having the maximum description degree in the predicted ALD, μ is the mean apparent age and σ denotes the standard deviation. For ϵ performance metric, the lower the better.

4.2 Comparative Evaluation with State-of-the-Arts

In this section, we evaluate and compare the proposed LALD framework with four state-of-the-art algorithms in two data-split settings, which are shown in Table 1. Besides IPAL [21], SVR [34] CPNN [4], and BFGS-LLD [3] were designed for chronological age estimation and are applied to apparent age estimation by using the mean pseudo age generating from apparent age annotations. Evidently, our method achieves significantly better performance over the rest four comparative algorithms, in details, consistently at least 7.04% better for leave-one-person-out protocol and at least 4.01% for randomly 80% training data-split. The direct competitor of LALD is BFGS-LLD. Both methods employ the identical low-level features and optimisation method BFGS. Consequently, the performance gain achieved by LALD over BFGS-LLD can only be explained by the superiority of the proposed ambiguous label distribution. In addition, we conduct a t-test on the predictions of the state-of-the-art methods and ours. The results of both data-split protocol show consistently statistical significance (rejection on null hypothesis at the 5% significance level).

Table 1. Comparative evaluation with state-of-the-art methods.

Methods	Leave-One-Person-Out	Randomly 80% Training
IPAL [21]	0.648 ± 0.132	0.574 ± 0.020
SVR [34]	0.611 ± 0.135	0.588 ± 0.027
CPNN [4]	0.676 ± 0.126	0.660 ± 0.043
BFGS-LLD [3]	0.618 ± 0.111	0.587 ± 0.019
LALD (ours)	**0.568 ± 0.118**	**0.551 ± 0.021**

4.3 Evaluation on Pseudo Age Acquisition

In this section, three types of strategies, mean apparent age annotations, maximum majority voting and mean age classes without counting the repeatability of annotations presented in Sect. 3.1 have been employed to determine pseudo age labels, which is an important factor for constructing the ambiguous label distribution. The results are illustrated in Table 2. Surprisingly, the strategy of mean classes beats the intuitive solutions of mean annotations and maximum majority voting, which are considered more robust against annotation outliers. The rational for such a phenomenon lies in the following observation. For age annotations having a roughly normal distribution (a peak in the middle), mean annotations, maximum majority voting and mean classes usually have the identical pseudo age. For tailed annotation density (a peak skewed to the boundary

Table 2. Evaluation on the strategies of pseudo age acquisition.

Methods	Leave-One-Person-Out	Randomly 80% Training
Mean annotations	0.579 ± 0.128	0.579 ± 0.026
Maximum majority voting	0.592 ± 0.121	0.575 ± 0.030
Mean classes	$\mathbf{0.568 \pm 0.118}$	$\mathbf{0.551 \pm 0.021}$

of annotated age range), mean classes is more robust against annotation outliers than the others, as the drift by repeated annotations leads to higher weights for outliers.

4.4   Weighted vs. Non-weighted Combination

This section compares the results of annotation reliableness weighted and non-weighted combination for generating the ambiguous label distribution, which are given in Table 3. It is evident that weighted combination of individual label distribution from apparent age annotations can benefit to construct a more informative label representation, which can verify the motivation of the introduction of reliableness weighting. It is worth mentioning here that even the results without adopting annotation reliableness as weights can also beat all four state-of-the art algorithms, which further demonstrates the effectiveness of the proposed LALD framework for apparent age estimation.

Table 3. Weighted vs. non-weighted combination to construct the ambiguous label distribution.

Methods	Leave-One-Person-Out	Randomly 80% Training
Non-weighted	0.595 ± 0.125	0.572 ± 0.025
Weighted	$\mathbf{0.568 \pm 0.118}$	$\mathbf{0.551 \pm 0.021}$

4.5   Evaluation on Reliableness Utilisation

This section evaluates on what kind of information is more favourable for capturing annotation reliableness, with the results illustrated in Table 4. We compare the first- and second- order distance from apparent age annotations to pseudo age, and find out that the first order statistics can consistently perform better on both data-split settings. Such an observation indicates that the choice of reliableness weighting is sensitive to estimation performance. Nevertheless, the performance achieved by employing second order statistics is still superior to all four comparative methods in Table 1.

4.6   Illustrative Samples

In this section, we illustrate a number of successful sampled facial images belonging to young, mid-age and old age groups respectively from the FG-NET benchmark in Fig. 4, with the generated ambiguous label distribution from annotations and the predicted label distribution. Evidently, compared to single peak and symmetric distribution in original label distribution, the ambiguous label distribution could have multiple peaks and asymmetric distribution, which is more informative and robust for apparent age estimation to capture the label ambiguity.

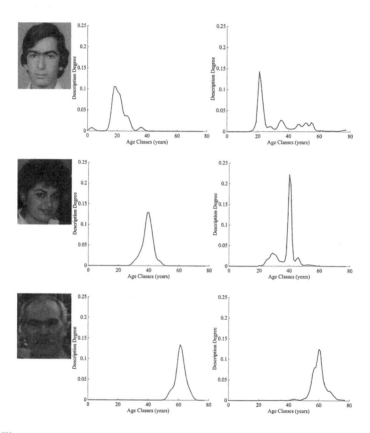

Fig. 4. Illustrative examples about the proposed LALD framework. In each subfigure, left is the original facial image, the middle is the generated ambiguious label distribution from apparent age, the right is the predicted distribution.

Table 4. Evaluation on the choices of reliableness utilisation.

Methods	Leave-One-Person-Out	Randomly 80% Training
First Order	**0.568 ± 0.118**	**0.551 ± 0.021**
Second Order	0.595 ± 0.117	0.565 ± 0.032

5   Conclusion

This paper proposes a novel concept of ambiguous label distribution designed for apparent age estimation problem, whose instances have multiple unreliable age annotations. Owing to discovering latent label correlation inherited from original label distribution learning framework and utilising annotation reliableness for weighting apparent age, the proposed learning with ambiguous label distribution method can achieve better and robust performance. We experimentally evaluate and analyse the variants of our algorithm, and notice that combining distributions adopted in this paper is important yet naive. In future, introducing a powerful combination method could be a promising research direction.

Acknowledgement. This work was funded by Academy of Finland under the Grant No. 267581 and 298700, and D2I SHOK project funded by Digile Oy and Nokia Technologies (Tampere, Finland). The authors wish to acknowledge CSC-IT Center for Science, Finland, for generous computational resources.

References

1. Chen, K., Gong, S., Xiang, T., Loy, C.C.: Cumulative attribute space for age and crowd density estimation. In: CVPR (2013)
2. Fu, Y., Guo, G., Huang, T.S.: Age synthesis and estimation via faces: a survey. TPAMI **32**(11), 1955–1976 (2010)
3. Geng, X., Ji, R.: Label distribution learning. In: ICDMW (2013)
4. Geng, X., Yin, C., Zhou, Z.H.: Facial age estimation by learning from label distributions. TPAMI **35**(10), 2401–2412 (2014)
5. Luu, K., Ricanek Jr., K., Bui, T.D., Suen, C.Y.: Age estimation using active appearance models and support vector machine regression. In: BTAS (2009)
6. Pontes, J.K., Britto, A.S., Fookes, C., Koerich, A.L.: A flexible hierarchical approach for facial age estimation based on multiple features. Pattern Recogn. **54**, 34–51 (2015)
7. Zhang, Y., Yeung, D.: Multi-tasks warped Gaussian process for personalized age estimation. In: CVPR (2010)
8. Han, H., Otto, C., Liu, X., Jain, A.K.: Demographic estimation from face images: human vs. machine performance. TPAMI **37**(6), 1148–1161 (2015)
9. Yang, X., Gao, B.B., Xing, C., Huo, Z.W., Wei, X.S., Zhou, Y., Wu, J., Geng, X.: Deep label distribution learning for apparent age estimation. In: CVPR Workshops (2015)
10. Rothe, R., Timofte, R., Gool, L.: DEX: deep expectation of apparent age from a single image. In: ICCV Workshops (2015)
11. Kuang, Z., Huang, C., Zhang, W.: Deeply learned rich coding for cross-dataset facial age estimation. In: ICCV Workshops (2015)
12. Zhu, Y., Li, Y., Mu, G., Guo, G.: A study on apparent age estimation. In: ICCV Workshops (2015)
13. Antipov, G., Baccouche, M., Berrani, S.A., Dugelay, J.L.: Apparent age estimation from face images combining general and children-specialized deep learning models. In: ICCV Workshops (2015)

14. Geng, X., Zhou, Z.H., Smith-Miles, K.: Automatic age estimation based on facial aging patterns. TPAMI **29**(12), 2234–2240 (2007)
15. Lanitis, A., Draganova, C., Christodoulou, C.: Comparing different classifiers for automatic age estimation. TSMC **34**(1), 621–628 (2004)
16. Geng, X., Wang, Q., Xia, Y.: Facial age estimation by adaptive label distribution learning. In: ICPR (2014)
17. Guo, G., Fu, Y., Huang, T.S., Dyer, C.R.: Image-based human age estimation by manifold learning and locally adjusted robust regression. TIP **17**(7), 1178–1188 (2008)
18. Guo, G., Mu, G., Fu, Y., Huang, T.S.: Human age estimation using bio-inspired features. In: CVPR (2009)
19. Chang, K.Y., Chen, C.S., Hung, Y.P.: Ordinal hyperplanes ranker with cost sensitivities for age estimation. In: CVPR (2011)
20. Wang, S., Tao, D., Yang, J.: Relative attribute SVM+ learning for age estimation. TC **46**(3), 827–839 (2015)
21. Zhang, M.L., Yu, F.: Solving the partial label learning problem: an instance-based approach. In: IJCAI (2015)
22. Cour, T., Sapp, B., Taskar, B.: Learning from partial labels. JMLR **12**, 1501–1536 (2011)
23. Cour, T., Sapp, B., Jordan, C., Taskar, B.: Learning from ambiguously labeled images. In: CVPR (2009)
24. Escalera, S., Fabian, J., Pardo, P., Baro, X., Gonzalez, J., Escalante, H., Guyon, I.: ChaLearn 2015 apparent age and cultural event recognition: datasets and results. In: ICCV, ChaLearn Looking at People Workshop (2015)
25. Krizhevsky, A., Sutskever, I., Hinton, G.E.: Imagenet classification with deep convolutional neural networks. In: NIPS (2012)
26. MacKay, D.J.: Information Theory, Inference and Learning Algorithms. Cambridge University Press, Cambridge (2003)
27. Berger, A.L., Pietra, V.J.D., Pietra, S.A.D.: A maximum entropy approach to natural language processing. Comput. Linguist. **22**(1), 39–71 (1996)
28. Pietra, S.D., Pietra, V.D., Lafferty, J.: Inducing features of random fields. TPAMI **19**(4), 380–393 (1997)
29. Nocedal, J., Wright, S.: Numerical Optimization. Springer Science & Business Media, New York (2006)
30. Malouf, R.: A comparison of algorithms for maximum entropy parameter estimation. In: The 6th Conference on Natural Language Learning (2002)
31. Cootes, T.F., Edwards, G.J., Taylor, C.J.: Active appearance models. TPAMI **23**(6), 681–685 (2001)
32. Yan, S., Wang, H., Huang, T.S., Yang, Q., Tang, X.: Ranking with uncertain labels. In: ICME (2007)
33. Yan, S., Wang, H., Tang, X., Huang, T.S.: Learning auto-structured regressor from uncertain nonnegative labels. In: ICCV (2007)
34. Smola, A.J., Schölkopf, B.: A tutorial on support vector regression. Stat. Comput. **14**(3), 199–222 (2004)

Prototype Discriminative Learning for Face Image Set Classification

Wen Wang[1,2], Ruiping Wang[1,2,3(✉)], Shiguang Shan[1,2,3], and Xilin Chen[1,2,3]

[1] Key Laboratory of Intelligent Information Processing of Chinese Academy
of Sciences (CAS), Institute of Computing Technology, CAS, Beijing 100190, China
wen.wang@vipl.ict.ac.cn, {wangruiping,sgshan,xlchen}@ict.ac.cn
[2] University of Chinese Academy of Sciences, Beijing 100049, China
[3] Cooperative Medianet Innovation Center, Beijing, China

Abstract. This paper presents a novel Prototype Discriminative Learning (PDL) method to solve the problem of face image set classification. We aim to simultaneously learn a set of prototypes for each image set and a linear discriminative transformation to make projections on the target subspace satisfy that each image set can be optimally classified to the same class with its nearest neighbor prototype. For an image set, its prototypes are actually "virtual" as they do not certainly appear in the set but are only assumed to belong to the corresponding affine hull, i.e., affine combinations of samples in the set. Thus, the proposed method not only inherits the merit of classical affine hull in revealing unseen appearance variations implicitly in an image set, but more importantly overcomes its flaw caused by too loose affine approximation via efficiently shrinking each affine hull with a set of discriminative prototypes. The proposed method is evaluated by face identification and verification tasks on three challenging and large-scale databases, YouTube Celebrities, COX and Point-and-Shoot Challenge, to demonstrate its superiority over the state-of-the-art.

1 Introduction

As one of the most important problems in the field of computer vision, traditional face recognition is usually posed as a single image classification problem. With development of imaging technology, multiple images can be available for one person in many real-world application scenarios such as video surveillance, multi-view camera photos or online photo albums, etc. Since multiple images usually incorporate dramatically large variations in pose, illumination, expression and other factors, it is no longer sufficient for traditional face recognition to handle such scenarios, which leads to a new research focus on face image set classification. Compared with a single image, a set of images can provide more information to describe the subjects of interest, hence image set classification

Electronic supplementary material The online version of this chapter (doi:10.1007/978-3-319-54187-7_23) contains supplementary material, which is available to authorized users.

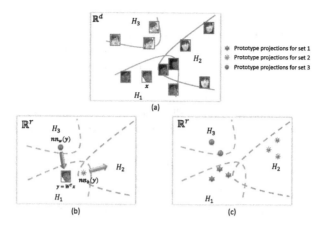

Fig. 1. Conceptual illustration. Different colors denote different subjects. $\mathbb{R}^d$ and $\mathbb{R}^r$ are respectively the original sample space and the projected subspace. (a) shows the affine hulls H_1, H_2 and H_3 of three image sets where H_1 and H_2 are overlapped which leads to a failed match. (b) illustrates the training process by taking any sample x in image set 1 as an example. The arrows imply the training objective, which is to make the projections in $\mathbb{R}^r$ satisfy that for a projection $y = W^T x$, its nearest neighbor in a prototype set from its own class (i.e., $nn_w(y)$) is closer than any other from different classes (i.e., $nn_b(y)$). (c) is an illustration of the learned target subspace and prototype sets.

is expected to achieve more appealing performance than single image classification. Generally speaking, existing image set classification methods mainly focus on how to model the image set and how to measure the dissimilarity between two sets.

In recent years, a simple but efficient affine hull model [1] is proposed to model the image set. The affine hull model tends to complement the unseen appearance variations that even do not appear in the image set via covering the affine combinations of sample images in this set. Thus the affine hull is quite appealing due to its favorable property of characterizing the implicit semantic relationship between the sample images in the set. Nevertheless, there are some fatal limitations. On the one hand, the affine hull matching fails when two hulls overlapped. This is usually caused by the over-large affine hull which usually occurs if the image set contains outliers such as incorrect or low-quality images. An illustration of such case is shown in Fig. 1(a). For shrinking the affine approximation, later methods attempt to artificially impose a tighter constraint (such as convex [1], sparse [2], regularized [3] or probabilistic [4] constraint) which, however, is a brute-force way and may lead to high time cost or missing of some representative candidate points. On the other hand, the discriminative information is ignored, while the affine hulls modeled based on original feature may not suffice to be discriminated linearly, which is iteratively learned in the form of discriminative metric in [5,6].

To address these limitations and explore a totally different and novel solution, this paper presents a Prototype Discriminative Learning (PDL) method for face image set classification. Our goal is to simultaneously learn a set of representative points (i.e. prototypes) for each image set and a linear discriminative projection. Thereinto, the learned prototypes of an image set are actually "virtual", that is, they do not certainly appear in the set but are assumed to belong to the corresponding affine hull, which aims to inherit the merit of affine hull in revealing unseen appearance variations. We expect that in the target projected subspace each image set can be optimally classified to the same class with its nearest neighbor prototype set. Figure 1(b) is an illustration of the training objective, while Fig. 1(c) shows the finally learned target subspace and prototype sets. Thus for an image set, its prototype set can be considered to shrink the corresponding affine hull discriminatively. Specifically, we estimate the loss function for classifying any image in an image set into the same class with its nearest prototype in the projected target subspace. Then by minimizing such loss, we can optimize the prototype sets and the linear projection simultaneously through gradient decent.

The rest of the paper is organized as follows: In Sect. 2, we review some existing works for face image set classification. Section 3 describes the proposed Prototype Discriminative Learning (PDL) method and then gives a discussion about some related works in the literature. In Sect. 4, we demonstrate the experimental evaluation on three challenging databases and analyze the comparison results with other works. Finally, Sect. 5 summarises the conclusion.

2   Related Work

In this section, we briefly introduce the existing works for face image set classification. To represent semantic relationship implicit in the image set, a lot of methods are proposed by exploring different kinds of image set models, for instance, one or several linear subspaces, statistical information, reconstruction model and affine hull.

In the literature, some methods tend to represent an image set as one or several linear subspaces. For example, Mutual Subspace Method (MSM) [7] and Discriminant analysis of Canonical Correlations (DCC) [8] model the image set with a single linear subspace and the difference between two subspaces is measured by principal angles. Grassmann Discriminant Analysis (GDA) [9] and Grassmann Embedding Discriminant Analysis (GEDA) [10] model the image set similarly but perform kernel discriminative learning on the Grassmann manifold where each point is a linear subspace. Besides, a series of works after Manifold-Manifold Distance (MMD) [11] propose to characterize an image set by multiple linear subspaces. Among them, MMD computes the distance between image sets by using the nearest distance between pair-wise local linear models and then Manifold Discriminant Analysis (MDA) [12] extends MMD by learning a discriminative feature subspace. Then an image set alignment method [13] is proposed to match the local linear subspaces more precisely. A later work of [14] proposes to search joint Sparse Approximated Nearest Subspaces (SANS) and employ their distance to measure the image set dissimilarity. A Robust Structured Subspace Learning (RSSL) method [15] is proposed for data representation, which respects the locally smooth property of visual geometric structure.

Some methods propose to extract the statistical information, such as mean vector, covariance matrix, probability distribution, or a combination of them, to describe the data structure in the image set. Some earlier methods, e.g., [16,17], exploit some parametric distribution, such as Gaussian, to represent each image set and compute the similarity by Kullback-Leibler Divergence (KLD). The Covariance Discriminative Learning (CDL) method [18] exploits the covariance matrix to represent the image set and conducts kernel discriminant analysis on the Symmetric Positive Definite (SPD) manifold. Harandi et al. [19] present an SPD Manifold Learning (SPDML) method to learn an orthonormal projection from the high-dimensional SPD manifold to a low-dimensional, more discriminative one. Then Huang et al. [20] propose to learn a tangent map from the original tangent space to a new discriminative tangent space. A later work of Wang et al. [21] proposes to model the image set with a GMM and derive a series of kernels for Gaussians to conduct Discriminant Analysis on Riemannian manifold of Gaussian distributions (DARG).

Different from the methods above, some methods attempt to employ a reconstruction model to learn the image set representation implicitly and then compute the dissimilarity between image set by corresponding reconstruction error. For instance, face dictionary is extended from still images to videos and the sparse representation of image set can be learned through a sparse reconstruction mechanism. Specifically, Chen et al. [22,23] present a video-based dictionary method and build one dictionary for each video clip. Cui et al. [24] propose a Joint Sparse Representation (JSR) method to adaptively learn the sparse representation of an video clip with consideration of the class-level and atom-level sparsity simultaneously. Further, a simultaneous feature and dictionary learning (SFDL) method is proposed in [25] so that discriminative information can be jointly exploited. Besides, Hayat et al. [26] present an Adaptive Deep Network Template (ADNT) to learn a deep reconstruction network for each class.

In addition to the above three categories, some works present an affine hull based model to reveal the unseen appearance within an image set and the implicit semantic relationship from the view of general data geometric structure. For example, Affine Hull based Image Set Distance (AHISD) [1] is proposed to model each image set by an affine hull model and thereby defines the dissimilarity between two hulls as the distance between a pair of nearest points belonging to either hull respectively. Aiming at overcoming the disadvantage that the affine hull may be too large and overlapped, a following trend of works attempt to add some constraints to avoid too loose affine approximation. For example, Convex Hull based Image Set Distance (CHISD) [1] adds a coefficient bound to control the looseness of the convex approximation. Then Sparse Approximated Nearest Points (SANP) [2] is proposed to introduce a sparse representation constraint to the candidate points that the selected nearest points are required to be sparsely represented by the original samples. More recently, Yang et al. [3] propose to approximate each image set by a regularized affine hull model, which exploits a constraint to regularize the affine hull. Further the work of [27] proposes to use multiple local convex hulls to approximate an image set. Chen et al. [28] propose to solve the matching problem by the minimal reconstruction error from a Dual Linear Regression Classification (DLRC) model. Wang et al. [4] propose to enhance the robustness against impure image sets by leveraging the statistical distribution of the involved image sets. To exploit the discriminative information, Zhu et al. [5] propose a Set-to-Set Distance Metric Learning (SSDML) method to learn proper metric between hulls iteratively. Besides, Leng et al. [6] extend SSDML with the strategy of prototype learning, which aims to iteratively filter out the outliers contained in original image set during SSDML.

3   Proposed Method

In this section, we first overview our proposed Prototype Discriminative Learning (PDL) method, followed by reviewing the affine hull model. Then we describe the details of the proposed method. Finally, we give a theoretical discussion about related methods.

3.1   Overview

This paper proposes a novel Prototype Discriminative Learning (PDL) method for face image set classification. As discussed in Sect. 1, it can promisingly improve the robustness of affine hull model to simultaneously learn prototypes and a linear discriminative projection, which are expected to satisfy the following two constraints.

(1) For an image set, its prototypes are a set of points belonging to the corresponding affine hull.
(2) Through the linear projection, the prototypes are mapped to a target subspace where for every sample image, its nearest neighbor in a prototype set from its own class is closer than any other from different classes.

With the first constraint, for an image set, its prototype set can be formulated as a set of combinations of the sample images in the set and to learn the prototypes, we just need to learn the corresponding affine coefficients. Hence, our proposed method inherits the favorable property of affine hull model that the unseen appearances can be revealed and employed to present the implicit semantic relationship by means of the general data geometric structure. The second constraint aims to drive that in the target subspace different image sets can be classified optimally to the same class with the nearest prototype set. We estimate the loss function similarly with the NN error estimation in [29–31]. Then by minimizing such loss function, we derive the corresponding gradients with respect to prototypes and the linear projection respectively, thus the optimized prototypes and linear discriminative projection can be learned simultaneously through gradient decent.

3.2 Affine Hull Model

Suppose there are a total of C image sets for training, the data matrix of the c-th image set is denoted by $X_c = \{x_{c,1}, x_{c,2}, ..., x_{c,n_c}\}$, where $x_{c,i}$ is a d-dimensional feature vector of the i-th image. The c-th image set can be approximated as the affine hull of the sample images [1].

$$H_c = \left\{ x = \sum_{i=1}^{n_c} \alpha_{c,i} \cdot x_{c,i} \,\middle|\, \sum_{i=1}^{n_c} \alpha_{c,i} = 1 \right\}, c = 1, ..., C. \tag{1}$$

By using the sample mean $\mu_c = \frac{1}{n_c} \sum_{i=1}^{n_c} x_{c,i}$ as a reference, we can rewrite the affine hull model as follows.

$$H_c = \{x = \mu_c + U_c v_c | v_c \in \mathbb{R}^l\}, c = 1, ..., C, \tag{2}$$

where U_c is an orthonormal basis and obtained by applying the Singular Value Decomposition (SVD) to the centered data matrix $[x_{c,1} - \mu_c, ..., x_{c,n_c} - \mu_c]$. Note that the directions corresponding to near-zero singular values are discarded, leading to l_c ($l_c < N_c$) singular vectors in U_c. As we have discussed in Sect. 1, the affine hull is a general geometric model containing all the affine combinations of sample images in the set, which can account for the unseen appearance, possible data variation, and further the semantic relationship between sample images. Nevertheless, such approximation is too loose and may lead to over-large affine hull. Therefore, in the following section we will introduce the proposed PDL method which simultaneously learns a prototype set to take place of the affine hull for each image set and a linear projection to make the prototype sets discriminative.

3.3 Prototype Discriminative Learning

Let $P = \{P_1, P_2, ..., P_C\}$ be a collection of the prototype sets to learn. Among them, for the c-th image set X_c, the prototype set can be denoted as $P_c = \{p_{c,1}, p_{c,2}, ..., p_{c,m_c}\} \subseteq H_c$, where

$$p_{c,i} = \mu_c + U_c v_{c,i}, \quad v_{c,i} \in \mathbb{R}^l. \tag{3}$$

Through a linear transformation W, we can obtain a projection in the target subspace which is denoted as

$$y = W^T x \in \mathbb{R}^r, \tag{4}$$

for each image data $x \in X_c, c = 1, ..., C$.

Our goal is to drive that for any image in each image set, it is closer to its nearest neighbor in any prototype set from the same class than that from different classes after mapped to the r-dimensional target subspace. Therefore, in reference of the NN error estimation in [29–31], we define a loss function as follows.

$$J(W, P_1, .., P_C) = \sum_{c=1}^{C} \sum_{x \in X_c} step(Q_x), \tag{5}$$

where $step(Q_x)$ is the step function, i.e.,

$$step(z) = \begin{cases} 0, & if \quad z < 1; \\ 1, & if \quad z \geq 1, \end{cases} \tag{6}$$

and

$$Q_x = \frac{d(y, nn_w^c(y))}{d(y, nn_b^c(y))}, \tag{7}$$

where $d(\cdot, \cdot)$ is the Euclidean distance. $nn_w^c(y)$ and $nn_b^c(y)$ are the nearest neighbors of y respectively from the projections of the same-class and different-class prototype sets, therefore we can formulate them as follows.

$$nn_w^c(y) = W^T a, \quad a = \operatorname*{argmin}_{\substack{a \in P \backslash P_c, \\ a \in Class(x)}} d(y, W^T a)$$

$$nn_b^c(y) = W^T b, \quad b = \operatorname*{argmin}_{\substack{b \in P \backslash P_c, \\ b \notin Class(x)}} d(y, W^T b) \tag{8}$$

Equation (5) denotes the total loss of classifying all sample data $x \in \forall X_c$, $c = 1, ..., C$. Specifically, after mapped though W, when sample data x is nearer to a prototype of its own class than any other from a different class, the loss for classifying x is zero. On the contrary, if in the projected subspace, x is nearer to a prototype from some different class than any other from its own class, the classification of x is mistaken and a large loss of 1 is imposed. Note that here the nearest neighbor of $x \in X_c$ is searched in prototype sets except for the one corresponding to X_c.

Considering the differential property, we employ a sigmoid function with slope to approximate the step function, i.e.,

$$\mathcal{S}_\beta(z) = \frac{1}{1 + e^{\beta(1-z)}}. \tag{9}$$

Note that when β is large, $\mathcal{S}_\beta(\cdot)$ is a smooth approximation of the step function. Then the objective function can be rewritten as follows.

$$J(W, P_1, .., P_C) = \sum_{c=1}^{C} \sum_{x \in X_c} \mathcal{S}_\beta(Q_x). \tag{10}$$

3.4   Optimization

For learning optimal prototype sets $P = \{P_1, P_2, ..., P_C\}$ and linear transformation W, we need to solve the optimization problem in the following.

$$\{W^*, P_1^*, P_2^*, ..., P_C^*\} = \operatorname*{argmin}_{W, P_1, P_2, ..., P_C} J(W, P_1, .., P_C). \tag{11}$$

In this paper, a gradient descent method is employed to solve such problem. Then we tend to derive the gradient of loss function J with respect to $W, P_1, P_2, ..., P_C$. Since the procedure to search the nearest prototype depends on the prototype sets and transformation matrix but is non-continuous and problematic, a simple approximation is usually exploited with such dependence ignored. That is to say, the same prototype neighbor is searched when the variation in the prototype sets and transformation matrix is sufficiently small [29]. Under such assumption, we can derive the gradient of J with respect to W approximately as follows:

$$
\begin{aligned}
\frac{\partial J}{\partial W_k} \approx & \sum_{c=1}^{C} \sum_{x \in X_c} \frac{S_\beta'(Q_x)Q_x}{d^2(y, nn_w^c(y))} \cdot (x - a)(y_k - nn_w^c(y)_k) \\
& - \sum_{c=1}^{C} \sum_{x \in X_c} \frac{S_\beta'(Q_x)Q_x}{d^2(y, nn_b^c(y))} \cdot (x - b)(y_k - nn_b^c(y)_k),
\end{aligned}
\tag{12}
$$

where $W_k \in \mathbb{R}^d$ denote the k-th column of W and y_k denote the k-th element of vector y. Note that denotations a and b have been defined in Eq. (8).

According to Eq. (3), for learning the prototype sets, we just need to learn the corresponding $\{v_{c,i}\}$, $i = 1, ..., m_c$ and $c = 1, ..., C$. Thus we derive the gradient of J with respect to each vector v_{ci} as follows.

$$
\begin{aligned}
\frac{\partial J}{\partial v_{ci}} \approx & \sum_{\substack{c=1 \\ p_{ci}=a}}^{C} \sum_{x \in X_c} \frac{S_\beta'(Q_x)Q_x}{d^2(y, nn_w^c(y))} \cdot U_c^T W W^T (a - x) \\
& - \sum_{\substack{c=1 \\ p_{ci}=b}}^{C} \sum_{x \in X_c} \frac{S_\beta'(Q_x)Q_x}{d^2(y, nn_b^c(y))} \cdot U_c^T W W^T (b - x),
\end{aligned}
\tag{13}
$$

For space limitation, the detailed derivations of Eqs. (12) and (13) are given in a supplementary material.

Based on the derived gradients above, we can update the prototype sets and the linear projection in an iterative procedure by using the limited-memory BFGS (L-BFGS) method [32].

3.5   Classification

After the training process, we have computed a optimal linear transformation W and prototype sets $P_1, P_2, ..., P_C$ for the total of C training image sets. Then given a total of K image sets as the gallery, we need to give a prediction of the label for a new test image set. First we optimize the prototype set for each gallery image set with W fixed by solving Eq. (11). Then we compute the projection of these gallery prototype sets and the test image set through W. Finally, the distance between the test image set and a gallery image set can be computed as the minimal distance between samples in the test image set and prototypes corresponding to the gallery image set. Thus, the test image set can be classified into the same class with its nearest gallery prototype set in the target subspace.

Algorithm 1. PDL-training

Input:

Data matrices of C image sets for training: $\{X_1, X_2, ..., X_C\}$ and their labels;

the slope for sigmoid function: β;

the initial prototype sets: $P = \{P_1, ..., P_C\}$;

the initial transformation matrix: W.

Output:

The optimal P and W

1: Initialize the value of J as zero;

2: **while** not converged **do**

3:     **for** $c = 1$ to C **do**

4:         **for all** x such that $x \in X_c$ **do**

5:             Compute the projection y by Eq. (4);

6:             Solve the optimization problem in Eq. (8) to compute $nn_w^c(y)$, $nn_b^c(y)$;

7:             Compute Q_x by Eq. (7);

8:             Add $\mathcal{S}_\beta(Q_x)$ to the value of J;

9:             Add to the gradient with respect to W and P respectively by Eqs. (12) and (13);

10:         **end for**

11:     **end for**

12:     Compute the step length and seeking direction by the L-BFGS algorithm;

13:     Update P and W;

14: **end while**

15: **return** P^*, W^*;

The Algorithms 1 and 2 summarize the training and testing process of our proposed PDL method respectively.

3.6   Discussion About Related Works

Firstly, we analyze the differences between our proposed PDL method and the unsupervised affine hull methods, such as AHISD [1], CHISD [1], SANP [2], RNP [3], DLRC [28] and ProNN [4], etc. (1) For AHISD, the affine hull may suffer from the issue of intersection, which makes the subsequent distance computation incorrect. Later CHISD, SANP, RNP and ProNN all attempt to solve such issue by imposing a constraint (such as convex, sparse, regularized, or probabilistic constraint) to the geometry structure of affine hull or the selection criteria of nearest points. These constraints are artificially set and based on additional assumption, which may lead to high time cost or missing of some useful information. On the contrary, our method efficiently ameliorates this issue by learning more representative and discriminative prototypes from the affine hull adaptively. (2) These methods are all unsupervised, while the discriminative information has been widely considered to be very important for the object classification.

Secondly, we figure out the differences from the supervised affine hull methods SSDML [5] and SPML [6]. (1) SSDML and SPML both follow a metric learning

---

Algorithm 2. PDL-testing

Input:
    Data matrices of K image sets as gallery and their labels $\{L_1, ..., L_K\}$;
    Data matrix of an image set for test: $T = \{t_1, ..., t_{n_t}\}$, $t_i \in \mathbb{R}^d$;
    the slope for sigmoid function: β;
    the initial prototype sets for the gallery image sets: $G = \{G_1, ..., G_K\}$;
    the initial transformation matrix: W.
Output:
    The label of the test image set L^*
1: Learn the prototype sets $G^* = \{G_1^*, ..., G_K^*\}$ with W fixed similarly with Alg.1.
2: Compute the projection $\widehat{T}$ by applying Eq. (4) to each sample vector in T;
3: Compute the projection $\widehat{G^*} = \{\widehat{G_1^*}, ..., \widehat{G_N^*}\}$ by applying Eq. (4) to each sample
    vector in G_i, $i = 1, ..., K$;
4: $k^* = \mathrm{argmin}_{j,k}\, d(\widehat{T}, \widehat{G_{j,k}^*})$
5: **return** $L^* = L_{k^*}$;

---

framework, while our PDL proposes a different strategy of learning a linear discriminative projection. (2) They both exploit a global discriminative learning, while we conduct PDL from a local view of nearest neighbor (NN), that is, only to penalize a larger distance between nearest neighbors from different classes than that from the same class. From the local view, the optimization objective is more consistent with the final NN-based classification, thus can facilitate more precise classification. (3) To solve the optimization problem, they both adopt a strategy of alternately optimizing. On the contrary, PDL presents a joint optimization mechanism, which can favorably reduce time complexity and avoid trapping in local optimum to some extent. (4) SPML can be considered as iteratively filtering out outlier samples (the remaining real samples are their so-called "prototypes") in the image set while learning discriminative metric. In contrast, our PDL aims to learn discriminative virtual prototypes, which do not necessarily appear in the original set as in SPML but are just required to belong to the corresponding affine hull. Based on this different problem formulation, the learning strategy in PDL is believed to be more direct and efficient.

Thirdly, we give a discussion about comparison with the prototype selection methods based on single sample/image [33,34]. These methods usually propose to select prototypes from the existing samples as a reference for nearest neighbor classifier, which is confined only to the existing samples. However, we argue that for the image set classification problem, the appearance variations within an image set may be too large to be matched only based on existing samples. On the contrary, our PDL first employs the affine hull to complement the unseen data variations and subsequently learns prototypes from these affine combinations, which is more specifically suited to the classification of image sets containing complex data variations.

4   Experiments

4.1   Databases and Settings

For evaluating our proposed PDL method, we used three challenging and large-scale databases: YouTube Celebrities (YTC) [35], COX [36] and Point-and-Shoot Challenge (PaSC) [37]. Examples in the three databases are shown in Fig. 2.

The YTC database is collected from YouTube and consists of 1,910 highly compressed and low-resolution video sequences belonging to 47 subjects. The face region in each image was resized into 20×20 intensity image, and was processed with histogram equalization to eliminate lighting effects. Following the similar protocols of [18,25], we conducted ten-fold cross validation experiments and randomly selected three clips for training and six for testing in each of the ten folds. This enables the whole testing sets to cover all of the 1,910 clips in the database.

The COX database is a large-scale video database and contains 3,000 video sequences from 1,000 different subjects which are captured by different camcorders. In each video, there is around 25–75 frames of low resolution and low quality, with blur, and captured under poor lighting. The face in each image was resized into 32×40 intensity image and histogram equalized. Since the database contains three settings of videos captured by different cameras, we conducted ten-fold cross validation respectively with one setting of video clips as gallery and another one as probe.

The PaSC database consists of 2,802 videos of 265 people carrying out simple actions. Half of these videos are captured by a controlled video camera, the rest are captured by one of five alternative hand held video cameras. It has a total of 280 sets for training and 1401 sets for testing. Verification experiments were conducted using control or handheld videos as target and query respectively. Since the database is relatively difficult, we followed a work of [38] to extract the state-of-the-art Deep Convolutional Neural Network (DCNN) features rather

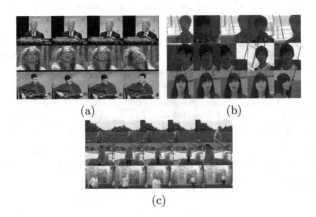

(a)   (b)

(c)

Fig. 2. Some examples of the databases. (a) YTC (b) COX (c) PaSC

than original gray features. Here the DCNN model is pre-trained on the CFW database [39] and subsequently fine-tuned on the training data of PaSC and COX database by using the Caffe [40].

4.2 Comparative Methods

To study the effectiveness of our proposed PDL method, we compared with several state-of-the-art image set classification methods. Among them, there are several affine hull based methods, including Mutual Subspace Method (MSM) [7], Affine Hull based Image Set Distance (AHISD) [1], Convex Hull based Image Set Distance (CHISD) [1], Sparse Approximated Nearest Point (SANP) [2], Regularized Nearest Points (RNP)[3], Dual Linear Regression Classification (DLRC) [28] and Set-to-Set Distance Metric Learning (SSDML) [5]. In addition, we also gave the comparison results with some other state-of-the-art supervised methods, such as Discriminative Canonical Correlations (DCC) [8], Manifold Discriminant Analysis (MDA) [12], Grassmann Discriminant Analysis (GDA) [9] and Grassmannian Graph Embedding Discriminant Analysis (GEDA) [10].

The source code of all comparative methods released by the original authors were used except that of DLRC. We carefully implemented the DLRC algorithm since downloading of its code is not available on its website now. For fair comparison, the important parameters of all the methods were carefully tuned following the recommendations in the original works: For AHISD, we retained 95% energy when learning the orthonormal basis. For CHISD, the error penalty was set to be $C = 100$ as in [1]. For SANP, the parameters were the same as [2]. Considering the high time cost in SANP, we only compared with it on YTC and COX. Note that since the SANP method is too time-consuming to run under the setting of COX, which contains large-scale data, we alternately took the image sets of 100 persons rather than all the 700 persons for testing. For RNP and DLRC, all the parameters were configured according to [3, 28] respectively. For SSDML, we set $\lambda_1 = 0.001$, $\lambda_2 = 0.5$, the numbers of the positive pairs and the negative pairs per set are set to 10 and 20. For DCC, corresponding 10 maximum canonical correlations were used. For MDA, the parameters were configured according to [12]. For GDA/GEDA, the dimension of Grassmannian manifold was set to 10. For our proposed PDL, we used the PCA transformation matrix as an initialization of W and employed unit vectors to initialize the coefficients in P.[1]

4.3 Results and Analysis

The identification experiments were conducted on the YTC and the COX database. Table 1 tabulates the rank-1 identification rates on the YTC and the COX databases, where each reported rate is a mean accuracy over the ten-fold trials. Then we used the PaSC database to evaluate our performance on the verification task and Table 2 lists the verification rate at a false accept rate (FAR) of 0.01.

---

[1] The source code of PDL is available at http://vipl.ict.ac.cn/resources/codes.

Table 1. Identification rates on YTC and COX. Here, "COX-ij" represents the experiment using the i-th set of videos as gallery and the j-th set of videos as probe.

Method	YTC	COX-12	COX-13	COX-23	COX-21	COX-31	COX-32
DCC [8]	0.668	0.625	0.661	0.506	0.561	0.638	0.452
MDA [12]	0.670	0.658	0.630	0.362	0.554	0.432	0.297
GDA [9]	0.659	0.723	0.807	0.744	0.714	0.820	0.776
GEDA [10]	0.668	0.767	0.838	0.766	0.726	0.828	0.800
AHISD [1]	0.637	0.530	0.361	0.175	0.435	0.350	0.188
CHISD [1]	0.665	0.569	0.301	0.148	0.444	0.264	0.137
SANP [2]	0.684	0.541	0.360	0.156	0.396	0.271	0.148
RNP [3]	0.703	0.525	0.333	0.148	0.581	0.379	0.146
DLRC [28]	0.692	0.492	0.379	0.155	0.441	0.361	0.175
SSDML [5]	0.689	0.601	0.531	0.287	0.479	0.444	0.273
PDL	**0.743**	**0.796**	**0.869**	**0.822**	**0.760**	**0.871**	**0.824**

Table 2. Verification rates on PaSC when false accept rate is 0.01 on PaSC dataset. Note that the control/handheld indicates the experiments with control/handheld videos.

Method	Control	Handheld
DCC [8]	0.389	0.375
GDA [9]	0.397	0.375
GEDA [10]	0.406	0.390
AHISD [1]	0.219	0.143
CHISD [1]	0.261	0.210
RNP [3]	0.274	0.198
DLRC [28]	0.242	0.171
SSDML [5]	0.292	0.229
PDL	**0.415**	**0.396**

As can be seen in the results, our method performs the best on all of the three databases. Firstly, our PDL achieves an impressively better result than the unsupervised affine hull based methods, such as AHISD, CHISD, SANP, RNP and DLRC. Specifically, on YTC, our method performs higher than a baseline method AHISD by 11%. On PaSC, our PDL outperforms AHISD by 19.6% for the control videos and 25.3% in the handheld scenario respectively. This supports the discussions in Sect. 3.6 that our PDL improves the affine hull model by learning prototypes discriminatively and adaptively, which is more flexible and robust than artificially imposing a tighter constraint. Secondly, our PDL is also superior over the supervised affine hull based method SSDML. As discussed in

Sect. 3.6, it mainly attributes to our innovation in learning virtual prototypes, the local discriminative learning strategy and the joint optimization mechanism. Besides, it can be generally observed that the supervised methods outperform the unsupervised methods more obviously on COX than the other two databases, due to the particularly large within-class variations on COX and the similar motions between different faces captured by the same camera.

4.4 Time Comparison

In addition, we compared the computational complexity of different methods on an Intel i7-3770, 3.40 GHz PC. Table 3 lists the time cost for the comparative methods for training and testing respectively on the YTC database. Note that only supervised methods need the training time. In practice, test time is more important for the efficiency of a method, as the training process can be conducted offline. From the table, we can see that our proposed method is very efficient and is faster than other affine hull based methods. Since For testing, our method only need to compute the projections and their distance, it is relatively efficient and is faster than other affine hull based methods.

Table 3. Time comparison (seconds) of different methods on YTC for training and testing.

Method	MSM	AHISD	CHISD	SANP	RNP	DLRC	SSDML	PDL
Training	N/A	N/A	N/A	N/A	N/A	N/A	346.33	75.30
Testing	1.31	1.58	1.71	56.77	1.56	1.91	2.35	1.15

5   Conclusions

This paper has proposed a novel Prototype Discriminative Learning method for face image set classification. We represented an image set by a prototype set learned from its basic affine hull model to shrink the loose affine hull effectively while inheriting the merit of affine hull in complementing the unseen appearance with affine combinations. Meanwhile, a linear projection was learned to drive that in the target projected subspace, the learned prototypes can be used to discriminate image sets of different classes. Our experimental evaluation has demonstrated that the proposed method can lead to state-of-the-art recognition accuracies on several challenging databases for face image set identification/verification.

In the future, we will study the regularization of W as well as prototypes more comprehensively. Further, we will explore the effect of learning representative prototypes in the large-scale and unclean image sets and study to construct dense and effective prototypes which can be easily adapted to other typical well established image set models, such as, linear subspace based set models, manifold based set models or statistical set models.

Acknowledgement. This work is partially supported by 973 Program under contract No. 2015CB351802, Natural Science Foundation of China under contracts Nos. 61390511, 61379083, 61272321, and Youth Innovation Promotion Association CAS No. 2015085.

References

1. Cevikalp, H., Triggs, B.: Face recognition based on image sets. In: IEEE Conference on Computer Vision and Pattern Recognition (CVPR) (2010)
2. Hu, Y., Mian, A.S., Owens, R.: Sparse approximated nearest points for image set classification. In: IEEE Conference on Computer Vision and Pattern Recognition (CVPR) (2011)
3. Yang, M., Zhu, P., Gool, L.V., Zhang, L.: Face recognition based on regularized nearest points between image sets. In: IEEE Conference on Automatic Face and Gesture Recognition (FG) (2013)
4. Wang, W., Wang, R., Shan, S., Chen, X.: Probabilistic nearest neighbor search for robust classification of face image sets. In: IEEE Conference on Automatic Face and Gesture Recognition (FG) (2015)
5. Zhu, P., Zhang, L., Zuo, W., Zhang, D.: From point to set: extend the learning of distance metrics. In: IEEE International Conference on Computer Vision (ICCV) (2013)
6. Leng, M., Moutafis, P., Kakadiaris, I.A.: Joint prototype and metric learning for set-to-set matching: application to biometrics. In: IEEE Conference on Biometrics Theory, Applications and Systems (BTAS) (2015)
7. Yamaguchi, O., Fukui, K., Maeda, K.: Face recognition using temporal image sequence. In: IEEE Conference on Automatic Face and Gesture Recognition (FG) (1998)
8. Kim, T.K., Kittler, J., Cipolla, R.: Discriminative learning and recognition of image set classes using canonical correlations. IEEE Trans. Pattern Anal. Mach. Intell. (TPAMI) **29**, 1005–1018 (2007)
9. Hamm, J., Lee, D.D.: Grassmann discriminant analysis: a unifying view on subspace-based learning. In: International Conference on Machine Learning (ICML) (2008)
10. Harandi, M.T., Sanderson, C., Shirazi, S., Lovell, B.C.: Graph embedding discriminant analysis on grassmannian manifolds for improved image set matching. In: IEEE Conference on Computer Vision and Pattern Recognition (CVPR) (2011)
11. Wang, R., Shan, S., Chen, X., Gao, W.: Manifold-manifold distance with application to face recognition based on image set. In: IEEE Conference on Computer Vision and Pattern Recognition (CVPR) (2008)
12. Wang, R., Chen, X.: Manifold discriminant analysis. In: IEEE Conference on Computer Vision and Pattern Recognition (CVPR) (2009)
13. Cui, Z., Shan, S., Zhang, H., Lao, S., Chen, X.: Image sets alignment for video-based face recognition. In: IEEE Conference on Computer Vision and Pattern Recognition (CVPR) (2012)
14. Chen, S., Sanderson, C., Harandi, M.T., Lovell, B.C.: Improved image set classification via joint sparse approximated nearest subspaces. In: IEEE Conference on Computer Vision and Pattern Recognition (CVPR) (2013)
15. Li, Z., Liu, J., Tang, J., Lu, H.: Robust structured subspace learning for data representation. IEEE Trans. Pattern Anal. Mach. Intell. (TPAMI) **37**, 2085–2098 (2015)

16. Shakhnarovich, G., Fisher, J.W., Darrell, T.: Face recognition from long-term observations. In: Heyden, A., Sparr, G., Nielsen, M., Johansen, P. (eds.) ECCV 2002. LNCS, vol. 2352, pp. 851–865. Springer, Heidelberg (2002). doi:10.1007/3-540-47977-5_56

17. Arandjelović, O., Shakhnarovich, G., Fisher, J., Cipolla, R., Darrell, T.: Face recognition with image sets using manifold density divergence. In: IEEE Conference on Computer Vision and Pattern Recognition (CVPR) (2005)

18. Wang, R., Guo, H., Davis, L.S., Dai, Q.: Covariance discriminative learning: a natural and efficient approach to image set classification. In: IEEE Conference on Computer Vision and Pattern Recognition (CVPR) (2012)

19. Harandi, M.T., Salzmann, M., Hartley, R.: From manifold to manifold: geometry-aware dimensionality reduction for SPD matrices. In: Fleet, D., Pajdla, T., Schiele, B., Tuytelaars, T. (eds.) ECCV 2014. LNCS, vol. 8690, pp. 17–32. Springer, Heidelberg (2014). doi:10.1007/978-3-319-10605-2_2

20. Huang, Z., Wang, R., Shan, S., Li, X., Chen, X.: Log-Euclidean metric learning on symmetric positive definite manifold with application to image set classification. In: International Conference on Machine Learning (ICML) (2015)

21. Wang, W., Wang, R., Huang, Z., Shan, S., Chen, X.: Discriminant analysis on Riemannian manifold of Gaussian distributions for face recognition with image sets. In: IEEE Conference on Computer Vision and Pattern Recognition (CVPR) (2015)

22. Chen, Y.-C., Patel, V.M., Phillips, P.J., Chellappa, R.: Dictionary-based face recognition from video. In: Fitzgibbon, A., Lazebnik, S., Perona, P., Sato, Y., Schmid, C. (eds.) ECCV 2012. LNCS, vol. 7577, pp. 766–779. Springer, Heidelberg (2012). doi:10.1007/978-3-642-33783-3_55

23. Chen, Y.C., Patel, V.M., Shekhar, S., Chellappa, R., Phillips, P.J.: Video-based face recognition via joint sparse representation. In: IEEE Conference on Automatic Face and Gesture Recognition (FG) (2013)

24. Cui, Z., Chang, H., Shan, S., Ma, B., Chen, X.: Joint sparse representation for video-based face recognition. Neurocomputing **135**, 306–312 (2014)

25. Lu, J., Wang, G., Deng, W., Moulin, P.: Simultaneous feature and dictionary learning for image set based face recognition. In: Fleet, D., Pajdla, T., Schiele, B., Tuytelaars, T. (eds.) ECCV 2014. LNCS, vol. 8689, pp. 265–280. Springer, Cham (2014). doi:10.1007/978-3-319-10590-1_18

26. Hayat, M., Bennamoun, M., An, S.: Learning non-linear reconstruction models for image set classification. In: IEEE Conference on Computer Vision and Pattern Recognition (CVPR) (2014)

27. Chen, S., Wiliem, A., Sanderson, C., Lovell, B.C.: Matching image sets via adaptive multi convex hull. arXiv preprint arXiv:1403.0320 (2014)

28. Chen, L.: Dual linear regression based classification for face cluster recognition. In: IEEE Conference on Computer Vision and Pattern Recognition (CVPR) (2014)

29. Paredes, R., Vidal, E.: Learning prototypes and distances: a prototype reduction technique based on nearest neighbor error minimization. Pattern Recogn. **39**, 180–188 (2006)

30. Paredes, R., Vidal, E.: Learning weighted metrics to minimize nearest-neighbor classification error. IEEE Trans. Pattern Anal. Mach. Intell. (TPAMI) **28**, 1100–1110 (2006)

31. Villegas, M., Paredes, R.: Simultaneous learning of a discriminative projection and prototypes for nearest-neighbor classification. In: IEEE Conference on Computer Vision and Pattern Recognition (CVPR) (2008)

32. Le, Q.V., Ngiam, J., Coates, A., Lahiri, A., Prochnow, B., Ng, A.Y.: On optimization methods for deep learning. In: International Conference on Machine Learning (ICML) (2011)

33. Garcia, S., Derrac, J., Cano, J.R., Herrera, F.: Prototype selection for nearest neighbor classification: taxonomy and empirical study. IEEE Trans. Pattern Anal. Mach. Intell. (TPAMI) **34**, 417–435 (2012)

34. Ma, M., Shao, M., Zhao, X., Fu, Y.: Prototype based feature learning for face image set classification. In: IEEE Conference on Automatic Face and Gesture Recognition (FG) (2013)

35. Kim, M., Kumar, S., Pavlovic, V., Rowley, H.: Face tracking and recognition with visual constraints in real-world videos. In: IEEE Conference on Computer Vision and Pattern Recognition (CVPR) (2008)

36. Huang, Z., Wang, R., Shan, S., Chen, X.: Learning euclidean-to-riemannian metric for point-to-set classification. In: IEEE Conference on Computer Vision and Pattern Recognition (CVPR) (2014)

37. Ross, B., Phillips, J., Bolme, D., Draper, B., Givens, G., Lui, Y.M., Teli, M.N., Zhang, H., Scruggs, W.T., Bowyer, K., Flynn, P., Cheng, S.: The challenge of face recognition from digital point-and-shoot cameras. In: IEEE Conference on Biometrics Theory, Applications and Systems (BTAS) (2013)

38. Beveridge, J.R., Zhang, H., Draper, B.A., Flynn, P.J., Feng, Z., Huber, P., Kittler, J., Huang, Z., Li, S., Li, Y., Kan, M., Wang, R., Shan, S., Chen, X.: Report on the FG 2015 video person recognition evaluation. In: IEEE Conference and Workshops on Automatic Face and Gesture Recognition (FG) (2015)

39. Zhang, X., Zhang, L., Wang, X.J., Shum, H.Y.: Finding celebrities in billions of web images. IEEE Trans. Multimedia **14**, 995–1007 (2012)

40. Jia, Y., Shelhamer, E., Donahue, J., Karayev, S., Long, J., Girshick, R., Guadarrama, S., Darrell, T.: Caffe: convolutional architecture for fast feature embedding. In: ACM International Conference on Multimedia (2014)

Collaborative Learning Network for Face Attribute Prediction

Shiyao Wang, Zhidong Deng$^{(\boxtimes)}$, and Zhenyang Wang

State Key Laboratory of Intelligent Technology and Systems,
Tsinghua National Laboratory for Information Science and Technology,
Department of Computer Science, Tsinghua University, Beijing 100084, China
michael@tsinghua.edu.cn

Abstract. This paper proposes a facial attributes learning algorithm with deep convolutional neural networks (CNN). Instead of jointly predicting all the facial attributes (40 attributes in our case) with a shared CNN feature extraction hierarchy, we cluster the facial attributes into groups and the CNN only shares features within each group in later feature extraction stages to jointly predicts the attributes in each group respectively. This paper also proposes a simple yet effective attribute clustering algorithm, based on the observation that some attributes are more collaborated (their prediction accuracy improve more when jointly learned) than others, and the proposed deep network is referred to as the collaborative learning network. Contrary to the previous state-of-the-art facial attribute recognition methods which require pre-training on external datasets, the proposed collaborative learning network is trained for attribute recognition from scratch without external data while achieving the best attribute recognition accuracy on the challenging CelebA dataset and the second best on the LFW dataset.

1 Introduction

Face attributes are deemed to be middle-level concepts which human use to describe a certain person. Empirically, when we describe a person, we may choose some characteristics on his/her attributes, such as gender, age, color, hair style(also see Fig. 1). Apart from many entertainment applications inspired by attribute prediction, it contributes to improve the performance of relative tasks as well, including face verification [1–4], face retrieval [3,5,6], suspect search according to eyewitness descriptions [7], and some relevant work about object recognition or classification [8–10]. All above previous work has proved that attributes act as powerful representations of images, extracting discriminative features benefit the following process and boost the final performance.

In order to accurately and robustly predict attributes from images, a general process is divided into four standard steps: (1) detect faces in a picture [11–13]; (2) face alignment [12,14,15]; (3) feature extraction; (4) classification about the presence/absence of each attribute. In this paper, we will focus on the latter two steps that given a face after alignment, and then predict the presence or absence of each required attributes which is similar to describ aspects of

© Springer International Publishing AG 2017
S.-H. Lai et al. (Eds.): ACCV 2016, Part III, LNCS 10113, pp. 361–374, 2017.
DOI: 10.1007/978-3-319-54187-7_24

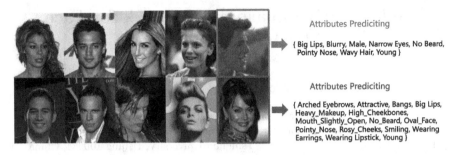

Attributes Prediciting

{ Big Lips, Blurry, Male, Narrow Eyes, No Beard, Pointy Nose, Wavy Hair, Young }

Attributes Prediciting

{ Arched Eyebrows, Attractive, Bangs, Big Lips, Heavy_Makeup, High_Cheekbones, Mouth_Slightly_Open, No_Beard, Oval_Face, Pointy_Nose, Rosy_Cheeks, Smiling, Wearing Earrings, Wearing Lipstick, Young }

Fig. 1. Illustration of the facial attribute prediction on CelebA.

visual appearance. Some papers suggest extracting hand-crafted features from the entire images or several specified local parts of the subject. Although well-designed features could sometime achieve appreciable results, it may lack of robustness in images with large variations or deformations of objects. On the other hand, the deep learning framework especially deep convolutional neural network (CNN) [16–18] is widely used to extract the features recently due to its superb ability to learn effective feature representations. It is a promising idea to use this deep network, nevertheless, given millions of labeled persons, the existing methods still need extra labels such as identity labels or other labeled data to pre-train or fine-tuning their nets which seems quite unefficient. Therefore, we instead present a new framework to address this issue by excavating valuable information behind the given data and providing priori knowledge benefits for designing a most suitable network.

In this paper, we propose a collaborative learning network to simultaneously predict 40 face attributes via a single model. Intuitively, modeling each attribute independently would be the best way to gain the superior accuracy and maximize the extraction of knowledge in the data whereas we consider it a heavy computational engineering more than a suitable model. Hence, we put forward a CNN structure equipped with well-designed both width and depth to jointly predict all the facial attributes with a shared feature extraction hierarchy and independent classification layers. Noticeably, our end-to-end model outperforms existing methods, gaining significant improvement in terms of the predicting accuracy evaluated on the CelebA [19]. On this basis, we further optimize our model inspired by [20], which described the hierarchical organization of face processing in the ventral stream. Similar to the visual cortical area organization which transforms the same low-level input into different representations in different visual cortical areas, we cluster the facial attributes into groups and the CNN only shares features within each group in later feature extraction stages to jointly predicts the attributes in each group respectively like the different visual cortical areas. This novel structure helps to further improve the predicting performance that significantly advances state-of-the-art attribute recognition on the CelebA. In summary, the contributions of our paper include:

- A well-designed convolutional neural network for predicting all the attributes both end to end and simultaneously.
- A simple yet useful algorithm to explore a pair-wise relationship between attributes and cluster them into different groups so that a new CNN structure can be built refer to above clusters.
- A novel framework that attributes share low-level feature extraction in the earlier stages of CNN, while focus on the relative tasks within a group in later stages that we called collaborative learning framework;
- Experiments which demonstrate that the collaborative learning framework significantly advances state-of-the-art attribute predicting on the CelebA dataset and the second best on the LFW dataset.

2   Relative Work

Attribute recognition methods are generally categorized into two groups: hand-crafted features and deep learning methods. Extracting hand-crafted features at pre-defined land-marks is mentioned in [1, 4, 8, 21]. [1] extracted hand-crafted features from different regions as "low-level" features, and then compute visual traits for attribute classification and face verification. [8] adopted color, texture, visual words, and edges as a base feature, shape, part and material as semantic attributes, embedding auxiliary discriminative attributes for describing objects. [4, 21] improve the discriminativeness of hand-crafted features via specializing a particular domain and set of parts or stronger classifier such as three-level SVM system. Deep learning method including [19, 22–27] has achieved great success in attribute predicting and any other vision tasks. Among these significant previous work, [19, 22] are the most relevant work with ours. The method in [19] cascades two CNNs, LNet and ANet, where LNet locates the entire face region and ANet extracts high-level face representation from the located region. The FC layer of Anet, namely, high-level representation is then adopted to train forty SVM classifiers for each attribute prediction. During the training process, LNet is pre-trained with one thousand object categories of ImageNet [28], while ANet is pre-trained by distinguishing massive face identities. In other word, both CNNs need large extra datasets in order to achieve good initialization. Besides, the overall method can be called Features+Classifier that motivated by the AdaBoost algorithm taken by Kumar et al. [29] which seems like optimizing each attributes model independently rather than end-to-end training simultaneously. Instead of relying on manually annotated images, Wang et al. [22] proposed a feature learning method relies on processing extra identity-unlabeled data and embeddings from a few supervised tasks. Actually, they still need extra datasets.

To sum up, all above significant prior work provided a variety of useful methods, especially [19, 22] which equipped with the competitive results and we will present their results as our strong baseline. The evaluation of above methods will be conducted on the benchmarks CelebA [30] and LFW datasets [31]. The attributes labels of these two datasets is provided by [19].

3   Our Approach

Framework Overview. Figure 2 illustrates our pipeline where a CNN model predicts all the facial attributes end-to-end as show in (a), while attributes clustering and the improved model as show in (b) and (c).

In the first step, we design a convolutional neural network learns multiple attribute labels simultaneously and predicts every attributes end to end.

In the second step, we exploit some latent information among the attributes. If the dataset consists of n people equipped with m attributes, it can be regard as n samples to explore the relevance to each pair of m attributes. If two attributes always appear to be the same output, namely presence or absence, we can consider them as positive correlation whereas one attribute is presence/absence, but the other one always appears to be the opposite state, we call them negative correlation. Mining from these large samples, we can draw a similarity matrix of these m attributes. And then, all the attributes can be clustered into k groups base on above matrix which is of great importance for the following step.

In the third step, differ from the general CNN structure, we present a novel *collaborative network*: on the bottom five stages, the net shares weights globally that works on extracting the rough features from the entire image; in the later feature extraction stage, the net is split to k branches based on the second step with each branch sharing weights within their own group that focuses on the correlated tasks. At last, the net break into m mini-branches corresponding to m attributes classifiers. Benefit from this architecture, the bottom stage can generate better features because of more supervised labels and the top stage would promote each other when they have correlated tasks, that we called collaborative network.

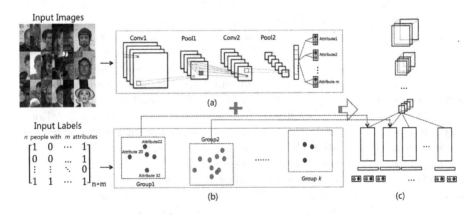

Fig. 2. The proposed pipeline of attribute prediction.

3.1   A Well-Designed Convolutional Neural Network

To predict the m attributes of a given image end-to-end, we adopt CNN as our feature extractor and classifier while the network in [19] is learned to extract features and SVM classifiers are used to predict attributes. We design a CNN structure as follows: the whole net can be divided into six stages with five max-pooling layers and each stage consists of four general convolution layers, batch normalization layers, non-linear(ReLU) layers whose depth is designed such that neurons in the highest layer have the effective receptive fields of approximately 1.5 - 2 times of the input sizes. And the number of filters in convolution layer preceding each pooling layer is always doubly expanded which used to be an empirical trick. After the last stage, all the feature maps are flattened into a vector, and connected to a FC layer whose size is 256 that is suitable for acting as a high-level feature. The classification layer is made up of m mini-classifiers corresponding to m attributes. Although the ANet in [19] was trained by classifying massive face identities in the pre-train stage, and there are some tricks during training in order to learn discriminative features with a large number of identities, our network is only fed by given dataset from scratch without pre-training. Our model achieves an impressive performance in CelebA and LFW datasets with the training process straight and concise. The detail of our network can be seen in Fig. 3.

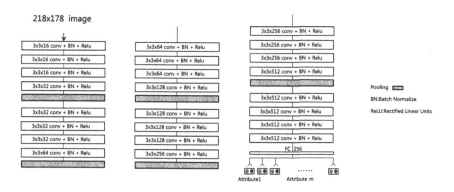

Fig. 3. The structure of our well-designed convolutional neural network.

3.2   Modeling the Relationship Among Attributes

Inspired by [20], we expect to build a model works as the visual cortical areas that receiving the low-level input, selective attention is paid to a particular attribute. Consequently, a pair-wise relationship among attributes should be established based on the relationship. With the face dataset consisted of n images, each labeled by m attributes, we will provide a simple yet useful algorithm to model the relationship among attributes. In CelebA, n denotes 162770 training samples while m indicating 40 attributes. As each attribute is represented as a binary code, 1 denoting presence while 0 denoting absence of a certain attribute, we can generate a $n * m$ matrix to present these data:

$$
\begin{pmatrix}
 & A_1 & A_2 & \cdots & A_j & \cdots & m \\
P_1 & I(P_1, A_1) & I(P_1, A_2) & \cdots & I(P_1, A_j) & \cdots & I(P_1, A_m) \\
P_2 & I(P_2, A_1) & I(P_2, A_2) & \cdots & I(P_2, A_j) & \cdots & I(P_2, A_m) \\
\vdots & \vdots & \vdots & \vdots & \vdots & \vdots & \vdots \\
P_i & I(P_i, A_1) & I(P_i, A_2) & \cdots & I(P_i, A_j) & \cdots & I(P_i, A_m) \\
\vdots & \vdots & \vdots & \vdots & \vdots & \vdots & \vdots \\
P_n & I(P_n, A_1) & I(P_n, A_2) & \cdots & I(P_n, A_j) & \cdots & I(P_n, A_m)
\end{pmatrix}
\tag{1}
$$

Where P_i presents the i-th person and A_j donotes the j-th attribute. For a given sample P_i, let $I(P_i, A_j) : P_i, A_j \rightarrow \{0,1\}$ be a function yielding the binary ground truth label for P_i and A_j, where $i \in \{1, \cdots, n\}$ is the people index and $j \in \{1, \cdots, m\}$ is the attribute index.

After this, all of the samples have been organized as above matrix (Eq. (1)) and each column vector belongs to a specified attribute. In order to explore a pair-wise relationship among attributes, we can calculate the co-occurrence of each two attributes. If they always appear to be the same output, we can consider them as positive correlation. Hence, we define the following equation as their similarity:

$$
s(A_j, A_{j'}) = p(A_j = 1, A_{j'} = 1) + p(A_j = 0, A_{j'} = 0)
\tag{2}
$$

$$
p(A_j = 1, A_{j'} = 1) = \frac{N(I(P_i, A_j) = I(P_i, A_{j'}) = 1)}{n}
\tag{3}
$$

$$
p(A_j = 0, A_{j'} = 0) = \frac{N(I(P_i, A_j) = I(P_i, A_{j'}) = 0)}{n}
\tag{4}
$$

A_j and $A_{j'}$ are two attributes, and $N(I(P_i, A_j) = I(P_i, A_{j'}) = 1/0)$ corresponds to the num of both A_j and $A_{j'}$ are equal to $1/0$.

After computing the similarity between two attributes nodes through above measures, we may model the relationship as a similar matrix:

$$
\begin{pmatrix}
 & A_1 & A_2 & \cdots & A_m \\
A_1 & s(A_1, A_1) & s(A_1, A_2) & \cdots & s(A_1, A_m) \\
A_2 & s(A_2, A_1) & s(A_2, A_2) & \cdots & s(A_2, A_m) \\
\vdots & \vdots & \vdots & \vdots & \vdots \\
A_m & s(A_m, A_1) & s(A_n, A_2) & \cdots & s(A_m, A_m)
\end{pmatrix}
\tag{5}
$$

We cluster these m attributes according to the similar matrix in a way motivated by K-means.

Step 1: We choose k attributes as initial cluster centers;

Step 2: Each attribute is assigned to its closest cluster center base on the similar matrix;

Step 3: Each cluster center is updated according to the above new groups;

Step 4: Back to step2 until no further change in cluster centers.

In the later experiments analysis, we choose $k = 6$. In fact, attribute prediction accuracy is improved consistently for different numbers of clusters (K), i.e., K $= 4, 5, 6, 7, 8$. Besides, we find that the attribute clusters have carried semantic concepts.

3.3   Collaborative Learning Network

After exploring the relationship between attributes, a new CNN architecture is presented in this section. We still believe that end-to-end is a better way exploiting all correlation instead of separating of feature extraction and attribute classification. Therefore, a collaborative learning network is presented in Fig. 4.

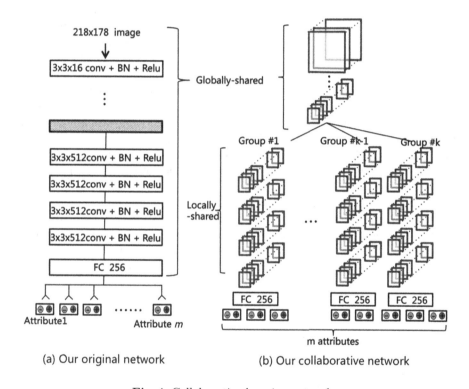

Fig. 4. Collaborative learning network.

During the first five stages, all the filters are globally-shared like our original net which aim at extracting the rough features over the entire image. And then, our net is divided into k sub-nets based on the clustering results provided by Sect. 3.1. In other word, in the last stage, our net generates k branches corresponding to the k clusters that including three standard units (a convolution layer, a batch normalization layer and a non-linear layer) and the filters are shared within their own branch respectively. In this way, each branch focuses on the specified and discriminative features extracting for relative attributes recognition and collaboratively learning by sharing the filters. Benefit from above stages, the features are rich and specialized enough for classification, so that we can apply the m classifiers after that process. Our collaborative network is still fed by the given attribute dataset and initialized with random weights. Contrary to our original net, the final accuracy is improved a lot compared with the strong baseline in [19,22] which well demonstrated our assumption.

4   Experiment

In this section we evaluate the effectiveness of collaborative learning network with quantitative results on two standard facial attribute datasets called CelebA and LFWA datasets. CelebA contains ten thousand identities, each of which has twenty images. For this dataset, there are eight thousand identities with 162770 images for training, another one thousand identities with 19867 images for validation and the remaining 19962 images for testing. LFWA has 13,233 images of 5,749 identities. Each image in CelebA and LFWA is annotated with forty face attributes. Our network is implemented based on the framework of caffe [32], and conducted on two GPUs with data parallelism. In all experiments, our models are trained using stochastic gradient descent (SGD) algorithm with a mini-batch of 64. The learning rate is initialized to 0.001 and repeatedly decreased 3 times, until it arrives at 1e−6. We run our net through the whole training data with only 8 epochs and a momentum of 0.9 is used in the entire training process to make SGD stable and fast. In the following sections, we conduct three kinds of experiments. First, we present a comparison between an end-to-end convolutional neural network and an approach separates the feature extraction and classification in [19]; Second, we show the pairwise relationship between attributes in Sect. 4.2 and a new CNN structure based on the above relationship is also presented; Last but not the least, some existing approaches including the sate-of-the-art are reported in Sect. 4.3 that proves our approach has advanced the other methods.

4.1   An End-to-End Convolutional Neural Network for Attribute Prediction

In this section, we compare the method in [19] and our well-designed convolutional neural network in order to demonstrate a way of end-to-end is able to achieve the higher accuracy. The Although Liu et al. [19] suggested to extract features via network and feed to independent linear SVMs for final attribute classification, we believe that our network has equipped with the ability of classification over 40 attributes labels as supervisory signals. Besides, contract to some suggestion that independently optimize each attributes by different networks, we tend to gather all the attributes together account for the collaborative learning. Actually, the comparison in Fig. 5 shows the average accuracy of LNets+ANet [19] is

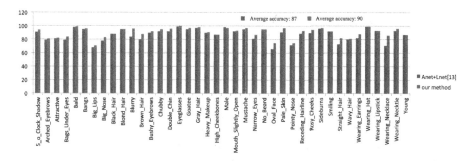

Fig. 5. Attribute prediction results of our method and [19].

87% while our net can achieve 90%. Almost every attributes have been improved, and the maximum improvement can obtain 15% (Wearing_Necklace). Consequently, we suggest that predicting face attributes though an end-to-end convolutional neural network would be better and all the attributes can be set as supervised signals in the meanwhile. When [19] prefers pre-taining by massive face identities for attribute prediction, we tend to train from scratch. Recognizing identities and attributes are two intuitively contradictive tasks. The former suppresses face variations unrelated to identities, such as pose, expression, age, wearing hat etc. while the latter reserves such variations. For comparison, we pre-train our model for face recognition on the Celeb face dataset and fine-tune it for attribute recognition. It gets 89.35% accuracy, inferior to 90.41% of the model without pre-training.

4.2 A Collaborative Learning Network Based on the Pair Wise Relationship of Attributes

We compute the correlation between two attributes as described in Sect. 3.2. Note that the relationship between any two sets of attributes can be computed in the same way. The correlation value is in the range of [0, 1], meaning positive correlation, respectively. And then we cluster all the attributes into six groups which presented as:

Cluster #1: High Cheekbones, Mouth Slightly Open, Smiling;

Cluster #2: No Beard, Young;

Cluster #3: Arched Eyebrows, Attractive, Heavy Makeup, Pointy Nose, Wavy Hair, Wearing Lipstick;

Cluster #4: Bags Under Eyes, Big Nose, Male;

Cluster #5: Black Hair, Oval Face, Straight Hair;

Cluster #6: 5 o Clock Shadow, Bald, Bangs, Big Lips, Blond Hair, Blurry, Brown Hair, Bushy Eyebrows, Chubby, Double Chin, Eyeglasses, Goatee, Gray Hair, Mustache, Narrow Eyes, Pale Skin, Receding Hairline, Rosy Cheeks, Sideburns, Wearing Earrings, Wearing Hat, Wearing Necklace, Wearing Necktie;

It seems that the clusters have carried some semantics concepts: Mouth Slightly Open has highly positive correlation with Smiling and they are in cluster

#1; No Beard is clustered with Young; Attractive, Heavy Makeup and Wearing are in cluster #3. We adopt these group information into our collaborative network by dividing our net into six branches on the last stage of original structure. The average accuracy is further improved from 90% to 91%. All groups get consistent improvement. 0.5%, 0.3%, 0.4%, 0.1%, 0.4%, and 0.2% average improvement is acquired for each of the six groups, respectively. We make a comparison between our original net and collaborative net for each attributes in Fig. 6(a). Why collaborative network helps to further improve the predicting accuracy? We consider that although there are 160 thousand images for training, the labels are not balanced such as Gray Hair which we only have 4% samples of positive labels.When add several relative attributes, the model will be capable of generating better features. In addition, we conduct anther experiment to prove the effectiveness of our framework. We designed three models taking different input sizes and network depth, all of which are based on the design principles in Sect. 3.1. As show in Fig. 6(b), all these three models achieve higher accuracy compared to our original network. Furthermore, we even applied our proposed method to GoogLeNet by branching its later feature extraction stages and using the clustered attributes for supervision. The accuracy is further improved from 90.9% to 91.24%. Therefore our method generally improves the attribute prediction accuracy irrespective of the network architectures that effectively demonstrate our collaborative learning network.

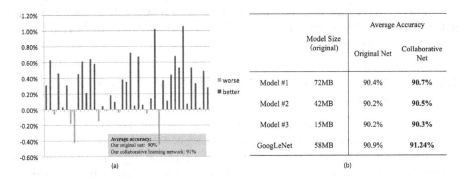

	Model Size (original)	Average Accuracy	
		Original Net	Collaborative Net
Model #1	72MB	90.4%	**90.7%**
Model #2	42MB	90.2%	**90.5%**
Model #3	15MB	90.2%	**90.3%**
GoogLeNet	58MB	90.9%	**91.24%**

(a)   (b)

Fig. 6. (a) Comparison between our original net and collaborative net for each attributes. (b) Performances of three different models.

4.3   Performance Comparison with Relative Work

In this section, our method is compared with five competitive approaches, i.e. FaceTracer [29], PANDA-w [26], and PANDA-l [26], Lnets+Anet [19] and Walk and Learn [22]. External data are needed during their training process while our net is trained from scratch without extra data. The average accuracy of all above methods in CelebA is 81%, 79%, 85%, 87%, 88% and 91%, respectively(also see Fig. 7). We also test these methods on LFWA dataset, and we get the second best. We check labels of LFWA to find some of images are mis-labeled. For example, some images labeled "female' is shown in Fig. 8. In order to further evaluate our model, we cleanse bad labels about "Male" and test our net on that cleaned labels again. The accuracy of the sex predicting can achieve 98% while it is 94% in the previous mis-labeled data that meet our expectations. In particular, our model is even not fine-tuned in LFWA data, and directly tested using the model trained over CelebA.

		5_o_Clock_Shadow	Arched Eyebrows	Attractive	Bags Under Eyes	Bald	Bangs	Big Lips	Big Nose	Black Hair	Blond Hair	Blurry	Brown Hair	Bushy Eyebrows	Chubby	Double Chin	Eyeglasses	Goatee	Gray Hair	Heavy Makeup	High Cheekbones	Male
	FaceTracer[29]	85	76	78	76	89	88	64	74	70	80	81	60	80	86	88	98	93	90	85	84	91
	PANDA-w [26]	82	73	77	71	92	89	61	70	74	81	77	69	76	82	85	94	86	88	84	80	93
CelebA	PANDA-l [26]	88	78	81	79	96	92	67	75	85	93	86	77	86	86	88	98	93	94	90	86	97
	Lnets+Anet [19]	91	79	81	79	98	95	68	78	88	95	84	80	90	91	92	99	95	97	90	87	98
	Work anf learn [22]	84	87	84	87	92	96	78	91	84	92	97	81	93	89	93	97	92	95	96	95	96
	Our method	94	82	82	85	99	96	71	83	89	96	96	88	93	95	96	100	97	98	91	87	98
	FaceTracer[29]	70	67	71	65	77	72	68	73	76	88	73	62	67	67	70	90	69	78	88	77	84
	PANDA-w [26]	64	63	70	63	82	79	64	71	78	87	70	65	63	65	64	84	65	77	86	75	86
LFWA	PANDA-l [26]	84	79	81	80	84	84	73	79	87	94	74	74	79	69	75	89	75	81	93	86	92
	Lnets+Anet [22]	84	82	83	83	88	88	75	81	90	97	74	77	82	73	78	95	78	84	95	88	94
	Work anf learn [22]	76	82	82	91	82	93	75	92	93	97	86	83	78	79	81	94	80	91	96	96	93
	Our method	74	80	79	83	91	88	79	84	92	96	85	81	81	78	79	91	81	86	96	90	94

		Mouth_Slightly_Open	Mustache	Narrow Eyes	No Beard	Oval Face	Pale Skin	Pointy nose	Receding Hairline	Rosy Cheeks	Sideburns	Smiling	Straight Hair	Wavy Hair	Wearing Earrings	Wearing Hat	Wearing Lipstick	Wearing Necklace	Wearing Necktie	Young	Average
	FaceTracer[29]	87	91	82	90	64	83	68	76	84	94	89	63	73	73	89	89	68	86	80	81
	PANDA-w [26]	82	83	79	87	62	84	65	82	81	90	89	68	76	72	91	88	67	88	77	79
CelebA	PANDA-l [26]	93	93	84	93	65	91	71	85	87	93	92	69	77	78	96	93	67	91	84	85
	Lnets+Anet[19]	92	95	81	95	66	91	72	89	90	96	92	73	80	82	99	93	71	93	87	87
	Work anf learn [22]	94	83	79	75	84	87	91	86	81	77	97	76	89	96	96	97	95	80	89	'88
	Our method	94	97	87	96	75	97	76	93	95	98	92	82	81	89	99	94	86	96	87	91
	FaceTracer[29]	77	83	73	69	66	70	74	63	70	71	78	67	62	88	75	87	81	71	80	74
	PANDA-w [26]	74	77	68	63	64	64	68	61	64	68	77	68	63	85	78	83	79	70	76	71
LFWA	PANDA-l [26]	78	87	73	75	72	84	76	84	73	76	89	73	75	92	82	93	86	79	82	81
	Lnets+Anet [19]	82	92	81	79	74	84	80	858	78	77	91	76	76	94	88	95	88	79	86	84
	Work anf learn [22]	98	90	79	90	79	85	77	84	96	92	98	75	85	91	96	92	77	84	86	87
	Our method	78	92	79	81	71	89	83	88	74	81	91	82	80	96	89	96	90	84	85	85

Fig. 7. Performance comparison with state of the art methods on 40 binary facial attributes. (Note that all the methods expect ours need external datasets for training.)

Fig. 8. Some mis-labeled images in LFWA. (Labed "female").

5   Conclusion

This paper have proposed a facial attribute learning algorithm with deep convolutional neural networks (CNN). First, we provide a well-designed convolutional neural network to predict attributes end to end with noticeable improvement. Second, to take a deep look into all the given attributes, we present a useful algorithm to learn a pairwise relationship between them; Third, based on the above relationship, unlike the previous facial attribute recognition studies which treated all the attributes equally, this paper discovered the grouping properties of the facial attributes and designed the deep network which explored the correlation of the attribute labels. The proposed method consistently improves the attribute prediction accuracy of very competitive baseline networks such as GoogLeNet and our designed deep networks. We have demonstrated the effectiveness of this framework on two challenging face datasets, CelebFaces and LFW datasets. Instead of pre-training on large-scale face recognition datasets like the previous state-of-the-art facial attribute recognition methods, the proposed collaborative learning network is trained from scratch without external data while achieving the best attribute recognition accuracy on the challenging CelebA face dataset and the second best on the LFW dataset.

Acknowledgement. The authors would like to thank the anonymous reviewers for their valuable comments that considerably contributed to improving this paper. This work was supported in part by the National Science Foundation of China (NSFC) under Grant Nos. 91420106, 90820305, and 60775040, and by the National High-Tech R&D Program of China under Grant No. 2012AA041402.

References

1. Kumar, N., Berg, A.C., Belhumeur, P.N., Nayar, S.K.: Attribute and simile classifiers for face verification. In: 2009 IEEE 12th International Conference on Computer Vision, pp. 365–372. IEEE (2009)
2. Song, F., Tan, X., Chen, S.: Exploiting relationship between attributes for improved face verification. Comput. Vis. Image Underst. **122**, 143–154 (2014)
3. Kumar, N., Berg, A.C., Belhumeur, P.N., Nayar, S.K.: Describable visual attributes for face verification and image search. IEEE Trans. Pattern Anal. Mach. Intell. **33**, 1962–1977 (2011)
4. Berg, T., Belhumeur, P.: Poof: part-based one-vs.-one features for fine-grained categorization, face verification, and attribute estimation. In: Proceedings of the IEEE Conference on Computer Vision and Pattern Recognition, pp. 955–962 (2013)
5. Siddiquie, B., Feris, R.S., Davis, L.S.: Image ranking and retrieval based on multi-attribute queries. In: 2011 IEEE Conference on Computer Vision and Pattern Recognition (CVPR), pp. 801–808. IEEE (2011)
6. Douze, M., Ramisa, A., Schmid, C.: Combining attributes and fisher vectors for efficient image retrieval. In: 2011 IEEE Conference on Computer Vision and Pattern Recognition (CVPR), pp. 745–752. IEEE (2011)
7. Feris, R., Bobbitt, R., Brown, L., Pankanti, S.: Attribute-based people search: lessons learnt from a practical surveillance system. In: Proceedings of International Conference on Multimedia Retrieval, p. 153. ACM (2014)

8. Farhadi, A., Endres, I., Hoiem, D., Forsyth, D.: Describing objects by their attributes. In: 2009 IEEE Conference on Computer Vision and Pattern Recognition, CVPR 2009, pp. 1778–1785. IEEE (2009)

9. Wang, Y., Mori, G.: A discriminative latent model of object classes and attributes. In: Daniilidis, K., Maragos, P., Paragios, N. (eds.) ECCV 2010. LNCS, vol. 6315, pp. 155–168. Springer, Heidelberg (2010). doi:10.1007/978-3-642-15555-0_12

10. Yu, F., Cao, L., Feris, R., Smith, J., Chang, S.F.: Designing category-level attributes for discriminative visual recognition. In: Proceedings of the IEEE Conference on Computer Vision and Pattern Recognition, pp. 771–778 (2013)

11. Sun, Y., Liang, D., Wang, X., Tang, X.: Deepid3: Face recognition with very deep neural networks. arXiv preprint arXiv:1502.00873 (2015)

12. Xiong, X., Torre, F.: Supervised descent method and its applications to face alignment. In: Proceedings of the IEEE Conference on Computer Vision and Pattern Recognition, pp. 532–539 (2013)

13. Ding, C., Xu, C., Tao, D.: Multi-task pose-invariant face recognition. IEEE Trans. Image Process. **24**, 980–993 (2015)

14. Zhu, S., Li, C., Change Loy, C., Tang, X.: Face alignment by coarse-to-fine shape searching. In: Proceedings of the IEEE Conference on Computer Vision and Pattern Recognition, pp. 4998–5006 (2015)

15. Tzimiropoulos, G.: Project-out cascaded regression with an application to face alignment. In: 2015 IEEE Conference on Computer Vision and Pattern Recognition (CVPR), pp. 3659–3667. IEEE (2015)

16. Sun, Y., Wang, X., Tang, X.: Hybrid deep learning for face verification. In: Proceedings of the IEEE International Conference on Computer Vision, pp. 1489–1496 (2013)

17. Zhu, Z., Luo, P., Wang, X., Tang, X.: Deep learning identity-preserving face space. In: Proceedings of the IEEE International Conference on Computer Vision, pp. 113–120 (2013)

18. Taigman, Y., Yang, M., Ranzato, M., Wolf, L.: Deepface: Closing the gap to human-level performance in face verification. In: Proceedings of the IEEE Conference on Computer Vision and Pattern Recognition, pp. 1701–1708 (2014)

19. Liu, Z., Luo, P., Wang, X., Tang, X.: Deep learning face attributes in the wild. In: Proceedings of the IEEE International Conference on Computer Vision, pp. 3730–3738 (2015)

20. Ungerleider, L.G., Haxby, J.V.: 'What' and 'where' in the human brain. Curr. Opin. Neurobiol. **4**, 157–165 (1994)

21. Bourdev, L., Maji, S., Malik, J.: Describing people: a poselet-based approach to attribute classification. In: 2011 IEEE International Conference on Computer Vision (ICCV), pp. 1543–1550. IEEE (2011)

22. Wang, J., Cheng, Y., Feris, R.S.: Walk and learn: facial attribute representation learning from egocentric video and contextual data. In: 2016 IEEE Conference on Computer Vision and Pattern Recognition (CVPR) (2016)

23. Luo, P., Wang, X., Tang, X.: Hierarchical face parsing via deep learning. In: 2012 IEEE Conference on Computer Vision and Pattern Recognition (CVPR), pp. 2480–2487. IEEE (2012)

24. Zhu, Z., Luo, P., Wang, X., Tang, X.: Multi-view perceptron: a deep model for learning face identity and view representations. In: Advances in Neural Information Processing Systems, pp. 217–225 (2014)

25. Zhang, N., Donahue, J., Girshick, R., Darrell, T.: Part-based R-CNNs for fine-grained category detection. In: Fleet, D., Pajdla, T., Schiele, B., Tuytelaars, T. (eds.) ECCV 2014. LNCS, vol. 8689, pp. 834–849. Springer, Heidelberg (2014). doi:10.1007/978-3-319-10590-1_54

26. Zhang, N., Paluri, M., Ranzato, M., Darrell, T., Bourdev, L.: Panda: pose aligned networks for deep attribute modeling. In: Proceedings of the IEEE Conference on Computer Vision and Pattern Recognition, pp. 1637–1644 (2014)

27. Zhang, Z., Luo, P., Loy, C.C., Tang, X.: Learning social relation traits from face images. In: Proceedings of the IEEE International Conference on Computer Vision, pp. 3631–3639 (2015)

28. Deng, J., Dong, W., Socher, R., Li, L.J., Li, K., Fei-Fei, L.: Imagenet: a large-scale hierarchical image database. In: 2009 IEEE Conference on Computer Vision and Pattern Recognition, CVPR 2009, pp. 248–255. IEEE (2009)

29. Kumar, N., Belhumeur, P., Nayar, S.: FaceTracer: a search engine for large collections of images with faces. In: Forsyth, D., Torr, P., Zisserman, A. (eds.) ECCV 2008. LNCS, vol. 5305, pp. 340–353. Springer, Heidelberg (2008). doi:10.1007/978-3-540-88693-8_25

30. Sun, Y., Chen, Y., Wang, X., Tang, X.: Deep learning face representation by joint identification-verification. In: Advances in Neural Information Processing Systems, pp. 1988–1996 (2014)

31. Huang, G.B., Ramesh, M., Berg, T., Learned-Miller, E.: Labeled faces in the wild: a database for studying face recognition in unconstrained environments. Technical report 07–49, University of Massachusetts, Amherst (2007)

32. Jia, Y., Shelhamer, E., Donahue, J., Karayev, S., Long, J., Girshick, R., Guadarrama, S., Darrell, T.: Caffe: convolutional architecture for fast feature embedding. In: Proceedings of the ACM International Conference on Multimedia, pp. 675–678. ACM (2014)

Facial Expression-Aware Face Frontalization

Yiming Wang[1], Hui Yu[1(✉)], Junyu Dong[2], Brett Stevens[1], and Honghai Liu[3]

[1] School of Creative Technologies, University of Portsmouth, Portsmouth, UK
hui.yu@port.ac.uk
[2] Ocean University of China, Qingdao, China
[3] School of Computing, University of Portsmouth, Portsmouth, UK

Abstract. Face frontalization is a rising technique for view-invariant face analysis. It enables a non-frontal facial image to recover its general facial appearances to frontal view. A few pioneering works have been proposed very recently. However, face frontalization with detailed facial expression recovering is still very challenging due to the non-linear relationships between head-pose and expression variations. In this paper, we propose a novel facial expression-aware face frontalization method aiming at reconstructing the frontal view while maintaining vivid appearances with regards to facial expressions. First of all, we design multiple face shape models as the reference templates in order to fit in with various shape of facial expressions. Each template describes a set of typical facial actions referred to Facial Action Coding System (FACS). Then a template matching strategy is applied by measuring a weighted Chi Square error such that the input image can be matched with the most approximate template. Finally, Robust Statistical face Frontalization (RSF) method is employed for the task of frontal view recovery. This method is validated on a spontaneous facial expression database and the experimental results show that the proposed method outperforms the state-of-the-art methods.

1 Introduction

Facial expression recognition (FER) forms the essence of human-machine system, and therefore is one of the most active research topic in the field of human-computer interaction, computer vision and machine intelligence [1,2]. For over two decades, much effort has been made to improve the FER system. However, most existing FER methods were still focusing on the recognition of frontal or near-frontal facial images with posed facial expression [3]. The performance of these methods will drop dramatically in the uncontrolled real-world environment, especially when there are large occlusion and head-pose variations. With the quick development of human-machine system, there has been a continuously increasing demand for effective spontaneous FER methods.

One of the key challenges for in-the-wild facial expression recognition is how to tackle the variations of out-of-plane head rotation. There are only a few approaches proposed for view-invariant FER. Up to now, these methods

© Springer International Publishing AG 2017
S.-H. Lai et al. (Eds.): ACCV 2016, Part III, LNCS 10113, pp. 375–388, 2017.
DOI: 10.1007/978-3-319-54187-7_25

can be divided into two categories: view-based methods and dictionary learning methods. In [4], a typical multi-view approach is applied, in which multiple discrete yaw angles are predefined. For each face image, the head pose is estimated and matched to the closest predefined angle. Then a view-specific facial expression classifier is trained in each discrete angle. It is obvious that a pose estimation step must be performed first, and the whole system has to be trained per viewpoint/person/expression. Similar situation exists in [5,6]. In [6,7], two approaches of view normalization are described. Rudovic et al. [7] present a pair-wise viewpoint normalization in which Coupled Gaussian Processes regression are used to model the pair-wise facial geometric features (facial points) between non-frontal face and its corresponding frontal counterpart. View normalization can tackle an input face image with unseen viewpoint, which is superior to pose-wise FER approaches. However, the pose estimation step is still inevitable. Furthermore, the pair-wise view modelling is time consuming and requires large amount of training data to ensure accuracy. For dictionary learning methods, include [8,9], both of them use SIFT [10] as low-level feature descriptor and encode SIFT feature pose-wise. Specifically, [8] uses generic sparse coding and [9] uses supervised super-vector encoding for high-level feature learning. After learning the high-level features from different views, a single classifier will be trained. Dictionary learning approaches still need to learn the features per viewpoint/person/expression in order to achieve promising results.

Apparently, current view-invariant FER methods highly rely on the quality of training data. These approaches require a large data volume in terms of different expressions and poses, and some of them even need to be trained person-specifically. The satisfied database is often not readily available.

As a newly rising research topic in view-invariant face analysis, face frontalization can overcome the drawbacks mentioned above. It aims to recover the frontal face from unconstrained image. Until now, the approaches and contributions in this field are very limited. The key idea of 2D face frontalization is estimating frontal facial shapes and compensating the missing part of frontal facial textures. Face frontalization starts from facial landmark localization, then performs shape matching and texture fitting schemes to reconstruct frontal face. Existing methods can be divided into two categories: 3D assisted methods and 2D methods. There are many 3D-assisted approaches that synthesize the 3D facial model from several images of one person and result in person-specific face frontalization [11,12]. These methods will need a large amount of facial images captured from poses and expressions. Furthermore, person-specific face frontalization approaches are often impractical since they cannot deal with a totally new face from a single image. Therefore, we will focus more on generic face frontalization. Hassner et al. [13] present a generic face frontalization method which attempt to approximate 3D facial shapes from a single 2D facial image. Given a single reference 3D surface, the landmarks of input image is projected to their 3D positions of reference 3D surface and compute a projection matrix. By generating frontal view using the estimated 3D surface, 3D coordinates will be projected back to the input image. The missing part due to pan angles of head pose is compensated by the corresponding symmetric parts of the face. This method is a breakthrough of generic face frontalization. However, the non-frontal face with tilt angles cannot be recovered, and the symmetry-based face compensation is not always reasonable, especially when there are occlusions or non-symmetric facial expressions. In general, 3D assisted method can achieve high performance of face frontalization. But 3D facial shape estimation from 2D is a hard problem and may potentially lead to misalignment. There are only a few works that has been done on 2D methods. In [14], the face image is divided into several overlapping blocks and Markov Random Field is employed to optimize patch-based local warps based on an assumption that the mean value of intensity of each block is the same. The global frontal face can be predicted by optimizing each local warp. In [15], Sagonas et al. presents Robust Statistical face Frontalization (RSF) based on the fact that frontal facial images have a smaller value of nuclear norm when compared with non-frontal facial images. RSF firstly warps a non-frontal face into a reference frame (base mesh). Then an optimization problem is solved by iteratively minimizing the nuclear norm of warped image.

Compared with the traditional view-invariant FER methods that is sensitive to the quality of training data, 2D face frontalization methods is potentially a better alternative for FER in the wild. Considering that 3D assisted face frontalization approaches has many limitations, we mainly focus on 2D methods. Thus, the purpose of this work is to derive a generic facial expression-aware face frontalization which is robust to individual differences, occlusions and head pose variations in whatever pan and tilt angles.

No work has been done to apply face frontalization to view-invariant FER so far due to the problem of identity bias [2] in 2D FER task. It has been proved that facial features are always biased to identity-related interpretation but poor in describing facial expressions. Current 2D face frontalization methods tend to aggravate this problem since they always need a shape template to perform rigid shape matching. The expression-related information of both facial shapes and textures may be lost after shape matching. This situation, therefore, makes it more difficult to apply face frontalization to FER.

In this research, we present a novel view-invariant FER method based on face frontalization. Knowing that changes in non-rigid expression and changes in head pose in 2D cannot be linearly decomposed [6,7], it is challenging perform a non-rigid frontal facial shape recovery from non-frontal images. Current 2D face frontalization approaches often present a rigid shape matching that matches the shape of input image with a single reference template, which often leads to a huge loss of facial expression cues. The non-linear problem mentioned above is very difficult to solve, but we can present a approximated solution when there are multiple templates instead of one single template, and non-frontal face shape can be matched with the most appropriate templates. Inspired by this idea, we developed a Facial Expression-Aware face Frontalization (FEAF) method using multi-template model. In this method, five different templates are developed. Among them, one face mesh is neutral and the other four are expression-enriched templates. These templates are designed by a combination of Action Units (AUs) described in Facial Action Coding System (FACS) according to the commonly accepted expert knowledge of the relations between six universal emotions and AUs [16].

With the five templates, a template matching step will be undertaken in order to match the input image with the most appropriate template. The landmarks of the input image must be detected first, but there is no need to detect all facial points. Only the landmarks located in the expression-enriched facial regions are needed. The satisfactory landmark detection method is Supervised Decent Method (SDM) [17] which can correctly localize 49 points in the facial regions of eyes, eyebrow, nose and mouth. Then the similarity between the input face shape and the five templates can be measured by using weighted Chi square error. After template matching, RSF is employed to fit the texture of input image with corresponding template. By observing that the running time of RSF can be significantly reduced if the initialization highly approximates to the optimal value, we use a landmark detector to localize facial landmarks and use its output as the initialization of RSF so that the efficiency can be improved prominently. Finally, some commonly used feature extraction methods and machine learning techniques will be employed to recognize the six universal facial emotions (happy, sad, surprise, fear, angry and disgust). The experimental validation shows that FEAF achieves an effective face frontalization with powerful description of facial expressions. FEAF outperforms the state-of-the-art methods for FER in the wild. The main contributions of this research include:

(1) We develop a multi-template model for facial expression-aware face frontalization in which detailed facial expression appearances are effectively reconstructed. As far as we know, this research is the first of its kind to recover full frontal facial expressions from unconstrained facial images.
(2) Whatever pan or tilt angles of head pose in test images (even the untrained ones) can be recovered in this work.
(3) The efficiency of RSF is improved by providing an appropriate initialization.

2   Multi-template Based Face Frontalization

2.1   Template Design and Template Matching

As previously discussed, 2D face frontalization methods may lose the expression-related cues during shape matching. Especially for RSF, the face shape will be arbitrarily warped to a neutral template whatever the shape of input image is. This warping strategy will cause a huge loss of expression-related information. In order to overcome this problem, we present a novel method by designing multiple templates which involve variant face shapes in terms of expressions, such that the non-frontal facial images can be warped to its most approximate template.

It is commonly accepted that the facial areas around eyebrow, eyes and mouth are the most enriched facial region related to facial emotions [18]. So we empirically design five templates which include five apparent characteristics: eyes wide open, eyebrow lower, lips apart, mouth wide open and neutral, as is exhibited in

Fig. 1. These facial behaviours are the most significant shape features in expressions of human beings.

The idea of the template design is inspired from FACS. FACS defines many AUs that encodes the basic unit of facial configurations. It is generally accepted that combination of different AUs can generate meaningful facial emotions. For example, the emotion of sad is often a combination of AU1, AU4 and AU15, meanwhile fear is usually a combination of AU1, AU2, AU4, AU5, AU7 and AU26. Apparently, Au1 and AU4 are basic component of both sad and fear. So we design the template t2 which includes AU1 and AU4 so that an input facial image with the emotion of sad or fear will be easier to be classified into t2. The other templates are designed in the similar way. Considering that many AUs, as well as their combinations, share the same shape and may not important in determine emotional categories, there is no need to design more templates. Five template are enough to meet the basic demand of expression-aware face frontalization.

Figure 1(a) means eyes wide open which is often related to surprise and fear. Figure 1(b) indicates lowering the eyebrow, which often appears in sad, angry and disgust. Figure 1(c) suggests lips apart that is relevant to smile and fear. Figure 1(d) means mouth wide open that often exists in surprise. Figure 1(e) is neutral face shape. As is shown in the figure, these five templates are very different from each other and can be immediately distinguished only by shapes. So the probability of mismatch can be reduced in the template matching step.

The procedure of of template design is as follow: (1) Given the training data, we manually select the frontal faces without occlusion. (2) Manually classifying the clean frontal face to five categories in terms of their most significant shape features. For those images that are suitable to more than one templates, a priority

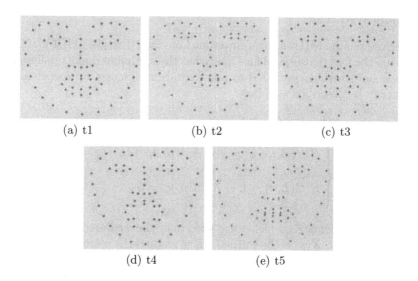

(a) t1   (b) t2   (c) t3

(d) t4   (e) t5

Fig. 1. Five templates of face shape

queue is listed, in which changes in eyebrow are priority, followed by eyes and finally the mouth. For example, a facial image with both eyes and mouth wide open will be assigned to template t1, but not t4, according to this priority list. (3) Five templates are obtained by computing the mean shape of each category.

Template matching will match the position of landmarks of input face with an appropriate template. It, thus, start from facial landmark detection. By observing that importance of different facial landmarks varies, there is no need to localize all these points. SDM is the best choice for this task due to its small computational cost and high accuracy in face alignment. It can effectively localize 49 fiducial landmark points in the region of eyebrow, nose, eyes and mouth. Then the facial points are normalized using Procrustes analysis.

Then, a similarity measure is undertaken to perform template matching. We select Chi square test statistic (χ^2) for similarity assessment. χ^2 statistic is a test of goodness of fit. It evaluates how well a statistic model fits with sets of observations. In this application, it has been discussed that some landmark points contribute more than others regarding facial expression variations. Therefore, the facial landmarks are weighted based on the importance. The weighted χ measure is then given as

$$\chi_w^2(S,T) = \sum_{i,j} w_j \frac{(S_j - T_{i,j})^2}{S_j + T_{i,j}} \tag{1}$$

where S and T are the positions of landmarks of input image and template, respectively. $T_{i,j}$ means the jth landmarks in template t_i. The function is weighted by w, which full considers the importance of the eyes and eyebrow. Figure 2 shows the weighting scheme.

2.2 Improved RSF

RSF is close related to Transform Invariant Low-rank Texture (TILT) [19]. RSF is based on the fact that frontal face image has the minimum rank (smallest value of nuclear norm) when compared to non-frontal face images. So a optimization problem can be described as follow:

$$\underset{L,e,c,\Delta p}{\operatorname{argmax}} \|L\|_* + \lambda \|E\|_1$$

$$s.t. \begin{cases} H^{(1)}(\Delta p, c, e) = x(p) + J(p)\Delta p - Uc - e = 0 \\ H^{(2)}(L, c) = L - \sum_{i=1}^{k} R(u_i)c_i = 0 \end{cases} \tag{2}$$

where L is low-rank matrix which is expected to be the recovered frontal face. E is sparse error matrix. $X(p)$ is the warped image and p is the parameter of its shape referred to the equation $s = s_0 + U_s p$ defined by Active Shape Model (ASM) [20] where s_0 is the reference template. $J(p) = x(p)\frac{\partial W}{\partial p}$ is the Jacobian matrix. $U = [u_1|u_2|\cdots|u_k]$ is the pre-computed appearance model (eigen faces computed

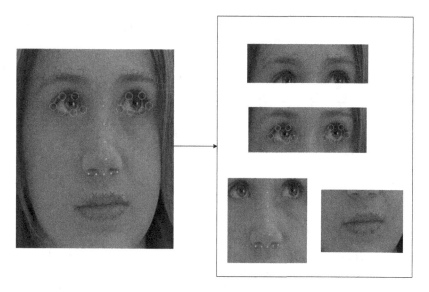

Fig. 2. Weighted Chi square measure where the weighted values are set by red '□' 2.0, '○' 1.5, '×' 1.0 and '•' 0.0. (Color figure online)

on only clean frontal faces). Equation $H^{(1)}$ indicates that the addition of low rank texture L and sparse error E agrees with the warped image, such that $X(p) = L + E$. In equation $H^{(2)}$, the low rank matrix is represented as a linear combination of U where c is its parameter. $R(\bullet)$ is an operator that reshape a vector to its corresponding matrix. By introducing augmented Lagrangian method (ALM) and alternating directions method of multipliers (ADMM). The parameters can be optimized iteratively.

The whole RSF includes outer loop and inner loop. Inner loop solves the above optimization problem and returns Δp. Outer loop updates p by $p = p + \Delta p$ and then use the new parameter to compute warp image $X(p)$ and Jacobian matrix. Both variables and parameter p will be inputs for the new round of inner loop.

In [15], the author does not provide the source code of RSF, so we implement it independently. One of the key techniques of RSF is image warping. It aims to warp the position of each pixel of input image to its corresponding location in base mesh (reference template). We employed piecewise affine warping method [21] for this task. Piecewise affine warping is based on an assumption that image warping on a small local region is linear although whole face warping is nonlinear. Given a base shape, Delaunay triangulation is used to create multiple non-overlapping triangles whose vertices are facial landmark points. Delaunay triangulation ensures that the circumcircle associated with each triangle contains no other points within it. Each triangle accounts for a fairly small region such that linear affine warping is reasonable. All these triangles make up the mesh.

For a pixel within triangle (x_i^0, y_i^0), (x_j^0, y_j^0) and (x_k^0, y_k^0) in base mesh, its location (x^0, y^0) can be expressed as:

$$(x^0, y^0) = (x_i^0, y_i^0) + \alpha[(x_j^0, y_j^0) - (x_i^0, y_i^0)] + \beta[(x_k^0, y_k^0) - (x_i^0, y_i^0)]$$

$$\begin{cases} \alpha = \dfrac{(x^0 - x_i^0)(y_k^0 - y_i^0) - (y^0 - y_i^0)(x_k^0 - x_i^0)}{(x_j^0 - x_i^0)(y_k^0 - y_i^0) - (y_j^0 - y_i^0)(x_k^0 - x_i^0)} \\[3mm] \beta = \dfrac{(y^0 - y_i^0)(x_j^0 - x_i^0) - (x^0 - x_i^0)(y_j^0 - y_i^0)}{(x_j^0 - x_i^0)(y_k^0 - y_i^0) - (y_j^0 - y_i^0)(x_k^0 - x_i^0)} \end{cases} \tag{3}$$

The warped image will follow the same deformation. Its position (x, y) in triangle (x_i, y_i), (x_j, y_j) and (x_k, y_k) is given as:

$$(x, y) = (x_i, y_i) + \alpha[(x_j, y_j) - (x_i, y_i)] + \beta[(x_k, y_k) - (x_i, y_i)] \tag{4}$$

With the location of each pixel warped in the base mesh, the pixel value will be computed by using linearly interpolation.

The time complexity of RSF is relatively high. Both optimization in inner loop and piecewise affine warping in outer loop are time consuming. In practical, there will be more than 120 iterations in each inner loop and no less than 40 iterations in outer loop. This situation makes RSF computationally expensive.

Considering that RSF optimization is unsupervised local direction search, it will lead to very slow convergence and expensive time cost. If the initialization is localized near optimal value, the number of outer loop will drop significantly. In order to achieve the expected initialization, a accurate landmark detector is inevitable. SDM is adequate, but this time 66 points is needed to synthesize the face mesh instead of the original 49 points. The authors of SDM only provide a 49-points detection model and the training part is totally hidden. So we implement SDM and train the 66-points landmark model. The 66 points computed using SDM is not so accurate compared with its 49-points counterpart, but it is still very close to the optimal values. In this strategy, outer loop of RSF can converge within five rounds. By introducing this simple step, this method become much more efficiency. The whole approach is shown in Algorithm 1.

3   Experiment

3.1   Database and Evaluation Protocol

We evaluate the proposed method on Statistical Facial Expression in the Wild (SFEW) [22]. SFEW contains 700 images with spontaneous facial expressions labelled by seven categories: six universal emotions and neutral. The images are captured from movies, which covers different real-world conditions such as occlusion, low resolution and variations in illumination and head pose. SFEW provide a clear evaluation protocol. Each category include 2 image sets: one for training and the other for testing. The experiment is strictly person-independent in which the images of one specific person with one specific emotion can only

---

Algorithm 1. Facial Expression-Aware face Frontalization

Input:

Test image X, orthonormal appearance model $U = [u_1|u_2|\cdots|u_k]$, orthonormal shape model $U_s = [u_{s1}|u_{s2}|\cdots|u_{sm}]$, five templates $T = [t_1|t_2|t_3|t_4|t_5]$

1: Detect facial shape S with 49 landmarks using SDM
2: Compute Chi-square error and find the most appropriate template t_i

$$\chi_w^2(S,T) = \sum_{i,j} w_j \frac{(S_j - T_{i,j})^2}{S_j + T_{i,j}}$$

$$t_i = \arg\min\{\chi_w^2(S,T)\}$$

3: initial 66 landmarks s using SDM again and compute $p = U_s^T(s - t_i)$.
4: **while** not converged **do**
5:    $X(p) \leftarrow$ Warp X to t_i
6:    $J \leftarrow$ compute Jacobian matrix
7:    **while** not converged **do**
8:       inner RSF loop
9:    **end**
10:   $p = p + \Delta p$
11: **end**

Output:

frontal face L, parameter p and sparse error E

---

exist in one image set, whatever training or testing set, in order to evaluate the generalization performance on totally new faces.

We perform face frontalization on both training and testing data. Then the uniformed Local Binary Pattern (LBP) features are extracted and finally, Support Vector Machine (SVM) with Radial Basis Function (RBF) kernel is utilized for emotion classification.

3.2   Face Frontalization

In this section, we will show visual results of face frontalization. In Fig. 3, the subfigures visually display the face frontalization result on the emotions of angry, disgust, fear, happy, sad and surprise, respectively. It can be seen that various head-poses in pan and tilt angles are recovered to their frontal views. The robustness against occlusion can be demonstrated by observing that all the subjects who wear glasses can be amended. Meanwhile, some images with low resolution can be processed as well. Figure 3 shows the robustness of FEAF with regards to occlusion, low resolution and head-pose variations. More importantly, the displayed frontal faces maintain detailed information of facial expressions.

Figure 4 shows the comparison of RSF and proposed method. We can intuitively see that both method achieve successful frontalization. But RSF lose much information regarding facia expressions. Whilst our method recovers the frontal view with more details of expressions. It is generally believed that texture cues are more important than shape cues in face analysis. But from this

Fig. 3. Face frontalization

comparison, it is obvious that shape features are also important in synthesizing the frontal faces and FEAF has a superior performance in the maintenance of facial expressions.

Another application of FEAF is face alignment. The position of facial landmark points can be computed by $s = t_i + U_s p$ where t_i is the shape of selected template and p is the shape parameter obtained by the output of our method. Landmark detection is not the main task of this work, but it still achieves considerable results, shown in Fig. 5.

3.3   Facial Expression Recognition in the Wild

The FER performance of FEAF is evaluated on SFEW database. With the derived frontal faces, we normalize the pixel values to the range of 0–255 and extract LBP features [23]. Each frontal face image is divided into 8 by 7 overlapping blocks with 70% overlapping rate and a spatially enhanced LBP feature representation is obtained by concatenating the histogram computed from each

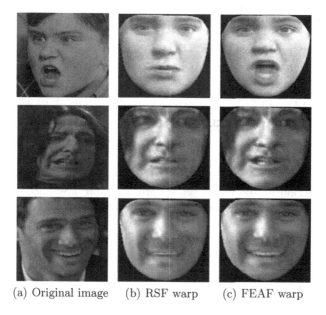

(a) Original image   (b) RSF warp   (c) FEAF warp

Fig. 4. Comparison of expression reconstruction

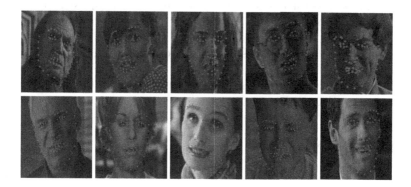

Fig. 5. Facial landmark localization

local block. The problem of facial expression recognition is a 7-class classification task, so RBF kernel based SVM classifier with one-vs-rest strategy is used for multi-class classification.

Table 1 illuminates the comparison with the state-of-the-art approaches for FER in the wild. The result of baseline is obtained from database creators [22]. It neither employs view-invariant approaches nor addresses the problem of discriminative feature learning. So baseline shows inferior performance than most view-invariant FER methods. In [6], the latest approach of view-normalization model is presented. It achieves relatively high accuracy in Angry, Disgust and Sadness. Nevertheless, its recognition rates of Neutral and Surprise are extremely poor,

Table 1. Recognition rate (%) of different methods on SFEW databse

	Angry	Disgust	Fear	Happy	Neutral	Sadness	Surprise	Total
Baseline	23.00	13.00	13.90	29.00	23.00	17.00	13.50	18.90
[6]	**25.89**	**28.24**	17.17	42.98	14.00	**33.33**	10.99	24.70
[24]	24.11	14.12	20.20	50.00	23.00	23.23	21.98	26.14
FEAF	23.21	18.82	**23.23**	**50.88**	**40.00**	26.26	**29.67**	**30.86**

even lower than random guess. This situation reveals the unstable performance of [6]. As previously discussed, view-normalization approaches needs large amount of training data to ensure accuracy. Considering SFEW is a small-scale database, the result of this method is consequently unstable. In [24], a state-of-the-art approach of dictionary-learning model is displayed and its result is a little better than [6]. Dictionary-learning methods, thus, are more stable than view-based methods for small sample experiment. The performance of FEAF can be summarized as below:

- FEAF outperforms all the other methods in terms of the overall recognition rate.
- FEAF achieves a considerable improvement in Fear, happy, Neutral and Surprise. Especially for the Neutral, our method achieves significantly higher accuracy than others.
- Unlike [6] which is bias against small sample data analysis, there is no obvious bias of FEAF.
- FEAF shows superior performance when compared with the state-of-the-art view-invariant FER.

4   Conclusions

In this paper, to the best of our knowledge, we present the first facial expression-aware face frontalization (FEAF) method for FER in the wild. FEAF includes three main step: multi-template design, template matching and improved Robust Statistical face Frontalization (RSF). The proposed method can successfully achieve generic face frontalization in 2D. The detailed information of facial expression, meanwhile, is maintained. The experimental validation shows that FEAF outperforms the state-of-the-art view-invariant FER methods. In the future, we are expected to further improve the accuracy of FEAF by developing a more effective template matching technique.

Acknowledgement. This work was supported by EU seventh framework programme under grant agreement No. 611391, DREAM, and the EPSRC project, 4D Facial Sensing and Modelling (EP/N025849/1).

References

1. Valstar, M., Jiang, B., Mehu, M., Pantic, M., Scherer, K.: The first facial expression recognition and analysis challenge. In: IEEE International Conference and Workshops on Automatic Face and Gesture Recognition, pp. 921–926 (2011)
2. Sariyanidi, E., Gunes, H., Cavallaro, A.: Automatic analysis of facial affect: a survey of registration, representation and recognition. IEEE Trans. Pattern Anal. Mach. Intell. **37**, 1113–1133 (2015)
3. Zeng, Z., Pantic, M., Roisman, G.I., Huang, T.S.: A survey of affect recognition methods: Audio, visual, and spontaneous expressions. IEEE Trans. Pattern Anal. Mach. Intell. **31**, 39–58 (2009)
4. Moore, S., Bowden, R.: Local binary patterns for multi-view facial expression recognition. Comput. Vis. Image Underst. **115**, 541–558 (2011)
5. Hesse, N., Gehrig, T., Gao, H., Ekenel, H.K.: Multi-view facial expression recognition using local appearance features. Int. J. Comput. Vis. **83**, 178–194 (2011)
6. Eleftheriadis, S., Rudovic, O., Pantic, M.: Discriminative shared Gaussian processes for multiview and view-invariant facial expression recognition. IEEE Trans. Image Process. **24**, 189–204 (2015)
7. Rudovic, O., Pantic, M., Patras, I.: Coupled Gaussian processes for pose-invariant facial expression recognition. IEEE Trans. Pattern Anal. Mach. Intell. **35**, 1357–1369 (2013)
8. Tariq, U., Yang, J., Huang, T.S.: Multi-view facial expression recognition analysis with generic sparse coding feature. In: Fusiello, A., Murino, V., Cucchiara, R. (eds.) ECCV 2012. LNCS, vol. 7585, pp. 578–588. Springer, Heidelberg (2012). doi:10.1007/978-3-642-33885-4_58
9. Tariq, U., Yang, J., Huang, T.S.: Supervised super-vector encoding for facial expression recognition. Pattern Recogn. Lett. **46**, 89–95 (2014)
10. Lowe, D.G.: Object recognition from local scale-invariant features. In: Proceedings of 7th IEEE International Conference on Computer Vision, vol. 2, pp. 1150–1157 (1999)
11. Jeni, L.A., Cohn, J.F., Kanade, T.: Dense 3D face alignment from 2D videos in real-time. In: IEEE International Conference and Workshops on Automatic Face and Gesture Recognition, pp. 1–8 (2015)
12. J. Roth, Y.T., Liu, X.: Unconstrained 3D face reconstruction. In: IEEE International Conference Computer Vision Workshops, pp. 2606–2615 (2015)
13. Hassner, T., Harel, S., Paz, E., Enbar, R.: Effective face frontalization in unconstrained images. In: IEEE Conference on Computer Vision and Pattern Recognition, pp. 4295–4304 (2015)
14. Ho, H.T., Chellappa, R.: Pose-invariant face recognition using markov random fields. IEEE Trans. Image Process. **22**, 1573–1584 (2013)
15. Sagonas, C., Panagakis, Y., Zafeiriou, S., Pantic, M.: Robust statistical face frontalization. In: Proceedings of IEEE International Conference on Computer Vision, pp. 3871–3879 (2015)
16. Taheri, S., Qiu, Q., Chellappa, R.: Structure-preserving sparse decomposition for facial expression analysis. IEEE Trans. Image Process. **23**, 3590–3603 (2014)
17. Xiong, X., la Torre, F.D.: Supervised descent method and its applications to face alignment. In: IEEE Conference on Computer Vision and Pattern Recognition, pp. 532–539 (2013)
18. Xue, M., Liu, W., Li, L.: Person-independent facial expression recognition via hierarchical classification. In: IEEE International Conference on Intelligent Sensors, Sensor Networks and Information Processing, pp. 449–454 (2013)

19. Zhang, Z., Ganesh, A., Liang, X., Ma, Y.: TILT: transform invariant low-rank textures. Int. J. Comput. Vis. **99**, 1–24 (2012)
20. Cootes, T.F., Taylor, C.J., Cooper, D.H., Graham, J.: Active shape model-their training and application. Comput. Vis. Image Underst. **61**, 38–59 (1995)
21. Metthews, I., Baker, S.: Active appearance model revisited. Int. J. Comput. Vis. **60**, 135–164 (2004)
22. Dhall, A., Goecke, R., Lucey, S., Gedeon, T.: Static facial expression analysis in tough conditions: Data evaluation protocol and benchmark. In: IEEE International Conference Computer Vision Workshops, pp. 2106–2112 (2011)
23. Huang, D., Shan, C., Ardabilian, M., Wang, Y., Chen, L.: Local binary patterns and its application to facial image analysis: a survey. IEEE Trans. Syst. Man Cybern. Part C Appl. Rev. **41**, 765–781 (2011)
24. Liu, M., Li, S., Chen, X.: AU-aware deep networks for facial expression recognition. In: IEEE International Conference and Workshops on Automatic Face and Gesture Recognition, pp. 1–6 (2013)

Eigen-Aging Reference Coding for Cross-Age Face Verification and Retrieval

Kaihua Tang[1,2(✉)], Sei-ichiro Kamata[1], Xiaonan Hou[2], Shouhong Ding[2], and Lizhuang Ma[2]

[1] Graduate School of Information, Production and Systems,
Waseda University, Shinjuku, Japan
kam@waseda.jp
[2] Shanghai Jiao Tong University, Shanghai, China
tkhchipaomian@gmail.com, xnhou1989@gmail.com, dingsh1987@gmail.com,
ma-lz@cs.sjtu.edu.cn

Abstract. Recent works have achieved near or over human performance in traditional face recognition under PIE (pose, illumination and expression) variation. However, few works focus on the cross-age face recognition task, which means identifying the faces from same person at different ages. Taking human-aging into consideration broadens the application area of face recognition. It comes at the cost of making existing algorithms hard to maintain effectiveness. This paper presents a new reference based approach to address cross-age problem, called Eigen-Aging Reference Coding (EARC). Different from other existing reference based methods, our reference traces eigen faces instead of specific individuals. The proposed reference has smaller size and contains more useful information. To the best of our knowledge, we achieve state-of-the-art performance and speed on CACD dataset, the largest public face dataset containing significant aging information.

1 Introduction

Growing number of corporations and organizations use face recognition algorithms to realize interaction and verification applications in recent years. In spite of the high recognition accuracy for well-captured images, it's still a tough task to maintain the effectiveness under various real-world deformation factors. Among all the factors that may cause the reduction of accuracy, PIE (pose, illumination, expression) and facial aging overwhelmed others. Compared with PIE, the facial aging is more complicated. Even for human beings, to identify the same person under different ages is not an easy job (see Fig. 1). Meanwhile, cross-age face recognition is so crucial for a real-world face recognition system, for example, it can be used to find escaped prisoners or missing people. Without cross-age face recognition ability, face information has to be updated frequently to keep the recognition system effective. Most of the previous researches achieve near or over human performance only under PIE variation [2,3,12,13]. Facial aging researches still have a lot to improve.

© Springer International Publishing AG 2017
S.-H. Lai et al. (Eds.): ACCV 2016, Part III, LNCS 10113, pp. 389–403, 2017.
DOI: 10.1007/978-3-319-54187-7_26

(a) Examples of face pairs with PIE variation.

(b) Examples of face pairs with facial aging variation.

Fig. 1. These image pairs show the difference between traditional PIE variation and facial aging variation. Facial aging variation is much more complicated. (a) The above image pairs come from LFW [21] dataset, a commonly used human face dataset, which doesn't contain aging information. (b) The bottom image pairs come from the largest public facial aging dataset CACD [1].

Among all the age related researches, age estimation [4,14,15] and aging simulation [5,16,17] researches have narrow application area, so we mainly investigate cross-age face recognition. As far as we know, the existing works can be roughly divided into three categories: the modeling approaches [6,7], the discriminative approaches [8,9,18,19] and the reference-based approaches [1]. Modeling approaches use aging simulation model to change query faces into the same age as gallery ones. Discriminative approaches try to separate or eliminate the age-sensitive features to increase the recognition ability. Reference-based approaches achieve age-invariance by comparing the face features with reference individuals in different ages. Figure 2 provides an example of how reference-based approaches achieve age-invariance.

The reference-based cross-age face recognition is first proposed by Bor-Chun Chen et al. [1], which is called Cross-Age Reference Coding (CARC). It achieves remarkable improvement compared to previous researches. But it still has some drawbacks. First, the reference set is extremely large, since it has to cover the diversity among race, gender and so on. Second, due to the expensive computational cost, it can't make full use of all training individuals. At last, in sparse coding, CARC only adds locality constraint to ensure the smoothness, but the global distribution may be changed.

To tackle these drawbacks, this paper introduces a new reference-based method called Eigen-Aging Reference Coding (EARC). Although eigen face isn't the first choice of recent face recognition researches, we find that it's effective in building eigen face reference in our task. Actually, it's the first time that eigen

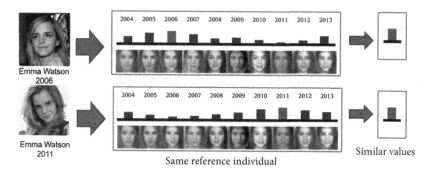

Fig. 2. After a pair of faces are encoded by a reference individual, they pool out their absolute maximums. If two faces come from the same person, the activated values are supposed to be similar.

face is used to build reference set. We use PCA to select some eigen components of face features just like how we calculate eigen faces [22]. Instead of tracing the aging process for specific individuals, the proposed reference set traces the aging process of these eigen face components as eigen-aging reference. We also add distribution constraint in our sparse coding to guarantee the encoded results follow good distributions. Contributions of this paper can be concluded into following four parts:

- Using PCA to remove the redundancy and cover the diversity among training individuals, which makes each of our eigen-aging reference contain more useful information.
- The number of reference and result dimension of our method are dramatically reduced compared with previous reference-based methods. EARC can make full use of all training individuals without increasing the computational cost.
- The proposed distribution constraint ensures that our encoded result is of good distributions. It improves the performance of sparse coding.
- The proposed method achieves state-of-the-art performance in both retrieval and verification, which is better than human average performance and very close to human voting result.

The rest of this paper is organized as follows. In Sect. 2, we introduce some related works. In Sects. 3 and 4, we separate our Eigen-Aging Reference Coding into reference construction part and encoding part. In Sect. 5, we provide some experiment results running on CACD. This paper will be concluded in Sect. 6.

2   Related Work

2.1   Cross-Age Face Recognition

Most of the highly qualified researches in cross-age face recognition start after MORPH [20] dataset is published. It keeps being the largest facial aging dataset until CACD [1] occurs. Except these two, the other facial aging datasets either contain small data size or have low quality. Limited by the rareness of datasets, there are still few researchers focusing on this field. To the best of our knowledge, we divide existing cross-age face recognition methods into three categories: the modeling approaches, the discriminative approaches and the reference-based approaches. The modeling approaches [6, 7] change the query faces into the same age as gallery one. Although it removes some variations caused by facial aging, the main problem is that the diversity of aging process between different race or gender can't be covered, which makes most of the modeling approaches hard to be general. The discriminative approaches have become popular in recent years. Most of them seem to be very effective in improving the recognition ability by eliminating age-sensitive features. Zhifeng Li et al. [8] build Local Patterns Selection feature descriptor to achieve age-invariance. It applies clustering encoding tree on feature space and removes facial aging variation by minimizing intra-user dissimilarity among different ages. Dihong Gong et al. [9] use hidden factor analysis to separate the features into age-sensitive factors and age-invariant factors. However, aging process is not just a process in chaos. Age-sensitive features can also be used for cross-age face recognition, if we take advantage of their inner regularity: similar faces are supposed to have similar aging process. That's why two twins look alike all through their life. Reference-based cross-age approaches make use of this regularity. They trace aging processes of some standard reference individuals.

Before reference set is used in cross-age face recognition, it has already been applied by lots of methods to improve traditional face recognition systems [10, 11] and achieves quite good results. Kumar et al. [11] present attribute classifier and simile classifier, the simile classifier uses reference people to do classification. Qin Yin et al. [10] propose an associate-predict model based on a 200 identities reference set. However, their reference sets don't contain aging variation. Bor-Chun Chen et al. [1] further find the value of reference in solving facial aging problem and use it to achieve age-invariance. They build a reference based on 600 individuals to deal with cross-age face recognition, which is called cross-age reference coding (CARC). CARC assumes that similar faces should still look alike when they both get older. It's an obvious phenomenon if we think about the above twins example. Its reference traces aging processes of reference individuals, then it encodes the faces by pooling out the maximum similarity between input face and reference individuals at different ages. Both younger faces or elder faces are supposed to activate their own corresponding age at reference and have similar activated values (see Fig. 2), so the age-invariant encoded feature can be obtained. Compared with other approaches, CARC shows much better performance and robustness.

However, all the existing reference methods mentioned above [1, 10, 11] use specific individuals as reference. Inspired by eigen face [22], the proposed eigen-aging reference first uses eigen faces as reference. It achieves much better performance and less computational consumption.

2.2 High-Dimensional LBP

Among all the feature descriptors applied in face recognition applications, we select high-dimensional LBP (HD-LBP) [2] to extract face features. It has been proved to have near human performance in traditional face recognition.

High-dimensional LBP features will be extracted on multi-scale patches around each landmark. Landmarks are some fixed points on human face like centers of eyes, corners of the mouth, tip of the nose and so on. Each landmark will have an independent high-dimensional feature vector. Because the multi-scale sampling extracts too much redundant information, we further use PCA to reduce its dimension. It will maintain the performance and reduce the computational consumption for further processing.

3 Eigen-Aging Reference

3.1 Form Training Individual Sets

Before calculating aging processes of eigen components, we have to obtain the training individual representations using Eq. (1). Average HD-LBP features of those images from same person at same age are used by:

$$R_{i,j}^k = \frac{1}{N_{ij}} \sum_{individual(x^k)=i, year(x^k)=j} x^k,$$
$$\forall i = 1, 2, ..., n; \forall j = 1, 2, ..., m; \forall k = 1, 2, ..., q; \tag{1}$$

where $x^k \in R^d$ is the HD-LBP feature at landmark k, $R_{i,j}^k$ means the average face feature of individual i in year j at landmark k, n is the number of training individuals, m is range of ages, and q is the number of landmarks. In our experiments, $n = 1200$, $m = 10$, $q = 16$. N_{ij} is the number of all images from individual i in year j. In CACD dataset, N_{ij} is nonzero for any i and j.

3.2 Train Eigen-Aging Reference

The Eigen-Aging Reference will be trained according to eigen face algorithm [22]. We make use of the above $R_{i,j}^k$ as training representations and calculate eigen components of them. For the convenience of formulization, we concatenate training representations into vectors by year as:

$$\hat{R}_i^k = [(R_{i,1}^k)^T, (R_{i,2}^k)^T, ..., (R_{i,m}^k)^T]^T \in R^{md}. \tag{2}$$

After that, PCA is applied to obtain eigen components of n training feature vectors $\hat{R}_i^k$, then we can get the average vector M^k and difference vectors Φ_i^k in the following:

$$M^k = \frac{1}{n} \sum_{i=1}^n \hat{R}_i^k, \qquad \Phi_i^k = \hat{R}_i^k - M^k. \tag{3}$$

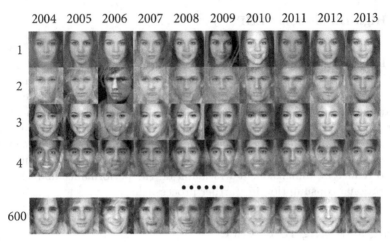

(a) The examples of Cross-Age Reference.

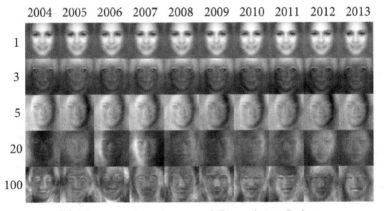

(b) The examples of proposed Eigen-Aging Reference.

Fig. 3. The above illustrations use raw images to make reference set easier to understand. (a) Cross-Age Reference is based on specific individuals. (b) Eigen-Aging Reference is based on eigen faces. The number on the left of each row is the rank of the corresponding eigen face.

Next we construct matrixes $\Phi^k = [\Phi_1^k, \Phi_2^k, ..., \Phi_n^k] \in R^{md \times n}$, and calculate eigenvalues λ_l^k and eigenvectors u_l^k of $(\Phi^k)^T \Phi^k$. Top-p λ_l^k and u_l^k are used, where $l = 1, 2, ..., p$. According to Sect. 5, the best performance can be achieved when $p = 50$ and $n = 1200$, so we come up with eigen-aging vector $\hat{E}_l^k$ for each eigen face component l by:

$$\hat{E}_l^k = \frac{1}{sqrt(\lambda_l^k)} \Phi^k u_l^k, \qquad l = 1, 2, ..., p. \tag{4}$$

We further separate $\hat{E}_l^k \in R^{md}$ into sub-vectors $E_{l,j}^k \in R^d$ by year as follows:

$$\hat{E}_l^k = [(E_{l,1}^k)^T, (E_{l,2}^k)^T, ..., (E_{l,m}^k)^T]^T. \tag{5}$$

$E_{l,j}^k$ is our proposed reference set representation, which is smaller and contain more useful information.

Figure 3 shows the difference between EARC and CARC. We use raw images to make reference set easier to understand. As we can see, specific individual based reference contains too much noise and has redundancy among different individuals while the proposed eigen component based reference is more representative and clear. With the above advantage, EARC only use tens of eigen face references but achieve better performance than CARC, the method based on hundreds of specific individuals. In the example of eigen aging reference, we display 1st, 3rd, 5th, 20th and 100th eigen face references. They are ranked according to their eigenvalues. We also find that the higher ranked components contain more structure information while the lower ranked ones may have more noise, so if we use too much eigen faces, the result could become worse. In Sect. 5, we will discuss how many eigen faces we should use.

4   Coding and Pooling

4.1   Sparse Coding

Before we apply sparse coding, We need to define relationship values $\alpha_{l,j}^k$ between input feature and l^{th} eigen face reference in year j at landmark k (we denote matrix $\check{E}_j^k = [E_{1,j}^k, E_{2,j}^k, ..., E_{p,j}^k] \in R^{d \times p}$ and vector $\check{\alpha}_j^k = [\alpha_{1,j}^k, \alpha_{2,j}^k, ..., \alpha_{p,j}^k] \in R^p$). Our reference set is used as dictionary, so it can be considered as solving a Tikhonov regularization problem:

$$\underset{\alpha_{i,j}^k}{\text{minimize}} ||x^k - \check{E}_j^k \check{\alpha}_j^k||^2 + \lambda ||\check{\alpha}_j^k||^2, \forall j, k. \tag{6}$$

An additional locality constraint [1] will also be used to improve performance. It will guarantee the smoothness of encoded results, which means that similar high dimensional points in original feature space still have similar value in encoded results. In other word, the relationship values between input face and eigen face reference l in year j should be similar to the values between input face and the same reference face in year $j + 1$ and $j - 1$, so for any k and l, $\alpha_{l,j}^k$ will always be similar to $\alpha_{l,j+1}^k$ and $\alpha_{l,j-1}^k$.

This locality smoothness constraint is defined as $\lambda ||LA^k||^2$. We let $A^k = [(\check{\alpha}_1^k)^T, (\check{\alpha}_2^k)^T, ..., (\check{\alpha}_m^k)^T]^T \in R^{mp}$ and matrix L:

$$L = \begin{bmatrix} I & -2I & I & 0 & \cdots & 0 & 0 & 0 \\ 0 & I & -2I & I & \cdots & 0 & 0 & 0 \\ \vdots & \vdots & \vdots & \vdots & \vdots & \vdots & \vdots & \vdots \\ 0 & 0 & 0 & 0 & \cdots & I & -2I & I \end{bmatrix} \in R^{(m-2)p \times (mp)}. \tag{7}$$

4.2   Distribution Constraint

In spite of that locality constraint $\lambda||LA^k||^2$ ensures the smoothness of encoded results and improves the coding performance, it can't keep the distribution still being similar to original features. Mathematically, the statement that near points maintain closer is not equal to that far points keep far away. Only when both of these two statements are satisfied, the global distribution can be maintained after changing feature spaces. So we propose a new constraint to guarantee the encoded features follow the distributions we need.

With smoothness constraint, we may have four possible distributions of encoded $\alpha_{l,j}^k$ sequences (see Fig. 4). Based on common sense, if we compare a face with a series of faces from identical individual in different ages, there should be only one most similar face. It can locate either at the boundary of age sequence or at one certain age inside the sequence, so there is supposed to be one and only one extreme point. The locality smoothness constraint may lead to distribution 4 in Fig. 4, which is not a good distribution in our work, So we propose two kinds of constraint terms that may force the encoded features follow the first three distributions.

Maximize boundary difference: We assume that the extreme point will occur at the boundary in most of the cases. It's quite rare for the extreme point located at exact center of the reference age. So we can simply try to maximize the difference of two boundary ages. The new constraint will be $-\beta||DA^k||^2$. The minus transfers the maximize problem to minimization by:

$$D = \begin{bmatrix} I\ 0\ \cdots\ 0\ -I \end{bmatrix} \in R^{p \times (mp)}. \tag{8}$$

Additional cost for extreme point: We also can force the encoded results into first three distributions by giving additional cost for extreme point. The good distributions should have only one extreme point while the bad distribution may have several ones. So we give the extreme point additional cost for occurrence. This constraint can also be written as $-\beta||DA^k||^2$ with new D:

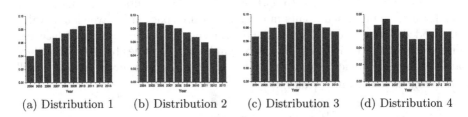

(a) Distribution 1   (b) Distribution 2   (c) Distribution 3   (d) Distribution 4

Fig. 4. With smoothness constraint, we may get above 4 distributions, but only the first three are good distributions in our work. Distribution 4 means all the possible distributions with more than one extreme point.

$$D = \begin{bmatrix} I & 0 & -I & 0 & \cdots & 0 & 0 & 0 \\ 0 & I & 0 & -I & \cdots & 0 & 0 & 0 \\ \vdots & \vdots & \vdots & & \vdots & \vdots & \vdots & \vdots \\ 0 & 0 & 0 & 0 & \cdots & I & 0 & -I \end{bmatrix} \in R^{(m-2)p \times (mp)}. \tag{9}$$

We should also notice that the locality constraint and distribution constraint both represent the distance relationship between original feature space and encoded feature space. For the convenience of learning parameters, they are supposed to be learned together. The combined constraint is $\lambda(||LA^k||^2 - \beta||DA^k||^2)$.

4.3 Optimization

To combine all the constraints, we denote $X^k = [(x^k)^T, ..., (x^k)^T]^T \in R^{mp}$ and matrix F:

$$F^k = \begin{bmatrix} \check{E}_1^k & 0 & \cdots & 0 \\ 0 & \check{E}_2^k & \cdots & 0 \\ \vdots & \vdots & \vdots & \vdots \\ 0 & 0 & \cdots & \check{E}_m^k \end{bmatrix} \in R^{(md) \times (mp)}, \tag{10}$$

so the final optimization function is:

$$\underset{A^k}{\text{minimize}} ||X^k - F^k A^k||^2 + \lambda_1||A^k||^2 + \lambda_2(||LA^k||^2 - \beta||DA^k||^2), \forall k. \tag{11}$$

It's easy to obtain $A^k = ((F^k)^T F^k + \lambda_1 I + \lambda_2 L^T L - \lambda_2 \beta D^T D)^{-1}(F^k)^T X^k, \forall k$. We define P^k to be a projection matrix, $P^k = ((F^k)^T F^k + \lambda_1 I + \lambda_2 L^T L - \lambda_2 \beta D^T D)^{-1}(F^k)^T$. HD-LBP face features can be easily transferred into reference space by multiplying with P^k.

4.4 Max Pooling

After sparse coding is applied, we use maximum pooling to achieve age-invariance. If a face is young, it should be more similar to the younger part of each eigen face reference, which means that the corresponding younger age has a little bit larger $\alpha_{l,j}^k$ than elder one. The maximum pooling will pool out this value. Two faces from same person at different age will pool out different ages but their maximum values are similar compared with the faces from different people. This is the reason why we can use maximum pooling to achieve age-invariance.

$A^k \in R^{mp}$ is a vector containing $\alpha_{l,j}^k$ as elements. We calculate the absolute maximum of each eigen face l on different ages, $max(|\alpha_{l,1}^k|, |\alpha_{l,2}^k|, ..., |\alpha_{l,m}^k|), \forall l, k$. The final age-invariant face feature is a p dimensional vector, p is the number of eigen faces we use in constructing eigen-aging reference.

5   Experiments

5.1   Cross-Age Face Dataset

Compared with traditional human face datasets, the cross-age datasets are extremely rare because it needs to track individuals over decades. The most popular datasets in this field are FG-NET, MORPH [20] and CACD [1]. CACD is published in recent years but its quality has already been proved. All the images of CACD are celebrity images captured in various unconstrained environments, compared with other cross-age datasets, images in CACD are more close to the complicated real-world environment.

The difference among FG-NET, MORPH and CACD is shown in Table 1. FG-NET is an early dataset. It contains only one thousand images from no more than one hundred individuals. Because of its limitation of data size, it's hard to support a general method. MORPH contains 55,134 images of 13,618 people with age range from 16–77, but they are in clear background environment. Another demerit of MORPH is that there are about 4 images in average for each individual and only one image in each certain age. It's hard to cover different PIE conditions. The CACD contains 163,446 images of 2,000 celebrities with age ranging from 16 to 62. All the collected images are from 2004 to 2013, and each individual has 80 images in average, which means about 8 images in every certain age per individual. It ensures that face information of each single individual in each age can be fully extracted, so CACD is the best choice for us to build a robust and effective reference. In our experiment, all the training and testing images come from CACD.

Table 1. The difference among FG-NET, MORPH and CACD.

Dataset	# of images	# of individuals	# images/individual	Age gap
FG-NET	1,002	82	12.2	0–45
MORPH [20]	55,134	13,618	4.1	0–5
CACD [1]	163,446	2,000	81.7	0–10

5.2   Similarity Measurement

We use cosine similarity to measure the difference between two age-invariant face features in our experiment. Because the feature from each landmark is calculate independently, we add up these similarities from all the landmarks, and use the sum to represent the total similarity between two faces.

5.3   Training Data and Parameters Selection

To compare our experiment results to the previous state-of-the-art methods in CACD, we organize our training data and test data in the same way. There are

Table 2. The coding parameters of EARC in three cases.

Case	λ_1	λ_2	β
EARC	10^0	10^1	-
EARC-maximize boundary difference	10^{-2}	10^4	10^{-1}
EARC-additional extreme point cost	10^{-1}	10^4	10^3

2,000 celebrities in total. 1,800 of them come from internet without annotation. (1) 1,200 of these 1,800 individuals are used to calculate reference representations; (2) 600 are used to calculate PCA subspace in High-Dimensional LBP. The rest images of 200 individuals have already been manually annotated, which means that their quality can be guaranteed. (3) We use 80 of them to learn parameters. (4) 120 of them test our experiment results.

To learn our parameters, we use 80 qualified celebrities. The parameters include the dimension of PCA subspace in high-dimensional LBP, the regularization parameters $\lambda_1, \lambda_2, \beta$, and the number of eigen faces we use. The images captured in 2004–2006 are collected as gallery set while those captured in 2013 are query set. The reason why we don't use images from 2007–2012 as gallery set to learn parameters is that a larger gap of age between gallery and query set is more meaningful for a cross-age face recognition system.

- In high-dimensional LBP, we use PCA to reduce the dimension while remove some variations. The original landmark dimension is 4,720. We try to reduce it to the range from 100 to 1,500. According to their performance and computational cost, we choose 800 dimension in our experiment. 900 and 1,000 dimension can increase a little bit performance, but the required computer memory and computational cost are extremely expensive.
- To learn regularization parameters λ_1, λ_2 and β, we greedily train them one by one from value 10^{-6} to 10^6. We record the parameters in three cases (see Table 2): EARC without distribution constraint, EARC with maximize boundary difference constraint, EARC with additional cost for extreme point constraint.
- In order to find a proper number of eigen faces, We first use 600 training individuals to train and select eigen face components from 10 to 100. It shows that the best performance is achieved at 70 eigen face components, since the lower ranked eigen components may have more noise and less structure information. Then we change the number of training individuals from 600 to 1200. The best performance achieves at 70, 60, 60, 50 when training individuals are 600, 800, 1000, 1200 (see Fig. 5(a)). It seems like the larger training data we use, the less eigen components we will need. So we make use of all the 1200 training individuals and choose top 50 eigen components as reference. Under this condition, it will have the best performance and the lowest computational expense of projection. CARC can't make full use of all 1200 training individuals due to its expensive projection cost.

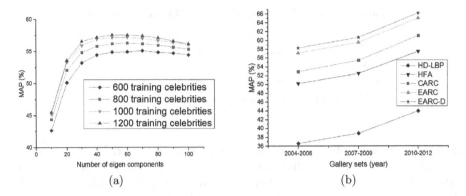

Fig. 5. (a) How does the retrieval performance change with the number of training individuals and number of eigen components. (b) The retrieval results of the proposed methods and previous state-of-the-art approaches. It shows that EARC-D achieves best performance at all the three age gaps.

5.4   Retrieval Experiments

In order to test retrieval performance of our proposed method, the rest annotated 120 celebrities are used. Because all the images are labelled the captured year from 2004–2013, we organize the gallery sets and query set by year. The images captured in 2013 will be gathered as query set while images captured in 2004–2006, 2007–2009 and 2010–2012 will be collected into 3 gallery sets.

The Mean Average Precision (MAP) is used to evaluate retrieval performance. It is widely used in information and image retrieval. If there is a query set contains Q query images, for each query image, it will compute its averaged precision (AP) of retrieval results at every recall level. MAP is the final average of these APs.

We compare our retrieval results (with and without distribution constraint) with the state-of-the-art methods CARC [1] and HFA [9]. They are proved to be very effective in CACD dataset. The results are shown in Fig. 5(b). We find that both maximize boundary difference constraint and additional cost for extreme point constraint have similar results, so we only use one EARC-D to represent distribution constraint. And it indeed improves the performance of original EARC.

5.5   Verification Experiments

Verification experiments are conducted under a verification subset of CACD called CACD-VS. It contains 4,000 image pairs, which have been manually checked to guarantee the quality. Half of these 4,000 image pairs are positive (come from the same person), the rest are negative (come from different people).

The CACD-VS contains two human performance benchmarks: average human performance and human voting performance. The former is the average of human accuracy, which is 85.7%. The human voting has a much better result, which is 94.2% accuracy. It uses decisions from 9 human beings to do each verification. The majority decision will be considered as final decision. For now, human voting result still shows the best accuracy over all the published methods.

In the proposed method, we separate data into 10 folds, 200 positive pairs and 200 negative pairs for each. 9 of them are used to learn PCA subspace and 1 to test. It will run 10 times and compute average results. After calculating cosine similarities, a simple threshold is used to classify. Both two kinds of distribution constraints have similar results, so we use EARC-D and EARC to represent the result with or without distribution constraint. We compare our methods with CARC [1], HFA [9] and HD-LBP [2] (see Table 3). It achieves the best accuracy, 91.2%, which is very close to human voting result.

Table 3. The verification results.

Method	Verification accuracy
High-dimensional LBP [2]	81.6%
Hidden factor analysis [9]	84.4%
Average human	85.7%
Cross-age reference coding [1]	87.6%
Eigen-aging reference coding	90.6%
Eigen-aging reference coding-D	91.2%
Human voting	94.2%

5.6   Computational Cost

Beside of good performance EARC and EARC-D achieve, they also significantly speed up the computation by reducing the encoded dimension. For a real-world face recognition system, it always contains huge amount of face information in the dataset. A computational expensive method has less practical application value. Compared with CARC, we only use 1/12 of the dimension to represent an age-invariant face and significantly improve the efficiency. The comparison of encoded dimension and computation time between CARC and EARC will be shown in Table 4. To measure the retrieval efficiency, we retrieve 100 face images in a gallery set with 10,000 images. The retrieval speed shows the computation

Table 4. The encoded dimension and computation time of CARC, EARC and EARC-D.

Method	Dimension of landmark	Dimension of face	Retrieval speed (ms)
CARC [1]	600	9600	9237
EARC	50	800	1352
EARC-D	50	800	1347

time of each method. This experiment is running under a computer with Intel(R) Core(TM) i7-4720HQ CPU @ 2.60 GHz 2.60 GHz, 16.0 GB RAM and MATLAB R2013a.

6   Conclusions

In this paper, we mainly propose an eigen face component based reference to encode the faces into an age-invariant space. It performs better than specific individual based reference and requires less computation time. We also present the distribution constraint to improve sparse coding. It further optimizes our method without costing additional computational consumption. Although the proposed two kinds of distribution constraint terms are based on different assumption, their mathematical similarity results in similar results.

In spite of the state-of-the-art result we achieve in CACD, it's still not as good as the human voting result. We suppose cross-dataset voting could improve our performance, because it might increase the stability of our system. We assume cross-dataset voting can lead to higher accuracy, because human voting is better than human average. Limited by the rareness of public cross-age dataset, we only try separating CACD into several small datasets. Although it improves a little bit, it doesn't make a mentionable difference. This is because the same dataset doesn't have enough appearance difference. If there are several large facial aging datasets with good quality and different appearance distributions, a more robust cross-dataset voting method may exceed human voting result.

Acknowledgement. We thank Xuchao Lu for his inspiring ideas and patient help on paper modification. This work was partially supported by JSPS KAKENHI Grant Number 15K00248, NSFC Grant Number 61133009 and fund of Shanghai Science and Technology Commission Grant Number 16511101300.

References

1. Chen, B.-C., Chen, C.-S., Hsu, W.H.: Cross-age reference coding for age-invariant face recognition and retrieval. In: Fleet, D., Pajdla, T., Schiele, B., Tuytelaars, T. (eds.) ECCV 2014. LNCS, vol. 8694, pp. 768–783. Springer, Heidelberg (2014). doi:10.1007/978-3-319-10599-4_49
2. Chen, D., Cao, X., Wen, F., Sun, J.: Blessing of dimensionality: high-dimensional feature and its efficient compression for face verification. In: Proceedings of the IEEE Conference on Computer Vision and Pattern Recognition, pp. 3025–3032 (2013)
3. Barkan, O., Weill, J., Wolf, L., Aronowitz, H.: Fast high dimensional vector multiplication face recognition. In: Proceedings of the IEEE International Conference on Computer Vision, pp. 1960–1967 (2013)
4. Zhu, K., Gong, D., Li, Z., Tang, X.: Orthogonal Gaussian process for automatic age estimation. In: Proceedings of the ACM International Conference on Multimedia, pp. 857–860. ACM (2014)
5. Lanitis, A., Taylor, C.J., Cootes, T.F.: Toward automatic simulation of aging effects on face images. IEEE Trans. Pattern Anal. Mach. Intell. **24**, 442–455 (2002)

6. Du, J.X., Zhai, C.M., Ye, Y.Q.: Face aging simulation and recognition based on NMF algorithm with sparseness constraints. Neurocomputing **116**, 250–259 (2013)
7. Park, U., Tong, Y., Jain, A.K.: Age-invariant face recognition. IEEE Trans. Pattern Anal. Mach. Intell. **32**, 947–954 (2010)
8. Li, Z., Gong, D., Li, X., Tao, D.: Aging face recognition: a hierarchical learning model based on local patterns selection. IEEE Trans. Image Process. **25**, 2146–2154 (2016)
9. Gong, D., Li, Z., Lin, D., Liu, J., Tang, X.: Hidden factor analysis for age invariant face recognition. In: Proceedings of the IEEE International Conference on Computer Vision, pp. 2872–2879 (2013)
10. Yin, Q., Tang, X., Sun, J.: An associate-predict model for face recognition. In: 2011 IEEE Conference on Computer Vision and Pattern Recognition (CVPR), pp. 497–504. IEEE (2011)
11. Kumar, N., Berg, A.C., Belhumeur, P.N., Nayar, S.K.: Attribute and simile classifiers for face verification. In: 2009 IEEE 12th International Conference on Computer Vision, pp. 365–372. IEEE (2009)
12. Li, Z., Gong, D., Li, X., Tao, D.: Learning compact feature descriptor and adaptive matching framework for face recognition. IEEE Trans. Image Process. **24**, 2736–2745 (2015)
13. Li, Y., Meng, L., Feng, J., Wu, J.: Downsampling sparse representation and discriminant information aided occluded face recognition. Sci. Chin. Inf. Sci. **57**, 1–8 (2014)
14. Fu, Y., Huang, T.S.: Human age estimation with regression on discriminative aging manifold. IEEE Trans. Multimedia **10**, 578–584 (2008)
15. Geng, X., Zhou, Z.H., Smith-Miles, K.: Automatic age estimation based on facial aging patterns. IEEE Trans. Pattern Anal. Mach. Intell. **29**, 2234–2240 (2007)
16. Suo, J., Chen, X., Shan, S., Gao, W.: Learning long term face aging patterns from partially dense aging databases. In: 2009 IEEE 12th International Conference on Computer Vision, pp. 622–629. IEEE (2009)
17. Tsumura, N., Ojima, N., Sato, K., Shiraishi, M., Shimizu, H., Nabeshima, H., Akazaki, S., Hori, K., Miyake, Y.: Image-based skin color and texture analysis/synthesis by extracting hemoglobin and melanin information in the skin. ACM Trans. Graph. (TOG) **22**, 770–779 (2003)
18. Ling, H., Soatto, S., Ramanathan, N., Jacobs, D.W.: Face verification across age progression using discriminative methods. IEEE Trans. Inf. Forensics Secur. **5**, 82–91 (2010)
19. Klare, B., Jain, A.K.: Face recognition across time lapse: on learning feature subspaces. In: 2011 International Joint Conference on Biometrics (IJCB), pp. 1–8. IEEE (2011)
20. Ricanek Jr., K., Tesafaye, T.: MORPH: a longitudinal image database of normal adult age-progression. In: 2006 7th International Conference on Automatic Face and Gesture Recognition, FGR 2006, pp, 341–345. IEEE (2006)
21. Huang, G.B., Ramesh, M., Berg, T., Learned-Miller, E.: Labeled faces in the wild: a database for studying face recognition in unconstrained environments. Technical report 07–49, University of Massachusetts, Amherst (2007)
22. Heseltine, T., Pears, N., Austin, J.: Evaluation of image preprocessing techniques for eigenface-based face recognition. In: Second International Conference on Image and Graphics, pp. 677–685. International Society for Optics and Photonics (2002)

Consistent Sparse Representation
for Video-Based Face Recognition

Xiuping Liu[1], Aihong Shen[1], Jie Zhang[2(✉)], Junjie Cao[1,3], and Yanfang Zhou[1]

[1] School of Mathematical Sciences, Dalian University of Technology,
Dalian, People's Republic of China
[2] School of Mathematical Sciences, Liaoning Normal University,
Dalian, People's Republic of China
Jiezxl1985@gmail.com
[3] School of Mathematics and Information Science, Nanchang Hangkong University,
Nanchang, People's Republic of China

Abstract. This paper presents a novel method named Consist Sparse Representation (CSR) to solve the problem of video-based face recognition. We treat face images from each set as an ensemble. For each probe set, our goal is that the non-zero elements of the coefficient matrix can ideally focus on the gallery examples from a few/one subject(s). To obtain the sparse representation of a probe set, we simultaneously consider group-sparsity of gallery sets and probe sets. A new matrix norm (*i.e.* $l_{F,0}$-mixed norm) is designed to describe the number of gallery sets selected to represent the probe set. The coefficient matrix is obtained by minimizing the $l_{F,0}$-mixed norm which directly counts the number of gallery sets used to represent the probe set. It could better characterize the relations among classes than previous methods based on sparse representation. Meanwhile, a special alternating optimization strategy based on the idea of introducing auxiliary variables is adopted to solve the discontinuous optimization problem. We conduct extensive experiments on Honda, COX and some image set databases. The results demonstrate that our method is more competitive than those state-of-the-art video-based face recognition methods.

1 Introduction

The recognition of human faces is one of the most important problems in the communities of computer vision and pattern recognition. Traditionally, the face recognition [1] is usually formulated as a problem of identifying a human face from a single image. However, it remains a big challenge to correctly identify a person from only a single face image in less controlled/uncontrolled environments since the facial appearance changes dramatically due to different variation in pose, illumination, expression, disguise, etc. Recently, the task of video-based

Electronic supplementary material The online version of this chapter (doi:10.1007/978-3-319-54187-7_27) contains supplementary material, which is available to authorized users.

S.-H. Lai et al. (Eds.): ACCV 2016, Part III, LNCS 10113, pp. 404–418, 2017.
DOI: 10.1007/978-3-319-54187-7_27

face recognition attracts more and more attention. In contrast to traditional face recognition task based on single-shot images, video-based face recognition can well overcome the dramatic appearance changes, since the video clips usually accompany with these changes. Moreover, video-based face recognition is more suitable for some practical applications, such as tracking a person by matching his/her video sequences taken somewhere against the surveillance videos recorded elsewhere.

There are a considerable mount of works on video-based face recognition which have achieved pretty good performance in term of different aspects. Sparse representation-based approaches are more attractive among them, since the elegant theory and the excellent performance in image and video processing [3–6]. In particular, the sparse representation (SR) [3] method sparsely represents each probe face image with a dictionary which is learned from all gallery data. Then the probe image can be classified into the one whose reconstruction error is the smallest. SR has shown favorable performance in term of face recognition, even if examples are partially occlusion [6].

However, SR treats every example in the dictionary equally and does not consider the structure of gallery data. Normally, images belong to the same subject should be treated as an ensemble, which illustrates that we should use the structure of group to describe the dictionary. There are some methods [7,8] emerged to address this problem. Elhamifar *et al.* [7] casted the classification as a structured sparse representation (SSR) problem in which they regard these images from the same subject in gallery set as a group. Majumdar *et al.* [8] proposed two alternate regularization methods, Elastic Net and Sum-Over-l_2-norm. Both of them favor the selection of multiple correlated training samples to represent the test sample. However, these methods only calculate the representation coefficient of one example in probe set at a time. When the scale of the image data is relatively large, the calculation and running time of these methods will increase rapidly. This degrades the recognition performance. And more importantly, they ignore the label consistency information in the probe set, *i.e.* all images in the probe set should be represented by examples from the same subject. Therefore, although each image is represented by a few/one subject(s), the whole probe set may be rebuilt by the whole dictionary, as shown in the Fig. 1. That is, the representation matrix of the whole probe set is not sparse enough.

Recently, Chen *et al.* [6] proposed a novel multivariate sparse representation method which takes the label consistency information into account. However, the structure of the dictionary is ignored and each face image in dictionary learned from each partition is treated equally. Cui *et al.* [9] proposed a Joint Sparse Representation method (JSR) which assumes that a probe set can be represented by a few gallery sets. This assumption is characterized by the $l_{2,1}$−mixed norm which puts l_2-norm on those coefficients to which each class corresponding and then sums up the coefficients of all classes (i.e. l_1-norm). However, the $l_{2,1}$-mixed norm is just a convex relaxation of the number of gallery sets selected. Sometimes, the coefficient matrix obtained by minimizing the mixed norm is not sparse enough, as shown in Fig. 1.

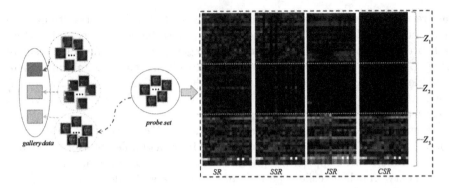

Fig. 1. The coefficient matrix obtained by SR, SSR, JSR, and CSR. We extract gallery data and probe set from COX database [2]. Gallery data consists of face images which belong to three subjects, and the probe set belongs to the third class. The coefficient matrix obtained by our method is more accurate and clear.

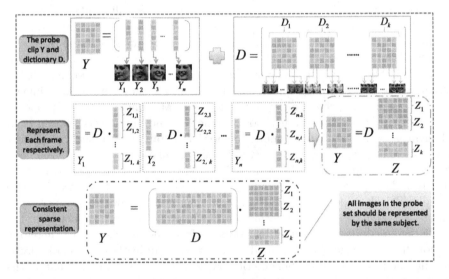

Fig. 2. Label consistency information. Given a probe clip Y and dictionary D, representation coefficients of all frames in probe set are combined to form the representation matrix Z (the second row). A more sparse coefficient matrix is obtained by using label consistency information (the last row). Where the red block represents non-zero and blue block is zero.

In this paper, we accept the fact that a probe set should be represented by a few/one correct gallery set(s) and formulate this fact into a consistent sparse representation (CSR) problem, as shown in Fig. 2. In CSR, a new matrix norm (*i.e.* $l_{F,0}$-mixed norm) is designed to describe the number of gallery sets selected. The $l_{F,0}$-mixed norm computes F-norm on those coefficient matrices to which each class corresponding and then counts the number of non-zeros coefficients of all classes. Compared with the $l_{2,1}$-mixed norm, $l_{F,0}$-mixed norm directly counts the number of gallery sets which is used to represent the probe set. Therefore, the coefficient matrix obtained by minimizing the $l_{F,0}$-mixed norm is more sparse, as shown in Fig. 1. However, the $l_{F,0}$-mixed norm is difficult to solve because it is discontinuous and traditional gradient decent method or other continuous optimization methods are not usable. We adopt a special alternating optimization strategy, which is based on the idea of introducing auxiliary variables to expand the original terms and update them iteratively. We conduct extensive experiments on several video databases and image set data. The results demonstrate that CSR is more competitive than the state-of-the-art methods for video-based face recognition.

The remainder of this paper is organized as follows. In Sect. 2, related work is introduced. In Sect. 3, we present the overview of the label consistency sparse. The optimization details of our algorithm is stated in Sect. 4. Experimental results are presented in Sects. 5 and 6 concludes the paper with a brief summary and discussion.

2   Related Work

Now many popular algorithms are the explosive development of Image Set based Classification (ISC) techniques and have achieved pretty good performance. To summarize, these techniques can be categorized according to the two key issues of the image set classification problem. The one is how to model image sets, the other is how to measure the similarity between two image sets, the similarity function usually varies with image set modeling or representation methods.

From the perspective of set modeling, existing methods can be divided into two categories: parametric and nonparametric representations. Parametric model-based representations use some parametric distributions to represent an image set, meanwhile, relative parameters can be estimated from the data. Specially, single Gaussian [1] or Gaussian mixture models (GMM) [10,11] is common parametric distribution function, and the similarity between two distributions can be measured by the Kullback-Leibler Divergence (KLD). For example, Wang et al. [10] presented a method named Discriminant Analysis on Riemannian manifold of Gaussian distributions (DARG) to solve the problem of face recognition with image sets. DARG represents the image set with Gaussian Mixture model (GMM), the dissimilarity between two distributions is then measured by the classical Kullback-Leibler Divergence (KLD) or Hellinger Distance (HD). The disadvantage of these methods is that a difficult parameter estimation problem should be solved and when the gallery sets and probe set have weak statistical correlation which may makes performance degradation.

In contrast, nonparametric model-free methods attempt to relax the assumptions on distributions of the data and utilize a more effective manner to model image set. In most cases, they represent an image set as a linear subspace [12–14] or nonlinear manifolds. For method based subspace, Discriminant-analysis of Canonical Correlations (DCC) [13] represented each image set as a single linear subspace, and usually principal angle is used to measure the similarity between two subspaces. Further Chen *et al.* [15] modeled image sets similarly, but they proposed a Dual Linear Regression Classification (DLRC) method to perform classification by a regression technique. While the linear subspace modeling cannot well conduct the case when the set is small but has large and complex data variations.

In order to address the limitation of subspace modeling, the more sophisticated nonlinear manifold has been used to model image set in the literature. In Manifold-Manifold Distance (MMD) [16], each image set is assumed to span a nonlinear manifold that can be partitioned into several local linear models and the distance between manifolds is converted into integrating the distances between pair-wise subspaces. Manifold Discriminant Analysis (MDA) [17] further extended MMD to work in a discriminative feature space rather than the original image space. Cui *et al.* [18] attempted to align all image sets to a pre-specified reference set and then measured the corresponding subspaces, which inevitably leads to the dependence on the choice of the reference set for the classification accuracy. However, these methods usually require a large data set with dense sampling to obtain appropriate manifold modeling. Generally speaking, nonparametric methods have shown favorable performance because a uniform prior is imposed on data variations in different image sets.

More recently, a new type of nonparametric methods [3–6,9] based on sparse representation has been introduced. These methods can represent each probe face image sparsely with a dictionary which is learned from all gallery data, and then the probe image can be classified into the one which has the smallest reconstruction error. In [7], they treated the classification task as a structured sparse recovery problem, and they regarded these images belonging to the same subject in gallery set as a group, then utilized the sparse coefficient of these groups. Chen *et al.* [6] proposed a novel multivariate sparse representation method in which the information among the video frames is inferred by seeking a row-sparse representation matrix. Cui *et al.* [9] proposed a Joint Sparse Representation method to handle the video-based face recognition problem. JSR treats multiple frames of a probe clip as an ensemble, and jointly recovers those face images in the clip. JSR solves the face recognition problem with $l_{2,1}$-mixed norm which is a convex relaxation of the number of selected gallery sets. While the method considers the label consistency of gallery sets, sometimes the coefficient matrix is not sparse enough.

3   Label Consistency

3.1   Problem Formulation

Suppose there are m video clips from k different subjects in gallery data. For each video clip, we crop original face images into images with fixed size and convert each frame to vector representation. We denote the i-th clip as $X_i = [X_{i,1}, X_{i,2}, \ldots, X_{i,n_i}] \in R^{d \times n_i}$, where $X_{i,j}$ is the vectorization of the j-th face image in i-th video clip. Each clip has a label denoted as l_i, where

$l_i \in [1, 2, \ldots, k]$. Note that all images of the same subject might come from multiple clips and these images should be an assemble. For each subject, we can obtain a subdictionary $D_i = [X_j | l_j = i]$, where $i = 1, 2, \ldots, k$. Then we combine these subdictionary D_i to form the dictionary $D = [D_1, D_2, \ldots, D_k] \in R^{d \times N}$, therefore, the dictionary can be used to present the whole gallery data.

Given a probe video clip, we make the same operation on the clip as those in gallery set, so the matrix $Y = [Y_1, Y_2, \ldots, Y_n] \in R^{d \times n}$ can be used to represent the clip, where Y_i is the vectorization of the i-th face image and n is the number of face images in the clip. Based on the theory of sparse representation, it can be represented by the dictionary linearly. We can recover the probe clip from the gallery dictionary D as follows:

$$Y = \sum_{i=1}^{k} D_i Z_i + E = DZ + E, \tag{1}$$

where Z_i is the coefficient matrix associated with the i-th subdictionary D_i. E is the residual term, and the coefficient matrix $Z = [Z_1; Z_2; \ldots; Z_k]$ is stacked block-wisely in height. Once the coefficient matrix Z obtained, we can identify the category of the probe clip Y according to the reconstruction error. The key of the problem is how to select a suitable coefficient matrix which contains the true class relation. In this paper, we formulate this problem as a $l_{F,0}$-mixed normal minimization problem.

3.2   Label Consistency

In video-based face recognition, since previous sparse representation [6] or group-level sparse representation [7] only consider that each frame in the probe clip should be represented by few subdictionaries, but not consider the consistency of these subdictionaries. Therefore, the coefficient matrix may not sparse enough.

For example, given a probe clip Y, each frame of the video can be represented by the dictionary D as shown in Fig. 2. Y_1 is represented by the first, second and k-th subjects. Y_2 is represented by the second and k-th subjects. Y_3 is represented by the first and k-th subjects, and the like. Although each image is represented by a few/one subject(s), the whole probe set may be rebuilt by the whole dictionary, as shown in the second row of Fig. 2. We accept the fact that all images from the probe clip should belong to the same subject and incorporate this fact into the choice of coefficient matrix.

Joint Sparse Representation: In JSR [9], they adopt $l_{2,1}$-mixed norm to characterize the fact that a probe clip should be represented by a few/one subject(s). Their model is described as following:

$$\begin{aligned} &\min_W F(W) = f(W) + \lambda_1 \zeta(W) + \lambda_2 \phi(W), \\ &W \geq 0(optional), \end{aligned} \tag{2}$$

where

$$\begin{aligned} f(W) &= \tfrac{1}{2} \|Y - DW\|_F^2, \\ \zeta(W) &= \sum_{i=1}^{M} \|W^i\|_1 = \|W\|_1, \\ \phi(W) &= \sum_{i=1}^{M} \|W^i\|_F. \end{aligned} \tag{3}$$

In the above model, f is the loss function of reconstruction error measured by the square of F-norm, ζ is the l_1 sparse function of the recovery coefficients W with $\|W\|_1 = \sum_{i,j} |W_{ij}|$, i.e., sum all the absolute values of each item in the representation matrix. They use $l_{2,1}$-norm to mark the structure sparse function ϕ. Since the function f is a smooth and convex function and two sparse regularization terms are convex, the model is solved by APG method. However, as described in the Sect. 1, the $l_{2,1}$-mixed norm is just a convex relaxation of the number of selected gallery sets. In some cases, the coefficient matrix obtained by minimize $l_{2,1}$-mixed norm is not sparse enough, as shown in Fig. 1.

Consistent sparse representation: We design a new matrix norm, $l_{F,0}$-mixed norm to describe the number of gallery sets which is used to represent the probe set. The $l_{F,0}$-mixed norm is defined as the number of non-zeros coefficient matrix Z_i and the model can be written as:

$$\min_{Z,E} \|Z\|_{F,0} + \lambda\|E\|_{2,1}, \\ s.t. Y = \sum_{i=1}^{k} D_i Z_i + E = DZ + E, \tag{4}$$

where

$$\|Z\|_{F,0} = \sum_{i=1}^{k} \|Z_i\|_{F,\#}, \tag{5}$$

$$\|Z_i\|_{F,\#} = \begin{cases} 1, if & \|Z_i\|_F \neq 0 \\ 0, otherwise \end{cases}.$$

The parameter λ is used to balance the effects of two parts and $\|\cdot\|_{2,1}$ is the $l_{2,1}$-norm defined as the sum of l_2-norm of the columns of a matrix. Compared with JSR, the coefficient matrix obtained by Eq. (4) is more suitable for the face recognition, as shown in Fig. 1.

4   Solver

In the literature, Eq. (4) is difficult to be solved directly, because the $l_{F,0}$-mixed norm is discontinuous. The alternating direction strategy ADM [19] can be adopted to solve model (4). We adopt the alternating optimization strategy of ADM, based on the idea of introducing only one auxiliary variable to expand the original terms and update them iteratively. Our algorithm contains three subproblems. All of them find their closed-form solutions.

We introduce an auxiliary variable J, then the object function is equivalent to the following problem:

$$\min_{Z,J,E} \|J\|_{F,0} + \lambda\|E\|_{2,1}, s.t. Y = DZ + E, Z = J. \tag{6}$$

We can solve the problem with the Augmented Lagrange Multiplier Method. The corresponding Lagrange function for the above model is:

$$L(J, Z, E, Y_{1,2}) = \|J\|_{F,0} + \lambda\|E\|_{2,1} + \frac{\mu}{2}(\|Y - DZ - E\|_F^2 \\ + \|Z - J\|_F^2) + \langle Y_1, Y - DZ - E \rangle + \langle Y_2, Z - J \rangle, \tag{7}$$

where Y_1 and Y_2 are Lagrange multiplier, μ is augmented parameter. Obviously, the augmented function $L(\cdot)$ are separable for each variable. The problem can be solved by ADM strategy. So we can optimize each variable and fix other variables to achieve optimization.

Subproblem 1: computing Z: When J and E are fixed, the estimation subproblem of Z corresponds to minimizing the following function:

$$\tfrac{\mu}{2}(\|Y - DZ - E\|_F^2 + \|Z - J\|_F^2) + \langle Y_1, Y - DZ - E \rangle + \langle Y_2, Z - J \rangle. \tag{8}$$

It is obtained by omitting the terms not involving Z in (7). The function is quadratic and thus has a global minimum. We differentiate the function (8) w.r.t Z and set it to zero. Then we have the analytical solution follows:

$$Z = (D^T D + I)^{-1}(D^T(Y - E) + J + \tfrac{1}{\mu}(D^T Y_1 - Y_2)). \tag{9}$$

Subproblem 2: computing J: The objective function for J is

$$\min_J \|J\|_{F,0} + \tfrac{\mu}{2}(\|Z - J\|_F^2) + \langle Y_2, Z - J \rangle. \tag{10}$$

However, the solving method of J is not the same as above. First, we divide J into several blocks J_i $(i = 1, 2, \ldots, k)$. J_i is composed of the rows associated with the Z_i. We define $W_i = Z_i + \frac{Y_{2,i}}{\mu}$ and the Eq. (10) can be divided into the following subproblem:

$$\min_{J_i} \|J_i\|_{F,\#} + \tfrac{\mu}{2}\|J_i - W_i\|_F^2. \tag{11}$$

Inspired by the idea of [20], we get the following theorem and the problem (11) can be solved by it. The proof of the theorem can be found in supplementary material.

Theorem 1. *Let $\lambda > 0$. Given A, B have the same width and height, the following objective function:*

$$\min_A \|A\|_{F,\#} + \lambda\|A - B\|_F^2, \tag{12}$$

has an optimal solution of the form

$$A = \begin{cases} B, if & \|B\|_F^2 > \tfrac{1}{\lambda} \\ 0, otherwise \end{cases}.$$

Subproblem 3: computing E: The objective function for E is

$$\min_E \lambda\|E\|_{2,1} + \tfrac{\mu}{2}\|E - P\|_F^2, \tag{13}$$

where $P = Y - DZ + \frac{Y_1}{\mu}$. The closed-form solutions of this problem can be solved via existing algorithm [21]. We outline the procedure in Algorithm 1.

Algorithm 1. Solving Equation (4)

Input: matrix $\mathbf{Y}, \mathbf{D}$, parameter λ
Initialize: $\mathbf{Z} = \mathbf{J} = 0$, $\mathbf{E} = 0$, $\mathbf{Y}_1 = \mathbf{Y}_2 = 0$, $\mu = 10^{-6}$, $max_\mu = 10^6$, $\rho = 1.1$,
$\varepsilon = 10^{-8}$.
while not converged **do**
    1. Fix the others and update $\mathbf{Z}$
    According to *Equation* (9) and obtain $\mathbf{Z}$.
    2. Fix the others and update $\mathbf{J}$
    According to *Equation* (11) and obtain $\mathbf{J}$.
    3. Fix the others and update $\mathbf{E}$
    Solving *Equation* (13) by existed algorithm.
    4. Update the multipliers
    $\mathbf{Y}_1 = \mathbf{Y}_1 + \mu(\mathbf{Y} - \mathbf{DZ} - \mathbf{E})$, $\mathbf{Y}_2 = \mathbf{Y}_2 + \mu(\mathbf{Z} - \mathbf{J})$,
    5. Update the parameter μ
    $\mu = min(\rho\mu, max_\mu)$.
    6. Check the convergence conditions
    $\|\mathbf{Y} - \mathbf{DZ} - \mathbf{E}\|_\infty < \varepsilon$, $\|\mathbf{Z} - \mathbf{J}\|_\infty < \varepsilon$,
end while

---

5   Experiments

We use two public databases (Honda [22], COX [2]) and some image sets, such as
AR Face, FERET, PIE, Yale Face Database B. Some examples from these databases are shown in Fig. 3. Images in these databases contain complex appearance
variations in poses, expressions, illuminations, etc. Below we first introduce these
databases and the experimental setup, and then evaluate our method by comparing several state-of-the-art methods. For all these databases, we conduct five
random experiments, i.e., five randomly selected training and testing combination, and then report the average accuracy.

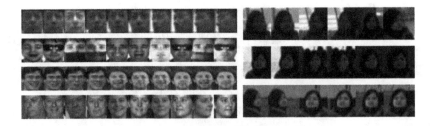

Fig. 3. Examples from video face databases: Honda [22] (the first row on the left), and
AR database (the second row on the left), ATT Face database (the third row on the
left), FERET face database (the last row on the left). COX [2] on the right, images
captured by camera one in the first row, the second row represents images obtained by
camera two, the last row images captured by camera three.

5.1 The Databases and Experimental Setup

Honda Database. Honda [22] was collected for video-based face recognition. In this paper, we use its subset which contains 59 videos of 20 subjects and each subject has at least 2 videos. The length of video clips varies from 12 to 645 frames. Different poses and expressions usually appear across different clips of each subject. We apply a online face recognition method to detect face from each video clip, and then resize all face images to gray-scale images with 20×20 pixels as used in [18]. We also employ histogram equalization in the pre-processing step to eliminate the lighting effects. For the varying set size problem, we uniformly down-sample each video clip (both gallery and probe) and use the obtained subsets for recognition. We test serval cases by extracting 30, 40, 50, and until 100 samples. If a set contains fewer images than the specified number, the original set is used. We select one video clip as training set from those video clips belong to the same subject and others is testing sets, as a result, we obtain 20 training sets and 39 testing sets totally.

COX Database. COX [2] is a public still and video face database from 1000 subjects, it contains 3000 video clips and has a training set contains 3 video clips for each subject. These three video clips captured by three different cameras respectively, the three videos have different illumination, pose and expression variation. We apply a online face recognition method to detect face from each video clip and resize the face to 32×40. Similar to the operation on Honda, we use its subsets and choose 5, 10, 15, and until 40 subjects to use instead of varying set size. One video clip can be selected as training set from three video clips of same subject, and other two video clips are testing sets.

AR Face Database. We select 100 subjects from AR Face database, each subject has 25 images. Images feature frontal view faces with different facial expressions, illumination conditions, and occlusions (sun glasses and scarf).

ATT Face Database. For ATT database, ten different images of each of 40 distinct subjects. For some subjects, the images were taken at different times, varying the lighting, facial expressions (open/closed eyes, smiling/not smiling) and facial details (glasses/no glasses). All the images were taken against a dark homogeneous background with the subjects in an upright, frontal position (with tolerance for some side movement).

FERET and PIE Face Database. We select 7 subjects from FERET database. Each subject has vary images. For PIE database, we use a subdatabase of 11,554 images of 68 people, each person under 13 different poses, 43 different illumination conditions, and with 4 different expressions.

YaleB Face Database. For YaleB database, we select 640 single light source images of 10 subjects each seen under different viewing conditions. For every subject in a particular pose, an image with ambient (background) illumination was also captured.

For these image sets data, histogram equalization is employed in the pre-processing step to eliminate the lighting effects. We resize all face images to gray-scale images with 30×30 pixels. In PIE database, images are resized with 32×32 pixels. Images from YaleB database are resized with 42×48 pixels. Images from the same subject are divided into three sets, one set of which is chosen for training and the rest sets for testing. Specially, images from the same subject are divided into two sets in ATT database, one set of which is chosen for training and another for testing.

5.2   The Comparison Methods

We compare our performance to several groups of state-of-the-art methods for video based face recognition. They include Manifold-Manifold Distance (MMD) [16], Manifold Discriminant Analysis (MDA) [17], Covariance Discriminative Learning (CDL) [23], Discriminant Analysis on Riemannian manifold of Gaussian distributions (DARG) [10], Joint sparse representation (JSR) [9]. In addition, we also compare with the baseline SR [3] by using all training data directly as the dictionary. The source codes of MMD, MDA, CDL and DARG are downloaded from author websites. For CDL [23], we use Linear Discriminant Analysis (LDA) and kernel formulation to discriminative learning. For DARG, we use kernel based on Mahalanobis distance (MD) and Log-Euclidean distance (LED) to measure the dissimilarity between two distributions.

The important parameters of different methods are carefully optimized as follows: For MMD and MDA, the parameters are configured according to [16,17]. Specifically, the ratio between euclidean distance and geodesic distance is 2. The number of connected nearest neighbors for computing geodesic distance both MMD and MDA is fixed to its default values, i.e., 12. For DARG and JSR, the important parameters were empirically tuned according to the recommendations in the original references [9,10].

5.3   Experiment Results and Analysis

First, we do experiments on several image set databases. The correct rate of these methods is shown in Table 1. Since the properties of each image set database are not the same, the performance of MMD, MDA, CDL and DARG on different databases is difference. Moreover, these image set databases are small. The number of face images in each training set and test set is few (*i.e.* 5–30 face images) which has a negative effect on performance of MMD, MDA, CDL and DARG. Among all these methods, our proposed method shows high stability and achieves better performance.

We summarize the recognition results of all methods on the Honda datasets in Fig. 4. Comparing the seven methods on Honda database, the image set based methods show distinct performances according to their properties. Among them, MDA better performs than MMD. This may be because MMD directly use image data in original space. DARG and CDL achieve better performance than MDA. This is because MDA learn the discriminant metrics in Euclidean space, whereas it classify the sets in non-Euclidean spaces. In contrast, these kernel-based methods utilize the statistics in Riemannian space and match them in the

Table 1. Correct rate of different methods on several image set databases.

Methods	MMD	MDA	CDL	DARG	SR	JSR	CSR
AR	0.587	0.545	0.531	0.592	1	0.956	1
ATT	0.980	0.755	0.835	0.955	0.995	0.995	1
FERET	0.812	0.386	0.786	0.975	1	1	1
PIE	1	0.673	0.872	1	1	1	1
YaleB	0.563	0.644	1	0.623	1	1	1

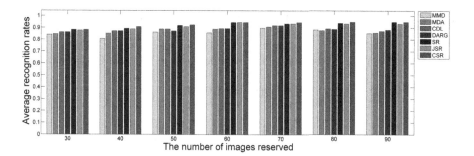

Fig. 4. Comparisons on Honda [22]. The horizontal axis represents the number of frames in each video reserved. The vertical axis represents the average accuracy. Comparisons on Honda [22]. The horizontal axis represents the number of frames in each video reserved. The vertical axis represents the average accuracy.

same space, which is more favorable for recognition problem. Compared with these classic video based methods, CSR, SR [3] and JSR [9] are more appealing than other methods. This finding is similar to that in [9]. Moreover, CSR achieves best performance among these three methods.

We also conduct comparison experiment on a more challenging database COX [2]. From the comparison in Fig. 5, the conclusion is consistent with perivous results on Honda database. Although CSR achieves better performance than other methods on Honda database, the superiority of our method is not obvious. However, CSR is superior to other methods distinctly on COX database. This may be contributed to that COX database is more challenging and the role of label consistency information is more remarkable on complicated and challenging database.

As shown in Table 2, we compared running time of CSR with other six methods on a PC with 4 GHz. In general, running time of our method is faster than SR and JSR. Compared with MMD, MDA, CDL and DARG, our method is faster than them on small database.

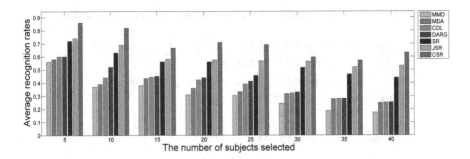

Fig. 5. Comparisons on COX [2]. The horizontal axis represents the number of subjects selected. The vertical axis represents the average accuracy.

Table 2. Running time (seconds) of different methods on PIE and COX databases

Methods	MMD	MDA	CDL	DARG	SR	JSR	CSR
PIE	1.59	6.67	8.45	9.56	29.82	21.94	1.13
COX	0.64	27.34	51.41	232.73	3211.52	458.02	52.21

6   Conclusion

In this paper, we propose a consistent sparse representation method to handle the video-based face recognition problem. In order to better characterize the number of gallery sets selected, we design a new matrix norm (*i.e.* $l_{F,0}$-mixed norm). A few/one correct gallery sets are utilized to reconstruct probe set by minimizing the $l_{F,0}$-mixed norm. A special alternating optimization strategy, which is based on the idea of introducing auxiliary variables to expand the original terms and update them iteratively, is adopted to solve the non-convex optimization problem. Experimental results on Honda database, COX database, and some image set face databases, demonstrate that our proposed method is more competitive than those state-of-the-art methods for video-based face recognition.

Acknowledgement. The authors would like to thank all the reviewers for their valuable comments. Thanks to Shiguang Shan, Zhen Cui and Ruiping Wang provide the data and code for us. Xiuping Liu is supported by the NSFC Fund (No. 61370143) and NEP Fund (No. f61632006). Junjie Cao is supported by the NSFC Fund (Nos.61363048 and 61262050).

References

1. Shakhnarovich, G., Fisher, J.W., Darrell, T.: Face recognition from long-term observations. In: Heyden, A., Sparr, G., Nielsen, M., Johansen, P. (eds.) ECCV 2002. LNCS, vol. 2352, pp. 851–865. Springer, Heidelberg (2002). doi:10.1007/3-540-47977-5_56

2. Huang, Z., Wang, R., Shan, S., Chen, X.: Learning Euclidean-to-Riemannian metric for point-to-set classification. In: Computer Vision and Pattern Recognition (CVPR), pp. 1677–1684 (2014)
3. Wright, J., Yang, A.Y., Ganesh, A., Sastry, S.S., Ma, Y.: Robust face recognition via sparse representation. IEEE Trans. Pattern Anal. Mach. Intell. **31**, 210–227 (2009)
4. Cui, Z., Shan, S., Chen, X., Zhang, L.: Sparsely encoded local descriptor for face recognition. In: Automatic Face and Gesture Recognition (FG), pp. 149–154. IEEE (2011)
5. Cui, Z., Shan, S., Zhang, H., Lao, S., Chen, X.: Structured sparse linear discriminant analysis. In: Image Processing (ICIP), pp. 1161–1164. IEEE (2012)
6. Chen, Y.C., Patel, V.M., Shekhar, S., Chellappa, R., Phillips, J.: Video-based face recognition via joint sparse representation. In: Automatic Face and Gesture Recognition (FG), pp. 1–8. IEEE (2013)
7. Elhamifar, E., Vidal, R.: Robust classification using structured sparse representation. In: Computer Vision and Pattern Recognition (CVPR), pp. 1873–1879 (2011)
8. Majumdar, A., Ward, R.K.: Classification via group sparsity promoting regularization. In: Acoustics, Speech and Signal Processing (ICASSP), pp. 861–864. IEEE (2009)
9. Cui, Z., Chang, H., Shan, S., Ma, B., Chen, X.: Joint sparse representation for video-based face recognition. Neurocomputing **135**, 306–312 (2014)
10. Wang, W., Wang, R., Huang, Z., Shan, S., Chen, X.: Discriminant analysis on Riemannian manifold of Gaussian distributions for face recognition with image sets. In: Computer Vision and Pattern Recognition (CVPR), pp. 2048–2057 (2015)
11. Arandjelović, O., Shakhnarovich, G., Fisher, J., Cipolla, R., Darrell, T.: Face recognition with image sets using manifold density divergence. In: Computer Vision and Pattern Recognition (CVPR), vol. 1, pp. 581–588 (2005)
12. Nishiyama, M., Yamaguchi, O., Fukui, K.: Face recognition with the multiple constrained mutual subspace method. In: Kanade, T., Jain, A., Ratha, N.K. (eds.) AVBPA 2005. LNCS, vol. 3546, pp. 71–80. Springer, Heidelberg (2005). doi:10.1007/11527923_8
13. Kim, T.K., Kittler, J., Cipolla, R.: Discriminative learning and recognition of image set classes using canonical correlations. IEEE Trans. Pattern Anal. Mach. Intell. **29**, 1005–1018 (2007)
14. Kim, T.K., Kittler, J., Cipolla, R.: Incremental learning of locally orthogonal subspaces for set-based object recognition. In: British Machine Vision Conference (BMVC), pp. 559–568 (2006)
15. Chen, L.: Dual linear regression based classification for face cluster recognition. In: Computer Vision and Pattern Recognition (CVPR), pp. 2673–2680. IEEE (2014)
16. Wang, R., Shan, S., Chen, X., Gao, W.: Manifold-manifold distance with application to face recognition based on image set. In: Computer Vision and Pattern Recognition (CVPR), pp. 1–8 (2008)
17. Wang, R., Chen, X.: Manifold discriminate analysis. In: Computer Vision and Pattern Recognition (CVPR), pp. 429–436. IEEE (2009)
18. Cui, Z., Shan, S., Zhang, H., Lao, S., Chen, X.: Image sets alignment for video-based face recognition. In: Computer Vision and Pattern Recognition (CVPR), pp. 2626–2633 (2012)
19. Lin, Z., Chen, M., Ma, Y.: The augmented Lagrange multiplier method for exact recovery of corrupted low-rank matrices, arXiv preprint arXiv:1009.5055 (2010)
20. Xu, L., Lu, C., Xu, Y., Jia, J.: Image smoothing via l 0 gradient minimization. ACM Trans. Graph. (TOG) **30**, 174 (2011). ACM

21. Tang, K., Liu, R., Su, Z., Zhang, J.: Structure-constrained low-rank representation. IEEE Neural Netw. Learn. Syst. **25**, 2167–2179 (2014)
22. Lee, K.C., Ho, J., Yang, M.H., Kriegman, D.: Video-based face recognition using probabilistic appearance manifolds. In: Computer Vision and Pattern Recognition (CVPR), vol. 1, p. I-313. IEEE (2003)
23. Wang, R., Guo, H., Davis, L.S., Dai, Q.: Covariance discriminative learning: a natural and efficient approach to image set classification. In: Computer Vision and Pattern Recognition (CVPR), pp. 2496–2503. IEEE (2012)

Unconstrained Gaze Estimation Using Random Forest Regression Voting

Amine Kacete$^{(\boxtimes)}$, Renaud Séguier, Michel Collobert, and Jérôme Royan

Institute of Research and Technology B-com, Cesson-Sévigné, France
amine.kacete@b-com.com

Abstract. In this paper we address the problem of automatic gaze estimation using a depth sensor under unconstrained head pose motion and large user-sensor distances. To achieve robustness, we formulate this problem as a regression problem. To solve the task in hand, we propose to use a regression forest according to their high ability of generalization by handling large training set. We train our trees on an important synthetic training data using a statistical model of the human face with an integrated parametric 3D eyeballs. Unlike previous works relying on learning the mapping function using only RGB cues represented by the eye image appearances, we propose to integrate the depth information around the face to build the input vector. In our experiments, we show that our approach can handle real data scenarios presenting strong head pose changes even though it is trained only on synthetic data, we illustrate also the importance of the depth information on the accuracy of the estimation especially in unconstrained scenarios.

1 Introduction

Automatic gaze estimation is the process of determining where the user is looking which can be represented as the point-of-regard or the visual axis. In recent years, gaze estimation has become the focus of several computer vision research according to the importance of this component in understanding the human behavior. Determining this information can be used in different areas such as Human Computer Interaction (HCI) systems, psychological and cognitive process understanding, security and monitoring systems and marketing research.

Many existing industrial solutions are commercialized and provide an acceptable accuracy in gaze estimation. These solutions often use a complex hardware such as range of infrared cameras (embedded on a head mounted or in a remote system) making them intrusive, very constrained by the user's environment and inappropriate for a large scale public use.

Current research focus on estimating gaze using low-cost devices such as a simple monocular camera relying on the analyze of the eye features and sometimes, head features extraction to infer the head pose parameters. [1] gives a

Electronic supplementary material The online version of this chapter (doi:10.1007/978-3-319-54187-7_28) contains supplementary material, which is available to authorized users.

© Springer International Publishing AG 2017
S.-H. Lai et al. (Eds.): ACCV 2016, Part III, LNCS 10113, pp. 419–432, 2017.
DOI: 10.1007/978-3-319-54187-7_28

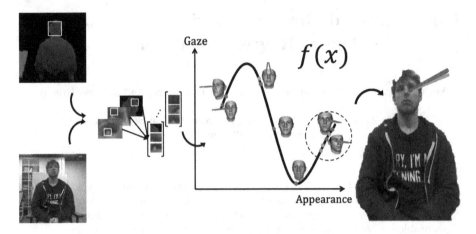

Fig. 1. Example of automatic gaze estimation based on our approach. We build a 3-channel global vector represented by the two RGB eye images and the face depth information using the depth sensor multimodal data, we extract a set of patches and project it through the forest represented here as the mapping function $f(x)$ (the learned gaze sample clusters are defined as the red centroid points). Each single tree casts votes for each patch (defined as the green points). By performing a non-parametric clustering technique, a final estimation is calculated (represented as the green line, the red one defines the ground truth). (Color figure online)

very comprehensive survey about it. In this paper we present an approach based on an ensemble of trees grouped in a single forest to learn the highly non-linear mapping function between the gaze information and the RGB eye image appearances including depth cues. To train our trees, we generate important training data with high head pose, illumination and scale variability using a statistical morphable model with an integrated parametric gaze model. At testing phase, we exploit the multimodal data of the Kinect sensor to grab the RGB and depth cues. By performing a face detection, we extract the two RGB eye images (converted to grayscale space) and the face depth information organized as a 3-channel global vector. Then we extract a set of patches so that each patch contains 3-channels extracted randomly from the global vector. We project all the extracted patches through the learned regression forest which casts votes for each patch. By clustering all the votes using a non-parametric technique, a final gaze estimate is calculated as illustrated in Fig. 1.

In our experiments we evaluate the robustness and accuracy of our approach on real data and we measure the importance of the depth cues in gaze estimation especially in highly unconstrained scenarios. The obtained results demonstrate the potential of our approach.

In the rest of the paper, we describe the related work in Sect. 2. In Sect. 3, we detail our approach and show the experimental results in Sect. 4. Section 5 concludes our work.

2   Related Work

In this section, we first present the existing work related to the automatic gaze estimation then we present a brief survey of the use of synthetic data to solve computer vision problems.

2.1   Automatic Gaze Estimation

The recent gaze estimation approaches can be divided into two global categories: feature-based and appearance-based approaches.

Based on geometrical assumptions, feature-based approaches rely on extracting some discriminative and invariant facial features from the eye image such as corneal infrared reflexion, pupil center and eye corners. Using these features, a user-specific 3D eyeball model is calculated to infer the visual axis information, [2] gives details about this model. [3] used the shape of the estimated pupil through an elliptic fitting. In addition to the pupil location information, [4] used the corners locations estimated through an AAM [5] fitting and by combining the two pieces of information, calculated the center and the radius of the eyeball giving the two angles of the visual axis. To get a direct access to the 3D information of the eyeball, [6] used a stereo setup. [7] performed the same strategy with a single camera by adding a calibration step. Some work uses a depth sensor, [8] estimated the head pose parameters using a multi-template ICP, based on these parameters and a template matching approach based on elliptical fitting, the eyeball parameters can be fixed. [9] used a flexible model fitting approach to compute the head pose parameters, coupled to the pupil location information estimated using the method from [10] and a calibration step by gazing a known fixed 3D points, the visual axis can be inferred. [2,11] used the corneal reflexion information based on one or multiple IR light sources. These methods still require sometimes complex devices such as Infrared cameras with a very heavy constrained calibration process, and sometimes a very high resolution imaging to extract accurately the facial eye points making them difficult to use in arbitrary environment.

Our method belongs to the appearance-based approach. Unlike feature-based approach, these methods aim to learn a direct mapping function from the high dimensional eye image appearances to the low space of the gaze information. [12] trained a neuronal network using $2k$ labeled training samples. [13] collected 252 training samples to build a manifold of the local linearity of the eye appearances and estimate an unknown sample using a linear interpolation. [14] exploited the Markov model interpolation to enhance the generalization of the mapping function over unseen data such as gaze sample under head movement. [15] introduced sparse semi-supervised Gaussian process to complete the training set with unlabeled samples. [16] proposed a visual saliency maps strategy to generate training data through a video stream and used a Gaussian process regression to determine the mapping function. [17] introduced the adaptative linear regression to learn on a very sparse training set. These methods perform on frontal head pose configuration and their accuracy decrease significantly with head pose changes.

[18] proposed to separate head pose component from the global gaze estimation system by performing an initial estimation under frontal configuration assumption then compensated with the head pose parameters for the final estimation geometrically. Using the same paradigm, [19] projected the training gaze sample in frontal manifold using a frontalization step based on the head pose parameters. These last two methods solved the problem of head changes successfully but still working under low user-camera distances. To cover all the eye image appearance variability, [20] recorded around $200k$ training samples and used a deeper strategy using a convolutional neuronal network to learn a very robust mapping function achieving a high gaze estimation accuracy but still very constrained by an important computational time.

2.2 Synthetic Data in Computer Vision

This last decade, machine learning techniques are considered as a very elegant way to tackle many problems in computer vision. They demonstrated a great potential in terms of efficiency and robustness. Nevertheless to achieve a high generalization across unseen scenarios, these methods often require a very representative training data set. Thus, the building of high amount of labeled data is a very tedious process and synthetic data represent a promising solution as the annotation is performed automatically instead of manual labeling. [21] developed an iterative model based on Gabor-filters applied on an empty image containing some seed points to render a fingerprint training samples. [22] rendered iris image samples obtained from a 2D polar projection of a cylindrical representation of continuous fibers. [23] improved face authentication by generating multiple virtual images using simple geometric transformations. [24] used a motion capture strategy to record RGB and depth cues of the body part movements, by varying body size and shape, scene position, camera position and mirroring the recorded data. They synthesize a highly varied training allowing a robust body part pose estimation. [25] tackled the head pose estimation problem with synthetic depth images by rendering an important amount of training data using a 3D statistical morphable model (3DMM).

In this work, we exploit the high generalization ability of the randomized regression trees by learning on a very representative rendered training data using the same 3D statistical morphable model as [25], and perform the gaze estimation.

3 Automatic Gaze Estimation with Regression Forest

We use randomized regression trees to estimate the two angles (θ, γ) of the gaze vector $\overrightarrow{g}$ from the RGB and depth cues combined on 3-channel patches. In Sect. 3.1, we provide some background on regression trees, then we detail the training and testing step in Sects. 3.2 and 3.3 respectively. In Sect. 3.4 we illustrate how we generated data for trees learning.

3.1   Random Regression Forest

Recently, many applications in computer vision have used Random forest to achieve the mapping from complex input spaces into discrete or continuous output space. Introduced by [26], randomized trees deal with different tasks such as classification [27–29], regression [24,30,31] and density estimation [32,33].

Regression forest is an ensemble of trees predictors which splits the initial problem into two low complex problems in a recursive way. At each node, a simple binary test is performed. According to the result of the test, a data sample is directed towards the left or the right child. The tests are selected to achieve an optimal clustering. The terminal nodes of the tree called leaves, store the estimation models approximating the best the desired output. To achieve high generalization, the trees are trained in a decorrelated way by introducing randomness in both the training data provided for each tree and the set of the binary tests.

3.2   Training

We trained each tree T in the forest $\mathcal{T} = \{T_k\}_{k=1:N_T}$ in a supervised way using a set of annotated patches $\{\mathcal{P}_i = (\mathcal{I}_i^c, g_i)\}_{i=1:N_P}$ randomly selected from the training data where:

- $\mathcal{I}_i^c$ represents the extracted visual features vector from a given patch $\mathcal{P}_i$, c defines the feature channel, we used 3 channels namely the two grayscale intensities extracted from the two eyes images, and the depth values extracted from the face.
- g_i represents the output gaze vector represented with two component (θ, γ).

Starting from the root, at each non-leaf node, we define a simple binary test t:

$$t_{x_1,y_1,x_2,y_2,c,\tau} = \begin{cases} 1, & if \ \mathcal{I}_i^c(x_1,y_1) - \mathcal{I}_i^c(x_2,y_2) \leq \tau \\ 0, & otherwise \end{cases}$$

where $(\mathcal{I}_i^c(x_1,y_1) - \mathcal{I}_i^c(x_2,y_2))$ represents the difference of intensity between two locations (x_1,y_1) and (x_2,y_2) in the channel c. Supervising the training consists in finding at each non-leaf node the optimal binary test t^* thats maximizes the purity of the data clustering. Maximizing the clustering purity is achieved by maximizing the information gain defined as the differential entropy of the set of patches at parent node $\mathcal{P}$ minus the weighted sum of the differential entropies computed at the children $\mathcal{P}_{\mathcal{L}}$ and $\mathcal{P}_{\mathcal{R}}$ defined as:

$$E = H(\mathcal{P}) - (w_{\mathcal{L}} H(\mathcal{P}_{\mathcal{L}}) + w_{\mathcal{R}} H(\mathcal{P}_{\mathcal{R}})) \tag{1}$$

The weights $w_{j \in \{R,L\}}$ are defined as the ratio between the number of patches reaching the parent node and the number of patches reaching the left node (or the right node respectively).$i.e.,$ $\frac{|\mathcal{P}_{j \in \{\mathcal{L},\mathcal{R}\}}|}{|\mathcal{P}|}$. Assuming that the gaze vector g at each node is a random variable with a multivariate Gaussian distribution such as $p(g) = \mathcal{N}(g, \bar{g}, \Sigma)$, it allows us to rewrite Eq. 1 as follows:

$$E = \log |\Sigma(\mathcal{P})| - (w_L \log |\Sigma(\mathcal{P_L})| + w_R \log |\Sigma(\mathcal{P_R})|) \tag{2}$$

where $|\Sigma(\mathcal{P})|$ represents the determinant of the covariance matrix Σ of the random variable g.

The learning process finishes when the data reach a predefined maximum depth value of the tree or the number of patches let down a threshold value yielding the creation of the leaves. A leaf l stores the mean of all the gaze vectors which reached it with the corresponding covariance.

3.3   Testing

To estimate the gaze vector from an unseen instance, we extract a set of patches from the RGB eye regions and the face depth information after a face detection step. Each patch is passed through all the learned trees in the forest. Using the optimal stored binary test each tree processes the patch until reaching a leaf. The gaze vector estimation according to a single tree is given by the reached leaf l in terms of the stored distribution $p(g|l) = \mathcal{N}(g, \bar{g}, \Sigma)$. The gaze vector estimation for a given patch $\mathcal{P}_i$ over all the trees is calculated as follows:

$$p(g|\mathcal{P}_i) = \frac{1}{N_T} \sum_t p(g|l_t(\mathcal{P}_i)) \tag{3}$$

All the estimations corresponding to the extracted patches are regrouped in votes. Before performing the clustering of these votes, we discard the estimations from the leaves with high variance considered as non-informative. To locate the centroid of the cluster of the votes, we perform 5 mean-shift iterations using a Gaussian kernel. Figure 2 shows an example of the final estimation, the green ones represent the votes casted by the forest which are selected by the mean-shift. The red lines corresponds some casted votes with a high variance discarded by the mean-shift. The final estimate is given by the blue line corresponding to the centroid of the selected votes.

3.4   Training Data Generation

To provide a very representative training dataset, we use the 3DMM from [34] to render the samples. This model is built from around 200 scans of human faces, it contains a very high mesh density including the face, frontal neck and ears. The shape and texture of the model are represented as a linear combinations of 199 components. They can be deformed according to the following equations:

$$\mathcal{A} = \mathcal{A}_0 + \mathcal{M}_\mathcal{A} \alpha \tag{4}$$

where $\mathcal{A}$ can denote the generated texture or shape respectively. $\mathcal{A}_0$ denotes the mean, $\mathcal{M}$ represents the basis components perturbed with parameters α.

Fig. 2. test instance example: the selected votes by mean-shift filtering are represented in green, the non-informative leaves responses in red and the final estimation in blue. (Color figure online)

To generate variability in the face identity we perturb the first 50 basis components of the shape and texture by ±1.5 of the standard deviation of each mode. To render images in different head pose configurations, we apply random rigid transformation on the model, the rotations spans ±60° for yaw and ±40° for pitch. For the scale, we translate the model along the z axis within 200 cm range.

Unfortunately, the basis components related to the shape and the texture of this model do not monitor the gaze direction. To integrate a parametric gaze model to the 3DMM able to generate different gaze direction instances, we decided to remove all the vertices related to the eye regions, and we place two spheres as eyeballs instead. We fix the diameters to the human average eyeball namely 25 mm. We use different textures for the eyeballs to handle the iris appearance variability. Moreover, to control the eyelids movements resulting from the gazing up and down, we introduce a linear translation for each vertex surrounding the eye regions. By defining the starting and the ending position in the global mesh, all the coefficients of the linear translations can be calculated. Thanks to the topology of the model, all these modifications keep the same behavior under identity variation. To generate gaze sample, we generate a virtual 3D point on which the two eyeballs turn toward, the gaze information angles can be easily calculated knowing the location of the eyeballs centers. Figure 3 shows the different steps applied to generate gaze samples.

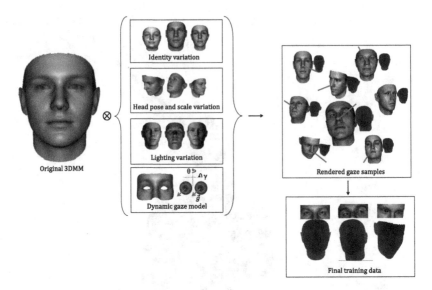

Fig. 3. Training data generation. We based our data generation on the 3D morphable model from [34], by introducing some variabilities such as identity (using shape and texture principal components respectively), head pose changes (using OpenGL camera with different rigid transformations), illumination (using different light intensities and directions) and an integrated parametric gaze model (represented by two global textured spheres). We obtain the final training data with the correspondent RGB-D images and gaze sample as annotation illustrated in red line. (Color figure online)

4   Experimental Results

Training dataset. To train the regression trees, we used $200k$ synthetic RGB-D samples. We extracted 30 patches from each sample giving $6M$ training data. After scaling the face depth image to (150×150) and the eyes rgb images to (80×70), the size of each channel of the extracted patches is fixed to (16×16). The trees parameters are fixed according to some empirical observation, *e.g.*, the maximum depth to 18 and at each node we randomly generate 400 splitting candidates with 50 thresholds giving a total number of $20k$ binary tests.

Testing dataset. To evaluate the performance of our algorithm on realistic data, we built our own gaze database using Kinect sensor. The database contains $17k$ RGB-D images of 42 people (15 females and 27 males, 4 with glasses and 38 without glasses) gazing different targets displayed on the screen. The subject performed 4 scenarios, gazing with a fixed head about $d_0 = 150\,\mathrm{cm}$ from the sensor, gazing with same distance d_0 under head pose changes and the two others scenarios are performed about $d_1 = 200\,\mathrm{cm}$ from the sensor. Knowing the Kinect intrinsic parameters and its rigid transformation to the screen, the displayed gaze points can be projected to the Kinect world space. The gaze vector is represented as vector stretching the head gravity point (computed using face

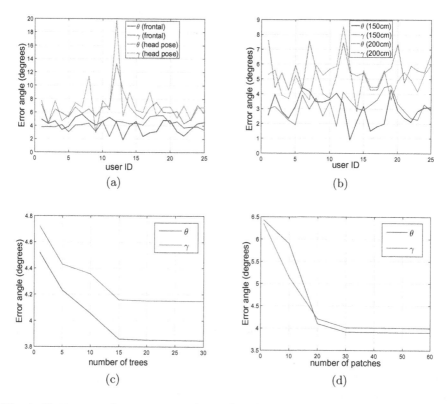

Fig. 4. Testing our forest accuracy learned with synthetic data on real annotated images. (a) The mean error for the two gaze directions under frontal and head pose changes. (b) The mean error of the two gaze directions under two distances from the sensor. (c) The global mean error (over 25 participants as a function of the number trees when the number of patches is fixed to 30). (d) The global mean error as a function of the number of patches extracted when the number of trees is fixed to 15.

detection area) and the 3D gazed point. The RGB-D images have a resolution of (1280×960) and (320×240) pixels respectively recorded at 15 fps.

Testing results. Some parameters control the performance of our method at the test time. Figure 4a represents the global error of the estimation (for both horizontal θ and vertical γ gaze angles) over 25 users from the database discussed previously under frontal and head changes configurations. For each user, a mean error across different gaze samples performed under two distances is calculated. In frontal case, the mean error over all the users is less than 3° for the two directions respectively whereas the error is less than 6.5° for head pose changes case. This difference in accuracy between the two configurations is directly linked to the high eye image appearances variability across head pose configuration making the trees prediction less accurate. In Fig. 4b we report the error as a function of distance from the sensor for a frontal configuration. The experiments show a mean error of 2.9° and 3.1° for θ and γ respectively at 150 cm from the sensor.

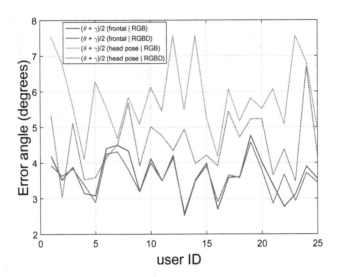

Fig. 5. The importance of depth channel in gaze estimation accuracy with our approach.

At 200 cm, we notified a slightly higher errors, 4.8° and 5.0° for the two directions respectively. The difference in accuracy between the two distances is related to the RGB eye images and face depth appearances which are significantly variable depending on the distance from the sensor. Figure 4c and d illustrate the variation of the mean errors over all the testing user under forest size and extracted patches number variation. In Fig. 4c the errors decrease by increasing the number of trees, they are reduced by approximately 15% compared to the initial value (from 4.5° to 3.8° and 4.7° to 4.1° for the two directions receptively) which is the result of output smoothing by different trees. We noticed that, using more than 15 trees does not perform more precision, so we fix the optimal forest size to 15. The number of patches extracted in the testing step is fixed to 30 according to Fig. 4d showing that the errors decrease approximatively by 40% (from 6.4 to 3.8 and 6.4 to 3.9 for the two directions). This behavior can be explained by the fact that trees get more information about the input which consequently gives more accurate estimations. To evaluate the importance of depth cues in our gaze estimation system, we performed our estimation with and without this information during the test under frontal and head pose changes, Fig. 5 shows the result. For the frontal configuration, we noticed that the depth doesn't enhance the estimation accuracy while difference in errors reach approximatively 2° under head pose changes. This result is expected since the depth cues, intrinsically, encodes more information related to the head pose variations than RGB cues giving better results.

Figure 6 illustrates the distribution of the mean square error of two gaze directions across θ and γ variations. We can distinguish 3 regions as follows:

– $\gamma < -20°$ represents the highest error range). These γ values correspond to the eyes closure making the eye image appearances very similar even if θ is

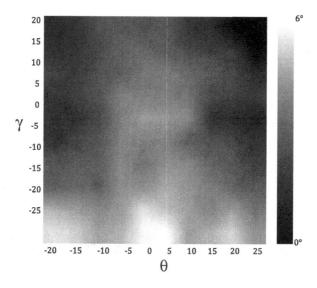

Fig. 6. Mean gaze error distribution over 25 participants across different gaze directions.

varying which produces bad gaze estimations. Furthermore, our parametric gaze model performs a linear shifting on the eyelid vertices to cover the new eye shape and stretches the original eyelid texture to cover the new texture giving a rough approximation of the real eye appearance. Our choice of such gaze model is strongly constrained by the 3DMM topology.

- $|\theta| < -7°$ describes a region with a relatively important error. Our forest is weakly discriminative with straight gazing samples under large distances. In addition, we noticed, for some users, an important error for upward gazing configuration ($\gamma > 10°$ and $\theta < 5°$) which can be explained by an elliptical deformation of the high part of the eyes. The fact that, this deformation is very person-specific and our parametric model performs the same deformation over the different face shapes generated by the 3DMM, the forest gives less accurate results.

- $\gamma > -20°$ and $|\theta| > 5°$ covers the range of good gaze estimation (error less than $4°$) which represents more than 50% of the total area. The appearance of the patches extracted from these gaze samples are very discriminative, in addition, for these configurations, our synthetic training data present a very high realism.

Table 1 compares our approach to the method from [9]. In the experiments conducted in [9] the gaze errors are computed for each eye, to get a direct comparison, we reported the mean of these errors over the two eyes. Note the improvement of our approach in accuracy overs 25 participants. Figure 7 shows some qualitative examples for successful and failure estimations.

Table 1. Comparison of our approach to the method in [9].

Method	θ error	γ error
Jianfeng et al.	5.53°	4.51°
Our method (worst participant result)	5.78°	5.81°
Our method (best participant result)	2.76°	2.93°
Our method (mean over 25 participants)	3.65°	3.88°

Fig. 7. Some example of automatic gaze estimation based on our approach. First and second row illustrate some successful estimation under frontal and head changes configurations respectively. Third row shows good estimation in a multi-users scenario. Last row describes some gaze estimations failure using our approach.

5   Conclusion

In this paper, we have proposed an approach based on regression forest trained on important amount of synthetic data to handle unconstrained gaze estimation (under head pose changes, illumination variation and large user-sensor distances). To generate the training data, we used a 3D morphable model with an integrated parametric gaze model allowing us to generate different gaze sample under identity, head pose, scale and illumination variations. We demonstrated that adding depth information performs better results for gaze estimation under high head pose changes and large user-sensor distance configurations. By establishing the gaze errors distribution we validate our integrated gaze model used for training data generation despite its linear aspect achieving state-of-the-art performance.

References

1. Hansen, D.W., Ji, Q.: In the eye of the beholder: a survey of models for eyes and gaze. In: TPAMI (2010)
2. Guestrin, E.D., Eizenman, M.: General theory of remote gaze estimation using the pupil center and corneal reflections. IEEE Trans. Biomed. Eng. **53**, 1124–1133 (2006)
3. Wang, J.G., Sung, E.: Study on eye gaze estimation. IEEE Trans. Syst. Man Cybern. Part B Cybern. **32**, 332–350 (2002)
4. Ishikawa, T.: Passive driver gaze tracking with active appearance models (2004)
5. Cootes, T.F., Edwards, G.J., Taylor, C.J.: Active appearance models. In: TPAMI (2001)
6. Matsumoto, Y., Zelinsky, A.: An algorithm for real-time stereo vision implementation of head pose and gaze direction measurement. In: Proceedings of the Fourth IEEE International Conference on Automatic Face and Gesture Recognition, pp. 499–504. IEEE (2000)
7. Chen, J., Ji, Q.: 3D gaze estimation with a single camera without IR illumination. In: 19th International Conference on Pattern Recognition, ICPR 2008, pp. 1–4. IEEE (2008)
8. Bär, T., Reuter, J.F., Zöllner, J.M.: Driver head pose and gaze estimation based on multi-template ICP 3-D point cloud alignment. In: 2012 15th International IEEE Conference on Intelligent Transportation Systems (ITSC), pp. 1797–1802. IEEE (2012)
9. Jianfeng, L., Shigang, L.: Eye-model-based gaze estimation by RGB-D camera. In: Proceedings of the IEEE Conference on Computer Vision and Pattern Recognition Workshops, pp. 592–596 (2014)
10. Timm, F., Barth, E.: Accurate eye centre localisation by means of gradients. In: VISAPP (2011)
11. Zhu, Z., Ji, Q.: Novel eye gaze tracking techniques under natural head movement. IEEE Trans. Biomed. Eng. **54**, 2246–2260 (2007)
12. Baluja, S., Pomerleau, D.: Non-intrusive gaze tracking using artificial neural networks. Technical report, DTIC Document (1994)
13. Tan, K.H., Kriegman, D.J., Ahuja, N.: Appearance-based eye gaze estimation. In: Proceedings of the Sixth IEEE Workshop on Applications of Computer Vision (WACV 2002), pp. 191–195. IEEE (2002)
14. Hansen, D.W., Hansen, J.P., Nielsen, M., Johansen, A.S., Stegmann, M.B.: Eye typing using Markov and active appearance models. In: Proceedings of the Sixth IEEE Workshop on Applications of Computer Vision (WACV 2002), pp. 132–136. IEEE (2002)
15. Williams, O., Blake, A., Cipolla, R.: Sparse and semi-supervised visual mapping with the S^3GP. In: 2006 IEEE Computer Society Conference on Computer Vision and Pattern Recognition, vol. 1, pp. 230–237. IEEE (2006)

16. Sugano, Y., Matsushita, Y., Sato, Y.: Calibration-free gaze sensing using saliency maps. In: 2010 IEEE Conference on Computer Vision and Pattern Recognition (CVPR), pp. 2667–2674. IEEE (2010)
17. Lu, F., Sugano, Y., Okabe, T., Sato, Y.: Inferring human gaze from appearance via adaptive linear regression. In: 2011 IEEE International Conference on Computer Vision (ICCV), pp. 153–160. IEEE (2011)
18. Lu, F., Okabe, T., Sugano, Y., Sato, Y.: A head pose-free approach for appearance-based gaze estimation. In: BMVC, pp. 1–11 (2011)
19. Mora, K.A.F., Odobez, J.M.: Gaze estimation from multimodal kinect data. In: 2012 IEEE Computer Society Conference on Computer Vision and Pattern Recognition Workshops (CVPRW), pp. 25–30. IEEE (2012)
20. Zhang, X., Sugano, Y., Fritz, M., Bulling, A.: Appearance-based gaze estimation in the wild. In: Proceedings of the IEEE Conference on Computer Vision and Pattern Recognition, pp. 4511–4520 (2015)
21. Cappelli, R., Erol, A., Maio, D., Maltoni, D.: Synthetic fingerprint-image generation. In: Proceedings of the 15th International Conference on Pattern Recognition, vol. 3, pp. 471–474. IEEE (2000)
22. Zuo, J., Schmid, N.A., Chen, X.: On generation and analysis of synthetic iris images. IEEE Trans. Inf. Forensics Secur. 2, 77–90 (2007)
23. Thian, N.P.H., Marcel, S., Bengio, S.: Improving face authentication using virtual samples. In: Proceedings of the 2003 IEEE International Conference on Acoustics, Speech, and Signal Processing (ICASSP 2003), vol. 3, p. III-233. IEEE (2003)
24. Shotton, J., Sharp, T., Kipman, A., Fitzgibbon, A., Finocchio, M., Blake, A., Cook, M., Moore, R.: Real-time human pose recognition in parts from single depth images. Commun. ACM 56, 116–124 (2013)
25. Fanelli, G., Gall, J., Van Gool, L.: Real time head pose estimation with random regression forests. In: CVPR (2011)
26. Breiman, L.: Random forests. Mach. Learn. 45, 2–32 (2001)
27. Marée, R., Wehenkel, L., Geurts, P.: Extremely randomized trees and random subwindows for image classification, annotation, and retrieval. In: Criminisi, A., Shotton, J. (eds.) Decision Forests for Computer Vision and Medical Image Analysis, pp. 125–141. Springer, London (2013)
28. Gall, J., Yao, A., Razavi, N., Van Gool, L., Lempitsky, V.: Hough forests for object detection, tracking, and action recognition. In: TPAMI (2011)
29. Lepetit, V., Lagger, P., Fua, P.: Randomized trees for real-time keypoint recognition. In: CVPR (2005)
30. Criminisi, A., Shotton, J., Robertson, D., Konukoglu, E.: Regression forests for efficient anatomy detection and localization in CT studies. In: Medical Computer Vision Workshop (2010)
31. Kacete, A., Seguier, R., Royan, J., Collobert, M., Soladie, C.: Real-time eye pupil localization using hough regression forest. In: Proceedings of the Sixth IEEE Workshop on Applications of Computer Vision (WACV 2016). IEEE (2016)
32. Moosmann, F., Triggs, B., Jurie, F.: Fast discriminative visual codebooks using randomized clustering forests. In: Twentieth Annual Conference on Neural Information Processing Systems (NIPS 2006), pp. 985–992. MIT Press (2007)
33. Ram, P., Gray, A.G.: Density estimation trees. In: Proceedings of the 17th ACM SIGKDD International Conference on Knowledge Discovery and Data Mining, pp. 627–635. ACM (2011)
34. Paysan, P., Knothe, R., Amberg, B., Romdhani, S., Vetter, T.: A 3D face model for pose and illumination invariant face recognition. In: Advanced Video and Signal Based Surveillance (2009)

A Novel Time Series Kernel for Sequences Generated by LTI Systems

Liliana Lo Presti[✉] and Marco La Cascia

DIID, Universitá degli studi di Palermo, V.le delle Scienze Ed. 6, Palermo, Italy
liliana.lopresti@unipa.it

Abstract. The recent introduction of Hankelets to describe time series relies on the assumption that the time series has been generated by a vector autoregressive model (VAR) of order p. The success of Hankelet-based time series representations prevalently in nearest neighbor classifiers poses questions about if and how this representation can be used in kernel machines without the usual adoption of mid-level representations (such as codebook-based representations). It is also of interest to investigate how this representation relates to probabilistic approaches for time series modeling, and which characteristics of the VAR model a Hankelet can capture. This paper aims at filling these gaps by: deriving a time series kernel function for Hankelets (TSK4H), demonstrating the relations between the derived TSK4H and former dissimilarity/similarity scores, highlighting an alternative probabilistic interpretation of Hankelets.

Experiments with an off-the-shelf SVM implementation and extensive validation in action classification and emotion recognition on several feature representations, show that the proposed TSK4H allows achieving state-of-the-art or even superior accuracy values in classification with respect to past work. In contrast to state-of-the-art time series kernel functions that suffer of numerical issues and tend to provide diagonally dominant kernel matrices, empirical results suggest that the TSK4H has limited numerical issues in high-dimensional spaces. On three widely used public benchmarks, TSK4H consistently outperforms other time series kernel functions despite its simplicity and limited time complexity.

1 Introduction

Time series arise naturally in several computer vision applications including tracking [25,26,42,56], action/motion modeling and classification [23,30,37,43], event causality [18,41,61], face emotion recognition [31], affective behavior [7,34], gait recognition [14,60], sequence alignment [64].

When dealing with time series, there is the need of formulating suitable kernel functions for adopting kernel methods [19] such as Support Vector Machine (SVM) [15]. Formerly proposed time series kernels are the Dynamic Time Warping (DTW) kernel [36] and the Global Alignment (GA) kernel [9,11]. While the DTW kernel considers data similarities along the optimal alignment path of the two time series, the GA kernel function takes into account all the possible alignments between two time series. The resulting kernel matrix is guaranteed to be

© Springer International Publishing AG 2017
S.-H. Lai et al. (Eds.): ACCV 2016, Part III, LNCS 10113, pp. 433–451, 2017.
DOI: 10.1007/978-3-319-54187-7_29

positive definite. GA kernel has shown promising results in face emotion classification given the non-rigid 2D deformations of facial landmarks [31]. With a focus on time series alignment, [40] proposes the temporal matching (TM) kernel to align videos efficiently. The TM kernel generalizes the circulant temporal encoding (CTE) [44] to consider the cross-correlation of two series of vectors in the Fourier domain.

In this paper, we assume that each time series is generated by a vector autoregressive model (VAR) of order p and unknown parameters, and formulate a time series kernel function to compare the generating VAR models without any costly system parameter identification [38] or sophisticated data embeddings [44].

The VAR(p) model assumption for data generation is not novel and has been adopted in several former works. In particular, in [4,47,52] VAR model parameters of each time series are explicitly estimated and used to discriminate between different classes within a SVM framework [4,52] or a NN classifier [47].

In contrast to these works, the autoregressive kernel (AR kernel) in [10] does not require of any system identification to compare time series. Under the VAR model assumption, the AR kernel is defined as the product kernel [21] of data posterior probability density functions of the two time series. However, the AR kernel suffers of numerical issues when dealing with high-dimensional time series and tends to produce diagonally dominant kernel matrices, which in turn may yield to serious difficulties during the learning stage of kernel machines.

In recent works [24,28,30], the observed time series are described by means of Hankel matrix-based representations denoted Hankelets. The main motivation behind the adoption of this representation is that Hankel matrices embed the system parameters and also represent the subspace where the trajectories lie [24]. The dissimilarity score in [24] is used to compare two Hankelets by approximating the cosine of the principal angles of the two different subspaces. Despite the dissimilarity score is not a distance, it has been successfully used for face emotion recognition within NN classifiers [28], and for action classification in [24,30] within a SVM framework and a discriminative HMM respectively. In order to adopt Hankelet-based representation of time series with SVM, bag-of-Hankelets and codebook-based representations are used in [24,28] respectively. At the best of our knowledge, Hankelets have never been used directly within a kernel machine due to the lack of a proper kernel function. This paper aims at filling this gap by defining a kernel function for Hankelets and, hence, a kernel for time series. We will discuss the relation between our proposed kernel and the dissimilarity score in [24], and the relation between our time series kernel function and the Matrix Cosine Similarity (MCS) in [50].

This paper will discuss, in order, the following main contributions:

- the interpretation of a (unnormalized) Hankelet in terms of precision matrix of the parameter posterior when assuming a Gaussian Autoregressive model for data generation;
- a time series kernel function for Hankelets (TSK4H). We formally show the relation between our kernel and formerly proposed scores;

– extensive validation of our approach in different settings to empirically show the generality of our approach.

In our experiments we focus on action classification and emotion recognition, and test our kernel within a standard SVM framework on publicly available benchmarks. In both application domains, we considered two different kinds of input data: (1) trajectories of 2D/3D landmarks, and (2) trajectories of visual features extracted from RGB videos. In our experiments, the adoption of our kernel function with SVM yields to comparable or superior performance with respect to other works at-the-state-of-the-art, but consistently outperforms GA and AR kernels.

2   Representing Time Series by Means of Hankelets

In recent years, there has been a growing interest into the representation of time series dynamics by Hankel matrices [24, 27, 28, 30]. In [24], a truncated block-Hankel matrix H represents the time series $Y = [y_1, \ldots, y_\tau]$ as follows:

$$
H = \begin{bmatrix}
y_1, & y_2, & y_3, & \ldots, & y_m \\
y_2, & y_3, & y_4, & \ldots, & y_{m+1} \\
\ldots & \ldots & \ldots & \ldots & \ldots \\
y_p, & y_{p+1}, & y_{p+2}, & \ldots, & y_\tau
\end{bmatrix}.
\tag{1}
$$

The implicit assumption is that the time series Y has been generated by a linear time invariant (LTI) system

$$
\begin{aligned}
x_k &= A \cdot x_{k-1} + \epsilon_{k-1}; \\
y_k &= C \cdot x_k,
\end{aligned}
\tag{2}
$$

where x_k represents the internal state of the system, the matrices A and C are the system and output matrices respectively, and ϵ_k is uncorrelated zero mean Gaussian noise. While the time series Y can be observed, x_k and ϵ_k are not, and A and C are unknown.

The main justification about the use of Hankel matrices as dynamics representation is that each Hankel matrix embeds the observability matrix $\Gamma = [CA^\tau, \ldots, CA, C]$ of the LTI system that has generated the time series. Indeed, $H = \Gamma \cdot X$, where $X = [x_1, \cdots, x_\tau]$ is the matrix formed by the sequence of internal states of the LTI system [30]. Former works such as [24, 30, 47] normalize the Hankel matrix H as follows:

$$
\hat{H} = \frac{H}{\sqrt{||H \cdot H^T||_F}}.
\tag{3}
$$

Finally, we note that HH^T (which is denoted Hankelet in [24]) plays a central role into the least square estimation of the AR model parameters [54].

2.1   Probabilistic Interpretation of Hankelets

Let us consider a vector autoregressive model of order p defined as follows:

$$y_k = \sum_{i=1}^{p} A_{p-i+1} \cdot y_{k-i} + \epsilon_k \qquad (4)$$

where $y_k \in \mathbb{R}^d$, $A_i \in \mathbb{R}^{d \times p}$ for $1 < i < p$, and $\epsilon_k \sim \mathcal{N}(0, V)$. Another equivalent formulation of the VAR model is

$$y_k = A \cdot x_k + \epsilon_k \qquad (5)$$

with $A = [A_1, \ldots A_p] \in \mathbb{R}^{d \times dp}$ and $x_k^T = [y_{k-p}^T, \ldots, y_{k-1}^T]$.

Due to the Gaussian VAR hypothesis, y_k follows a normal distribution, i.e. $p(y_k|A, x_k, V) = \mathcal{N}(A \cdot x_k, V)$. Let us denote the set of $m - 1$ vectors in a temporal window $Y = [y_{p+1}, \ldots, y_\tau]$ ($Y \in \mathbb{R}^{d \times m-1}$) and $X = [x_{p+1}, \ldots, x_\tau]$ ($X \in \mathbb{R}^{dp \times m-1}$). Due to the Markov property, the joint density of Y is

$$p(Y|A, X, V) = \frac{1}{(2\pi)^{\frac{md^2}{2}} |V|^{\frac{C}{2}}} e^{-\frac{1}{2} \mathrm{Trace}((Y-AX)^T V^{-1}(Y-AX))}. \qquad (6)$$

By taking a closer look at X and comparing it with Eq. 1, we find that X is the (unnormalized) Hankel matrix H of the sequence of predictors (i.e. past values), i.e. $[y_1, \cdots, y_{\tau-1}]$.

Furthermore, if we consider a normal matrix distribution prior for A with zero mean and V equals to the noise covariance matrix [10], $A \sim \mathcal{NM}_{d,dp}(0, \Sigma, V)$, then it is possible to express the posterior distribution over A as a normal matrix distribution, namely:

$$A|Y, X \sim \mathcal{NM}_{d,dp}(M, U, W); \qquad (7)$$
$$M = YX^T U; \qquad (8)$$
$$W = V; \qquad (9)$$
$$U = (XX^T + \Sigma^{-1})^{-1}. \qquad (10)$$

As a consequence, the precision matrix (or inverse covariance matrix) of the parameter posterior can be rewritten as follows:

$$U^{-1} = (HH^T + \Sigma^{-1}). \qquad (11)$$

If the prior for A has a precision matrix equals to $\Sigma^{-1} = \alpha I$ with $0 < \alpha << 1$ (i.e., the prior is mostly uninformative), U^{-1} can be interpreted as a regularized version of the matrix HH^T.

In this sense, the comparison of (unnormalized) Hankelets entails the comparison of precision matrices of the two parameter posterior densities of the underlying Gaussian Processes. It is well known that elements of a precision matrix represent partial covariances of pairs of variables, that is they measure how two variables covariate conditionally on the remaining ones.

In our application, each precision matrix refers to the parameter posterior and its elements reflect statistical links on the parameters of the VAR(p) model referring to different time lags and components of the time series vectors. Hence, given two time series, comparison of the corresponding Hankelets allows comparison of how the model parameters conditionally covariate in the two underlying Gaussian Processes.

3   Time Series Kernel for Hankelets

The comparison of two Hankelets aims to establish if the corresponding time series might have been generated by similar or even the same VAR model. The lack of a suitable kernel function has limited the adoption of Hankelets within kernel machine frameworks. This paper aims at filling this gap by deriving a suitable kernel function for Hankelets.

In a nutshell, we propose to adopt a very popular kernel function, the cosine similarity kernel [46] that, as we will show in Sects. 3.1 and 3.2, assumes a special meaning for Hankelets. Given two vectors u and v, the cosine similarity kernel is defined as follows:

$$K(u, v) = \frac{<u, v>}{\sqrt{<u, u>}\sqrt{<v, v>}}. \tag{12}$$

We now rely on a well-known relation between the vectorization operator of a matrix, $\text{vec}(A) : \mathbb{R}^{n \times m} \to \mathbb{R}^{nm \times 1}$, and the Frobenius dot product of matrices, $< \cdot, \cdot >_F$. For a matrix A, it holds that

$$< \text{vex}(A), \text{vex}(A) >= \text{Trace}(A^T A) =< A, A >_F = ||A||_F^2. \tag{13}$$

Given two matrices A and B both in $\mathbb{R}^{n \times m}$, their Frobenius dot product is defined as $< A, B >_F = \text{Trace}(A^T B)$.

Let us assume that $A = H_p H_p^T$ and $B = H_q H_q^T$ are two (unnormalized) Hankelets for the time series Y_p and Y_q respectively. In this special case:

$$< H_p H_p^T, H_q H_q^T >_F = \text{Trace}(H_p H_p^T H_q H_q^T) = ||H_p^T H_q||_F^2 \tag{14}$$

and it turns out that the cosine similarity kernel of two Hankelets is

$$K(H_p H_p^T, H_q H_q^T) = \frac{<H_p H_p^T, H_q H_q^T>_F}{\sqrt{<H_p H_p^T, H_p H_p^T>_F}\sqrt{<H_q H_q^T, H_q H_q^T>_F}} \tag{15}$$

$$= < \hat{H}_p \hat{H}_p^T, \hat{H}_q \hat{H}_q^T >_F = ||\hat{H}_p^T \hat{H}_q||_F^2 \tag{16}$$

which is a valid, separable, positive definite kernel function for Hankelets.

In the statistics literature, the measurement in Eq. 16 is known as RV-coefficient or vector correlation [1,50,53]. The RV-coefficient was proposed as a measure of similarity between positive semi-definite matrices and as a theoretical tool to analyze multivariate techniques [1]. The RV-coefficient measures the alignment of the subspaces represented by positive semi-definite matrices and is invariant to rotation transformations [53]. Given a data matrix X, by denoting with $\Sigma_{X_p X_q} = X_p^T X_q$, the RV-coefficient coincides with our kernel function when $X = H$, that is:

$$RV(X_p, X_q) = \frac{\text{Trace}(\Sigma_{X_p X_q} \Sigma_{X_q X_p})}{\sqrt{\text{Trace}(\Sigma_{X_p X_p} \Sigma_{X_p X_p})}\sqrt{\text{Trace}(\Sigma_{X_q X_q} \Sigma_{X_q X_q})}} = K(H_p H_p^T, H_q H_q^T).$$

Finally, our kernel relates to the one proposed in [16] where data points are compared on the Grassmannian manifold. In [16], it is also proposed to embed points X to XX^T, which is similar to the unnormalized Hankelet. This embedding suffers of numerical issues when high-dimensional time series are considered. In our kernel definition, each Hankelet is normalized by means of its Frobenius norm, and a different rescaling is applied to each pair of time series.

3.1 Relation with Dissimilarity and Similarity Scores

In [24], two normalized Hankelets $\hat{H}_p \hat{H}_p^T$ and $\hat{H}_q \hat{H}_q^T$ are compared by the dissimilarity score defined as follows:

$$d(\hat{H}_p \hat{H}_p^T, \hat{H}_q \hat{H}_q^T) = 2 - ||\hat{H}_p \hat{H}_p^T + \hat{H}_q \hat{H}_q^T||_F. \tag{17}$$

An equivalent form of the dissimilarity score that takes advantage of the normalization in Eq. 3 (see [24]) is:

$$d(\hat{H}_p \hat{H}_p^T, \hat{H}_q \hat{H}_q^T) = 2 - \sqrt{2 + 2||\hat{H}_p^T \hat{H}_q||_F^2} \tag{18}$$

and, by considering the SVDs of the two Hankelets, this score can be regarded as an approximation of the cosine of the principle angles between the two subspaces [24,30]. Based on Eq. 18, the work in [27] proposes a similarity score:

$$s(\hat{H}_p \hat{H}_p^T, \hat{H}_q \hat{H}_q^T) = ||\hat{H}_p^T \hat{H}_q||_F. \tag{19}$$

By comparing Eq. 19 to our kernel formulation in Eq. 16, we get that

$$K(H_p H_p^T, H_q H_q^T) = s(\hat{H}_p \hat{H}_p^T, \hat{H}_q \hat{H}_q^T)^2. \tag{20}$$

In practice, the cosine similarity kernel of two Hankelets is the squared similarity score. It measures the cosine of the angle between the two vectorized Hankelets and, since the same considerations in [24] hold also in our case, the cosine similarity kernel of two Hankelets might be regarded as an approximation of the cosine of the principal angles of the two subspaces.

3.2 Relation with the Matrix Cosine Similarity

The work in [50] proposes to compare matrices of features by means of the matrix cosine similarity (MCS) score. Features f_i are extracted from local neighborhood and stacked into columns of a feature matrix F. Two feature matrices F and Q (of same size) are compared by means of the MCS defined as

$$MCS(F, Q) = \frac{<F, Q>_F}{\sqrt{<F, F>_F}\sqrt{<Q, Q>_F}}. \tag{21}$$

MCS is a generalized version of the vector cosine similarity designed to compare matrices of same size.

Our time series kernel function differs from the MCS in several respects. (1) In contrast to [50], which deals with objects detection and stacks in a unique matrix a set of visual features, our kernel function applies to Hankelets computed upon vector time series (or any ordered sequence of features); (2) In [50], there is not an underlying model for the generation of the feature matrices F and Q. In contrast, in our formulation we assume that the two time series are generated by VAR(p) models, and the kernel function aims at measuring model similarities rather than feature similarities. (3) In [50], the MCS is a RV-coefficient if F and Q are positive semi-definite matrices, which might not be true in general. Hankelets are positive semi-definite symmetric matrices and our kernel estimates exactly the RV-coefficient of the two matrices. This is of interest because, in this case, the RV-coefficient measures communality between the two subspaces even in high-dimensional data [53]. (4) Finally, the MSC score can only compare sets of feature descriptors of same size (same number of feature vectors). Our TSK4H can compare vector time series of different lengths.

3.3 Time and Space Complexity

A Hankel matrix H of maximal order p of a d-dimensional vector time series is built by a simple reordering of the time series elements as shown in Eq. 1. Computing the Frobenius norm of HH^T when H has size $dp \times m$ with $dp >> m$, considering that $||HH^T||_F = ||H^T H||_F$, has a cost of $O(m^2(dp + 1))$. The time complexity of evaluating our kernel function is of about $O(m^2(dp + 1))$, that is linear in the dimension of the vector time series and quadratic in the number of columns of H. Finally, storing a single Hankel matrix H has a space complexity of $O(dpm)$ and is more convenient than storing HH^T ($O(\frac{d^2 p^2}{2})$).

4 Applications

Our kernel function may be adopted in any domain where time series (or any ordered sequence of features) arise. Here, we detail how our TSK4H can be used in two challenging applications: face emotion recognition and action classification. In each of these applications, we will extract a time series of per-frame feature vector to represent a video depicting a face emotion or an action respectively. A set of these time series will be used to train a standard SVM[1] by employing our TSK4H. To perform multi-class classification, we will consider a

---

[1] In our experiments, we used the publicly available library LIBSVM [3].

1-vs-all classification schema. Due to the VAR(p) model assumption, each time series has to be made zero mean before evaluating our kernel. Whenever the Hankel matrices have a dimensionality higher than the training set size, each SVM can be trained in the dual space. In the following, we briefly describe both the adopted visual features extracted to represent RGB videos, and features that are more application domain dependent.

Visual Features. We adopt two widely used per-frame descriptors: Haar-like features [57] and HoG features [12].

We extract the 6 basic Haar-like features[2] used in [27] from 13×13 non-overlapping regions of same size from each image. The extracted features are stacked into one vector of dimension 1014. A video of N frames is represented by a 1014-dimensional time series of length N.

We extract HoG features[3] from blocks of size 32 pixels, cells of size 16 pixels and a block stride (shift) of 16 pixels. Before extracting the HoG features, images are resized to a multiple of the block size. A video of N frames is represented by a M-dimensional time series of length N, with M dependent on the number of blocks.

Face Emotion Recognition deals with the problem of inferring the emotion (i.e., fear, anger, surprise, etc.) given a sequence of face images, and suffers of strong inter-subject variations, illumination changes, biometric differences, head pose changes, etc. Useful literature reviews on the topic are [48,62].

A common approach is that of representing a face expression through 2D face landmark coordinates estimated, for instance, by an active appearance model [8]. Following [28], we represent a sequence of face expressions by means of vector time series of: (1) concatenated 2D facial landmark coordinates (L); (2) pairwise landmark distances (D); (3) concatenation of pairwise landmark distances and landmark coordinates (L+D).

Since face landmark detection is still an open problem, it is appealing the adoption of visual features extracted from face regions, such as Haar-like and HoG features. In contrast to [27], which extracts Haar-like features from different spatial windows within the face image and builds a Hankelet for each kind of Haar-like template, we concatenate all Haar-like features in a single vector. When adopting HoG features, each face image is resized to 160×128 and the per-frame descriptor has a size of 2268.

Action Classification entails the problem of assigning an action label to a sequence of per-frame feature descriptors and, in general, it suffers of the following issues: difficulties in reliably describing human poses, biometric differences, subjective velocities and characteristics (i.e. different human gaits), illumination

---

[2] We used the public implementation available within the Struck tracking method [17], which is the one suggested in [27].

[3] We used the implementation available with the OpenCV library [2].

changes and clothing variation (especially in RGB videos), etc. More details on these challenges can be found in popular literature reviews [5,39,55].

In recent years, there has been a proliferation of works about action classification based on sequences of skeletons obtained from MoCap data [33] or estimated from depth data [51]. Our kernel function applies to this kind of application as well. We will consider the case in which action samples are described as sequences of 3D body joints, and the case in which action samples are described as time series of per-frame descriptors (Haar-like or HoG features) extracted from bounding boxes of the detected persons. Before extracting HoG features, each bounding box is resized to 128×64 to better preserve its aspect ratio. The resulting HoG descriptor is of 756 features.

5 Experimental Results

In the following, we will refer to our method as to time series kernel for Hankelets $TSK4H(f)$, where f is the feature type that has been considered.

We have compared our kernel function to the GA and the AR kernels [10][4] on equal terms of features. In our experiments, the AR kernel performed very poorly compared with both our kernel and the GA kernel and we decided to drop these results. We believe the poor performance of the AR kernel might be ascribable to the high data dimensionality that raises serious numerical issues. As for the GA kernel, we have found benefits in adopting the normalized GA kernel (NGAK) [9]. NGAK has two parameters: the bandwidth σ of the exponential kernel, and T that regulates the triangular weighting function within the kernel. This weighting function is necessary to take into account the level of warping needed to align the two time series. In our experiments we used brute force to set these parameters by testing various parameter combinations. Similarly to [31], we set $\sigma = 2^s$ and let s varying in $[0, 20]$ with step 2. Moreover, we let T varying in $[0, 14]$ with step 2 ($T = 0$ indicates that no triangular weighting function is used). As a result, a total number of 88 parameter combinations have been tested. For each experiment, we report the parameters and the accuracy values in classification corresponding to the best parameter combination. We also report the accuracy values achieved with NGAK combined with a NN classifier. We will use the notation $NGAK(f, \sigma, T)$ to indicate that the NGA kernel was computed on sequences of features f with parameters (σ, T).

We have further implemented a baseline method that, given a pair of time series, aligns them with DTW by maximizing the cosine similarity of pairs of vectors. The similarity value of the best alignment was normalized by the length of the aligned sequences. As explained in [11], a DTW kernel is not guaranteed to be positive definite. Therefore we adopt the NN classifier over this similarity score. We refer to this baseline method as to DTW-S(f), where f is the considered feature type.

---

[4] Both the code of the AR Kernel and of the (normalized) GA kernel are publicly available at Dr. Cuturi's website.

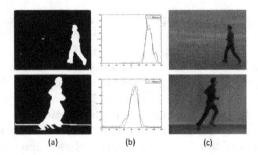

Fig. 1. Examples of bounding boxes extracted for two frames in the KTH dataset: (a) the mask obtained by applying a threshold on the gradient magnitude, (b) the Gaussian curve fitted on the sum of binary values along the columns, (c) the final bounding box centered on the mean of the Gaussian curve.

We will use $HH^T(f) + NN$ to indicate that Hankelets have been computed on the feature f and classified by the NN classifier. During training we set the parameter C of our SVM to 1000. In all of our experiments, when applying PCA, we skipped the first (less discriminative) component and retained 99% of the total variance. Coefficients of the PCA were estimated on the training set, then the test set was transformed accordingly.

Datasets. In experiments for face emotion recognition, we adopted the widely used Extended Cohn-Kanade dataset (CK+) [32], which provides 327 video sequences of 118 different individuals displaying 7 emotions: *angry (A), contempt (C), disgust (D), fear (F), happy (H), sadness (Sa), surprise (Su).* The number of frames of these sequences ranges in $[6, 71]$ with an average value of about 18 ± 8.6. Considering that some sequences have very few frames, to guarantee a fair comparison with other works and to use all the sequences in the dataset, the order p of the Hankelets was set to 3 (with $p = 3$, at least $2 \cdot p - 1 = 5$ frames are needed to build a Hankelet). The adopted validation protocol is leave-one-subject-out cross-validation. The CK+ dataset provides landmark tracking results estimated by an active appearance model, which we use in our experiments as 2D face landmarks as also proposed in [28,32].

In experiments for action classification based on 3D skeletons, we have adopted the UCF dataset [13], which provides skeletons of 15 joints for 16 actions performed 5 times by 16 subjects. It comprises 1280 action sequences of length in $[27, 229]$, (on average 6634 frames). The actions are: *balance (B), climbladder (CR), climbup (CP), duck (D), hop (H), kick (K), leap (L), punch (P), run (R), stepback (SB), stepfront (SF), stepleft (SL), stepright (SR), twistleft (TL), twistright (TR), and vault (V).* The adopted protocol on this dataset is 4-fold cross-subject validation: subjects are split in 4 subsets; at each run, 3 subsets are used for training the models, the remaining subset is used in test.

Finally, in experiments for action classification based on visual features extracted from RGB videos, we have adopted the KTH dataset [49], which con-

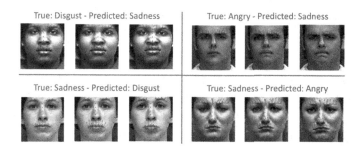

Fig. 2. Misclassified samples from the CK+ dataset. Blue dots represent the 2D facial landmarks. Red arrows show the velocities of each landmark. (Color figure online)

tains six types of human actions, *boxing (B), hand clapping (HC), hand waving (HW), jogging (J), walking (W), running*, performed several times by 25 subjects in four different scenarios: outdoors, outdoors with scale variation, outdoors with different clothes, and indoors. The dataset contains 2391 sequences with a spatial resolution of 160×120 pixels. The length of the sequences (after people detection) ranges in $[12, 362]$, with an average length of about 81.9 ± 43.1 frames. Considering the minimum length of the sequences, the order of the Hankel matrix was set to $p = 6$. The adopted protocol is leave-one-subject-out cross-validation. Bounding boxes for person detection (shown in Fig. 1) were computed with a simple detector that: computes the gradient magnitude at each frame, performs a morphological closing operation with a line structuring element (to highlight vertical edges), applies a threshold to the resulting image, computes the sum of the binary pixels across the rows, fits a Gaussian and takes the bounding box centered on the Gaussian mean and with a width proportional to the standard deviation (which corresponds to a Gaussian-based peak detector).

Face Emotion Recognition. On the CK+ dataset, when adopting 2D face landmarks, PCA was used to filter out noise and reduce data redundancy, and provided per-frame vectors of average size 88, 123, 127 for L, D, and L+D features respectively. We report our results and comparison with former works in Table 1. The first column describes the type of features used to represent the time series and the classifier. The other columns indicate the emotion labels, and the last column reports the average per-class accuracy value.

The first and second parts of the table reports the results described in [28] when adopting a NN classifier and a codebook-based SVM (CSVM). The third part of the table reports the results achieved when using the NGAK in SVM and NN classifiers. The fourth part of the table reports the accuracy values achieved with our baseline method DTW-S while the fifth part of the table shows the accuracy values achieved by adopting our time series kernel for Hankelets (TSK4H). Finally, the bottom part of the table reports the accuracy values of other works at the state-of-the-art that uses 2D facial landmark time series.

Table 1. Accuracy values (in %) for the Emotion Recognition task on the CK+ dataset when using 2D facial landmarks. In bold font the highest accuracy values.

Emotions:	A	C	D	F	H	Sa	Su	Avr.
$HH^T(L)$+NN [28]	82.2	77.8	94.9	80	**100**	64.3	97.6	85.3
$HH^T(D)$+NN [28]	88.9	**83.3**	**96.6**	84	**100**	67.9	98.8	88.5
$HH^T(L+D)$+NN [28]	91.1	**83.3**	94.9	84	**100**	71.4	98.8	89.1
$HH^T(L)$+CSVM [28]	86	75	92	85.6	98.3	74.3	95.9	86.7
$HH^T(D)$+CSVM [28]	89.1	72.8	92.4	**89.6**	97	80.7	97.2	88.4
$HH^T(L+D)$+CSVM [28]	89.8	73.9	90.8	89.2	97.4	81.8	97.7	88.7
NGAK(L, 2^6, 10)+NN	62.2	77.8	83	56	97.1	50	94	74.3
NGAK(D, 2^8, 10)+NN	73.3	88.9	88.1	68	97.1	75	94	83.5
NGAK(L, 2^8, 14)+SVM	86.7	72.2	94.9	64	95.6	78.6	96.4	84.1
NGAK(D, 2^{10}, 0)+SVM	88.9	77.8	91.5	72	97.1	85.7	95.2	86.9
DTW-S(L)+NN	86.7	72.2	93.2	68	**100**	53.6	97.6	81.6
DTW-S(D)+NN	88.9	83.3	96.6	68	**100**	60.7	97.6	85
TSK4H(L)+SVM [ours]	86.7	**83.3**	94.9	88	98.5	**82.1**	97.6	90.2
TSK4H(D)+SVM [ours]	91.1	**83.3**	**96.6**	88	**100**	78.6	98.8	**90.9**
CK+ [32]	35	25	68.4	21.7	98.4	4	**100**	50.4
CLM-based [6]	70.1	52.4	92.5	72.1	94.2	45.9	93.6	74.4
LRBM [35]	**97.8**	72.2	89.8	84	**100**	78.6	97.6	88.6
ITBN [58]	91.1	78.6	94	83.3	89.8	76	91.3	86.3

In contrast to [28], in our experiments we did not notice any significant difference when using the features D or $L+D$ and we dropped the results obtained with $L+D$. As the table shows, the adoption of a discriminative method such as SVM over Hankelets by means of our TSK4H allows us to get an increase in the emotion classification accuracy values.

By comparing the results achieved with the CSVM [28] and with our approach, the gain in accuracy value is of about 3.4% (average gain over L and D). We stress that our approach does not require a codebook learning stage. The gain in accuracy value is even higher when comparing to the NGAK and DTW-S (5.9% and 8.7% respectively – average gains over L and D).

Irrespectively of the adopted representations (L, D, L+D), most of the confusion in our experiments was between the classes *sadness* and *disgust*, and the classes *angry* and *sadness*. Some misclassified samples are shown in Fig. 2. This kind of mistakes was formerly observed in [29,30]. In practice, 2D landmark trajectories corresponding to raising and lowering eyebrows/lips tend to be mirrored versions of each other, and yield to similar Hankelet representations.

We report in Table 2 the experimental results obtained when adopting Haar-like and HoG features as per-frame face expression descriptors. Adoption of PCA resulted in 94 and 807 (average) dimensional per-frame vectors for Haar-like and HoG features respectively.

Table 2. Accuracy values in Emotion Recognition (CK+) with visual features. Bold font highlights the highest accuracy values.

Emotions:	A	C	D	F	H	Sa	Su	Avr.
Ens. of H(Haar-like)+NN [27]	86.7	83.3	96.6	52	**100**	71.4	97.6	83.9
HH^T(Haar-like)+NN	60	77.8	93.2	56	**100**	64.3	96.4	78.2
HH^T(HoG)+NN	57.8	61.1	**100**	72	97.1	64.3	98.8	78.7
NGAK(Haar-like, 1, 12)+NN	55.6	88.9	79.7	28	87	60.7	83.1	69
NGAK(HoG, 2^2, 12)+NN	26.7	88.9	54.2	12	49.3	85.7	59	53.7
NGAK(Haar-like, 2^2, 0)+SVM	80	88.9	88.1	56	95.7	75	97.6	83
NGAK(HoG, 2^4, 0)+SVM	84.4	72.2	94.9	64	98.6	85.7	98.8	85.5
DTW-S(Haar-like)+NN	71.1	77.8	96.6	52	**100**	71.4	98.8	81.1
DTW-S(HoG)+NN	71.1	50	**100**	64	98.5	53.6	98.8	76.6
TSK4H(Haar-like)+SVM [ours]	91.1	83.3	98.3	72	**100**	85.7	97.6	89.7
TSK4H(HoG)+SVM [ours]	86.7	83.3	98.3	88	**100**	75	98.8	90.01
CAPP+SVM [32]	70	21.9	94.7	21.7	**100**	60	98.7	66.7
LDN+RBF-SVM [45][a]	71.7	73.7	93.4	**90.5**	95.8	78.9	97.6	85.9
LBP+CC+SVM [20]	**93**	**89**	98	80	**100**	**86**	**100**	**92.3**

[a] 10-fold cross validation

As Table 2 shows, we achieve state-of-the-art results on the CK+ dataset when adopting visual features. When using Haar-like features, the adoption of SVM over Hankelets allows us to achieve a higher accuracy value with respect to the work in [27], and to NN classifier (the gain in accuracy value is of about 6.9% and 14.7%, respectively). With respect to the NGAK, the gain in accuracy values of our TSK4H is of 8.1% and 5.3% on Haar-like and HoG features respectively. Overall, the average accuracy values reached with our TSK4H on Haar-like features and HoG features are close each other.

Action Classification. On the UCF dataset, application of PCA over skeletal data resulted in 28-dimensional per-frame descriptors (on average). For comparison purposes, we test our approach with order $p = 4$ as used in [30], and further report results achieved with the highest possible value of $p = 14$ considering the minimum length of the sequences.

Table 3 compares the results of different approaches. The method in [30] adopts discriminative HMMs over small temporal windows described in terms of Hankelets of order 4. With respect to this method, our approach achieves a gain in accuracy values of 1% on equal terms of order. We stress here that, in contrast to [30], we calculate Hankelets over entire sequences. The gain in accuracy values when $p = 14$ is of about 1.8%. With respect to NGAK, on equal terms of classification framework (SVM), the gain in accuracy value of our TSK4H is of about 2.6%. Interesting, in this experiment, NGAK+NN performs better than NGAK+SVM. Differently than the experiments presented for emotion

Table 3. Accuracy values in Action Recognition (UCF dataset) with 3D body joints trajectories. Bold font highlights the highest values.

Actions:	B	CR	CP	D	H	K	L	P	R	SB	SF	SL	SR	TL	TR	V	Avr.
HH^T+ DHMM [30]	99.9	98.7	96.9	98.6	96.9	98.5	95	98.4	98.5	97.9	99.3	98.1	97.5	93.6	92.9	97.8	97.4
HH^T+NN (p = 4)	100	98.8	98.8	100	96.3	100	100	100	96.3	91.3	78.8	85	93.8	98.8	100	96.3	95.9
HH^T+NN (p = 14)	100	98.8	100	100	97.5	100	100	100	93.8	98.8	98.8	97.5	98.8	98.7	100	**98.8**	98.8
NGAK(1,0)+ NN	98.7	98.7	100	100	100	100	97.5	100	95	100	100	98.7	98.7	100	98.7	98.7	99
NGAK(4,0)+ SVM	83.7	98.7	96.2	100	98.7	100	97.5	97.5	88.7	100	100	98.7	97.5	96.3	97.5	96.3	96.7
DTW-S+NN	100	100	98.8	100	95	100	100	98.8	98.8	100	100	97.5	98.8	100	100	**98.8**	98.5
TSK4H (p = 4)	100	98.8	100	100	100	98.8	98.8	100	97.5	98.8	95	96.3	97.5	97.5	98.8	97.5	98.4
TSK4H (p = 14)	100	98.8	100	100	100	100	98.8	100	97.5	100	98.8	98.8	97.5	100	100	97.5	**99.2**
Log. Reg. [13]	97.5	93.8	98.8	100	96.2	98.8	100	95	97.5	97.5	97.5	96.2	98.8	88.8	86.2	92.5	95.9
LTBSVM [52][a]	100	100	93.3	100	93.3	100	96.7	100	100	93.3	100	100	100	100	93.3	96.7	97.9

[a] 70–30% cross validation protocol.

recognition, on the UCF dataset NGAK performs similarly to our kernel. We note that the vector dimensionality in this experiment is of about 28, and it is much lower than the feature representation dimensionality of the other experiments we present. This suggests that high-dimensionality may have a negative impact on NGAK while our kernel function seems to be less affected by the vector dimensionality. To verify this, we performed a further experiment on the KTH dataset.

On the KTH dataset, application of PCA resulted in 460 and 517 dimensional per-frame vectors for Haar-like and HoG features respectively. Table 4 reports the achieved accuracy values in classification. Overall, our method achieves the same accuracy value of [59]. The work in [59] proposes a shape-motion prototype-based approach. An action is represented as a sequence of prototypes. Prototypes are trained via K-means clustering and, at test time, are inferred by maximizing a model conditional probability. Given the sequence of prototypes, classification is performed by applying DTW and K-NN classifier. We note that the whole framework in [59], which is a composition of methods, has a higher time complexity than our approach due to the need of performing DTW and comparing with the sequences in the training set for applying KNN. With respect to the NGAK on equal terms of classification framework (SVM) and feature representations, our method allows to obtain a gain in the accuracy values of about 20.3% and 22.8% on Haar-like and HoG features respectively. The table also shows that, on this dataset, our method works better on HoG features rather than Haar-like features. By inspecting the confusion matrices, when adopting Haar-like features, most of the confusion is between the classes *jogging* and *running*: 14.75% of sequences in the class *jogging* are recognized as *running*, while 16.25% of sequences in the class *running* are classified as *jogging*. Such percentages reduce to 5.25% and 8.5% respectively when adopting HoG features. These results suggest that Haar-like features might not be suitable to represent fine-grained differences in the body poses of these two actions.

Table 4. Accuracy values in Action Classification on the KTH dataset when using all the scenarios in leave-one-subject-out cross-validation.

Actions:	B	HC	HW	J	R	W	Avg.
HH^T(Haar-like)+NN	90	90	93	61	53	89	79
HH^T(HoG)+NN	93	99	98	84	73	98	91
NGAK(Haar-like, 2^2, 0)+NN	24	73	50	70	73	74	61
NGAK(HoG, 2^2, 8)+NN	26	81	69	62	55	52	57
NGAK(Haar-like, 2^8, 0)+SVM	47	72	76	81	77	90	74
NGAK(HoG, 2^8, 0)+SVM	65	72	81	82	81	95	79
DTW-S(Haar-like)+NN	82	87	95	57	53	83	76
DTW-S(HoG)+NN	87	97	**99**	87	78	97	91
TSK4H(Haar-like)+SVM [ours]	97	95	89	77	84	94	89
TSK4H(HoG)+SVM [ours]	99	98	98	**95**	91	99	**97**
Descriptor-based [22]	96	99	86	91	85	92	92
MSRR [63]	98	97	**99**	90	90	**100**	96
SMP+DTW+KNN [59]	**100**	**100**	99	90	**93**	**100**	**97**

6   Conclusion

Recent works [24, 27, 28, 30] have successfully adopted Hankelets as a time series dynamics representation especially in NN classifiers. This paper discusses a probabilistic interpretation of Hankelets in terms of precision matrix of the Gaussian process that generates the time series. Based on this interpretation, comparison of Hankelets turns into the comparison of partial covariances of the VAR(p) model parameters.

Furthermore, this paper proposes a time series kernel function for Hankelets, which is the cosine similarity kernel function of the vectorized Hankelets. This paper shows that: (1) the proposed TSK4H measures the angle of the vectorized Hankelets, and hence of the vectors of partial covariances of the model parameters; (2) the proposed kernel coincides with the RV-coefficient used in statistics to measure the similarity between positive semi-definite matrices; (3) TSK4K has a relation with the dissimilarity score proposed in [24] but, in contrast to it, our TSK4H defines a valid positive definite kernel that allows the use of kernel machines directly over Hankelets; this offers the advantage of skipping the codebook generation step that was necessary in [24, 28] in order to adopt SVM. Finally, similarly to the score in [24], TSK4H approximates the cosine of the principal angles of the two subspaces.

Our extensive validation in action and emotion classification suggests that TSK4H is robust to numerical issues in high-dimensional spaces, and provides high accuracy values irrespectively of the adopted feature representation. In our experiments, TSK4H consistently outperforms other time series kernels such as GA and AR kernels. Time complexity of the GA kernel is claimed to be of about

$O(n_p n_q d)$ with n_i indicating the length of the i-th time series [9], and d is the vector dimension. Time complexity of the AR kernel is $O((p + 1)dN^2 + N^3)$ with $N = \max(n_p, n_q)$, and p the order of the assumed VAR model [10]. Our TSK4H and the GA kernel have comparable time complexity (see Sect. 3.3 for the complexity of TSK4H). However, in practice we have noticed that computation of the TSK4H (implemented in Matlab) seems faster than the computation of the GAK (publicly available C++ implementation). The main reason might be that the claimed complexity for the GA kernel does not consider the local kernel computation that, in high dimensions, may greatly affect the complexity.

References

1. Abdi, H.: RV coefficient and congruence coefficient. In: Encyclopedia of Measurement and Statistics, pp. 849–853. Sage, Thousand Oaks (2007)
2. Bradski, G.: The OpenCV library. Dr. Dobb's J. Softw. Tools **25**(11), 120–126 (2000)
3. Chang, C.C., Lin, C.J.: LIBSVM: a library for support vector machines. ACM Trans. Intell. Syst. Technol. (TIST) **2**(3), 27–37 (2011). ACM
4. Chaudhry, R., Ofli, F., Kurillo, G., Bajcsy, R., Vidal, R.: Bio-inspired dynamic 3D discriminative skeletal features for human action recognition. In: Proceedings of the IEEE Conference on Computer Vision and Pattern Recognition Workshops (CVPRW 2013), pp. 471–478. IEEE (2013)
5. Chen, L., Wei, H., Ferryman, J.: A survey of human motion analysis using depth imagery. Pattern Recogn. Lett. 34(15), 1995–2006 (2013). Elsevier
6. Chew, S., Lucey, P., Lucey, S., Saragih, J., Cohn, J., Sridharan, S.: Person-independent facial expression detection using constrained local models. In: Proceedings of Conference and Workshop on Automatic Face and Gesture Recognition (FG), pp. 915–920. IEEE (2011)
7. Cohn, J., Schmidt, K.: The timing of facial motion in posed and spontaneous smiles. Int. J. Wavelets Multiresolut. Inf. Process. **2**(2), 121–132 (2004). World Scientific
8. Cootes, T., Edwards, G., Taylor, C.: Active appearance models. IEEE Trans. Pattern Anal. Mach. Intell. (PAMI) **23**(6), 681–685 (2001). IEEE
9. Cuturi, M.: Fast global alignment kernels. In: Proceedings of International Conference on Machine Learning (ICML), pp. 929–936 (2011)
10. Cuturi, M., Doucet, A.: Autoregressive kernels for time series. arXiv preprint arXiv:1101.0673 (2011)
11. Cuturi, M., Vert, J., Birkenes, O., Matsui, T.: A kernel for time series based on global alignments. In: Proceedings of International Conference on Acoustics, Speech and Signal Processing (ICASSP), vol. 2, pp. 413–420. IEEE (2007)
12. Dalal, N., Triggs, B.: Histograms of oriented gradients for human detection. In: Proceedings of Conference on Computer Vision and Pattern Recognition (CVPR 2005), vol. 1, pp. 886–893. IEEE (2005)
13. Ellis, C., Masood, S.Z., Tappen, M.F., Laviola Jr., J.J., Sukthankar, R.: Exploring the trade-off between accuracy and observational latency in action recognition. Int. J. Comput. Vis. **101**(3), 420–436 (2013). Springer
14. Frank, J., Mannor, S., Precup, D.: Activity and gait recognition with time-delay embeddings. In: Conference on Artificial Intelligence (AAAI) (2010)
15. Gehler, P.V.: Kernel learning approaches for image classification. Ph.D. thesis, Universitat des Saarlandes (2009)

16. Harandi, M.T., Salzmann, M., Jayasumana, S., Hartley, R., Li, H.: Expanding the family of Grassmannian Kernels: an embedding perspective. In: Fleet, D., Pajdla, T., Schiele, B., Tuytelaars, T. (eds.) ECCV 2014. LNCS, vol. 8695, pp. 408–423. Springer, Heidelberg (2014). doi:10.1007/978-3-319-10584-0_27

17. Hare, S., Saffari, A., Torr, P.H.S.: Struck: structured output tracking with kernels. In: Proceedings of International Conference on Computer Vision (ICCV 2011), pp. 263–270. IEEE (2011)

18. Haufe, S., Nolte, G., Mueller, K., Krämer, N.: Sparse causal discovery in multivariate time series. arXiv preprint arXiv:0901.2234 (2009)

19. Hofmann, T., Schölkopf, B., Smola, A.: Kernel methods in machine learning. Ann. stat. **36**(3), 1171–1220 (2008). JSTOR

20. Huang, X., Zhao, G., Pietikainen, M., Zheng, W.: Robust facial expression recognition using revised canonical correlation. In: Proceedings of International Conference on Pattern Recognition (ICPR), pp. 1734–1739. IEEE (2014)

21. Jebara, T., Kondor, R., Howard, A.: Probability product kernels. J. Mach. Learn. Res. **5**, 819–844 (2004). JMLR.org

22. Jiang, Z., Lin, Z., Davis, L.S.: Recognizing human actions by learning and matching shape-motion prototype trees. Trans. Pattern Anal. Mach. Intell. **34**(3), 533–547 (2012). IEEE

23. Lehrmann, A., Gehler, P., Nowozin, S.: Efficient nonlinear Markov models for human motion. In: Proceedings of Conference on Computer Vision and Pattern Recognition (CVPR 2014), pp. 1314–1321. IEEE (2014)

24. Li, B., Camps, O., Sznaier, M.: Cross-view activity recognition using Hankelets. In: Proceedings of Conference on Computer Vision and Pattern Recognition (CVPR 2012), pp. 1362–1369. IEEE (2012)

25. Lin, R.-S., Liu, C.-B., Yang, M.-H., Ahuja, N., Levinson, S.: Learning nonlinear manifolds from time series. In: Leonardis, A., Bischof, H., Pinz, A. (eds.) ECCV 2006. LNCS, vol. 3952, pp. 245–256. Springer, Heidelberg (2006). doi:10.1007/11744047_19

26. Lo Presti, L., La Cascia, M.: An on-line learning method for face association in personal photo collection. Image Vis. Comput. **30**(4), 306–316 (2012). Elsevier

27. Lo Presti, L., La Cascia, M.: Ensemble of Hankel matrices for face emotion recognition. In: Murino, V., Puppo, E. (eds.) ICIAP 2015. LNCS, vol. 9280, pp. 586–597. Springer, Heidelberg (2015). doi:10.1007/978-3-319-23234-8_54

28. Lo Presti, L., La Cascia, M.: Using Hankel matrices for dynamics-based facial emotion recognition and pain detection. In: Proceedings of the IEEE Conference on Computer Vision and Pattern Recognition Workshops (CVPRW 2015), pp. 26–33. IEEE (2015)

29. Lo Presti, L., La Cascia, M., Sclaroff, S., Camps, O.: Gesture modeling by Hanklet-based hidden Markov model. In: Cremers, D., Reid, I., Saito, H., Yang, M.-H. (eds.) ACCV 2014. LNCS, vol. 9005, pp. 529–546. Springer, Heidelberg (2015). doi:10.1007/978-3-319-16811-1_35

30. Lo Presti, L., La Cascia, M., Sclaroff, S., Camps, O.: Hanklet-based dynamical systems modeling for 3D action recognition. Image Vis. Comput. **40**, 1–53 (2015). Elsevier

31. Lorincz, A., Jeni, L., Szabó, Z., Cohn, J., Kanade, T.: Emotional expression classification using time-series kernels. In: Proceedings of Conference on Computer Vision and Pattern Recognition Workshops (CVPRW), pp. 889–895. IEEE (2013)

32. Lucey, P., Cohn, J., Kanade, T., Saragih, J., Ambadar, Z., Matthews, I.: The extended Cohn-Kanade dataset (CK+): a complete dataset for action unit and emotion-specified expression. In: Proceedings of Conference on Computer Vision and Pattern Recognition Workshops (CVPRW), pp. 94–101. IEEE (2010)

33. Moeslund, T., Granum, E.: A survey of computer vision-based human motion capture. Comput. Vis. Image Underst. 81(3), 231–268 (2001). Elsevier

34. Nicolaou, M.A., Pavlovic, V., Pantic, M.: Dynamic probabilistic CCA for analysis of affective behaviour. In: Fitzgibbon, A., Lazebnik, S., Perona, P., Sato, Y., Schmid, C. (eds.) ECCV 2012. LNCS, vol. 7578, pp. 98–111. Springer, Heidelberg (2012). doi:10.1007/978-3-642-33786-4_8

35. Nie, S., Wang, Z., Ji, Q.: A generative restricted Boltzmann machine based method for high-dimensional motion data modeling. Comput. Vis. Image Underst. 136, 14–22 (2015). Elsevier

36. Noma, H., Shimodaira, K.: Dynamic time-alignment kernel in support vector machine. Adv. Neural Inf. Process. Syst. 14, 921–930 (2002)

37. Ofli, F., Chaudhry, R., Kurillo, G., Vidal, R., Bajcsy, R.: Sequence of the most informative joints (SMIJ): a new representation for human skeletal action recognition. J. Vis. Commun. Image Represent. 25(1), 24–38 (2014). Elsevier

38. Paoletti, S., Juloski, A., Ferrari-Trecate, G., Vidal, R.: Identification of hybrid systems a tutorial. Eur. J. Control 13(2), 242–260 (2007). Elsevier

39. Poppe, R.: A survey on vision-based human action recognition. Image and Vis. Comput. 28(6), 976–990 (2010). Elsevier

40. Poullot, S., Tsukatani, S., Phuong Nguyen, A., Jégou, H., Satoh, S.: Temporal matching kernel with explicit feature maps. In: Proceedings of Conference on Multimedia Conference, pp. 381–390. ACM (2015)

41. Prabhakar, K., Oh, S., Wang, P., Abowd, G., Rehg, J.M.: Temporal causality for the analysis of visual events. In: Proceedings on Computer Vision and Pattern Recognition (CVPR 2010), pp. 1967–1974. IEEE (2010)

42. Rahimi, A., Recht, B., Darrell, T.: Learning to transform time series with a few examples. Trans. Pattern Anal. Mach. Intell. 29(10), 1759–1775 (2007). IEEE

43. Raptis, M., Kokkinos, I., Soatto, S.: Discovering discriminative action parts from mid-level video representations. In: Proceedings of Conference on Computer Vision and Pattern Recognition (CVPR 2012), pp. 1242–1249. IEEE (2012)

44. Revaud, J., Douze, M., Schmid, C., Jégou, H.: Event retrieval in large video collections with circulant temporal encoding. In: Proceedings of Conference on Computer Vision and Pattern Recognition (CVPR 2013), pp. 2459–2466. IEEE (2013)

45. Ramirez Rivera, A., Castillo, R., Chae, O.: Local directional number pattern for face analysis: face and expression recognition. Trans. Image Process. (TIP) 22(5), 1740–1752. IEEE (2013)

46. Sahami, M., Heilman, T.D.: A web-based kernel function for measuring the similarity of short text snippets. In: Proceedings of International Conference on World Wide Web, pp. 377–386. ACM (2006)

47. Sankaranarayanan, A.C., Turaga, P.K., Baraniuk, R.G., Chellappa, R.: Compressive acquisition of dynamic scenes. In: Daniilidis, K., Maragos, P., Paragios, N. (eds.) ECCV 2010. LNCS, vol. 6311, pp. 129–142. Springer, Heidelberg (2010). doi:10.1007/978-3-642-15549-9_10

48. Sariyanidi, E., Gunes, H., Cavallaro, A.: Automatic analysis of facial affect: a survey of registration, representation and recognition. Trans. Pattern Anal. Mach. Intell. (PAMI) 37(6), 1113–1133 (2014). IEEE

49. Schüldt, C., Laptev, I., Caputo, B.: Recognizing human actions: a local SVM approach. In: Proceedings of International Conference on Pattern Recognition (ICPR 2004), vol. 3, pp. 32–36. IEEE (2004)
50. Seo, H.J., Milanfar, P.: Training-free, generic object detection using locally adaptive regression kernels. Trans. Pattern Anal. Mach. Intell. **32**(9), 1688–1704 (2010). IEEE
51. Shotton, J., Sharp, T., Kipman, A., Fitzgibbon, A., Finocchio, M., Blake, A., Cook, M., Moore, R.: Real-time human pose recognition in parts from single depth images. Commun. ACM **56**(1), 116–124 (2013). ACM
52. Slama, R., Wannous, H., Daoudi, M., Srivastava, A.: Accurate 3D action recognition using learning on the Grassmann manifold. Pattern Recognit. (PR) **48**(2), 556–567 (2015). Elsevier
53. Smilde, A.K., Kiers, H.A.L., Bijlsma, S., Rubingh, C.M., Van Erk, M.J.: Matrix correlations for high-dimensional data: the modified RV-coefficient. Bioinformatics **25**(3), 401–405 (2009). Oxford University Press
54. Songsiri, J., Dahl, J., Vandenberghe, L.: Graphical models of autoregressive processes. In: Convex Optimization in Signal Processing and Communications, pp. 89–116. Cambridge University Press. Cambridge (2010)
55. Turaga, P., Chellappa, R., Subrahmanian, V.S., Udrea, O.: Machine recognition of human activities: a survey. Trans. Circ. Syst. Video Technol. **18**(11), 1473–1488 (2008). IEEE
56. Urtasun, R., Fleet, D.J., Fua, P.: 3D people tracking with Gaussian process dynamical models. In: Proceedings of Conference on Computer Vision and Pattern Recognition (CVPR 2006), vol. 1, pp. 238–245. IEEE (2006)
57. Viola, P., Jones, M.J.: Robust real-time face detection. Int. J. Comput. Vis. **57**(2), 137–154 (2004). Springer
58. Wang, Z., Wang, S., Ji, Q.: Capturing complex spatio-temporal relations among facial muscles for facial expression recognition. In: Proceedings of Conference on Computer Vision and Pattern Recognition (CVPR), pp. 3422–3429. IEEE (2013)
59. Wu, B., Yuan, C., Hu, W.: Human action recognition based on context-dependent graph kernels. In: Proceedings of the IEEE Conference on Computer Vision and Pattern Recognition (CVPR 2014), pp. 2609–2616. IEEE (2014)
60. Xu, D., Yan, S., Tao, D., Zhang, L., Li, X., Zhang, H.: Human gait recognition with matrix representation. Trans. Circ. Syst. Video Technol. **16**(7), 896–903 (2006). IEEE
61. Yang, M.H., Ahuja, N., Tabb, M.: Extraction of 2D motion trajectories and its application to hand gesture recognition. Trans. Pattern Anal. Mach. Intell. **24**(8), 1061–1074 (2002). IEEE
62. Zeng, Z., Pantic, M., Roisman, G.I., Huang, T.S.: A survey of affect recognition methods: audio, visual, and spontaneous expressions. Trans. Pattern Anal. Mach. Intell. **31**(1), 39–58 (2009). IEEE
63. Zhang, X., Yang, Y., Jiao, L.C., Dong, F.: Manifold-constrained coding and sparse representation for human action recognition. Pattern Recogn. **46**(7), 1819–1831 (2013). Elsevier
64. Zhou, F., De la Torre, F.: Generalized canonical time warping. Trans. Pattern Anal. Mach. Intell. (PAMI) **38**(2), 279–294 (2016). IEEE

Hand Pose Regression
via a Classification-Guided Approach

Hongwei Yang and Juyong Zhang[(⊠)]

University of Science and Technology of China, Hefei, Anhui, China
juyong@ustc.edu.cn

Abstract. Hand pose estimation from single depth image has achieved great progress in recent years, however, up-to-data methods are still not satisfying the application requirements like in human-computer interaction. One possible reason is that existing methods try to learn a general regression function for all types of hand depth images. To handle this problem, we propose a novel "divide-and-conquer" method, which includes a classification step and a regression step. At first, a convolutional neural network classifier is used to classify the input hand depth image into different types. Then, an effective and efficient multiway cascaded random forest regressor is used to estimate the hand joints' 3D positions. Experiments demonstrate that the proposed method achieves state-of-the-art performance on challenging dataset. Moreover, the proposed method can be easily combined with other regression method.

1 Introduction

In recent years, the problem of pose estimation of 3D articulated objects such as human body [1–3] and hand [4–11] from markerless visual observations has been widely studied due to their wide applications on human-computer interaction, Augmented Reality (AR), motion sensing game, robotic control, etc. In the earlier period, researchers estimated gestures from the 2D RGB images [12–14] or video [15]. Along with the development of hardware technology, in particular, low-cost commodity depth cameras like MicroSoft Kinect, PrimeSense and Intel RealSense have emerged in recent years. Human body and hand pose estimation from RGB-D data [7,16–26] have noticeably progressed after the introduction of depth sensors.

Since human hand has large viewpoint variance, self-occlusion and similarity between fingers, hand pose estimation based on markerless visual observations is still a challenging task. Although it is an extremely difficult problem, a lot of literatures about hand pose estimation have been proposed in recent years. In the survey [27], vision-based markerless hand tracking algorithms were roughly classified into two types, model-based and appearance-based approaches.

Electronic supplementary material The online version of this chapter (doi:10. 1007/978-3-319-54187-7_30) contains supplementary material, which is available to authorized users.

© Springer International Publishing AG 2017
S.-H. Lai et al. (Eds.): ACCV 2016, Part III, LNCS 10113, pp. 452–466, 2017.
DOI: 10.1007/978-3-319-54187-7_30

Model-based methods often fit a hand template to the input data to estimate hand poses. Pose estimation can be formulated as a optimization problem [4,16] or nearest-neighbor search problem [28]. Recently, Sridhar et al. [29,30] proposed an accurate model-based hand tracking method in a multiple camera setup, with the purpose of resolving serious self-occlusions. Although multiple camera setup can achieve more precise pose recovery, the complex acquisition setup and manual calibration is less suitable for the consumer-level applications. In [4,16,31,32], the 3D hand poses were reconstructed via inverse kinematics techniques, which optimize a nonlinear energy function that is extremely difficult to find global minimum. The numerical optimization algorithms of above methods usually are complex, time-consuming and easy to trap into local minima. Therefore, it limits their usages in real-time and accurate applications.

On the other hand, appearance-based approaches use direct mapping techniques which try to learn a direct mapping from the input image space to the output pose space. During the past few years, numerous appearance-based methods based on nearest neighbor search [20,28], decision forest [5,6,8,18,33,34] or convolutional networks [21,23,26,35] have been developed for hand pose estimation. In [5], Keskin et al. introduced a multi-layered randomized decision forest framework. They divided a whole classification task into two classification stages, which only focus on different learning task and make the whole learning task more efficient and accurate. Recently, Sun et al. [33] presented a cascaded hierarchical regression approach with 3D pose-indexed features. Although the 3D pose-indexed features achieve approximatively strict 3D invariance and cascaded framework reinforces learning ability. It is extremely difficult to handle complicated hand poses estimation by only one regression model. Therefore, it is a good choice to improve the whole learning ability by introducing multiway regression models.

In this paper we propose a novel "divide-and-conquer" classification-guided regression learning framework to estimate hand pose from single depth image. At first, in order to reduce the search space of regression, a convolutional network based classifier is introduced to predict the hand gesture type. Cascaded random forest regressors for different hand gesture types are trained on disjoint part of training dataset. Then based on the predicted class of classifier, a corresponding regressor is selected to estimate the final hand pose. It means that the classifier divides the learning task and the regressors conquer their own task. The algorithm pipeline is shown in Fig. 1.

Our main contributions are as follow: a new classification-guided hand pose regression framework is developed. Based on the training dataset of regression, we train a classifier to partition a complex and difficult regression learning task into several more easier subproblems. Then each regressor only focuses on a part of training data so that it is more professional and accurate. The classification-guided regression approach outperforms the state-of-the-art methods. More broadly speaking, other discriminative hand pose regression model can be used as the regressor module in our framework and further improve the pose estimation accuracy.

This paper is organized as follows: Sect. 2 describes our proposed framework, which includes the new convolutional networks hand pose classifier and

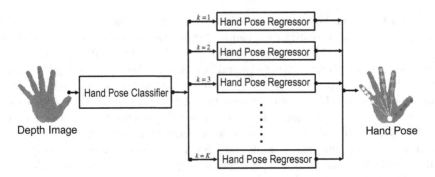

Fig. 1. Algorithm pipeline. We integrate a hand pose classifier and several hand pose estimation regressors into one framework. The hand pose classifier predicts the class of gesture from the depth image. Based on the predicted class, one of hand pose regressors is selected to estimate the final hand joints' 3D locations.

the cascaded random regression forest. After that, we introduce the experimental details and analyze quantitative and qualitative results of experiments in Sect. 3. Finally, we conclude this paper in Sect. 4.

2 Methodology

2.1 Hand Model and Method Overview

We use a hand skeleton model as illustrated in Fig. 2. The hand pose $\Theta = \{\mathbf{p}_i\}_{i=1}^{21}$, where $\mathbf{p}_i = (x_i, y_i, z_i)$, represents 21 kinematic joints' 3D positions. We divide the hand pose Θ into six parts including the palm Θ^p (6 joints of the palm) and five fingers Θ^f (each 3 joints of the maniphalanx), where $f \in F = \{1, 2, 3, 4, 5\}$.

An overview of our pipeline is showed in Fig. 1. We estimate the hand pose Θ in the form of the 3D locations of its joints from a single depth image I. And we denoted a training dataset by $\{(I_i, \Theta_i)\}_{i=1}^N$, each element of which is a depth image labeled with its corresponding ground truth joints' locations. Our proposed method integrates the hand pose classifier and hand pose regressor into one framework. In advance, the training dataset of regression is clustered into K subsets. And based on the clustering result, a hand pose classifier are trained on the entire dataset, but several hand pose regressors are respectively trained on each subset. At the testing stage, the hand pose classifier infers the hand pose class k from the depth image first. Then the hand pose regressor which is corresponding to k-class estimates the final hand pose Θ.

2.2 Clustering Training Data

To train the hand pose classifier, we utilize the existing dataset $\{(I_i, \Theta_i)\}_{i=1}^N$ of hand pose regression to generate the training dataset $\{(I_i, L_i)\}_{i=1}^N$ of classifier.

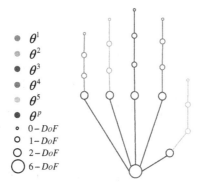

Fig. 2. A 21-joint representing of a canonical hand pose. The palm root (wrist joint) encodes 6 degrees of freedom (DoF) of the global rotation and translation. Each finger and its corresponding root point on the palm (4 joints in total) encode 4 degrees of freedom of finger articulation.

Thus, we cluster the hand joints' position vector Θ_i to generate corresponding target label L_i for depth image I_i. The K-Means clustering algorithm with rigid alignment is used to cluster the hand poses. Since Θ_i is related to camera viewpoint, the rigid registration is applied to the hand pose to remove the affects caused by the camera viewpoint. The rigid registration procedure is as follows:

$$\mathcal{T}_i = (\mathbf{R}_i, \mathbf{t}_i) = RigidAlig(\Theta_C^p, \Theta_i^p), \quad i = 1, 2, ..., N,$$

where N is the number of the training set, Θ_C is a canonical hand pose, which is arbitrarily chosen from $\{(I_i, \Theta_i)\}_{i=1}^N$. $RigidAlig$ is refered as rigid registration and it is achieved by Iterated Closest Point algorithm [36], which is used to compute the rigid transformation between the palm joints of canonical hand pose and each other hand pose. And $\mathbf{R}_i, \mathbf{t}_i$ respectively represent rotation and translation. The rigid transformation $\mathcal{T}_i$ aligns each hand pose Θ_i of training dataset to a certain coordinate system that determined by canonical hand pose. Then we cluster the aligned training data $\{\mathbf{R}_i\Theta_i + \mathbf{t}_i\}_{i=1}^N$ into K classes by K-Means clustering. Therefore, the target label L_i of depth image I_i equals its corresponding class of hand pose Θ_i of K-Means cluster.

2.3 CNN Classifier

Hand pose classifier predicts the hand pose type k for each input depth image I. As the hand pose registration space and variation of camera viewpoints are very large and complex, it is difficult to directly classify hand poses based on depth images which are the only input information. Because of the excellent performance of Convolutional neural network(CNN) on complex and large-scale image classification task [37], we adopt CNN method in this work to classify the hand poses.

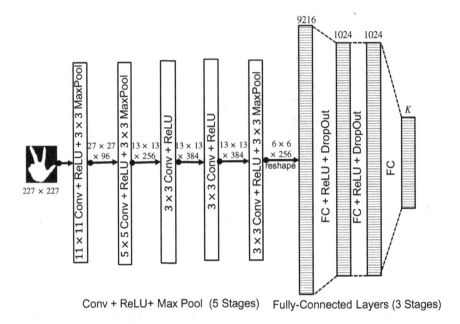

Fig. 3. The convolutional network architecture used in our paper. The network contains five convolutional layers and three fully-connected layers.

In this paper, the hand pose classifier is based on a standard CNN framework (Fig. 3). The CNN, similar to fully-connected neural networks, performs end-to-end feature learning and is trained with the back-propagation algorithm. However, they are different in many respects, most notably local connectivity, weight sharing, and local pooling. The first two properties significantly reduce the number of free parameters and the need to learn repeated feature detectors at different locations of the input. The third property makes the learned representation invariant to small translations of the input [38].

The CNN classifier is illustrated in Fig. 3. In the original depth image, the proportion of background pixels is much greater than hand pixels. Therefore, we crop the bounding box of the hand region from original depth image and resize it according to requirement of input data. The input is then processed by five stages of convolution and subsampling, which use rectified linear units (ReLUs) [39] and max-pooling. The convolution kernel stride of the first convolution layers is 4, and others are zero. The padding of the five convolution layers are respectively $0, 2, 1, 1$ and 1. Internal pooling layers contribute to reduce computational complexity and improve classification tolerance for small input image translations. Unfortunately, pooling also results in a loss of spatial precision. Since invariance to input translations can be learned with sufficient training exemplars, we only choose three stages of pooling where the stride is 2.

Following the five convolution and subsampling layers, the top-level pooled map is flattened to a vector and processed by three fully connected layers. Each of these output stages is composed of a linear matrix-vector multiplication with learned bias, followed by a point-wise non-linearity (ReLU). Dropout [40] is used on the input to each of fully-connected linear stages to reduce over-fitting for the restricted-size training set. There are two dropout layers with dropout ratio 0.5 that behind the first two fully connected layers. The output layer has a K-ways softmax unit which produces a distribution over K hand pose classes.

2.4   Hand Pose Regression

To estimate the hand pose Θ, we adopt the cascaded random regression forest method, which is a state-of-the-art hand pose estimation method presented in [33], as hand pose regressor. The final hand pose Θ^T is progressively estimated via a series of sequent random forest regressors $\{\mathcal{R}^t\}, t = 1, 2, ..., T$, with pose indexed features, which depend on the estimated pose Θ^{t-1} from the previous stage. In order to facilitate understanding, we will give a brief introduction of this method in the rest of this section.

Cascaded random regression forest needs a depth image I and an initial hand pose Θ^0 as input. In each stage t, it progressively updates the current pose estimation Θ^t as

$$\Theta^t = \Theta^{t-1} + \mathcal{R}^t(I, \Theta^{t-1}).$$

The above formula indicates that the hand pose is updated in the 3D camera coordinate system of its corresponding depth image I. When the 3D camera viewpoint is fixed, a specific pose hand model can be used to generate different depth images towards different 3D rigid transformations (corresponding to 3D camera coordinate systems). In the training stage, we compute the hand pose residual $\delta\Theta$ which is irrelevant to 3D camera coordinate system. Therefore, it is necessary to align the pose Θ to the canonical coordinate system which is determined by the canonical hand pose Θ_C. For a given hand pose Θ, we compute a 3D rigid transformation $\mathcal{T}_\Theta$ between itself and the canonical hand pose Θ_C.

In the training stage, the stage regressor $\mathcal{R}^t$ is learnt to approximate the current pose residual $\delta\Theta_i$, which is the difference between the ground truth pose and the previous pose estimation Θ_i^{t-1}, over all training samples $i(= 1, 2, ..., N)$. It's worth noting that the features of $\mathcal{R}^t$ depend on the estimated pose Θ_i^{t-1} from the previous stage. Similar to previous random forest methods for image processing [1,5,6,8,34], the pixel difference features, i.e., the difference of two random pixels, are also used. The 3D pose indexed features are constructed as follow:

1. In the canonical coordinate system, randomly select a point pair $(\mathbf{p}_1, \mathbf{p}_2)$ within a 3D sphere whose centre is the centroid of hand point cloud and radius is R, which is related to the size of a real 3D hand model.
2. The point pair $(\mathbf{p}_1, \mathbf{p}_2)$ is transformed to camera coordinate system using the inversed rigid transformation $\overline{\mathcal{T}_\Theta}$.
3. Transformed point pair is projected on depth image to get their corresponding pixels $(\mathbf{u}_1, \mathbf{u}_2)$, and then the pixel difference feature is computed.

The pose indexed feature is written as

$$I(\mathbf{u}_1) - I(\mathbf{u}_2),$$

where $\mathbf{u}_i = CamProj(\overline{\mathcal{T}_\Theta}(\mathbf{p}_i)), i = 1, 2$.

In this paper, we use holistic regression algorithm as our hand pose regressor which regresses the entire hand pose Θ at each stage. As for hierarchical regression algorithm of [33], it is completely feasible to directly replace holistic algorithm in our framework, and the accuracy can be further improved. Although we use the holistic regression algorithm in our framework, our approach also performs better than the hierarchical regression algorithm without classification step. The training algorithm for holistic cascaded regression is shown in Algorithm 1.

Input: depth image I_i, ground truth pose Θ_i, and initial pose Θ_i^0 for all training samples i

Output: regressors $\{\mathcal{R}^t\}_{t=1}^T$

1 **for** $t = 1$ **to** T **do**

2 $\quad \delta\Theta_i = \mathcal{T}_{\Theta_i^p, t-1}^p(\Theta_i) - \mathcal{T}_{\Theta_i^p, t-1}^p(\Theta_i^{t-1})$;

3 $\quad$ learn $\mathcal{R}^t$ to approximate $\delta\Theta_i$;

4 $\quad \Theta_i^t = \Theta_i^{t-1} + \overline{\mathcal{T}_{\Theta_i^p, t-1}^p}(\mathcal{R}^t(I_i, \Theta_i^{t-1}))$;

5 **end**

Algorithm 1. Training algorithm for holistic cascaded hand pose regression. Let $\overline{\mathcal{T}}$ represent the inverse of the rigid transformation $\mathcal{T}$.

3   Experiments

In this section we evaluate the proposed method on the MSRA Hand Pose Dataset [33] that is a real-world depth based dataset. We first describe the implementation details of the classifier and the regressors. Then we introduce the dataset and the evaluation metrics, and quantitatively and qualitatively evaluate the proposed method with the state-of-the art methods.

3.1   Implementation Details

The CNN classifier is implemented in CAFFE [41] framework and those parameters are optimized by using error back-propagation. We choose decay parameter as 0.2 and set batch size to 64, momentum to 0.9 and a weight decay to 0.0005. The learning rate decays over about 10 epochs and starts with 0.005, and the networks are trained for 50 epochs. We choose parameter K as 17 and the classifier is trained on the GPU mode.

The initial hand pose Θ^0 of hand pose regressor is similar to [33]. Each hand pose regressor consists of 6 cascaded stages and each random regression forest of hand pose regressor consists of 10 trees. Each split node of tree samples 540 random feature point pairs, and we pick one that gives rise to maximum variance reduction over all dimensions of the pose residual. The tree node splits until the node includes less than 10 samples.

In this paper, we propose a heuristic and effective left-right hand pose estimation method only based on the right hand training dataset. At first, a left-right hand binary classifier is used to predict the binary labels of hand. The binary classifier has a same architecture with the hand pose CNN classifier (in Sect. 2.3), except that the output layer is a 2-way softmax unit. The training dataset for binary classifier is constructed by flipping the right hand image of regression training dataset to generate the left hand depth image. In the testing stage, a depth image I is put into the binary classifier to predict left or right hand. If the predicted class is left hand, the original depth image is flipped horizontally. This means that the point clouds of hand are projected symmetrically about YOZ plane, i.e.

$$flip(\mathbf{p}) = flip((x,y,z)) = (-x,y,z), \quad I' = flip(I),$$

where $\mathbf{p}(x,y,z)$ is a point in the original point clouds. Then the pseudo-right depth image I' is put into the right hand pose classifier to get predicted hand pose class k. The k-th cascaded random forest regressor estimates the right hand joints' 3D positions Θ. Finally, we can flip back to get the left hand pose, that is $\Theta' = flip(\Theta)$.

3.2   Dataset and Evaluation Metric

There exist some public real-world depth based datasets for hand pose estimation. However, the depth images of the dataset [7] include the forearm that causes terrible initialization which usually produce large errors in pose estimation, and the dataset [34] has restricted range of viewpoints and large annotation errors of ground truth hand poses. The datasets [16,17,29] provide too little training data to train meaningful models. The above datasets are not suitable for our task. The MSRA Hand Pose Dataset [33] is a large-scale and challenging real-world benchmark for hand pose estimation. It consists of $76,500$ depth images with accurate ground truth hand poses. The depth images are captured from 9 subjects, and each subject contains 17 gestures. This dataset has larger viewpoint variations (yaw nearly spans the full $[-90,90]$ range and pitch within $[-10,90]$ degrees). Thus we select the MSRA Hand Pose Dataset to evaluate our proposed method.

Although the MSRA Hand Pose Dataset performs well for training hand pose regressor in [33], it is not sufficient to train a well-behaved CNN model to classify the complex hand poses that have large variation range of viewpoints. To avoid overfitting and improve the classification accuracy, data enhancement is applied to improve the diversity of dataset. In the MSRA Hand Pose Dataset, each depth image is rotated in $+10°$, $-10°$, $+20°$, $-20°$, $+30°$, $-30°$, $+40°$, $-40°$ to generate 8 depth images. Then the dataset is expanded 8 times from the previous one.

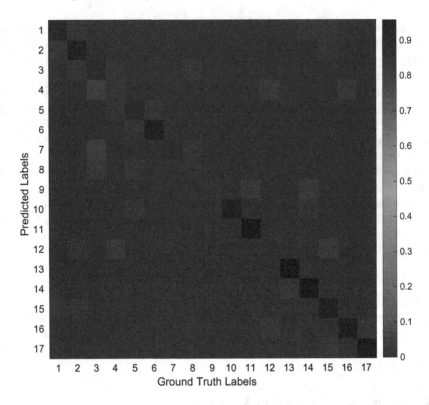

Fig. 4. The confusion matrix of hand pose classifier on the classification training dataset constructed in Sect. 2.2. The average accuracy of leave-one-subject-out cross-validation is 91.2%.

Similar to the previous work [33,34], there are two accuracy metrics for hand pose estimation. The first one is the averaged Euclidean distance of entire predicted joints from the ground truth across all the test samples, that is $M_1 = \frac{1}{N} \sum_{i=1}^{N} \sum_{j=1}^{21} \|\hat{\Theta}_{i,j} - \Theta_{i,j}\|$, where $\hat{\Theta}$ and Θ are respectively the predicted joints and the ground truth joints. The second metric is the success rate, i.e., the percentage of frames where all joints are within a maximum distance threshold ε, that is $M_2 = \frac{1}{N} \sum_{i=1}^{N} I(\max_{1 \leq j \leq 21} \|\hat{\Theta}_{i,j} - \Theta_{i,j}\| \leq \varepsilon) \times 100\%$, where $I(\cdot)$ is an indicator function. It is obvious that the second metric is more strict than the first one.

3.3   Quantitative Results

Our proposed method is evaluated by leave-one-subject-out cross-validation. For left-right hand classifier and hand pose classifier, the average classification accuracy is respectively 95.0% and 91.2%. The average confusion matrix of hand pose classifiers across all the subjects is shown in Fig. 4.

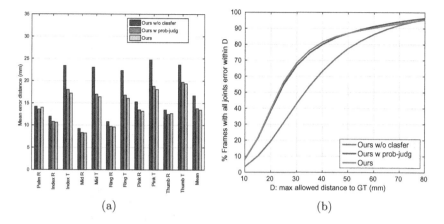

(a)   (b)

Fig. 5. Comparison our method with the two baselines. (a) Mean error distance. (b) Success rate with max allowed distance threshold ε. Compared to **ours w/o clasfer**, our method and the second baseline achieve a considerable improvement in the two error metrics.

In order to demonstrate the efficiency of our pipeline, we implement two baselines. The first baseline directly estimates the hand pose without hand pose classifier. We refer this baseline as **ours w/o clasfer**. Our proposed method largely outperforms the first baseline in both mean error distance and success rate metrics. But when the max distance threshold ε is larger than 75 mm, the success rate of our method is slightly less than the first baseline in Fig. 5(b). The reason is that the hand pose classifier predicts an incorrect class and thus results in a poor estimation of the incorrect regressor.

It is well known that a classifier has a receiver operating characteristic curve. The probability which the classifier predicts a true-positive result is higher, when the probability of top-1 label is higher. In order to resolve that incorrect predicted hand pose class results in poor estimation, we propose the second baseline, a classification-guided regression pipeline with predicted probability judgement, which means a judgement of predicted probability for hand pose classifier is added to decide whether to trust the predicted label. We refer this baseline as **ours w prob-judg**. If predicted probability is greater than a given probability threshold ϵ, we trust the predicted label and use the k-label regressor. Otherwise, we don't trust the predicted label and use the regressor trained on entirety training dataset.

In all the experiments, we choose $\epsilon = 99\%$ as the threshold of predicted probability. The average true-positive ratio for our hand pose classifier is 97.4%. As shown in Fig. 5, the second baseline has a better performance than our proposed method on large distance threshold $\varepsilon \in [50, 80]$. This is because we only trust samples that have high predicted probability and do not coercively execute classification-guided regression method for which has low predicted probability.

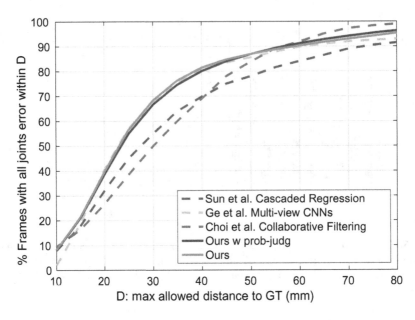

Fig. 6. Quantitative evaluation of hand pose estimation. The figure shows that the proportion of frames where all joints are within max allowed distance threshold ε. We compare our proposed approach to the second baseline and three state-of-the-art methods [10, 26, 33].

We compare our pipeline with state-of-the-art methods [10, 26, 33] on MSRA Hand Pose Dataset. As shown in Fig. 6, our method entirely and substantially performs great better than **Cascaded Hierarchical Regression** [33]. This is because the hand pose classifier divides the overall complicated and difficult learning task into several relatively easy-to-learn tasks, which are suited to random forest regressor. For **Collaborative Filtering** [10], we achieve superior accuracy predictions on most threshold interval, especially when distance threshold ε is within 50 mm. When the distance threshold ε is greater than 55 mm, the performance of our method get worse than [10]. This is because that our method suffers from poor hand pose estimation caused by the incorrect predicted label. Compared with **Multi-view CNNs** [26], our method performs better in almost all distance threshold interval, especially in the highest and lowest threshold interval. We just use a holistic hand pose regressor to achieve the state-of-the-art performance on accuracy. Furthermore, it will achieve better performance than [26] by replacing regressor with **Multi-view CNNs** in our framework.

We also compare the mean error distance metric over different viewpoint angles of our proposed method and two methods [26, 33] in Fig. 7. Our proposed method has smaller average errors than those of **Cascaded Hierarchical Regression** over all yaw and pitch viewpoint angles, and performs better than **Multi-view CNNs** over most yaw viewpoint angles and partial pitch viewpoint angles.

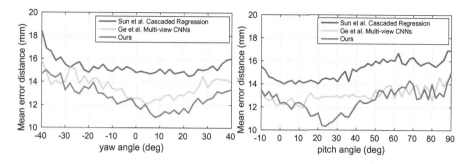

Fig. 7. Quantitative evaluation of hand pose estimation. We compare our approach and two state-of-the-art methods [26,33] with respect to the mean error distance metric. Left: the mean joint errors distributed over all yaw viewpoint angles. Right: the mean joint errors distributed over all pitch viewpoint angles.

3.4   Qualitative Results

Some qualitative results of the proposed method and the two baselines on several challenging examples are shown in Fig. 8 to further illustrate the superiority of our method over the other two baselines. And more qualitative examples are demonstrated in the accompanying supplementary demo video.

Our proposed algorithm is tested on Intel i5 3.3 GHz with NVIDIA GTX980 GPU running Ubuntu 14.04. The overall hand pose estimation pipeline runs on a single thread on CPU, except that the left-right hand classifier and hand pose classifier are tested on the GPU mode. The left-right hand classifier and hand pose classifier cost 7.1 ms in all, and the hand pose estimation regressor

Fig. 8. Qualitative results for dataset in [33] of three approaches. (a) The ground truth hand poses. (b) Regression without classifier. (c) Our classification-guided regression method. (d) Classification-guided regression method with predicted probability judgement.

costs 0.7 ms. Therefore, the overall computation time of our method is around 8 ms. Such high performance is sufficient for real-time applications. In terms of method efficiency, the proposed algorithm is faster than most existing methods [5,7,8,10,17,26,29].

4   Conclusions

In this paper, a classification-guided regression learning framework is presented to estimate hand joints' 3D locations from single depth image. In order to simplify the challenging task, a well-trained CNN classifier is applied to identify hand gesture types. Based on the predicted class of classifier, an accurate and efficient cascaded random forest regressor is used to estimate the final hand joints' positions. Our classifier reduces the search space of regression and speeds up hand pose estimation. Experiments demonstrate that the proposed method achieves state-of-the-art performance on challenging dataset. Our proposed method is effective and has strong extensibility that is easy to integrate classifier and regressor into single pipeline. More broadly speaking, any discriminative hand pose regression model can be used as the regressor module in our framework. Likewise, our classifier unit can be substituted by any kind of efficient classification models.

Acknowledgement. We thank Zishun Liu for his helpful suggestions on this paper. This work was supported by the National Key R&D Program of China (No. 2016YFC0800501), NSF of China (Nos. 61672481, 61303148), NSF of Anhui Province, China (No. 1408085QF119), Specialized Research Fund for the Doctoral Program of Higher Education under contract (No. 20133402120002).

References

1. Shotton, J., Sharp, T., Kipman, A., Fitzgibbon, A., Finocchio, M., Blake, A., Cook, M., Moore, R.: Real-time human pose recognition in parts from single depth images. In: CVPR (2011)
2. Ganapathi, V., Plagemann, C., Koller, D., Thrun, S.: Real-time human pose tracking from range data. In: Fitzgibbon, A., Lazebnik, S., Perona, P., Sato, Y., Schmid, C. (eds.) ECCV 2012. LNCS, vol. 7577, pp. 738–751. Springer, Heidelberg (2012). doi:10.1007/978-3-642-33783-3_53
3. Ye, M., Zhang, Q., Wang, L., Zhu, J., Yang, R., Gall, J.: A survey on human motion analysis from depth data. In: Grzegorzek, M., Theobalt, C., Koch, R., Kolb, A. (eds.) Time-of-Flight and Depth Imaging. Sensors, Algorithms, and Applications. LNCS, vol. 8200, pp. 149–187. Springer, Heidelberg (2013). doi:10.1007/978-3-642-44964-2_8
4. Oikonomidis, I., Kyriazis, N., Argyros, A.A.: Efficient model-based 3D tracking of hand articulations using kinect. In: BMVC (2011)
5. Keskin, C., Kıraç, F., Kara, Y.E., Akarun, L.: Hand pose estimation and hand shape classification using multi-layered randomized decision forests. In: Fitzgibbon, A., Lazebnik, S., Perona, P., Sato, Y., Schmid, C. (eds.) ECCV 2012. LNCS, vol. 7577, pp. 852–863. Springer, Heidelberg (2012). doi:10.1007/978-3-642-33783-3_61

6. Sun, M., Kohli, P., Shotton, J.: Conditional regression forests for human pose estimation. In: CVPR (2012)
7. Tompson, J., Stein, M., Lecun, Y., Perlin, K.: Real-time continuous pose recovery of human hands using convolutional networks. ACM Trans. Graph. **33**, 169 (2014)
8. Tang, D., Yu, T.H., Kim, T.K.: Real-time articulated hand pose estimation using semi-supervised transductive regression forests. In: ICCV (2013)
9. Tang, D., Taylor, J., Kohli, P., Keskin, C., Kim, T.K., Shotton, J.: Opening the black box: hierarchical sampling optimization for estimating human hand pose. In: ICCV (2015)
10. Choi, C., Sinha, A., Choi, J.H., Jang, S., Ramani, K.: A collaborative filtering approach to real-time hand pose estimation. In: ICCV (2015)
11. Tzionas, D., Ballan, L., Srikantha, A., Aponte, P., Pollefeys, M., Gall, J.: Capturing hands in action using discriminative salient points and physics simulation. IJCV, 1–22 (2015)
12. Erol, A., Bebis, G., Nicolescu, M., Boyle, R.D., Twombly, X.: Vision-based hand pose estimation: a review. CVIU **108**, 52–73 (2007)
13. Puwein, J., Ballan, L., Ziegler, R., Pollefeys, M.: Joint camera pose estimation and 3D human pose estimation in a multi-camera setup. In: Cremers, D., Reid, I., Saito, H., Yang, M.-H. (eds.) ACCV 2014. LNCS, vol. 9004, pp. 473–487. Springer, Heidelberg (2015). doi:10.1007/978-3-319-16808-1_32
14. Oikonomidis, I., Kyriazis, N., Argyros, A.A.: Markerless and efficient 26-DOF hand pose recovery. In: Kimmel, R., Klette, R., Sugimoto, A. (eds.) ACCV 2010. LNCS, vol. 6494, pp. 744–757. Springer, Heidelberg (2011). doi:10.1007/978-3-642-19318-7_58
15. de La Gorce, M., Fleet, D.J., Paragios, N.: Model-based 3d hand pose estimation from monocular video. PAMI **33**, 1793–1805 (2011)
16. Qian, C., Sun, X., Wei, Y., Tang, X., Sun, J.: Realtime and robust hand tracking from depth. In: CVPR (2014)
17. Xu, C., Cheng, L.: Efficient hand pose estimation from a single depth image. In: ICCV (2013)
18. Li, P., Ling, H., Li, X., Liao, C.: 3D hand pose estimation using randomized decision forest with segmentation index points. In: ICCV (2015)
19. Sharp, T., Keskin, C., Robertson, D., Taylor, J., Shotton, J., Kim, D., Rhemann, C., Leichter, I., Vinnikov, A., Wei, Y., Freedman, D., Kohli, P., Krupka, E., Fitzgibbon, A., Izadi, S.: Accurate, robust, and flexible real-time hand tracking. In: CHI (2015)
20. Supancic III., J.S., Rogez, G., Yang, Y., Shotton, J., Ramanan, D.: Depth-based hand pose estimation: methods, data, and challenges. In: ICCV (2015)
21. Oberweger, M., Wohlhart, P., Lepetit, V.: Hands deep in deep learning for hand pose estimation. In: CVWW (2015)
22. Poier, G., Roditakis, K., Schulter, S., Michel, D., Bischof, H., Argyros, A.A.: Hybrid one-shot 3D hand pose estimation by exploiting uncertainties. In: BMVC (2015)
23. Oberweger, M., Wohlhart, P., Lepetit, V.: Training a feedback loop for hand pose estimation. In: ICCV (2015)
24. Oberweger, M., Riegler, G., Wohlhart, P., Lepetit, V.: Efficiently Creating 3D training data for fine hand pose estimation. In: CVPR (2016)
25. Taylor, J., Bordeaux, L., Cashman, T., Corish, B., Keskin, C., Soto, E., Sweeney, D., Valentin, J., Luff, B., Topalian, A., Wood, E., Khamis, S., Kohli, P., Sharp, T., Izadi, S., Banks, R., Fitzgibbon, A., Shotton, J.: Efficient and precise interactive hand tracking through joint, continuous optimization of pose and correspondences. ACM SIGGRAPH **35**(4), 143 (2016)

26. Ge, L., Liang, H., Yuan, J., Thalmann, D.: Robust 3D hand pose estimation in single depth images: from single-view CNN to multi-view CNNs. In: CVPR (2016)
27. Mohr, D., Zachmann, G.: A survey of vision-based markerless hand tracking approaches. CVIU (2013)
28. Wang, R.Y., Popović, J.: Real-time hand-tracking with a color glove. ACM Trans. Graph. **28**(3), 63 (2009)
29. Sridhar, S., Oulasvirta, A., Theobalt, C.: Interactive markerless articulated hand motion tracking using RGB and depth data. In: ICCV (2013)
30. Sridhar, S., Rhodin, H., Seidel, H.P., Oulasvirta, A., Theobalt, C.: Real-time hand tracking using a sum of anisotropic gaussians model. In: 3DV (2014)
31. Sridhar, S., Mueller, F., Oulasvirta, A., Theobalt, C.: Fast and robust hand tracking using detection-guided optimization. In: CVPR (2015)
32. Tagliasacchi, A., Schröder, M., Tkach, A., Bouaziz, S., Botsch, M., Pauly, M.: Robust articulated-ICP for real-time hand tracking. Comput. Graph. Forum **34**, 101–114 (2015)
33. Sun, X., Wei, Y., Liang, S., Tang, X., Sun, J.: Cascaded hand pose regression. In: CVPR (2015)
34. Tang, D., Chang, H., Tejani, A., Kim, T.K.: Latent regression forest: structured estimation of 3D articulated hand posture. In: CVPR (2014)
35. Neverova, N., Wolf, C., Nebout, F., Taylor, G.: Hand pose estimation through weakly-supervised learning of a rich intermediate representation. Computer Science (2015)
36. Rusinkiewicz, S., Levoy, M.: Efficient variants of the ICP algorithm. In: 3DIM (2001)
37. Krizhevsky, A., Sutskever, I., Hinton, G.E.: Imagenet classification with deep convolutional neural networks. In: NIPS (2012)
38. Jain, A., Tompson, J., Andriluka, M., Taylor, G.W., Bregler, C.: Learning human pose estimation features with convolutional networks. Computer Science (2013)
39. Glorot, X., Bordes, A., Bengio, Y.: Deep sparse rectifier neural networks. In: AISTATS (2011)
40. Hinton, G.E., Srivastava, N., Krizhevsky, A., Sutskever, I., Salakhutdinov, R.R.: Improving neural networks by preventing co-adaptation of feature detectors. arXiv preprint arXiv:1207.0580 (2012)
41. Jia, Y., Shelhamer, E., Donahue, J., Karayev, S., Long, J., Girshick, R., Guadarrama, S., Darrell, T.: Caffe: convolutional architecture for fast feature embedding. arXiv preprint arXiv:1408.5093 (2014)

Who's that Actor? Automatic Labelling of Actors in TV Series Starting from IMDB Images

Rahaf Aljundi[✉], Punarjay Chakravarty, and Tinne Tuytelaars

KU Leuven, ESAT-PSI, iMinds, Leuven, Belgium
rahaf.aljundi@esat.kuleuven.be

Abstract. In this work, we aim at automatically labeling actors in a TV series. Rather than relying on transcripts and subtitles, as has been demonstrated in the past, we show how to achieve this goal starting from a set of example images of each of the main actors involved, collected from the Internet Movie Database (IMDB). The problem then becomes one of domain adaptation: actors' IMDB photos are typically taken at awards ceremonies and are quite different from their appearances in TV series. In each series as well, there is considerable change in actor appearance due to makeup, lighting, ageing, etc. To bridge this gap, we propose a graph-matching based self-labelling algorithm, which we coin HSL (Hungarian Self Labeling). Further, we propose a new metric to be used in this context, as well as an extension that is more robust to outliers, where prototypical faces for each of the actors are selected based on a hierarchical clustering procedure. We conduct experiments with 15 episodes from 3 different TV series and demonstrate automatic annotation with an accuracy of 90% and up.

1 Introduction

There has been an explosion of video data in the recent past. In addition to professionally shot movies and TV series, with the proliferation of smart phones and the advent of social media, users document more and more of their lives on home-video. To properly use this data, like the search engines of the world wide web, there is a need for archiving and indexing these videos for easy search and retrieval. To this end, we present an automatic person labelling system in video starting from only a few sample images, which could be actor images from IMDB (as in our experiments) or "tagged" images on social media. Our system can be used to archive, index and search large databases of video using just a few images of the main characters as starting point. Using our method, one can imagine a video search app that could for example search for the scene in the

---

R. Aljundi and P. Chakravarty—Equal contribution.

Electronic supplementary material The online version of this chapter (doi:10.1007/978-3-319-54187-7_31) contains supplementary material, which is available to authorized users.

© Springer International Publishing AG 2017
S.-H. Lai et al. (Eds.): ACCV 2016, Part III, LNCS 10113, pp. 467–483, 2017.
DOI: 10.1007/978-3-319-54187-7_31

TV series Breaking Bad, where Walter White first meets Jesse Pinkman, or for all videos of grandpa, from a collection of home videos.

The first major attempt at automatically labelling actors in a TV series was made by Everingham et al. [1], who in their seminal work demonstrated the training of classifiers for actor recognition using the weak supervision of subtitle and transcript files. However, subtitles and transcripts are not always available. They can be hard to find for some movies and TV shows, and are perhaps non-existent for TV news broadcasts, music-videos, documentaries and silent films. Even if they are available, transcripts, in most cases obtained from fan websites, vary in quality and reliability. They also come in different formats, making it hard to automate the process. In addition, there is the need to align the transcripts with the subtitles. In the absence of transcripts, [2–4] use supervision from hand-labelled ground-truth for some of the video data. Collecting such annotations is, however, a cumbersome and user-unfriendly process.

In this paper, we consider an alternative source for training appearance models for actors, the Internet Movie Database, or IMDB. Most, if not all TV series are listed on IMDB, along with photos of the major actors. Our method is simple to use. All it needs is the name of the show. If the title exists on IMDB, the downloading of actor images and their use in the labelling of actor appearances in the TV show or movie is completely automatic.

This may seem straightforward, but actually it is not. The images in IMDB are mostly taken at awards functions and ceremonies using flash-enabled DSLR cameras, or are from promotional material for the shows. An actor's appearance in a show is normally quite different from her appearance in IMDB. This is due to various factors like camera motion (and consequent blur), lighting changes, make-up of the actor, and in-show or real-life ageing of the star. In other words, there is a *domain shift* between the IMDB data and the data from the TV shows. In this work, we aim at overcoming this domain shift (see Fig. 1).

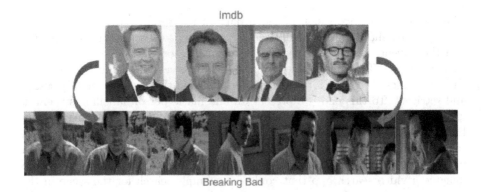

Fig. 1. The problem of actor labelling using IMDB images is one of accounting for the domain shift between photos in IMDB and frames from the TV show.

In a self-labelling inspired approach, we iteratively select actors' faces from the TV series that are closest to their counterparts in IMDB and use them to enrich the set of IMDB template images. As more and more actor faces from the show are added to the original set of IMDB templates, the actor's appearance in the show is better represented, the domain shift is accounted for, and the method is better able to label the actor inspite of major changes in appearance compared to the original IMDB templates.

To optimally select the set of images in each step of the self-labelling procedure, we propose to use bipartite graph matching and incorporate a new graph edge cost. In order to strengthen the method's stability in various challenging cases (pose change, lighting change, presence of side actors, etc.), we further add a clustering/outlier removal scheme resulting in an *actor profile* that is used for face labelling in later iterations. We utilize the power of the latest generation face recognition and verification techniques - Deep Face features [5–7]. These features are activations of Convolutional Neural Networks trained on vast face recognition databases. They are able to characterize a person under a wide range of poses and lighting conditions. We use a multi-target tracker to simultaneously track multiple people in video. Each track has many faces of the same person under different poses and our labelling algorithm takes advantage of the information available from all the faces in a track to identify it. We do not use any context information such as clothing.

We demonstrate actor labelling in the TV series Big Bang Theory, Breaking Bad and Mad Men. For each of these series, we manually download a small number of template images of the top 5/6 actors from IMDB and demonstrate an average of 91% accuracy in detecting those characters in these TV series.

In summary, our main contributions are: (1) a new graph edge cost that improves the selection of hitherto unlabelled faces in the self-labelling process; (2) the use of bipartite graph matching for optimal selection of unlabelled faces in each step of the self-labelling process; (3) a hierarchical clustering procedure yielding an actor profile and increasing the robustness of the method against outliers and side-actors; and (4) a dataset comprising of labelled faces in a total of 15 episodes from 3 TV series.

We detail prior related work in Sect. 2, describe our system in Sect. 3, report experiments in Sect. 4, discuss results in Sect. 5 and conclude with Sect. 6.

2   Related Work

Weak Supervision for Actor Labelling. The topic of labeling actors in movies and TV-series has been tackled mostly using weak supervision from subtitle and transcript files [1,8–13], building on the seminal work of Everingham et al. [1,8]. The transcript tells us what each character says in the show, but is not aligned with the video. On the other hand, the subtitle file is aligned with the video, but does not have speaker labels. [1,8] proposes to combine the two (using words in the dialogues) to get the label for an active speaker, if one exists, in each frame. Active speakers detected by lip movement detection in video can then be labelled from the aligned subtitle and transcript files. This weak supervision allows training face and clothing based classifiers for each person based on which they can be labelled even in the absence of speech.

Tapaswi et al. [10] build on this work, using additional cues to train classifiers in a joint optimization framework, such as the fact that no two people in the same scene can have the same label. Bauml et al. [9] employ an additional class for side-actors and take into account unlabelled data (faces without transcript associations) by using an entropy function that encourages classifier decision boundaries to lie along low-density areas of unlabelled data points. Both [9] and [10] report average accuracy results of around 83% on the first 6 episodes of the TV series The Big Bang Theory.

Building on the work of [14], the assignment of actor names in transcripts to one (of several) people in the frame has also been treated as a Multiple Instance Learning (MIL) problem [12,13]. They both use CNN features as face descriptors. [12] also shows that having a separate side-actor classifier boosts actor recognition performance. Finally, action recognition has been used as an additional cue as well, along with weak supervision from transcripts [11].

In contrast to this previous work on actor labelling, our supervision is not from transcripts and subtitle files, which can be difficult to obtain and align with video, but from IMDB actor images. The names of the main actors are used to obtain a few IMDB images for weak supervision, and then propagated through the video data via self-labelling.

Label Propagation in Domain Adaptation. Label propagation is widely used in semi-supervised learning, i.e. dealing with the case when there is a mix of labelled and unlabelled data. In one of the first papers on label propagation, [15] builds a graph with weighted edges indicating the similarity between nodes. Some of the nodes are labelled, and these labels are propagated to their neighbouring, unlabelled nodes according to some similarity measure. The labels propagate through the graph while preventing the initially given labels from changing.

Pham et al. [16] use label propagation to name people appearing in TV news broadcasts starting from transcripts for weak supervision. Kumar et al. [17] also use label propagation to propagate actor labels from a set of key frames that are matched to a manually-curated selection of template images in a movie. However, in their case, and in semi-supervised learning in general, it is supposed that the source and target data are from the same distribution. This is not the case in domain adaptation (DA, see [18] for a survey) where there is a domain shift between the labeled source data and unlabelled target data. We use self-labelling [19], which is a variant of label-propagation in the presence of domain shift, to deal with this. In our case, the domain shift is between the IMDB images that are used for supervision, and the actor faces in the TV series. Bruzzone et al. [19] suggest an iterative labelling strategy that uses the initial labelled source data to train an SVM model and then gradually add the target data obtained from self-labelling to adapt the decision function (i.e. the learned model), while simultaneously slowly removing the source labels. The final classifier is only learnt from the self-labelled target samples. A theoretical guarantee for the self-labelling DA to work has been provided in [20].

We incorporate bipartite graph matching to optimally select the best unlabelled faces to be added to the actor template images in each iteration of the self-labelling process. Graph matching has been used for DA before in remote sensing [21,22]. For example, in [22] the authors match the built graphs in the source and target domains after non linearly transforming (aligning) the source domain with the target domain. However, to the best of our knowledge, the use of graph matching within self-labelling has not been studied before.

3   System Description

We first describe the data preprocessing: face detection, description and tracking (Sect. 3.1). Then, we move on to the graph-matching based self-labelling, including the new graph edge cost and the hierarchical clustering extension (Sect. 3.2).

3.1   Preprocessing

Face Detection and Face Feature Extraction. We use a Deformable Parts Model (DPM) [23,24] for face detection and a pretrained CNN model (Deep Face VGG model [6]) for face description. The latter model has been trained for face recognition on a database of 2622 celebrities (with 375 face images each). We use the output of the last but one fully connected layer (after L2-normalization), which gives us a 4096 dimensional descriptor, used both for tracking faces and for matching actor face tracks to faces in the IMDB database.

Tracking. We employ a multi-target tracker for tracking faces in the TV series. The tracker receives as input the face detections in each frame, along with their CNN feature vectors. New tracks are initialized at detections that do not overlap with previously operating tracks. Existing tracks are updated as follows: if the bounding box of the track and a detection are overlapping, and the Euclidean distance between their feature vectors is below a threshold, the track's coordinates are updated to those of the bounding box of the detection. When there are multiple tracks and detections in close proximity, the Hungarian algorithm [25] is used to optimally associate tracks and detections based on the matching scores (Euclidean distances) of their feature vectors, avoiding multiple detections being assigned to the same track, or a single detection being assigned to multiple tracks. Only associations that are below the previously mentioned threshold distance are considered in the Hungarian optimization.

Tracks that are not updated for a threshold number of frames are deleted. Tracks are also deleted at shot boundaries. Shot boundary detection is done by a simple histogram comparison between frames.

Tracking gives us a sequence of faces in time, that are similar in face descriptor space (see Fig. 2). We adjust our Euclidean distance threshold (for similarity between face descriptors) so that we err on the side of more tracks of the same

Fig. 2. Different faces and poses in a single track (Breaking Bad, Episode 1).

actor in the same shot (early track termination) instead of merging tracks of different actors. The advantage with tracking is that the labelling algorithm has the information from all the faces in the track to make a decision about the track identity.

3.2   H(C)SL: Hungarian (Clustering-Based) Self-Labelling

Tracking the detected faces gives us a set of tracks for a given video/episode. These tracks belong to the main actors, whose images we have from IMDB, as well as side actors, for whom we do not have profile images. Each track is composed of a set of detected faces and we consider each track as one entity. As explained earlier, we also have from IMDB, a set of faces for each given main actor. We consider the set of faces of each actor from IMDB as a point cloud in feature space. The appearances of these faces are different from the actors' appearances in the tracks, because of reasons explained earlier. Because of this domain shift, a naive nearest neighbour matching of the tracks to the closest IMDB faces, will not give good results. We later illustrate this with a baseline experiment.

In spite of the variation between the tracks and the IMDB faces, there still exist some tracks that look sufficiently similar to the IMDB faces to be labelled correctly. To successfully label all the tracks, we make use of these best matching tracks, gradually adding tracks to the actor cloud in a strategy inspired by self-labelling.

Graph-Based Matching Using the Hungarian Algorithm. We now explain our method for selecting tracks and moving them to the actor cloud. Note that, in each iteration, at most one track is added to each actor's cloud (face set).

We first construct a bipartite graph $G = (V, E)$ where vertices represent sets of faces (actor clouds or tracks), and edges indicate an assignment (at a certain cost) of a track to an actor cloud. V consists of two sets: the actors set, $V_{Actors} = \{V_{Actors}^i\}$, and the set of tracks, $V_{Tracks} = \{V_{Tracks}^j\}$. The number of vertices in V_{Actors} corresponds to the number of main actors. The number of vertices in V_{Tracks} is much larger. Each IMDB actor vertex represents the set of faces that correspond to that actor:

$$V_{Actors}^i = \{f_1^i, \cdots, f_{im}^i, \cdots, f_{n_{im}}^i\}, \text{with } n_{im} \text{ the number of faces of actor } i. \tag{1}$$

Each vertex in V_{tracks} represents a track extracted from the TV series:

$$V_{Tracks}^j = \{f_1^j, \cdots, f_{tr}^j, \cdots, f_{n_{tr}}^j\}, \text{with } n_{tr} \text{ the number of faces in track } j. \tag{2}$$

Actor clouds

Tracks

Fig. 3. Hungarian graph matching between vertices in the actors set (top row) and tracks from the video (bottom rows). Matched tracks at this iteration are indicated with solid connections.

Each vertex in V_{Tracks} can have an assignment (i.e. an edge) to one and only one vertex in V_{Actors}. It is possible that a V_{Tracks} vertex remains unassigned, for example if the track belongs to a side actor. While in theory a V_{Actors} vertex can be linked to more than one vertex in V_{Tracks}, we only allow one edge to it at each step in our implementation. Each edge describes the cost of matching the track to the actor. Our task is to find the optimal matching of the tracks to the actors. To do so, we use the Hungarian method [25] that assigns to each actor V_{Actors}^i, a track V_{Tracks}^j, as illustrated in Fig. 3. We set up a cost matrix with each element in the matrix representing the matching cost between an actor and a track, $w(V_{Tracks}^j, V_{Actors}^i)$. This matching cost w is given by Eq. 4 below.

Graph Edge Cost. The choice of the cost measure $w(V_{Tracks}^j, V_{Actors}^i)$ in the Hungarian cost matrix, with V_{Tracks}^j and V_{Actors}^i sets of faces, is crucial as it is the main ingredient of the selection process. Rather than using the minimum or average distance as a cost, we introduce the *normalized edge cost* between a track and a set of faces. To this end, we first compute the average distance d between the faces in the track and the actor faces in the actor cloud:

$$d(V_{Tracks}^j, V_{Actors}^i) = \frac{1}{n_{tr}} \sum_{f_{tr}^j \in V_{Tracks}^j} \frac{1}{n_{im}} \sum_{f_{im}^i \in V_{Actors}^i} \| f_{tr}^j - f_{im}^i \|^2 \quad (3)$$

Then the cost of assigning a track j to an actor i is this average distance, followed by the subtraction of the mean of the average distance of the track j to all actors:

$$w(V_{Tracks}^j, V_{Actors}^i) = d(V_{Tracks}^j, V_{Actors}^i) - \frac{1}{n_{ac}} \sum_i d(V_{Tracks}^j, V_{Actors}^i) \quad (4)$$

where n_{ac} is the number of main actors. This encourages the selection of face tracks that are closer to a specific actor cloud compared to others, while tracks that are equidistant from all actor clouds, typically side-actors or faces recorded under atypical conditions, are less likely to be selected.

Hungarian Self-Labelling. Having explained the graph matching step, let us summarize the steps of the proposed method. We start with actor clouds containing only the IMDB faces. Then, we iteratively assign tracks from the TV series to their corresponding actor clouds using Hungarian graph matching (at most one track per actor at a time), and move the faces in these tracks to the corresponding actor clouds. We repeat this process until no more tracks can be assigned. The iterative assignment step reduces the distance between the actor clouds and the remaining TV series tracks and gradually removes the domain shift. As a result, with each successive iteration of graph-matching based self-labelling, the process is able to label tracks in the TV series that contain faces that are more and more different from the original IMDB images. If there is a mismatch in the number of tracks between actor A and actor B - say 500 tracks for actor A and 100 tracks for actor B - after all the tracks for actor B have been labelled, the Hungarian method will still need to assign tracks for actor B. These will be tracks that don't actually belong to actor B. To prevent this, we use a threshold (λ) for track assignment. Algorithm 1 shows the steps of the proposed graph-matching based self-labelling procedure.

Algorithm 1. Hungarian self-labelling $(V_{Tracks}, V_{Actors}, \lambda)$

1: $V_{Tracks} = V_{Tracks}^1, V_{Tracks}^2, .., V_{Tracks}^m$ ▷ the set of tracks, m: the number of tracks
2: $V_{Actors} = V_{Actors}^1, V_{Actors}^2, .., V_{Actors}^n$ ▷ the actor clouds, n: the number of actors
3: **do**
4:     $number\_of\_added\_tracks = 0$
5:     w=compute_cost(V_{Tracks}, V_{Actors}) ▷ the cost of assigning each track to each
6:   actor (equation 4)
7:     assignments=hungarian$(V_{Tracks}, V_{Actors}, w)$
8:     **for** V_{Actors}^i in V_{Actors} **do**
9:         best_track=assignments(V_{Actors}^i)
10:        **if** $w(V_{Tracks}^{best\_track}, V_{Actors}^i) < \lambda$ **then**
11:            $V_{Actors}^i = (V_{Actors}^i \cup V_{Tracks}^{best\_track})$
12:            $V_{Tracks} = V_{Tracks} \setminus \{V_{Tracks}^{best\_track}\}$
13:            $number\_of\_added\_tracks = number\_of\_added\_tracks + 1$
14:        **end if**
15:    **end for**
16: **while** $number\_of\_added\_tracks > 0$

It is worth noting that the domain shift between IMDB and TV series is not constant, even for the same actor. Hence, it is difficult to come up with a global transformation to overcome this shift. We performed some experiments with Transfer Joint Matching [26], which is a global optimization procedure, but it took 60× longer to run, and gave inferior results to our baselines.

Hierarchical Clustering/Outlier Removal. In the self-labelling process, tracks are matched to actors and all the faces in the track are moved to the corresponding actor cloud. However, some of these faces might be outliers: belonging to a false face detection, quite different from most other faces in the cloud, or resembling rare cases of the actor appearance that are extremely different, such as a face in a dimly-lit night scene, blurred or half-occluded. Including all the faces in the actor cloud (incl. those outliers) in the cost computation for self-labelling might result in false labelling of the remaining tracks. This motivates having a better representation for each actor rather than using all the faces in the actor cloud. To this end, we use a hierarchical clustering approach that **1.** removes the outlier faces, and **2.** selects representative sample points from all the faces representing the actor appearance to obtain a complete profile of the actor, to be used for the cost computation in the self-labelling. We achieve this in two steps: first we obtain big clusters using a relaxed clustering criterion that excludes all the outlying faces; subsequently, we use a more strict clustering criterion to obtain tighter sub-clusters within each big cluster. The centroids of these sub-clusters are chosen as our set of representative points/faces for the actor under consideration. Since we do not care about the exact boundaries between the clusters, and do not have to make a decision using points on these boundaries (we only use the centroids), a simple, yet efficient nearest neighbour clustering method [27,28] is sufficient for our problem. This hierarchical nearest neighbour clustering is done in an online fashion, during the addition of each track from the TV series to the actor cloud. Figure 4 illustrates our clustering procedure.

Online Nearest Neighbor Clustering. In our clustering method we leave the original IMDB images out of the calculation: they remain in the cost computation function as regularizers. So we start with empty clouds for each actor, and receive one track in each assignment step. Each track V^j_{Tracks} is composed of multiple faces $\{f^j_1, f^j_2, \ldots, f^j_n\}$, and we process each face independently. The first face f^j_1 initializes the first cluster $cluster_1$. The second face f^j_2 is then compared with the first cluster $cluster_1$ and if the distance is less than a threshold, it is merged with the first cluster. Otherwise, it is assigned to a new clus-

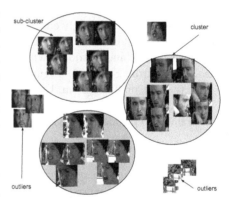

Fig. 4. Clustering of track faces into a cloud - comprising of clusters, sub-clusters and outliers for each actor.

ter. The same procedure is repeated, until all tracks have been clustered. Small clusters (with number of elements below a threshold) are considered as outliers. For all other clusters, a similar clustering scheme is applied within the clusters, to obtain sub-clusters. The centroids of these sub-clusters are the representative points/faces for each actor cloud. Figure 5 shows some actor profile examples.

Fig. 5. From top to bottom: actor profiles for actor Bryan Canston in Breaking Bad, actor Jon Hamm in Mad Men and actress Kaley Cuoco in Big Bang Theory, obtained from the centroids of the sub-clusters.

4   Experiments

Data Set Creation. We conduct a variety of experiments on three different TV series: Breaking Bad, Big Bang Theory and Mad Men. Each of these series exhibit different characteristics and challenges, from the low quality indoor scenes of Big Bang Theory to the crime drama in Breaking Bad with varied actors looks, light, etc. **Breaking Bad (BB)** is a crime drama with challenging indoor and outdoor scenes, shaky camera work and a reasonable number of side-actors. We process episodes 1–6 from Season 1. **Big Bang Theory (BBT)** is a sitcom, shot indoors, with a relatively constant cast, and a smaller number of side-actors. We process episodes 1–6 from Season 1. **Mad Men (MM)** is a period drama, shot both indoor and outdoors with a large number of side-actors. We process episodes 1–3 from Season 1. Figure 6a shows the percentage of the tracks that belong to each main actor and those belonging to the side actors in the three TV series.

We put all episodes through the following pipeline:

1. Face detection using the DPM model
2. CNN feature extraction
3. Tracking
4. Repeat 1 and 2 for IMDB face images
5. Label the tracks using IMDB images as supervision

We download a small number (15–20) of IMDB images as templates for each actor and ensure that no images from the TV series themselves are included. This mimics a fully automated web crawler that would use the actor name as search query and download the top images from IMDB. We used one of our unsupervised labelling methods to get an initial labelling, which we corrected ourselves manually to get ground truth labels. On average for each episode, we have 636 tracks with 20584 faces for BBT; 1113 tracks with 20044 faces for BB and 1590 tracks with 29020 faces for MM. These have been labelled as belonging

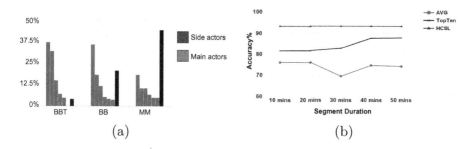

Fig. 6. (a) Percentage of each main actor and the side actors. (b) Baselines and our method compared on variable length segments in Breaking Bad.

to one of 5/6 main actors in the show, or to a generic side-actor class. This dataset is available online[1].

Baselines. In the next series of experiments we compare our Hungarian Clustering based self-labelling algorithm referred to as **HCSL** and its variant without clustering **HSL** with two direct matching baselines and one self-labelling approach:

1. Closest Average Actor (**AVG**): As explained before, we start from a set of faces for each of the main actors, downloaded from IMDB. The task is to match each track from the TV series to its corresponding actor. So, in the first direct matching baseline, we compute the average distance between the faces in the track and the set of faces that belongs to each actor. The track is then assigned to its closest actor according to this assignment cost, i.e. the corresponding average distance. If the average distance between the track and its closest actor cloud is bigger than a predefined threshold, the track is considered to belong to a side actor.
2. Nearest Neighbor Face (**1NN**): This baseline is similar to AVG, but instead of computing the assignment cost of a track to an actor as the average distance between the set of track faces and the set of IMDB actor faces, the cost is simply the minimum distance between a face in the track and its closest face in the set of actor faces. As with AVG, a track is assigned to the actor with the minimum cost and if that cost is bigger than a threshold it means it is a side actor.
3. Self-labelling with top 10 percentile (**TopTen**): In order to examine the added value of the Hungarian Matching coupled with the self-labelling strategy, we add another baseline which is a simple self-labelling approach. Instead of selecting the globally optimal tracks to be assigned, we select the tracks such that their assignment cost is in the top ten percentile of all track costs. We move them to their corresponding actors clouds and repeat the self-labelling process until no track is remaining. To determine the side actors, we examine

---

[1] http://www.jaychakravarty.com/research/whos-that-actor/.

the cost of assignment for each track at the end of the self-labelling process and if a track has an assignment cost higher than a threshold it is considered a side actor.

Note that we do not compare to other work because our framework is different (we use actor names for weak supervision - see Supplementary Material).

Choice of Graph Edge Cost. We compare our proposed normalized edge cost (**NC**) to the Euclidean Distance (**EUC**). The main difference is that NC favours the selection of tracks that are close to one actor compared to all others, while EUC looks at the absolute distances and thus could select a track that is close to more than one actor.

Accuracy results for all our experiments are in terms of track labels. The track is our measurement unit - all faces in each track have the same label.

Parameter Selection. The threshold for HSL is a tradeoff between assigning hard tracks to IMDB clusters and false assignment of side actor tracks. It is set to the average distance between main actor clusters and side actor tracks in the last iterations of self-labelling. All hyperparameters (see Supplementary Material) were selected once (while testing on the first episode of BB) and not changed for the duration of our experiments.

5   Results and Discussion

Table 1 shows that all methods including baselines achieve above 90% results averaged over 6 episodes of the **Bang Theory (BBT)** dataset. This is because all the main actors have distinctive appearances and also because of the existence of very few side-actors (Fig. 6a). In addition, there is very little change in each actor's appearance over the course of the TV series, except in episode 6, where the actors are wearing masks. Consequently, baseline methods' performance decrease noticeably in episode 6, while our HSL method retains a performance of 90%. In this case, the clustering HCSL scheme does not seem to improve on top of the Hungarian self-labelling HSL method, probably due to the limited change in actor appearance over the episode and the existence of very few side-actors.

The second TV series, **Breaking Bad (BB)**, has much more change in actor appearance over the course of the show. For example, the character Walter White develops cancer and shaves his head during chemotherapy. When the character Jesse Pinkman gets hit in the face, he gets a black eye. There is also more variety in the shots in this show, including indoors as well as outdoors, and dimly lit conditions. There is camera shake, leading to blurred faces and noisy tracks. This results in a drop in performance of the direct matching methods (1NN and AVG) from 92% in BBT to 77.3% in BB as reported in Table 2. The Hungarian Self-labelling (HSL) shows a good mean accuracy of 90.3%. The self labelling with top ten percentile (TopTen) performs slightly less with an

Table 1. Different methods performance on Big Bang Theory Dataset

Method	Metric	EP1	EP2	EP3	EP4	EP5	EP6	Avg
1NN	Euc	92.0	93.3	95.5	95.4	94.9	83.3	92.4
AVG	Euc	93.0	92.2	94.1	**97.5**	93.4	82.0	92.0
TopTen	Euc	94.8	**97.8**	98.1	91.2	91.9	84.0	92.9
TopTen	NC	94.2	**97.8**	**99.8**	93.7	96.6	85.8	94.6
HSL (Our)	Euc	**96.0**	97.6	98.8	90.8	96.1	88.1	94.5
HSL (Our)	NC	95.5	**97.8**	99.7	97.0	**97.1**	**90.0**	**96.1**
HCSL (Our)	NC	95.4	97.4	**99.8**	92.5	96.9	88.4	95.0

Table 2. Baselines and our method compared on Breaking Bad

Method	Metric	EP1	EP2	EP3	EP4	EP5	EP6	Avg
1NN	Euc	66.9	63.5	58.5	65.4	73.5	70.0	66.3
AVG	Euc	75.1	74.0	70.9	76.8	82.8	84.5	77.3
TopTen	Euc	93.4	90.3	66.9	92.5	90.7	94.5	88.0
TopTen	NC	95.2	93.8	69.7	92.2	**92.0**	94.0	89.4
HSL (Our)	Euc	94.7	94.3	56.4	**93.4**	91.6	**94.6**	87.5
HSL (Our)	NC	**95.9**	94.9	72.6	91.9	91.9	**94.6**	90.3
HCSL (Our)	NC	95.5	**95.1**	**88.8**	93.2	91.6	94.3	**93.0**

average of 89.4%. In contrast to BBT, the Hungarian Clustering based Self-labelling (HCSL) algorithm achieves better results (average accuracy of 93.0%) in the presence of above mentioned actor appearance change and false detections (outliers).

From Tables 1 and 2, we see that using our normalized measure (the choice of edge cost for each method is shown in column 2), helps improving the performance of the self-labelling methods by 2%–3% on average. This illustrates the need for such a metric to improve the selection procedure by taking into account the track distance to the other actors.

In the third dataset, **Mad Men** TV series, we only report the results of the best direct matching baseline (AVG) and the best self-labelling baseline (TopTen with NC edge cost). Mad Men presents a unique situation, where the number of side-actors is greater than the number of main actors (Fig. 6a). In addition, the women in the show are presented both with and without makeup, resulting in appearance changes that made it difficult even for the human annotator to identify them as the same person purely based on facial features. The self-labelling based methods show decreased performance compared to the previous shows, but still improve noticeably on the direct matching baseline that achieves an average of 74.8%. TopTen gives an average performance of 85.8%, while our proposed method (HCSL) achieves a performance of 86.1% on average.

Variable Length Movies. We examine the effect of variable length videos on the performance using segments of increasing length (in 10 min increments) in BB (Fig. 6b). The Hungarian self-labelling (HCSL) beats the other baselines, and it is important to note that the performance of HCSL remains relatively constant over the variable lengths of videos, in contrast with the TopTen baseline. This is because the selection of tracks in each iteration of the TopTen self-labelling process is dependent on the proportion of tracks belonging to each actor. This effect is more pronounced in segments of shorter duration. The effect of track imbalance between actors is overcome by a larger number of tracks in segments of longer duration. However, HCSL ensures the optimal selection of tracks for every actor in each iteration, regardless of the number of tracks per actor.

Time Complexity. Here's a breakdown of the computation time of our current system: face detection: 22 seconds (per frame); deep CNN face feature extraction per face: 0.5 seconds (per frame); Tracking: 0.1 second (per frame); HCSL: about the same as the running time of the episode. The complexity is of order $O(n^2)$ where n is the number of tracks. This is dominated by the pairwise distance calculation between the tracks. The main bottleneck is the DPM face detection, which could be sped up using GPU.

Qualitative Results. Figure 7 shows sample results of our HCSL algorithm from the Breaking Bad TV series. We present a mixture of randomly selected,

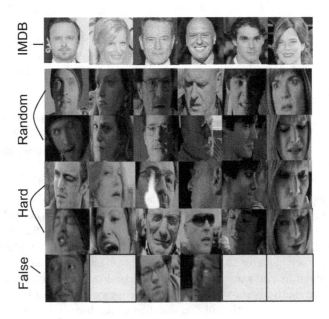

Fig. 7. Samples of labelled images for the main actors in the Breaking Bad dataset. Top row IMDB images, Rows 2–3: randomly selected images. Rows 4–5: hard cases. Row 6: false labels. Best viewed in pdf.

hard and wrongly labelled faces per actor. During the initial stages of self-labelling, the algorithm picks the easier faces (more similar to IMDB faces). After a few iterations, the algorithm acquires more faces from the dataset, is better able to bridge the domain shift between IMDB and the TV show, and is able to correctly label the more challenging faces. At the end of the self-labelling process, when the main actor tracks have been exhausted, the algorithm starts assigning the side-actor tracks to the actor clouds. During this stage of the self-labelling process, the edge cost threshold is the only thing controlling the assignment process. This threshold (as mentioned earlier) is fixed for all our experiments, and might not be the optimal one for the particular show/actor combination (Table 3).

Table 3. Baselines and our method compared on Mad Men

Method	Ep1	Ep2	Ep3	Avg
AVG	72.4	72.8	79.4	74.8
TopTen	**86.2**	86.1	85.3	85.8
HCSL (Our)	85.9	**86.6**	**86.1**	**86.1**

6   Conclusions and Future Work

In this work we utilize the power of deep CNN features to automatically label actors in TV series, using their photos from IMDB, starting only from the names of the lead actors as weak supervision. There is a domain shift between actor images in IMDB and their appearance in TV series. We overcome this problem by introducing a graph-matching based self-labelling approach that iteratively adds actor faces from the TV series to the original collection of actor photos from IMDB. We obtain an appearance profile for each actor based on clustering the self-labelled actor images. We believe that our method is generic enough to be applied to TV news broadcasts, documentary films, silent movies and home videos, given only a few representative images of the actors therein. In future work, we plan on including side actors in the graph matching procedure by automatically building side actor templates which should further decrease the number of false positives. We will also explore metric learning to learn a good similarity measure from the data.

Acknowledgment. This work was supported by the iMinds HiViz project. The first author's PhD is funded by an FWO scholarship.

References

1. Everingham, M., Sivic, J., Zisserman, A.: "Hello! My name is... Buffy"-automatic naming of characters in TV video. In: BMVC, vol. 2, p. 6 (2006)
2. Tapaswi, M., Bäuml, M., Stiefelhagen, R.: Knock! Knock! Who is it? Probabilistic person identification in TV-series. In: 2012 IEEE Conference on Computer Vision and Pattern Recognition (CVPR), pp. 2658–2665. IEEE (2012)
3. Hu, Y., Ren, J.S., Dai, J., Yuan, C., Xu, L., Wang, W.: Deep multimodal speaker naming. In: Proceedings of the 23rd Annual ACM Conference on Multimedia Conference, pp. 1107–1110. ACM (2015)
4. Ren, J., Hu, Y., Tai, Y.W., Wang, C., Xu, L., Sun, W., Yan, Q.: Look, listen and learn-a multimodal LSTM for speaker identification. arXiv preprint arXiv:1602.04364 (2016)
5. Taigman, Y., Yang, M., Ranzato, M., Wolf, L.: DeepFace: closing the gap to human-level performance in face verification. In: Proceedings of the IEEE Conference on Computer Vision and Pattern Recognition, pp. 1701–1708 (2014)
6. Parkhi, O.M., Vedaldi, A., Zisserman, A.: Deep face recognition. In: British Machine Vision Conference (2015)
7. Schroff, F., Kalenichenko, D., Philbin, J.: FaceNet: a unified embedding for face recognition and clustering. In: Proceedings of the IEEE Conference on Computer Vision and Pattern Recognition, pp. 815–823 (2015)
8. Sivic, J., Everingham, M., Zisserman, A.: Who are you? - Learning person specific classifiers from video. In: IEEE Conference on Computer Vision and Pattern Recognition, CVPR 2009, pp. 1145–1152. IEEE (2009)
9. Bauml, M., Tapaswi, M., Stiefelhagen, R.: Semi-supervised learning with constraints for person identification in multimedia data. In: Proceedings of the IEEE Conference on Computer Vision and Pattern Recognition, pp. 3602–3609 (2013)
10. Tapaswi, M., Bauml, M., Stiefelhagen, R.: Improved weak labels using contextual cues for person identification in videos. In: 2015 11th IEEE International Conference and Workshops on Automatic Face and Gesture Recognition (FG), vol. 1, pp. 1–8. IEEE (2015)
11. Bojanowski, P., Bach, F., Laptev, I., Ponce, J., Schmid, C., Sivic, J.: Finding actors and actions in movies. In: Proceedings of the IEEE International Conference on Computer Vision, pp. 2280–2287 (2013)
12. Parkhi, O.M., Rahtu, E., Zisserman, A.: It's in the bag: stronger supervision for automated face labelling. In: ICCV Workshop: Describing and Understanding Video & the Large Scale Movie Description Challenge. IEEE (2015)
13. Haurilet, M.L., Tapaswi, M., Al-Halah, Z., Stiefelhagen, R.: Naming TV characters by watching and analyzing dialogs. In: IEEE Winter Conference on Applications of Computer Vision (WACV) (2016)
14. Guillaumin, M., Verbeek, J., Schmid, C.: Multiple instance metric learning from automatically labeled bags of faces. In: Daniilidis, K., Maragos, P., Paragios, N. (eds.) ECCV 2010. LNCS, vol. 6311, pp. 634–647. Springer, Heidelberg (2010). doi:10.1007/978-3-642-15549-9_46
15. Zhu, X., Ghahramani, Z.: Learning from labeled and unlabeled data with label propagation. Technical report. Citeseer (2002)
16. Pham, P.T., Tuytelaars, T., Moens, M.F.: Naming people in news videos with label propagation. IEEE Multimedia 18, 44–55 (2011)
17. Kumar, V., Namboodiri, A.M., Jawahar, C.: Face recognition in videos by label propagation. In: 2014 22nd International Conference on Pattern Recognition (ICPR), pp. 303–308. IEEE (2014)

18. Patel, V.M., Gopalan, R., Li, R., Chellappa, R.: Visual domain adaptation: a survey of recent advances. IEEE Signal Process. Mag. **32**, 53–69 (2015)
19. Bruzzone, L., Marconcini, M.: Domain adaptation problems: a DASVM classification technique and a circular validation strategy. IEEE Trans. Pattern Anal. Mach. Intell. **32**, 770–787 (2010)
20. Habrard, A., Peyrache, J.P., Sebban, M.: Iterative self-labeling domain adaptation for linear structured image classification. Int. J. Artif. Intell. Tools **22**, 1360005 (2013)
21. Banerjee, B., Bovolo, F., Bhattacharya, A., Bruzzone, L., Chaudhuri, S., Buddhiraju, K.M.: A novel graph-matching-based approach for domain adaptation in classification of remote sensing image pair. IEEE Geosci. Remote Sens. Lett. **53**, 4045–4062 (2015)
22. Tuia, D., Muñoz-Marí, J., Gómez-Chova, L., Malo, J.: Graph matching for adaptation in remote sensing. IEEE Geosci. Remote Sens. **51**, 329–341 (2013)
23. Mathias, M., Benenson, R., Pedersoli, M., Gool, L.: Face detection without bells and whistles. In: Fleet, D., Pajdla, T., Schiele, B., Tuytelaars, T. (eds.) ECCV 2014. LNCS, vol. 8692, pp. 720–735. Springer, Heidelberg (2014). doi:10.1007/978-3-319-10593-2_47
24. Girshick, R.B., Felzenszwalb, P.F., McAllester, D.: Discriminatively trained deformable part models, release 5. http://people.cs.uchicago.edu/rbg/latent-release5/
25. Kuhn, H.W.: The Hungarian method for the assignment problem. Naval Res. Logistics Q. **2**, 83–97 (1955)
26. Long, M., Wang, J., Ding, G., Sun, J., Yu, P.S.: Transfer joint matching for unsupervised domain adaptation. In: The IEEE Conference on Computer Vision and Pattern Recognition (CVPR) (2014)
27. Ward Jr., J.H.: Hierarchical grouping to optimize an objective function. J. Am. Stat. Assoc. **58**, 236–244 (1963)
28. Bubeck, S., von Luxburg, U.: Nearest neighbor clustering: a baseline method for consistent clustering with arbitrary objective functions. J. Mach. Learn. Res. **10**, 657–698 (2009)

Author Index

Printed in the United States
By Bookmasters